GW01260675

Coastal and Estuarine Fine Sediment Processes

Companion books to this title in the ***Proceedings in Marine Science*** series are:

Solent Science – *A Review*
M. Collins and K. Ansell (Eds.)

Muddy Coast Dynamics and Resource Management
B.W. Flemming, M.T. Delafontaine and G. Liebezeit (Eds.)

Proceedings in Marine Science 3

Coastal and Estuarine Fine Sediment Processes

Edited by

William H. McAnally
Coastal & Hydraulics Laboratory
Engineering Research and Development Center
Vicksburg, MS, USA

Ashish J. Mehta
Department of Civil and Coastal Engineering
University of Florida
Gainesville, FL, USA

ELSEVIER

Amsterdam – Boston – Heidelberg – London – New York – Oxford – Paris – San Diego
San Francisco – Singapore – Sydney – Tokyo

ELSEVIER SCIENCE B.V.
Sara Burgerhartstraat 25
P.O. Box 211, 1000 AE Amsterdam, The Netherlands

First edition 2001
Second impression 2003

Library of Congress Cataloging in Publication Data
A catalog record from the Library of Congress has been applied for.

British Library Cataloguing in Publication Data
A catalogue record from the British Library has been applied for.

ISBN: 0 444 50463 X

♾ The paper used in this publication meets the requirements of ANSI/NISO Z39.48-1992 (Permanence of Paper). Printed in The Netherlands.

Preface

The contributions in this volume are based on submissions following the 5th International Conference on Nearshore and Estuarine Cohesive Sediment Processes (INTERCOH 1998) held in Seoul, Korea. The volume is dedicated to the memory of Ranjan Ariathurai (1945-85), a valued colleague who made seminal and lasting contributions to the field of fine sediment engineering.

Ranjan Ariathurai
1945-1985

Chita Ranjan Ariathurai was born in Jaffna, Sri Lanka in 1945. He received a Bachelor of Science degree in Civil Engineering from the University of Ceylon in 1967, a Master of Science in Civil Engineering from the University of Aston, Birmingham, UK, in 1970, and Doctor of Philosophy in Civil Engineering from the University of California, Davis, USA, in 1974. His doctoral thesis (Ariathurai, 1974) was the first comprehensive multi-dimensional numerical model for suspended fine sediment transport, deposition and erosion, and most subsequent models have built upon his work. He recognized the need to introduce a layered bed in the model, with each layer defined by density and bed shear strength. This scheme made it possible to model depth-varying bed properties, a key feature of fine sediment beds. His thesis also introduced an unusual "mis-attribution" of a popular equation.

In Ariathurai's Ph.D. thesis (1974) and some subsequent publications (e.g., Ariathurai et al., 1976), he presented the following equation for cohesive sediment erosion and attributed it to Partheniades (1962):

$$\frac{dm}{dt}\begin{cases} M\left(\frac{\tau_b}{\tau_{ce}}-1\right) & \tau_b > \tau_{ce} \\ 0 & \tau_b \le \tau_{ce} \end{cases} \tag{1}$$

where dm/dt = erosion rate in mass per time per unit area, M = empirical erosion rate constant, τ_b = shear stress exerted by the flow on the bed, and τ_{ce} = critical shear stress for erosion, or equivalently the bed shear strength.

Equation (1) has been widely and successfully used to characterize surface erosion of cohesive sediment, but it does not actually appear in the cited reference by Partheniades. When asked about this discrepancy, Ariathurai (Personal discussion with McAnally, about 1979) replied that he needed a simple equation for his model, so he fit a straight line through Professor Emmanuel Partheniades' experimental plots of erosion rate versus applied shear stress for remolded sediment samples. He further said that he credited Professor Partheniades with the equation as a courtesy and in respect for his pioneering work in cohesive sediment research. When he learned that Partheniades declined to take credit for the equation, Ariathurai subsequently presented the equation without citing a source (e.g., see Ariathurai and Arulanandan, 1978). Nevertheless, Partheniades (1962) is still often cited as the source of the equation.

The following form of the erosion equation appears, unattributed, in the Ph.D. thesis of Arumugam Kandiah (1974), who was, like Ariathurai, a student of Professor Ray Krone at the University of California, Davis.

$$\frac{dm}{dt} = M'(\tau_b - \tau_{ce}) \qquad \tau_b > \tau_{ce} \tag{2}$$

where M' = an empirical erosion rate coefficient which includes the critical shear stress.

Kandiah acknowledged in his dissertation the help and encouragement of "...my good friend and colleague, Dr. Ranjan Ariathurai...," which indicates that when his thesis was written Ariathurai's had already been submitted. That indication is supported by the sequence numbers assigned to their respective theses by University Microfilms International. This evidence establishes primacy of publication, but given the exchange of ideas among graduate students and faculty it seems likely that the equation was the product of several minds at the university.

The approximately linear relationship between erosion rate and applied shear stress became known at the Davis campus of the University of California during the late 1960's and early 1970's. Professor Kandiah Arulanandan, acting on the advice of Professor Krone, constructed a cylindrical erosion device based on a design developed at the University of Texas, and Arulanandan's student Patrick Riley performed experiments that demonstrated the linear relationship anticipated by Professor Krone (letter to McAnally dated July 21, 1999). Subsequent students of Arulanandan and Krone further explored that relationship, including A. Sargunam (Sargunam et al., 1973), P. Loganathan and the previously cited A. Kandiah. Kandiah's (1974) contributions were substantial in that he thoroughly explored the relationships for the critical shear stress, τ_{ce} and erosion rate constant, M', versus key variables such as clay mineralogy and pore fluid composition.

This evidence leads to the observation that (1) was the product of the fertile intellectual climates at the University of California campuses of Davis and Berkeley (where Professors Krone and Partheniades did their seminal work in the early 1960's) and involved several individual contributions. A proper attribution should include Ariathurai, for setting the form of the equation and publishing it first; Riley, for his experiments showing that the erosion rate was a linear function of the excess applied shear stress; Kandiah, for his experiments fully demonstrating linearity; Professor Partheniades, for his initial erosion experiments that served as the foundation; Professor Krone, for his guidance of Riley's work, and his supervision of the dissertation research of both Ariathurai and Kandiah; and Professor Arulanandan, for his supervision of the work of Riley. It would be more than awkward to use precious journal space with an appellation of the "Ariathurai-Kandiah-Riley-Partheniades-Krone-Arulanandan equation," so it is proposed that (1) be known as the Ariathurai-Partheniades equation, and with Ariathurai (1974) cited as the source.

Before we conclude this preface, we would like to extend our deepest gratitude to Cynthia Vey, without whose diligence and persistence, it would have been far more difficult to put this volume together. INTERCOH 1998 was sponsored and organized by the Korea Ocean Research and Development Institute (KORDI) under the initiative of Dr. Dong-Young Lee and his colleagues, who naturally deserve special thanks.

REFERENCES

Ariathurai, C. R., 1974. A finite element model for sediment transport in estuaries. *Ph.D. Thesis*, University of California, Davis.

Ariathurai, C. R., and Arulanandan, K., 1978. Erosion rates of cohesive soils. *Journal of the Hydraulics Division,* ASCE, 104(2), 279-282.

Kandiah, A. 1974. Fundamental aspects of surface erosion of cohesive soils. *Ph.D. Thesis*, University of California, Davis.

Partheniades, E., 1962. A study of erosion and deposition of cohesive soils in salt water. *Ph.D. Thesis*, University of California, Berkeley.

Sargunam, A., Riley, P., Arulanandan, K., and Krone, R. B., 1973. Physico-chemical factors in erosion of cohesive soils. *Journal of the Hydraulics Division*, ASCE, 99(3), 555-558.

William H. McAnally
Ashish J. Mehta

Contents

Contributing Authors

T. J. Andersen
Institute of Geography
University of Copenhagen
Øster Voldgade 10
DK 1350 Copenhagen K, Denmark
E-mail: tja@server1.geogr.ku.dk

P. Bassoullet
Institut Français pour la Recherche et Exploitation de la Mer (IFREMER)
Coastal Environment Division
DEL-EC-TP, B.P. 70, 29280, Plouzané, France
E-mail: philippe.bassoullet@ifremer.fr

P. Castaing
Université Bordeaux I
Département de Géologie et Océanographie
UMR 5805 EPOC
Avenue des Facultés
33405 Talence cedex, France
E-mail: weber@geocean.u-bordeaux.fr

H. Christiansen
Strom und Hafenbau
Hamburg Port Authority
Dalmannstrasse 1
20457 Hamburg II, Germany
E-mail: drchr@t-online.de

M. C. Christie
Institute of Marine Studies
The University of Plymouth
Plymouth, Devon PL4 8AA, United Kingdom
E-mail: m.christie@plymouth.ac.uk

J. M. de Kok
Rijkswaterstaat
RIKZ
The Hague, The Netherlands
E-mail: j.m.dkok@rikz.rws.minvenw.nl

K. R. Dyer
Institute of Marine Studies
The University of Plymouth
Plymouth, Devon PL4 8AA, United Kingdom
E-mail: kdyer@plymouth.ac.uk

R. Gu
Iowa State University
Department of Civil and Construction Engineering
Ames, IA 50011 USA
E-mail: roygu@iastate.edu

E. J. Hayter
U.S. Environmental Protection Agency
National Exposure Research Laboratory
Ecosystems Research Division
960 College Station Road
Athens, GA 30605 USA
E-mail: hayter.earl@.epa.gov

O. Higashi
City and Environment Planning, Research
EX Corporation
2-17-22 Takada, Toshima-ku
Tokyo, 171-0033, Japan
E-mail: (no e-mail address)

B. M. Janssen-Stelder
Institute for Marine and Atmospheric Research Utrecht
Department of Physical Geography
Faculty of Geographical Sciences
Utrecht University
P. O. Box 80.115
3508 TC, Utrecht, The Netherlands
E-mail: b.stelder@geog.uu.nl

H. Jestin
Institut Français pour la Recherche
et Exploitation de la Mer (IFREMER)
Coastal Environment Division
DEL-EC-TP, B.P. 70, 29280
Plouzané, France
E-mail: herve.jestin@ifremer.fr

J. Jiang
ASL Environmental Sciences Inc.
1986 Mills Road
Sidney, V8L 5Y3, BC, Canada
E-mail: jjiang@aslenv.com

J.-Y. Jin
Coastal and Harbor Engineering Research Center
Korea Ocean Research and Development Institute
Ansan P.O. Box 29
Seoul 425-600, Korea
E-mail: jyjin@kordi.re.kr

P. Jinchai
Hydrography Department
Royal Thai Navy
Bangkok, Thailand

C. Johansen
NIRAS
Vestre Havnepromenade 9
DK-9100 Aalborg, Denmark
E-mail: cjo@niras.dk

K. C. Jun
Korea Ocean Research and Development Institute
P.O. Box 29
Ansan, Kyunggi, Korea
E-mail: kcjun@kordi.re.kr

R. Kirby
Ravensrodd Consultants Limited
6 Queen's Drive
Taunton, Somerset TA1 4XW, United Kingdom
E-mail: robkirby@globalnet.co.uk

C. Kranenburg
Delft University of Technology
Faculty of Civil Engineering
P.O. Box 5048
2600 GA Delft, The Netherlands
E-mail: c.kranenburg@ct.tudelft.nl

T. Kusuda
Department of Urban and Environmental Engineering
Kyushu University
6-10-1 Hakozaki, Higashi-ku
Fukuoka, 812-8581, Japan
E-mail: kusuda@civil.kyushu-u.ac.jp

T. Larsen
Hydraulics & Coastal Engineering Laboratory
Aalborg University
Sohngaardsholmsvej 57
DK-9000 Aalborg, Denmark
E-mail: i5tl@civil.auc.dk

D. Y. Lee
Korea Ocean Research and Development Institute
P.O. Box 29
Ansan, Kyunggi, Korea
E-mail: dylee@kordi.re.kr

J. L. Lee
Sungkyungkwan University
Civil Engineering Department
Suwon, Kyunggi, Korea
E-mail: jllee@skku.ac.kr

W. T. B. van der Lee
Institute for Marine and Atmospheric Research Utrecht
Department of Physical Geography, Faculty of Geographical Sciences
Utrecht University
P. O. Box 80.115
3508 TC, Utrecht, The Netherlands
Present Affiliation: HKV Consultants
P.O. Box 2120
8203 AC Lelystad, The Netherlands
E-mail: w.vanderlee@hkv.nl

P. Le Hir
Institut Français pour la Recherche et l'Exploitation de la Mer (IFREMER)
Direction de l'Environnement Littoral
DEL-EC-TP, B.P. 70
29280 Plouzané, France
E-mail: plehir@ifremer.fr

LI Yan
Second Institute of Oceanography, SOA
P. O. Box 1207
Hangzhou, 310012, Peoples Republic of China
liyan@mail.hz.zj.cn

Y. Li
Tetra Tech, Inc./ Infrastructure Southwest Group
17770 Cartwright Road, Suite 500
Irvine, California 92614 USA
E-mail: wally.li@ttisg.com

P. Manojkumar
Department of Marine Geology and Geophysics
Cochin University of Science and Technology
Lakeside Campus
Cochin, 682 016, India
E-mail: manojkumar_@hotmail.com

W. H. McAnally
Coastal and Hydraulics Laboratory
U.S. Army Engineer Waterways Experiment Station
3909 Halls Ferry Road
Vicksburg, Mississippi 39180, USA
E-mail: mcanalw@wes.army.mil

A. J. Mehta
Civil and Coastal Engineering Department
345 Weil Hall, P.O. Box 116580
University of Florida
Gainesville, Florida 32611, USA
E-mail: mehta@coastal.ufl.edu

L. M. Merckelbach
Delft University of Technology
Faculty of Civil Engineering
P.O. Box 5048
2600 GA Delft, The Netherlands
E-mail: l.merckelbach@ct.tudelft.nl

H. J. Mitchener (Now: H. J. Lowe)
Defence Evaluation Research Agency
Underwater Sensors and Oceanography
Winfrith Technology Centre
Dorchester, Dorset DT2 8XJ, United Kingdom
Former address:
H. R. Wallingford Ltd., Marine Sediments Group
Howbery Park
Wallingford, Oxfordshire OX14 1XD United Kingdom
E-mail: hlowe@dera.gov.uk

M. Mory
Ecole Nationale Supérieure en Génie des Technologies Industrielles
BP 576
64012 Pau Cédex, France
E-mail: mathieu.mory@univ-pau.fr

A. C. Narayana
Department of Marine Geology and Geophysics
Cochin University of Science and Technology, Lakeside Campus
Cochin, 682 016, India
E-mail (2): a_c_narayana@yahoo.com or acn@cusat.ker.nic.in

D. J. O'Brien
Hyder Consulting Ltd.
P.O. Box 4, Pentwyn Road
Nelson, Mid Glam CF 46 6YA, United Kingdom
E-mail: damon.obrien@hyder-con.co.uk

J. S. Park
Coastal and Harbor Engineering Research Center
Korea Ocean Research and Development Institute
Ansan P.O. Box 29
Seoul 425-600, Korea
E-mail: jpark@kordi.re.kr

K. S. Park
Coastal and Harbor Engineering Research Center
Korea Ocean Research and Development Institute
Ansan P.O. Box 29
Seoul 425-600, Korea
E-mail: kspark@kordi.re.kr

I. Piedra-Cueva
Institute of Fluid Mechanics and Environmental Engineering (IMFIA)
Faculty of Engineering
C.C. 30
Montevideo, Uruguay
E-mail: ismaelp@aljofar.fing.edu.uy

W. Roberts
16 McMullan Close
Wallingford, Oxfordshire OX10 0LQ, United Kingdom
E-mail: bill.roberts@ndirect.co.uk

D. H. Schoellhamer
U.S. Geological Survey
Placer Hall, 6000 J Street
Sacramento, California 95819-6129 USA
E-mail: dschoell@usgs.gov

G. C. Sills
University of Oxford
Department of Engineering Science
Parks Road
Oxford OX1 3PJ, United Kingdom
E-mail: gilliane.sills@eng.ox.ac.uk

R. Silva Jacinto
Institut Français pour la Recherche et l'Exploitation de la Mer (IFREMER)
Direction de l'Environnement Littoral, DEL-EC-TP, B.P. 70
29280 Plouzané, France
E-mail: rsilva@ifremer.fr

T. J. Smith
Ravensrodd Consultants Limited
6 Queen's Drive
Taunton, Somerset TA1 4XW, United Kingdom
E-mail: tims@tsc.gov.uk

A. Sottolichio
Université Bordeaux I
Département de Géologie et Océanographie
UMR 5805 EPOC
Avenue des Facultés
33405 Talence cedex, France
E-mail: a.sottolichio@geocean.u-bordeaux.fr

K. Taki
Department of Civil Engineering
Chiba Institute of Technology
2-17-1, Tsudanuma, Narashino, 275-8588, Japan
E-mail: taki@ce.it-chiba.ac.jp

R. Tatavarti
Naval Physical and Oceanographic Laboratory
Thrikkahara, Cochin, 682 021, India
E-mail: tatavarti@vsnl.net.in

A. M. Teeter
Coastal and Hydraulics Laboratory
U.S. Army Engineer Waterways Experiment Station
3909 Halls Ferry Road
Vicksburg, Mississippi 39180, USA
E-mail: teeter@hl.wes.army.mil

E. A. Toorman
Hydraulics Laboratory
Civil Engineering Department
Katholieke Universiteit Leuven
De Croylaan 2,
B-3001 Heverlee, Belgium
E-mail: erik.toorman@bwk.kuleuven.ac.be

H. Torfs
Provincie Vlaams-Brabant
Directie Infrastructuur
Dienst Waterlopen
3010 Leuven, Belgium
E-mail: htorfs@vl-brabant.be

P. Turner
Institute of Marine Studies
The University of Plymouth
Plymouth, Devon PL4 8AA, United Kingdom
E-mail: dturner3944@talk21.com

R. E. Uittenbogaard
WL Delft Hydraulics
P.O. Box 177
2600 MH Delft, The Netherlands
E-mail: rob.uittenbogaard@wldelft.nl

S. B. Vinzon
Programa de Engenharia Oceânica
COPPE - Universidade Federal do Rio de Janeiro
CEP 21945-970
Rio de Janeiro, RJ, Brazil
E-mail: susana@peno.coppe.ufrj.br

R. Watanabe
Department of Civil Engineering
Fukuoka University
8-19-1 Nanakuma, Jonan-ku
Fukuoka, 814-0180, Japan
E-mail: wata@fukuoka-u.ac.jp

R. J. S. Whitehouse
HR Wallingford
Howbery Park
Oxfordshire OX10 8BA, United Kingdom
E-mail: r.whitehouse@hrwallingford.co.uk

J. C. Winterwerp
WL Delft Hydraulics
P.O. Box 177
2600 MH Delft, The Netherlands
E-mail: han.winterwerp@wldelft.nl

H. Yamanishi
Department of Urban and Environmental Engineering
Kyushu University
6-10-1 Hakozaki, Higashi-ku
Fukuoka, 812-8581, Japan
E-mail: yamanishi@civil.kyushu-u.ac.jp

K. Yamasaki
Department of Civil Engineering
Fukuoka University
8-19-1 Nanakuma, Jonan-ku
Fukuoka, 814-0180, Japan
E-mail: yama@fukuoka-u.ac.jp

K. D. Yum
Coastal and Harbor Engineering Research Center
Korea Ocean Research and Development Institute
Ansan P.O. Box 29
Seoul 425-600, Korea
E-mail: kdyum@kordi.re.kr

XIA Xiaoming
Second Institute of Oceanography, SOA
P. O. Box 1207
Hangzhou, 310012, Peoples Republic of China
E-mail: xiaxm@mail.hz.zj.cn

Coastal and Estuarine Fine Sediment Processes
W.H. McAnally and A.J. Mehta (Editors)

Cohesive sediment transport modeling: European perspective

E. A. Toorman

Hydraulics Laboratory, Civil Engineering Department, Katholieke Universiteit Leuven, De Croylaan 2, B-3001 Heverlee, Belgium

Ongoing and recently completed efforts in cohesive sediment transport research in Europe are described, with emphasis on the use of models as research tools. Numerical models are increasingly used as virtual laboratories to study complex phenomena in which sediment-turbulence interactions play a central role. A related research area is the study of the strength development in cohesive sediment beds. This paper attempts to indicate where European research stands and what it hopes to achieve in the coming years.

1. INTRODUCTION

Cohesive sediment transport (CST) processes in coastal and estuarine areas have important effects on the economy (large areas of the coast are developed, e.g., harbors, tourism, fisheries) and the environment (mudflats are the basis of ecologically valuable, but vulnerable coastal wetlands). Therefore, CST is an important interdisciplinary research field with a wide range of applications including morphodynamic changes due to construction works, dredging and dredged material disposal, wetland restoration etc., sediment transport in rivers, lakes, reservoirs and sewers, and land slides. The applications even cross the borders of hydraulic engineering as materials with similar properties occur in mining engineering (drilling muds and mine tailings), chemical process engineering (colloidal suspensions and ceramic materials) and civil engineering (concrete and grout).

This introductory paper aims at presenting a summary of fundamental CST research, highlighting the present horizons of our understanding, and showing directions on how to go beyond, particularly by the use of computer models as virtual laboratories. The paper is limited to the European context where several joint research projects, which aim at filling in some of the gaps in our knowledge, are presently being carried out or have been carried out in recent years.

2. FUNDAMENTAL RESEARCH

Until a few year ago, fundamental CST research was largely divided into separate studies of the various processes involved (particularly: erosion, deposition, consolidation, flocculation and rheology), in order to obtain a better understanding of each process. A brief historical review may help understand how CST research has evolved.

It appears that initially only geologists and geographers showed interest in cohesive sediment transport (e.g., Postma, 1967). Systematic engineering research on this topic seems to have started seriously in the 1960s. Most agree that it was pioneered by Krone (1962) in the U.S. Among the earliest engineering studies on CST in Europe one finds the Demerara River project by Delft Hydraulics (1962), including studies on sedimentation, flocculation and erosion using a paddle-driven annular flume, the Thames barrier project by HR Wallingford (Owen and Odd, 1972; Odd and Owen, 1972) and projects by NedeCo (1965, 1968). The studies at the Laboratoire Central d'Hydraulique de France (LCHF) have become especially well known, in particular the paper by Migniot (1968), which gives an overview of the mainly empirical study of the behavior of cohesive sediments (i.e., sedimentation, flocculation, consolidation, rheology, gravity flow, erosion under uniform and tidal currents and waves). Other early reviews on CST processes were published by Owen (1966) and Postma (1967).

However, the entire problem of estuarine morphodynamics and CST is one of the interaction of many processes, resulting in features such as the formation of turbidity maxima, density currents, lutoclines etc., which are the net result of all the forces and fluxes involved. It is practically impossible to reproduce these in the laboratory due to scaling problems and the high costs. Field work to obtain comprehensive data sets is even more expensive. These complex problems can only be studied with the help of mathematical models, complemented by experimental data for calibration and validation. Therefore, one has to give a correct mathematical description of all these forces and fluxes in the three dimensions, and, in principle, the features that occur in nature should be reproducible. This is the approach that has led to what some call "integrated modeling" (Le Hir, 1997), which requires a greater degree of detail than engineering models offer in general.

2.1. Numerical modeling as research tool

Application of CST models in engineering and environmental problems concern morphodynamic changes occurring in large areas and over a long period, requiring an enormous computer capacity because of the small length and time scales over which the processes occur. Therefore, until the present, engineering models contain many simplifications. According to a review by Odd (1988), the first CST model, in which the complete cycle of mud transport and siltation in an estuary was simulated, was the 1D(imensional)H(orizontal)-2L(ayer) model developed at HR Wallingford in 1968 for the Thames barrier investigation (Owen and Odd, 1972; Odd and Owen, 1972). Since then a large number of models have emerged (2DH, sometimes multi-layered, 2DV(ertical) and recently Quasi-3D). Full 3D modeling for engineering applications is still very limited, as no computer is powerful enough to do the task within a reasonable time. Therefore, some simplifications are necessary, i.e., vertical momentum exchange is neglected, reducing the vertical momentum balance to the hydrostatic pressure condition, and often no coupling between sediment transport and hydrodynamics is taken into account.

In order to study the shortcomings of the traditional applied models, more detailed 2DV research models have been developed in recent years at the Institut Français pour la Recherche et l'Exploitation de la Mer (IFREMER) (Le Hir, 1997) and the Laboratoire National d'Hydraulique (LNH) in France (Galland, 1996), the Katholieke Universiteit Leuven (K.U.Leuven) in Belgium (Toorman, 1997b) and Delft Hydraulics in the Netherlands (Winterwerp, 1998c), that are used as "virtual laboratories". These models solve the full hydrodynamics and sediment balance, together with a one- or two-equation turbulence closure model. At LNH even a Reynolds-stress turbulence model has been implemented (Teisson et al., 1992; Galland, 1996). Unfortunately, complete data sets to validate every aspect of these models (i.e., detailed distributions of all the field variables under well described conditions and material parameters), particularly with regard to the turbulence interactions, are still not available.

The results of the detailed 2DV models have to be parameterized, i.e., translated into a simplified form, in order to make them suitable for presently used engineering models. In particular, in Delft much effort is currently being done in this field regarding many processes and features (flocculation, entrainment, lutocline formation and fluid mud), by making extensive use of scaling laws and validation with a 1DV integrated model (Winterwerp, 1998a; Kranenburg and Winterwerp, 1997; Winterwerp and Kranenburg, 1997; Winterwerp, 1999). Introduction of the description of cohesive sediment flocs by a fractal structure has proven to be useful for parameterization (Kranenburg, 1994).

As current CST research increasingly uses these detailed research models to study complex interactions, the need in CST research is presently primarily determined by the model requirements. Of course, models are only approximations of reality and fail to reproduce some phenomena occurring in the field, which only measurements can reveal. Furthermore, models are necessarily oversimplified because of the limitations in computer capacity on the one hand, and on the other hand the impossibility to account for the complexity of the very non-homogeneous and variable composition of natural cohesive sediments. These factors contribute to the fact that estimated quantities of transported sediment by CST models can show large discrepancies from reality. Models, however, have the advantage over measurements in covering much larger areas and providing data at many more points. Complementary measurements are required for calibration, validation and verification of the model. Unfortunately, the collected data, either in the field or in the laboratory, are often incomplete, and do not allow calibration or validation of the models without making unverifiable additional assumptions, which contribute to a great deal to the inaccuracy of model predictions. A relevant example of the inadequacy of laboratory data was the inter-comparison exercise in the MAST II G8M project (see Section 2.2), in which flume tests were simulated with traditional depth-averaged models by various institutes with their own software (Hamm et al., 1996). Even the simple flat bottom flume erosion test could not be reproduced by any model due to a lack of a complete data set to calibrate the model (particularly the data to set up the erosion model, because the strength-density relationship was incomplete). Furthermore, a high degree of sensitivity was found with respect to certain model parameters, thus raising questions on the reliability of traditional CST models.

2.2. Collaborations

One of the major aims of CST research is to provide the authorities with predictive models to be used as a management tool. As CST is a wide and multi-disciplinary research field, the institutes involved have specialized in certain areas, and collaboration is necessary in order to provide the end-users with the necessary process modules and experimental or field data for calibration and validation.

It should be mentioned that CST research owes much to the Cohesive Sediment Workshops organized by Professor Mehta from the University of Florida (Gainesville) since 1980. These meetings have been followed up by the INTERCOH conferences. They have stimulated collaboration in Europe, because it was the first forum where the CST research community met. Another initiative was launched in 1974 by a few scientists to form the Estuary Study Group, which is a yearly meeting of (mainly European) scientists concerned with processes in the estuarine environment.

The funding by the European Commission through the Marine Science & Technology (MAST) research program and the Training & Mobility for Researchers (TMR) program of its Directorate General XII for Science, Research & Development allowed a collaboration between various institutes within the European Union (Table 1).

The COSINUS project in particular aims at improving applied CST models by studying processes that are not well understood with the help of basic research models, supplemented by dedicated field and laboratory experiments. Special attention is given to parameterization of the newly developed process modules. These parameterized formulations will be incorporated into various 2D and 3D models currently used by some of the major hydraulics institutes in Europe. Validation will be done by inter-comparison with the 2DV integrated research models. Finally the models will be applied to a few European estuaries.

2.3. Research priorities

Based on the concept that CST research needs should originate from model needs, one should start from the list of model parameters for which closure relationships are necessary (see Table 2).

Current fundamental CST research in Europe focuses on three major subjects: 1) sediment-turbulence interaction, 2) bed strength evolution and 3) intertidal mudflats. Interestingly, when these topics are compared with the list of major gaps in our understanding of CST processes, as identified a decade ago by Dyer (1989), it can be concluded that both are roughly equivalent. This indicates that the progress made over the last ten years was not as hoped, because with each step forward the understanding of CST processes has revealed new problems to be solved. In the following sections, attention will be focused on the first two topics.

Table 1
Joint cohesive sediment research projects funded by the European Commission

E.C. Program	Project Acronym	Title	Start	End	Reference
MAST I	G6M	Coastal Morphodynamics	1990	1992	de Vriend, 1992
MAST II	G8M	Coastal Morphodynamics	1993	1996	Stive et al., 1996
MAST II	-	3D Modeling of Estuaries	1993	1996	Peltier et al., 1996
MAST III	INTRMUD	The morphological development of Intertidal Mudflats	1997	1999	Dyer, 1998
MAST III	COSINUS	Prediction of Cohesive Sediment Transport and bed dynamics in Estuaries and coastal zones with Integrated Numerical Simulation models	1998	2000	Berlamont and Toorman, 1998
TMR	SWAMIEE	Sediment and Water Movement in Industrialized Estuarine Environments	1998	2000	

Table 2
Parameters required for a detailed 2DV model based on mixture theory

Sediment transport equation:
- settling velocity (flocculation model)
- mass diffusivity or Schmidt number (empirical)
- erosion flux (semi-empirical)
- deposition flux (empirical)

Momentum equation:
- suspension viscosity (rheological model)
- eddy viscosity (turbulence closure)
- suspension density (sediment transport equation)

Turbulence model:
- turbulence damping functions
- Schmidt number
- buoyancy factor in the ε-equation (k–ε model only)

3. SEDIMENT-TURBULENCE INTERACTION

Several important CST phenomena observed in nature can only be explained in terms of the interaction between the suspended particles and the turbulent fluctuations of flow. Sediment-turbulence interactions play an important role in the processes of settling and flocculation, stratification and erosion. This interaction is a complicated problem, which is finally getting attention now that the latest generation of CST models allows the implementation of relatively detailed turbulence models.

3.1. Settling and flocculation

The net settling velocity of mud aggregates is determined by the balance between buoyancy (or gravity), drag, turbulent mixing (diffusion) and inter-particle collisions. Buoyancy and vertical drag determine the Stokes' fall velocity, which depends on the floc structure (size, density and shape). Aggregation and break-up of suspended flocs is controlled by turbulent shear and inter-particle collisions. The effective settling rate (the average settling velocity) is the fall velocity multiplied by a hindrance factor representing the reduction due to the presence of other particles.

Flocculation has been studied for a long time, because it plays an important role in applications other than CST, particularly in chemical engineering (e.g., for water purification). Important research on flocculation in the context of estuarine CST has been carried out by Dyer (1989) and van Leussen (1994). A simple relationship between floc size and turbulent shear in equilibrium conditions has been proposed, and its incorporation in applied models has yielded a qualitative improvement (Malcherek, 1996). A more general flocculation model, based on the structural kinetics theory, which accounts for growth and break-up of flocs, has recently been developed by Winterwerp (1998a). A fractal dimension can be associated with the floc structure (Kranenburg, 1994).

3.2. Stratification

Due to the vertical distribution of the turbulent eddy diffusivity in the water column and the effect of gravity, the balance between the vertical forces on particles results in a non-homogeneous concentration distribution over the vertical, i.e., stratification. Stable stratification (when a negative vertical density gradient occurs) leads to damping of turbulence due to the buoyancy effect; turbulent mixing is reduced by gravity.

Certain time-dependent conditions lead to very sharp concentration gradients, known as lutoclines, resulting in the generation of a two-layer stratified medium. When the flux Richardson number Ri_f, the ratio of buoyancy damping to production of turbulent kinetic energy, exceeds an empirically determined critical value Ri_c of about 0.15 (a necessary but not a sufficient condition for damping; Toorman, 1999b) at a lutocline, turbulent mixing between the two layers is inhibited as vertical turbulent fluctuations are nearly completely damped due to the large density difference, like at the air-water interface. The collapse of turbulence above the lutocline may temporarily cause the top layer to exhibit a laminar flow regime (laminarization). This introduces a challenge for numerical modeling, and may need special so-called "low-Reynolds number" modifications (where the local turbulent Reynolds number $k^2/\nu\varepsilon$ is implied, with k = turbulent kinetic energy, ν = kinematic viscosity of the suspension and ε = turbulent dissipation rate).

The denser bottom layer has been labeled as "concentrated benthic suspension" (CBS) and forms the central theme in the MAST3 COSINUS research project, which investigates the conditions under which CBS layers occur, and how local laminarization around the lutocline can be modeled (Berlamont and Toorman, 1998). Some would call CBS "fluid mud", but CBS is a sub-type of fluid mud where the particles are kept in suspension by turbulence. Fluid mud can also exist as a dense non-Newtonian slurry, which develops structure, and thus strength (effective stress), at rest (e.g., Toorman, 1992). Important progress was made by the development of a general rheological model which accounts for the thixotropic behavior of fluid mud (Toorman, 1997b). There are several data sets which suggest that CBS layers occur in the field (e.g., Wolanski et al., 1988; Winterwerp, 1998b). Within the SILTMAN project concerning siltation of the Rotterdam Waterway (the Netherlands), this feature has been studied extensively (Winterwerp, 1998c). With the help of a 1DV integrated model a theoretical study of the conditions for the occurrence of CBS has been investigated. The study indicates that relatively well-mixed concentration profiles can become "saturated" and then rapidly collapse, resulting in a dense mud layer and causing rapid siltation by gravity flow (Winterwerp et al., 2000).

In analogy with temperature gradient-induced stratified flow, the Schmidt number, the ratio of the eddy viscosity to the sediment mixing coefficient, is affected by the buoyancy effect and the damping correction factors are empirically written as a function of Ri. Attempts have been made to apply an empirical damping function, determined for thermal stratification in oceans by Munk and Anderson (1948), but this formulation has not been found satisfactory for simulating CST (e.g., Galland, 1996). No conclusive alternative formulation has been found thus far. Attempts have been undertaken with one-equation (Le Hir, 1997; Kranenburg, 1998), two-equation (Galland, 1996; Toorman, 1997; Winterwerp, 1999) and even Reynolds stress turbulence closure models (Galland, 1996) to study this phenomenon numerically, thus far with qualitative results only. More institutes are currently working on this subject, particularly within the MAST3 COSINUS project.

When the flux Richardson number becomes increasingly smaller than Ri_c, the lutocline becomes more and more unstable. Internal waves are generated which can lead to Kelvin-Helmholtz vortices, which is the beginning of turbulent mixing of the two layers. This process has been studied extensively by Uittenbogaard (1995), and a mathematical formulation for use with a k-ε model has been derived and incorporated in the DELFT-3D model. Kelvin-Helmholtz instabilities in stratified cohesive sediment suspensions are currently being investigated experimentally in Edinburgh using particle image velocimetry (PIV) and ultrasonic techniques (Crapper, personal communication, 1998).

The damping of turbulence by density gradients is possibly also the reason why lid-generated turbulence in annular flumes in some cases does not reach the bottom and no erosion occurs. This hypothesis will be investigated numerically in Edinburgh (Crapper, personal communication, 1998). A new experiment will be carried out in a modified set-up of the annular flume at the Delft University of Technology within the COSINUS project. The modification consists in lowering the rotating lid to the bottom such that the bed can be rotated relative to the walls (Kranenburg and Bruens, 1999). Secondary currents due to wall friction however cannot be avoided.

3.3. Deposition and erosion

Thus far only sediment-turbulence interaction in the water column has been addressed. At the bottom, one needs to determine the shear stress for the estimation of deposition and erosion rates. However, near the bottom the problem becomes more complex because of the transient and viscous sub-layers, where traditional turbulence models are not valid. Implementation of a low-Reynolds turbulence model would solve the problem, but such models require a very fine mesh near the wall, which is too expensive. If one ever wishes to apply numerical models to real scale problems, one must try to keep them as simple as possible and the number of grid points as small as possible. Therefore, the law-of-the-wall is still applied to bridge the wall layer. However, the validity of this law in sediment-laden turbulent channel flow has been subject of debate since the first experimental data revealed some anomalies. Attempts have been made to solve this problem by modifying the von Karman constant κ (e.g., Vanoni, 1946). As this constant can be related to the statistical distribution of the turbulent fluctuations, i.e., $\kappa = (2\pi)^{-1/2} = 0.4$, assuming a normal distribution (Ni and Wang, 1991), it makes sense that deviations from a normal distribution occur because of the expected asymmetric damping of fluctuations where concentration gradients occur. Others, starting with Coleman (1981), have claimed that the deviation is due to the so-called wake effect, which can be explained, at least in part, by secondary currents generated in experimental flumes.

Another phenomenon which occurs in the benthic boundary layer of cohesive sediment laden turbulent flows is drag reduction; a reduction of the shear velocity is observed, resulting in a thickening of the viscous sub-layer. This subject was studied in the context of CST for the first time by Gust (1976) and is hypothetically attributed to energy dissipation by stretching of the flocs, in analogy with drag reduction by polymers. New experimental data (at concentrations of 4-8 g/l in salt and fresh water) have recently been published (Li and Gust, 2000). Similar work at higher, but uniform concentrations in Karlsruhe has been carried out by Wang et al. (1998).

These problems are crucial for determining the correct shear stress at the bottom, which controls erosion (although some argue that the bottom roughness by topography and bed forms may imply that drag reduction will not be important under field conditions). Erosion rates may thus decrease, not (only) because of saturation, but because the bottom shear stress may become smaller than in clear water, at least under certain conditions; these are still subjects of investigations. Numerical simulations (e.g., Toorman, 1997b) indicate that a standard modeling approach with the k-ε turbulence model applied to transient cases indeed computes a temporary reduction of the bottom shear stress due to the presence of suspended sediment.

Drag reduction also implies that, for the same energy input, the depth-averaged flow rate can increase and the total transport of sediment can be much higher than anticipated, certainly when concentration effects on the bottom shear stress are neglected (as in traditional models).

High concentrations near the bottom may also lead to laminarization over a greater depth, i.e., the formation of a non-Newtonian fluid mud layer. This process has been incorporated into IFREMER's continuous model (Le Hir et al., 2000), and will be incorporated into various other integrated models (e.g., at Delft Hydraulics within the framework of the SILTMAN project, e.g., Winterwerp et al., 2000, and at the K.U.Leuven). Nevertheless, the increase of the mixture viscosity with increasing concentration is not sufficient to explain laminarization.

3.4. Validation

A serious problem for the modelers at this stage is the lack of data to validate new process models. There is a particular need for fully-developed turbulent shear flow experiments with relatively high concentrations of suspended cohesive sediment where profiles of velocity, concentration and turbulent fluctuations are measured. Thus far, the only tests known to the author, where turbulent fluctuations have been measured, are those carried out by Crapper and Ali (1997). New annular flume test will be conducted at the Delft University of Technology within the COSINUS project. Possibly useful are the interesting experiments on settling on an inclined bottom, which have been carried out in Liverpool (Ali and Georgiadis, 1991).

4. BED DYNAMICS

The next problem is how to link the bed shear stress, generated by currents and waves, to the erosion pattern, as there are several modes of erosion (like surface and bulk erosion). Recent flume experiments by the author (Toorman and Luyckx, 1997) indicate that, even qualitatively, not all erosion phenomena have been described sufficiently (such as: internal waves of the fluffy layer prior to erosion, upturning and pealing of the oxidated top layer, and dominance of the bed load of relatively large pieces of ruptured bed material to suspended load after bulk erosion). As erosion changes the bed roughness at micro- and macro-scales, the estimation of the true bed shear stress can be troublesome, even in the laboratory.

The second problem is that of the estimation of the erosion strength of the bed, required to quantify the erosion rates. It is traditionally assumed that the critical fluid stress for erosion is directly related to bed strength, but it is still not known how (despite extensive experimental work), primarily because quantification of the bed erosion strength itself is a problem. Several definitions or characteristics can be used. Traditionally, it has been assumed that the erosion strength can be represented by an empirical relationship between shear strength, determined with a vane test or other appropriate devices, and bed density. In practice, it is nearly impossible to measure the bed strength without disturbing the original structure. Only at the surface of a flat bed can one directly determine the erosion threshold, if one knows the velocity profile. Special devices have been designed to measure surface erosion strength; see Cornelisse et al. (1997) and Gust and Müller (1997).

Furthermore, it has become clear that the bed strength is time dependent due to densification by self-weight consolidation, thixotropy, creep, dynamic loading by waves and tides, mass exchange with the water column due to erosion and deposition and biological activity.

A national experimental program on "strength evolution of the bed" is being conducted in Delft, funded by the Netherlands Technology Foundation (1996-2000), with additional funding through the COSINUS project (with some experiments carried out in Oxford, England). It aims at finding the relationship between laboratory and field conditions. Relationships are sought between bed strength determined with a penetration (sounding) and a vane test, using a controlled stress rheometer, and effective stresses (Merckelbach, 1998).

4.1. Consolidation

In numerical models, the erosion threshold is often computed using density profiles obtained from simplified consolidation models.

The systematic geotechnical study of the self-weight consolidation behavior of estuarine cohesive sediments was pioneered in Oxford by Sills and co-workers (e.g., Been and Sills, 1981; Sills, 1997). The initial conditions are governed by the bed formation history due to settling. Therefore a unifying theory for sedimentation and self-weight consolidation has been developed (Toorman, 1996 and 1999a).

Consolidation experiments have made it evident that the traditional empirical relationships for effective stress and even permeability as functions of density are incorrect. For effective stress this assumption is also inconsistent on theoretical grounds, as its value can never exceed the buoyant weight of the sediment layer above (which includes the condition that its value is always zero at the bed surface, independent on the surface density). Attempts to define new empirical relation-ships for effective stress (e.g., Toorman and Huysentruyt, 1997) do not seem to provide a good alternative. The non-uniqueness of the constitutive relationships is attributed to time-dependent effects. These effects are related to structural changes of the original building blocks of the bed, the flocs, which loose their identity as they break-up under the increasing load, partly releasing their previously immobilized interstitial pore water. Hence, with increasing depth, and thus self-weight load, it becomes less probable that immobilized pore water can remain captured within the floc structure and block pores. This can explain the non-uniqueness of the permeability (Toorman, 1999a). On the other hand, new bonds of bio-physico-chemical nature are formed. In geotechnics one distinguishes between thixotropy (strength variation at constant effective stress and constant volume) and creep (densification at constant effective stress) (Sills, 1995).

As the bed is formed by settling mud flocs, whose structure is controlled by the hydrodynamic conditions and concentration in the water column, the differences in obtained permeabilities and effective stresses for different initial conditions, including scale effects, are slowly being understood. Within the COSINUS project experiments will be conducted in Oxford where the floc structure (determined from video images) will be varied by modifying the floc formation conditions.

Finally, it can be expected that compaction in a shear flow will be higher than in quiescent water as particles are subjected to a horizontal drag which causes the particle to move to a site where it undergoes better interlocking. There is still little experimental evidence for this effect.

4.2. Generalized bed dynamics

Thus far, consolidation models, linked to transport models, are based on traditional geotechnical models using the Terzaghi or the more general Gibson approach. The traditional constitutive equations used in these models are empirical and inadequate. In order to compute the strength development under the various loading forces (currents, waves and gravity), the model should enable the simulation of not only consolidation (strength development), but also liquefaction (structural break-up under shear stresses) and/or fluidization (structural break-up under high excess pore pressures).

In principle, this can be done by using the generalized Biot theory for porous media (Zienkiewicz and Shiomi, 1984), where stress-strain relationships are used as constitutive equations for effective stress and shear stress to describe the soil behavior. It has the additional advantage that time-dependency is introduced in a completely different way, and is expected to yield insight into experimental data. This seems the only consistent way to construct a comprehensive model for the simulation of consolidation, liquefaction and fluidization. Until recently, this approach has only been used for the study of wave effects on an idealized poro-elastic bed (e.g., Yamamoto et al., 1978). A generally valid theory must account for large deformations (e.g., Chen and Mizuno, 1990; Fowler and Noon, 1999), which makes the solution very complex from a mathematical point of view. The construction of such a general bed dynamics model is under investigation at K.U.Leuven within the COSINUS project. Various soil models (visco-elastic-plastic, possibly with addition of creep) will be tested. Thixotropy can be introduced in analogy with thixotropic modeling of fluid mud.

The 1DV bed model, developed within the DYNASTAR project at Delft Hydraulics, has been used to derive erosion criteria (van Kesteren et al., 1997). This model assumes elastic behavior, amongst some other simplifications. However, a pure elastic description can be shown to be equivalent to an effective stress-concentration relationship, which is physically invalid, as mentioned above.

5. PIONEERING AREAS

Certain complexities, particularly the non-uniform grain size distribution of natural mud and biological effects, have hardly been accounted for. In particular, the behavior of the sand-mud mixtures and long-term morphodynamics are still pioneering areas where little progress has been made and which have received much less attention than desired.

5.1. Behavior of sand-mud mixtures

There have been some studies on sedimentation/consolidation and erosion behavior on mixtures at several institutes (e.g., Toorman and Berlamont, 1993; Torfs et al., 1996; Mitchener et al., 1997), but the complexity of this problem is very high and few results can be used in models. A few 2DH models have been developed in which the influence of the mud/sand ratio on the erosion characteristics is included (e.g., Chesher and Ockenden, 1997). A hybrid modeling technique, coupling deterministic models with empirical data, is currently being developed by Delft Hydraulics and Rijkswaterstaat (The Netherlands). The important effect of layering of the bed due to differential settling is not yet included, except for an attempt in the sedimentation/ consolidation model of K.U.Leuven (Toorman and Berlamont, 1993).

5.2. Mudflats dynamics and biological mediation

It has been recognized for a long time that biological factors play an important role in the evolution of floc and bed structure, affecting deposition/erosion behavior. Biological mediation is particularly important for understanding the evolution of mudflats (e.g., Paterson, 1997). This is an area within CST research where recently important initiatives on large joint projects have occurred. The study of mudflat dynamics is an interdisciplinary research (involving physicists, biologists, sedimentologists and engineers) in which much attention is given to the biological aspects and seasonal variations.

In the Netherlands, nine institutes have put their resources together into the "BOA Intertidal" project (1993-1999) funded by the Dutch Organization for Scientific Research (NWO) (Terwindt et al., 1992). Its main emphasis is on the collection and interpretation of data during a complete seasonal cycle. The studies comprise: strength evolution of the sediment by variations in water content (e.g., by exposure, rain, evaporation, ...), biogenic stabilization and meio- and macro-fauna, near-bed turbulence versus suspended concentration, flocculation (effects of biological processes, salinity, sediment concentration, current velocity and turbulence) and quantification of CST by numerical modeling of a tidal basin.

The MAST III INTRMUD project (Dyer, 1998) aims at investigating the characteristics of mudflats in order to establish a classification which reflects the morphological effects of variations in parameters as: tidal range and phase, wave climate, sediment physical and biological properties, biological community structure, etc. A GIS database system has been developed for this purpose. A series of conceptual models for mudflat development will be proposed. *In situ* data have been collected. The resulting algorithms will be implemented in computer models enabling consideration of the effects of climate change, sea level rise and anthropogenic stresses.

The universities of Hamburg and St. Andrews are collaborating in a study on biogenic stabilization of intertidal and subtidal mudflats, with special attention to the movements of bacteria within the porespace of permeable sediments under flow (Gust, personal communication,1998).

5.3. Long-term predictions

Limitations of computer capacity restrict the application of traditional CST models to relatively short periods up to a few days. Strategies for long-term predictions of morphodynamic variations over several years have been developed mainly for non-cohesive sediments. As to cohesive sediments, they can be studied approximately in physical scale models, as has been done for the preservation of the Mont-Saint-Michel island in France, at SOGREAH (1999). A preliminary investigation to develop a modeling methodology was carried out during the MAST II G8M project by Capobianco (1994), based on the analysis of flume experiments. Within the context of mudflat morphodynamics it has also received attention in the INTRMUD project (Roberts and Whitehouse, 2000).

6. CONCLUSIONS

Cohesive sediment transport research in Europe is currently dealing with some of the major problems for a fuller understanding of the behavior of this type of sediment in the aquatic environment. An inventory of the present work allows the classification of research into three major areas: sediment-turbulence interaction, bed dynamics and intertidal mudflats. Over the past few years one can notice a change in approach: there is now much more integrated research in which the interactions of processes are studied. A trend has been set to build detailed "integrated" models in which all physical aspects are described, which allows the study of all the processes in a vertical plane. The results of these virtual laboratories are expected to serve in the near future as data for the validation of "simpler" engineering models. Unfortunately, experimental data generally do not provide the same degree of detail, resulting in a lack of data for validation of these models. In field surveys more attention is given to the simultaneous measurement of a larger number of parameters. The first steps towards integration of biology into engineering research are set with the interdisciplinary studies of intertidal mudflats.

7. ACKNOWLEDGMENT

The author's post-doctoral research position is financed by the Flemish Fund for Scientific Research.

REFERENCES

Ali, K. H. M., and Georgiadis, K., 1991. Laminar motion of fluid mud. *Proceedings of the Institution of Civil Engineers, Part 2, Research and Theory*, Institution of Civil Engineers, London, 795-821.

Been, K., and Sills, G. C., 1981. Self-weight consolidation of soft soils: an experimental and theoretical study. *Géotechnique*, 31(4), 519-535.

Berlamont, J., and Toorman, E., 1998. Prediction of COhesive Sediment transport and bed morphodynamics in estuaries and coastal zones with Integrated NUmerical Simulation models (COSINUS). *Proceedings of the 3rd European Marine Science and Technology Conference* (Lisbon, May 1998), Project Synopses, Vol. II, 585-590, E.C., D.G. XII, Luxembourg.

Capobianco, M., 1994. Evidences of long term effects on flume measurements of mud processes under unsteady currents and proposal for a possible behaviour-oriented model. *MAST II G8M Overall Workshop*, Gregynog, Wales, UK.

Chen, W. F., and Mizuno, E., 1990. *Nonlinear analysis in soil mechanics.* Developments in Geotechnical Engineering No. 53, Elsevier, Amsterdam, 602p.

Chesher, T. J., and Ockenden, M. C., 1997. Numerical modeling of mud and sand mixtures. In: *Cohesive Sediments,* N. Burt, R. Parker and J. Watts eds., John Wiley, Chichester, UK, 395-415.

Coleman, N. L., 1981. Velocity profiles with suspended sediment. *Journal of Hydraulic Research,* 19(3), 211-229.

Cornelisse, J. M., Mulder, H. P., Houwing, E. J., Williamson, H.J., and Witte, G., 1997. On the development of instruments for *in situ* erosion measurements. In: *Cohesive Sediments,* N. Burt, R. Parker and J. Watts eds., John Wiley, Chichester, UK, 175-186.

Crapper, M., and Ali, K. H. M., 1997. A laboratory study of cohesive sediment transport. In: *Cohesive Sediments,* N. Burt, R. Parker and J. Watts eds., John Wiley, Chichester, UK, 197-211.

de Vriend, H. J., ed., 1992. *Abstracts-in-depth.* G6M coastal morphodynamics, final workshop. Delft Hydraulics, Emmeloord, The Netherlands.

Delft Hydraulics Laboratory, 1962. Demerara coastal investigation. *Report*, Delft Hydraulics, Delft, The Netherlands.

Dyer, K. R., 1989. Sediment processes in estuaries: future research requirements. *Journal of Geophysical Research*, 94(C10), 14327-14339.

Dyer, K. R., 1998. The morphological development of intertidal mudflats (INTRMUD). *Proceedings of the 3rd European Marine Science and Technology Conference,* (Lisbon, May 1998), Project Synopses, Vol.II:521-532, E.C., D.G. XII, Luxembourg.

Fowler, A. C., and Noon, C. G., 1999. Mathematical models of compaction, consolidation and regional groundwater flow. *Geophysical Journal International*, 136, 251-260.

Galland, J.-C., 1996. Transport de sédiments en suspension et turbulence. *Report HE-42/96/007/A,* LNH (EDF-DER), Chatou, France (in French).

Gust, G., 1976. Observations on turbulent drag reduction in a dilute suspension of clay in sea-water. *Journal of Fluid Mechanics*, 75(1), 29-47.

Gust, G., and Müller, V., 1997. Interfacial hydrodynamics and entrainment functions of currently used erosion devices. In: *Cohesive Sediments,* N. Burt, R. Parker and J. Watts eds., John Wiley, Chichester, UK, 149-174.

Hamm, L., Chesher, T. J., Jakobsen, F., Peltier, E., and Toorman, E., 1996. Data analysis for cohesive sediment transport modeling. *MAST II G8M Coastal*

Morphodynamics, Project 4 - Topic E: Mud Morphodynamic Modeling. Report No.52184R7, SOGREAH, Grenoble, France.

Kranenburg, C., 1994. The fractal structure of cohesive sediment aggregates. *Estuarine, Coastal and Shelf Science*, 39, 451-460.

Kranenburg, C., 1998. Saturation concentrations of suspended fine sediment. Computations with the Prandtl mixing-length model. *Report No.5-98*, Department of Civil Engineering, Delft University of Technology, Delft, The Netherlands.

Kranenburg, C., and Winterwerp, J. C., 1997. Erosion of fluid mud layers. I: entrainment model. *Journal of Hydraulic Engineering*, 123(6), 504-511.

Kranenburg, C., and Bruens, A., 1999. A flume experiment on CBS dynamics. *COSINUS Annual General Meeting*, Book of Abstracts, Grenoble, France, 21-22.

Krone, R. B., 1962. Flume studies of the transport of sediment in estuarial shoaling processes. *Final Report*, Hydraulic Engineering Laboratory and Sanitary Engineering Research Laboratory, University of California, Berkeley, 118p.

Le Hir, P., 1997. Fluid and sediment "integrated" modeling: application to fluid mud flows in estuaries. In: *Cohesive Sediments,* N. Burt, R. Parker and J. Watts eds., John Wiley, Chichester, UK, 417-428.

Le Hir, P., Bassoullet, P., Jestin, H., and Sottolichio, A., 2000. Application of the continuous modeling concept to simulate high-concentration suspended sediment in a macrotidal estuary. (this volume).

Li, M. Z., and Gust, G., 2000. Boundary layer dynamics and drag reduction in flows of high cohesive sediment suspensions. *Sedimentology*, 47, 71-86.

Malcherek, A., 1996. Numerical modeling of floc dynamics. In: *Three Dimensional Numerical Modeling of Cohesive Sediment Transport Processes in Estuarine Environments,* E. Peltier et al. eds., Final Report to the EC contract MAS2-CT92-0013, LNH (EDF-DER), Chatou, France, 34-45.

Merckelbach, L. M., 1998. Consolidation and strength evolution of Caland-Beer mud. Measurement report of laboratory experiments. *Report No. 7-98*, Department of Civil Engineering, Delft University of Technology, Delft, The Netherlands, 44p.

Migniot, C., 1968. Etude des propriétés physiques de différents sediments très fins et leur comportement sous des actions hydrodynamiques (A study of the physical properties of various forms of very fine sediment and their behavior under hydrodynamic actions). *La Houille Blanche*, 7, 591-620.

Mitchener, H., Torfs, H., and Whitehouse, R. J. S., 1997. Erosion of mud/sand mixtures. *Coastal Engineering*, 29 (1-2), 1-26.

Munk, W. H., and Anderson, E. A., 1948. Notes on a theory of the thermocline. *Journal of Marine Research*, 3(1), 276-295.

NedeCo, 1965. Siltation of the Bangkok port channel. Vol.11, NedeCo, Den Haag.

NedeCo, 1968. Surinam transportation study. Hydraulic investigation. *Report,* NedeCo, The Hague, The Netherlands.

Ni, J. R., and Wang, G. Q., 1991. Vertical sediment distribution. *Journal of Hydraulic Engineering*, 117(9), 1184-1194.

Odd, N. V. M., 1988. Mathematical modeling of mud transport in estuaries. In: *Physical Processes in Estuaries,* J. Dronkers and W. van Leussen eds., Springer Verlag, Berlin, 503-531.

Odd, N. V. M., and Owen, M. W. 1972. A two-layer model for mud transport in the Thames estuary. *Proceedings of the Institution of Civil Engineers*, Supplement paper 7517S, London, 175-205.

Owen, M. W., 1966. A study of the properties and behaviour of muds. Literature review. *Report INT61*, Hydraulics Research Station, Wallingford, UK.

Owen, M. W., and Odd, N. V. M., 1972. A mathematical model of the effect of a tidal barrier on siltation in an estuary. In: *Tidal Power*, T. J. Gray and O. K. Gashus eds., Plenum Press, New York, 456-485.

Paterson, D. M., 1997. Biological mediation of sediment erodibility: ecology and physical dynamics. In: *Cohesive Sediments,* N. Burt, R. Parker and J. Watts eds., John Wiley, Chichester, UK, 215-229.

Peltier, E., Le Normant, C., Teisson, C., Malcherek, A., Markofsky, M., Zielke, W., Cornelisse, J., Molinaro, P., Corti, S., and Greco, G., 1996. Three-dimensional numerical modeling of cohesive sediment transport processes in estuarine environments. *Final report to the EC MAST2 contract MAS2-CT92-0013*, LNH, EDF-DER, Chatou, France, 253p.

Postma, H., 1967. Sediment transport and sedimentation in the estuarine environment. In: *Estuaries,* G.H. Lauff ed., Publication No. 83, American Association for the Advancement of Science, Washington, DC, 158-179.

Roberts, W. R., and Whitehouse, R. J. S., 2000. Predicting the profile of intertidal mudflats formed by cross-shore tidal currents. (this volume).

Sills, G. C., 1995. Time dependent processes in soil consolidation. In: *Compression and Consolidation of Clayey Soils,* H. Yoshikuni and O. Kusakabe eds., Balkema, Rotterdam, 875-890.

Sills, G. C., 1997. Consolidation of cohesive sediments in settling columns. In: *Cohesive Sediments,* N. Burt, R. Parker and J. Watts eds., John Wiley, Chichester, UK, 215-229.

SOGREAH, 1999. Etude hydrosédimentaire & Interview. *La Baie, J. de l'Opération.* No. Mars 99:6-9, Syndicat Mixte pour le Rétablissement du Caractêre Maritime du Mont-Saint-Michel, Caen, France.

Stive, M. J. F., DeVriend, H. J., Fredsoe, J., Hamm, L., Soulsby, R. L., Teisson, C., and Winterwerp, J. C. eds., 1996. Advances in coastal morphodynamics. An overview of the G8 Coastal Morphodynamics Project. *Final report to the EC MAST2 contract MAS2-CT92-0027*, Delft Hydraulics, Delft, The Netherlands.

Teisson, C., Simonin, O., Galland, J. C., and Laurence, D., 1992. Turbulence and mud sedimentation: a Reynolds stress model and a two-phase flow model. *Proceedings of the International Conference on Coastal Engineering*, ASCE, New York, 2853-2866.

Terwindt, J. H. J., Heip, C. H. R., and Zimmerman, J. T. F., 1992. BOA research theme on tidal areas. The physical and biodynamical behaviour of mud and sand beds in tidal inlets and flats. *Report*, Netherlands Institute for Sea Research, Den Burg.

Toorman, E. A., 1996. Sedimentation and self-weight consolidation: general unifying theory. *Géotechnique*, 46(91), 103-113.

Toorman, E. A., 1997a. Modeling the thixotropic behaviour of dense cohesive sediment suspensions. *Rheologica Acta*, 36(1), 56-65.

Toorman, E. A., 1997b. Eddy viscosity models for sediment-laden turbulent flow. *Seminar on New Developments and Validation in Turbulence Modeling*, ERCOFTAC Belgian Pilot Centre, Gent, Belgium, 24p.

Toorman, E. A. 1999a. Sedimentation and self-weight consolidation: constitutive equations and numerical modelling. *Géotechnique,* 49(6), 709-726.

Toorman, E. A., 1999b. Sediment-induced damping in turbulent flow of suspensions. 1. The "Siltman" testcase and the concept of "saturation". *COSINUS Project Report HYD/ET99.02*, Hydraulics Laboratory, K.U.Leuven, Heverlee, Belgium.

Toorman, E. A., and Berlamont, J. E., 1993. Settling and consolidation of mixtures of cohesive and non-cohesive sediments. In: *Advances in Hydro-Science and Engineering,* S. Wang ed., University of Mississippi, University, Mississippi, 606-613.

Toorman, E. A., and Huysentruyt, H., 1997. Towards a new constitutive equation for effective stress in self-weight consolidation. In: *Cohesive Sediments,* N. Burt, R. Parker and J. Watts eds., John Wiley, Chichester, UK, 121-132.

Toorman, E. A., and Luyckx, G., 1997. Erosion sensitivity of bottom sediments. Laboratory experiments. *Container Dock West. Hydraulic and Sedimentological Research. Report HYD/ET97.3 to IMDC*, Hydraulics Laboratory, K.U.Leuven, Heverlee, Belgium, 39p (in Dutch).

Torfs, H., Mitchener, H., Huysentruyt, H., and Toorman, E., 1996. Settling and consolidation of mud/sand mixtures. *Coastal Engineering*, 29(1-2), 27-45.

Uittenbogaard, R. E., 1995. The importance of internal waves for mixing in a stratified estuarine tidal flow. *Ph.D. Thesis*, Technical University of Delft, Delft, The Netherlands.

van Kesteren, W., Cornelisse, J. C., and Kuijper, C., 1997. Dynastar bed model: bed strength, liquefaction and erosion. *Cohesive Sediment Report No.53* to Rijkswaterstaat, Delft Hydraulics.

van Leussen, W., 1994. Estuarine macroflocs and their role in fine-grained sediment transport. *Ph.D. Thesis*, University of Utrecht, Utrecht, The Netherlands, 488p.

Vanoni, V. A., 1946. Transportation of suspended sediment by water. *Transactions of ASCE*, 111, 67-133.

Wang, Z. Y., Larsen, P., Nestmann, F., and Dittrich, A., 1998. Resistance and drag reduction of flows of clay suspensions. *Journal of Hydraulic Engineering*, 124(1), 41-49.

Winterwerp, J. C., 1998a. A simple model for turbulence-induced flocculation of cohesive sediment. *Journal of Hydraulic Research*, 36(3), 309-326.

Winterwerp, J. C., 1998b. Sediment-fluid interactions: a literature survey. *Report* Z2386 (MAST 3 COSINUS project), Delft Hydraulics, Delft, The Netherlands, 24p.

Winterwerp, J. C., 1998c. SILTMAN: Analysis of field measurements. *Report* Z2263, Delft Hydraulics, The Netherlands, 40p + figs.

Winterwerp, J. C., 1999. On the dynamics of high-concentrated mud suspensions. *Ph.D. Thesis*, Technical University of Delft, Delft, The Netherlands, 180p + appendices.

Winterwerp, J. C., and Kranenburg, C., 1997. Erosion of fluid mud layers. II: Experiments and model validation. *Journal of Hydraulic Engineering*, 123(6), 512-519.

Winterwerp, J. C., Uittenbogaard, R. E., and de Kok, J. M., 2000. Rapid siltation from saturated mud suspensions. (this volume).

Wolanski, E., Chappell, J., Ridd, P., and Vertessy, R., 1988. Fluidization of mud in estuaries. *Journal of Geophysical Research*, 93(C3), 2351-2361.

Yamamoto, T., Koning, H. L., Sellmeijer, H., and van Hijum, E., 1978. On the response of a poro-elastic bed to water waves. *Journal of Fluid Mechanics*, 87(1), 193-206.

Zienkiewicz, O. C., and Shiomi, T., 1984. Dynamic behaviour of saturated porous media: the generalized Biot formulation and its numerical solution. *International Journal for Numerical and Analytical Methods in Geomechanics*, 8, 71-96.

Coastal and Estuarine Fine Sediment Processes
W.H. McAnally and A.J. Mehta (Editors)
2001 Elsevier Science B.V.

Collisional aggregation of fine estuarial sediment

W. H. McAnally[a] and A. J. Mehta[b]

[a]Coastal and Hydraulics Laboratory, U.S. Army Engineer Waterways Experiment Station, 3909 Halls Ferry Road, Vicksburg, Mississippi 39180, USA

[b]Department of Civil and Coastal Engineering, University of Florida, 345 Weil Hall, P.O. Box 116580, Gainesville, Florida 32611, USA

The rate of estuarial fine sediment aggregation is related to the frequency of particle collisions, which is a function of fluid, flow and particle characteristics plus the number concentration of particles. Most are two-body collisions, but three-body collisions can play a significant role in high-concentration suspensions and can be calculated from standard statistical theory. Experimental data suggest that the efficiency of aggregation, i.e., that fraction of collisions resulting in a bond between the colliding particles, spans five orders of magnitude; however, a new formulation replaces the traditional efficiency. This formulation partitions the efficiency through a non-dimensional analysis of the significant parameters in collision, aggregation, and disaggregation. Analysis shows that the new formulation successfully reproduces some of the limited aggregation rate data available.

1. INTRODUCTION

Fine sediments in estuaries are mixtures of inorganic minerals, organic materials, and biochemicals. Mineral grains consist of clays (e.g., montmorillonite, illite, and kaolinite) and non-clay minerals (e.g., quartz and carbonate). Organic materials include plant and animal detritus and bacteria. The relative organic/non-organic composition of estuarial sediments varies over wide ranges between estuaries and within the same estuary spatially and seasonally.

The characterizing feature of these sediments is that the individual primary grains are too small to settle under their own weight when suspended in water, since any disturbance, even Brownian motion, will tend to keep them suspended. Clay mineral grains exhibit cohesion, which causes them to bond if they collide, forming aggregates of grains. These aggregates may grow through continuing collision and aggregation to a size which enables them to settle. The process by which cohesive grains collide and bind to each other by physico-chemical forces into larger particles is termed aggregation.

This paper presents an analytic description for fine estuarial sediment interparticle collisions. The description relates the number of aggregating collisions, those that produce a larger particle by bonding colliding particles, to fundamental characteristics of fluid, flow and particles. The roles of biological organisms, organic matter, and biochemicals in aggregation are not addressed here.

2. COHESION

Cohesion describes the tendency of clayey sediment grains to bind together (aggregate) under some circumstances, which significantly affects sediment behavior, as described below. In general, smaller sized grains are more cohesive, with diameters greater than 40 μm essentially cohesionless, and cohesion becoming progressively more important as grain size decreases below 20 μm (Mehta and Lee, 1994).

Cohesion as a property of clay minerals arises from surface electrical charges on the grains, and one measure of cohesion is in terms of the ease with which cations held within the lattice can be exchanged for more active cations in the surrounding fluid--the cation exchange capacity (CEC) being expressed in milliequivalents per 100 g of clay. Table 1 lists the four most common clay minerals, their characteristic size, their CEC, and the salinity critical to aggregation (also called flocculation or coagulation). Aggregation of colliding clay mineral particles occurs when the sum of forces (including van der Waals attraction and the coulombic repulsion of the grains' ion clouds) results in a net attractive force.

Table 1
Common clay minerals and characteristics

Clay Mineral	Typical grain size (μm)	Equivalent (nominal) diameter (μm)	Cation exchange capacity (meq/100 g)	Critical salinity for aggregation (ppt)
Kaolinite	1 by 0.1	0.36	3 – 15	0.6
Illite	0.01 by 0.3	0.062	10 – 40	1.1
Smectite (Montmorillonite)	0.001 by 0.1	0.011	80 – 150	2.4
Chlorite	0.01 by 0.3	0.062	24 – 45	--

Sources: CTH, 1960; Grim, 1968; Ariathurai et al., 1977.

The degree of cohesion displayed by fine sediment is a function of the sediment CEC and fluid properties, most notably salinity, and cohesion of clayey sediments may be modulated by metallic or organic (biochemical) coatings on the particles (Gibbs, 1977; Kranck, 1980). However, those effects are not directly considered here.

3. AGGREGATE CHARACTERISTICS

The density of a fine sediment aggregate may diminish with increasing size (Krone, 1963; Gibbs, 1985; Kranck et al., 1993), and has been expressed by an equation of the form (McAnally, 1999):

$$\rho_j = \text{smaller of} \begin{cases} \rho_g \\ \rho + B_\rho \left(\dfrac{D_g}{D_j} \right)^{3-n_f} \end{cases} \tag{1}$$

where ρ_j = density of the jth size class, ρ = fluid density, ρ_g = sediment grain density, B_ρ = sediment-dependent coefficient, D_g = grain diameter, D_j = diameter of the jth size class, and n_f = fractal dimension, usually about 2.

While the form of (1) has been widely used (e.g., see Tambo and Watanabe, 1979; McCave, 1984; Kranenburg, 1994), the coefficients for each range vary from one data set to another; thus the equation must be fit to the sediment to be modeled.

Aggregate strength (resistance to disaggregation) is a function of grain-to-grain cohesion, size and orientation of particles within the aggregate, and organic content. Experimental results (e.g., Krone, 1963; Hunt, 1982; Krone, 1983; Mehta and Parchure, 2000) have shown that as aggregate size and organic content increase, both aggregate density and aggregate strength decrease. Partheniades (1993) reported that Krone's (1963) data for critical yield stress of San Francisco Bay sediment aggregates fit a simple power-law expression. Here a power-law employing the fractal approach (Kranenburg, 1994; Winterwerp, 1999) has been adopted for particle strength by size class:

$$\tau_j = B_\tau \left(\frac{\Delta\rho_j}{\rho} \right)^{\frac{2}{3-n_f}} \tag{2}$$

where $\tau_j = j^{th}$ size class aggregate strength, and B_τ = sediment-specific coefficient.

4. COLLISIONS

Given a suspension of cohesive grains with sufficient dissolved salts and enough grains to permit aggregation, three mechanisms are primarily responsible for collisions that can lead to aggregation: a) Brownian motion, which affects grains and small aggregates of only a few grains and is thus most important in the early stages of aggregation and nearly quiescent waters; b) viscous or turbulent fluid shear, which brings two particles together as the gradient in local velocity permits one particle to overtake and capture the other; and c) differential settling, which results in collisions as faster-settling particles overtake slower-settling ones and capture them. Before two suspended particles can make physical contact, fluid must flow out of the narrowing gap between the particles. The pressure increase required to force the fluid out exerts repelling forces on the particles and may or may not prevent a collision, depending on fluid viscosity and the particles' positions, porosities, masses, and relative velocity.

4.1. Two-body collisions

The frequency of collisions between two particles can be expressed by (Smoluchowski, 1917):

$$N_{im} = \alpha_a \beta_{im} n_i n_m \tag{3}$$

where N_{im} = number of collisions between i and m class particles per unit time per unit volume, α_a = aggregation efficiency factor, which is discussed further below, β_{im} = collision frequency function, dependent on particle diameters and system characteristics, i, m = indices for i and m size class particles, respectively, and n_i, n_m = number concentration of i and m class particles, respectively.

The collision frequency function, β_{im}, can be calculated by a simple analysis of particle motions under several modes of collision. The analysis begins with two idealized spherical particles as shown in Figure 1--one from the ith size class (the i particle) and one from the mth size class (the m particle). We surround the m particle with a collision sphere of diameter:

$$D_{c,im} = F_c (D_i + D_m) \tag{4}$$

where F_c = collision diameter function, with a value between 0 and 1, and D_i, D_m = diameter of the i and m particles, respectively. The two particles will experience a close encounter if their relative motion causes particle i to intrude within the collision sphere of particle m.

Brownian motion. The collision frequency function for Brownian motion can be treated as a case of Fickian diffusion (Smoluchowski, 1917):

$$\beta_{B,im} = 4\pi E_{im} F_c (D_i + D_m) \tag{5}$$

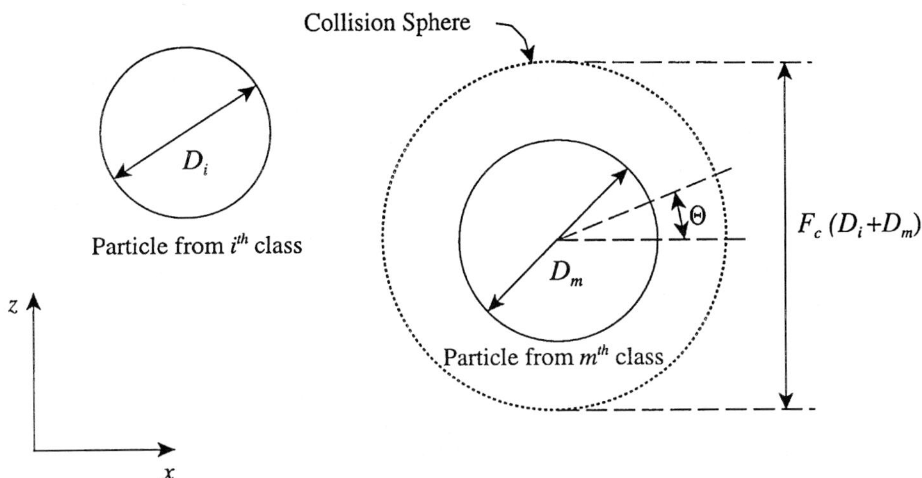

Figure 1. Definition of terms for two-body particle encounters.

where E_{im} = relative diffusion coefficient for the two particles, given by (Overbeek, 1952):

$$E_{im} = E_g D_g \left(\frac{1}{D_i} + \frac{1}{D_m} \right) \tag{6}$$

where E_g = Brownian diffusion coefficient of the primary grain = $\kappa T/3\pi\mu D_g$, κ = Boltzman constant, T = absolute temperature in deg. K, μ = dynamic viscosity of the fluid, and D_g = primary grain diameter. Substituting (6) and E_g into (5) yields the two-body collision frequency function for Brownian motion:

$$\beta_{B,im} = \left[\frac{2}{3} \; \frac{\kappa T F_c}{\mu} \right] \frac{(D_i + D_m)^2}{D_i D_m} \tag{7}$$

Flow shear. Taking the center of the m particle in Figure 2 as the coordinate origin moving at the flow speed in the x direction, the transport rate of i particles by turbulent flow into the m particle's collision sphere gives the rate of shear-induced collisions (Saffman and Turner, 1956):

$$N_{im} = 2n_i \int_0^{\pi/2} u_i \pi F_c^2 (D_i + D_m)^2 \cos\Theta d\Theta \tag{8}$$

where Θ = angle between x axis and a location on the sphere's surface, and u_i = velocity of the i particle relative to center of the m particle, given by:

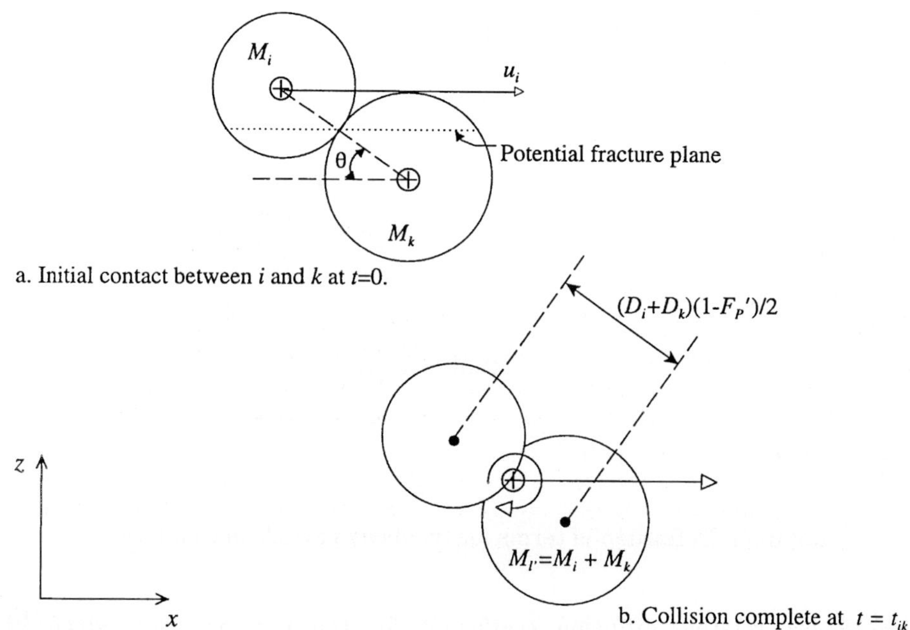

Figure 2. Schematic of two-body collision.

$$u_i = \frac{(D_i + D_m)}{2} \left|\frac{du}{dx}\right| \tag{9}$$

In (9) it is assumed that the two particles are approximately the same size, they do not influence each other's motion, and energy is isotropically dissipated through eddies much smaller than $D_i + D_m$. If du/dx is normally distributed and (9) is substituted into (8), the resulting flow shear collision frequency function is:

$$\beta_{S,im} = \left[\frac{\pi F_c^2}{4} \sqrt{\frac{2}{15\pi}}\right] \sqrt{\frac{\varepsilon}{\nu}}\,(D_i + D_m)^3 \tag{10}$$

where ε = turbulent energy dissipation rate, and ν = kinematic viscosity of the fluid.

Differential settling. If the i and m particles in Figure 2 have different settling velocities, the number of i particles passing into the m particle collision sphere by differential settling alone is described by (8) with the velocity replaced by the difference in settling velocities between the two particles (McCave, 1984).

Integrating that equation over the collision sphere surface yields the collision frequency function for differential settling:

$$\beta_{D,im} = \left[\frac{\pi F_c^2}{4}\right] (D_i + D_m)^2 \left|W_{s,i} - W_{s,m}\right| \tag{11}$$

where $W_{s,i}$, $W_{s,m}$ = settling velocity of the i and m particles, respectively.

4.2. Three-body collisions

The rate of three-body collisions is calculated in terms of the probability of two essentially simultaneous two-body collisions--an i particle collides with a k particle and during that collision an m particle also collides with the same k particle. The two collisions are independent events, where $EV(iK)$ is the event in which any i particle collides with a particular k particle, referred to as K, and $EV(Km)$ is the event in which the K particle collides with any m particle. The individual probabilities of those two events occurring in any given time interval t can be expressed as (Clarke and McChesney, 1964):

$$Pr[EV(iK)] = \frac{N_{ik}}{n_k}\ t \tag{12}$$

The probability of a three-body collision is then the probability that a Km collision occurs in the time interval over which an iK collision occurs. Since events $EV(iK)$ and $EV(Km)$ are statistically independent, the probability of their intersection, namely, the probability of an iKm collision during the same time interval, is:

$$\begin{aligned} Pr[EV(iKm)] &= Pr[EV(iK) \cap EV(Km)] \\ &= Pr[EV(iK)]Pr[EV(Km)] \\ &= \frac{N_{ik}t_{ik}}{n_k}\ \frac{N_{km}t_{km}}{n_k} \end{aligned} \tag{13}$$

where t_{ik}, t_{km} = duration of iK and Km collisions, respectively. Equation (13) can be recast into collision frequency by:

$$\begin{aligned} N_{ikm} &= Pr[EV(iKm)]\ \frac{n_k}{t_{ikm}} \\ &= N_{ik}N_{km}\ \frac{t_{ik}t_{km}}{n_k t_{ikm}} \end{aligned} \tag{14}$$

where t_{ikm} = total duration of the three-body collision.

If the particles were cohesionless inelastic bodies, the collision duration t_{ik} and thus the probability of a three-body collision would be essentially zero. However, porous cohesive aggregates will intermesh during a collision and thus experience a finite collision time between the moment of first contact and the moment at which enough grain-to-grain contact has been made to halt further interpenetration. Referring to Figure 2, if the two-body collision begins when the *i* particle first touches the *k* particle and ends when it has penetrated some distance into the *k* particle, then an *i-k* collision duration will be approximately:

$$t_{ik} = \frac{(D_i + D_k)F_p}{2\ u_i} \tag{15}$$

where $F_p = F_p' \cos\theta, F_p'$ = coefficient representing the relative depth of interparticle penetration, θ = angle between direction of u_i and the line connecting the *i* and *k* particle centers, and u_i = velocity of *i* particle relative to the *k* particle, given by:

$$u_i = \begin{cases} \dfrac{\kappa T}{3\pi\mu D_i D_k} & \text{Brownian Motion} \\ \dfrac{D_i + D_k}{2}\sqrt{\dfrac{2\varepsilon}{15\pi\nu}} & \text{Flow Shear} \\ \left|W_{s,i} - W_{s,k}\right| & \text{Differential Settling} \end{cases} \tag{16}$$

The value of F_p in (15) will be a function of particle momentum and the number and strength of intra-particle bonds of the colliding particles. For the present purposes, F_p is estimated to be about 0.1, following Krone (1963).

Considering now the three-body collision schematized in Figure 3, if the *ik* and *km* collisions are simultaneous, then the minimum value of t_{ikm} will be the larger of t_{ik} and t_{km} and will be best approximated over many collisions by $(t_{ik} + t_{km})/2$. The maximum possible value of t_{ikm} will occur if the two collisions are sequential, with the *km* collision beginning at the end of the *ik* collision, and $t_{ikm} = t_{ik} + t_{km}$. Assuming a normal distribution for both t_{ik} and t_{km}, the mean value of t_{ikm} can be estimated to be $3(t_{ik} + t_{km})/4$. Using these time values, the time ratio term of (14) becomes:

$$\frac{t_{ik}t_{km}}{t_{ikm}} = \frac{2F_p(D_i + D_k)(D_k + D_m)}{3[u_m(D_i + D_k) + u_i(D_k + D_m)]} \tag{17}$$

where u_m = velocity of the *m* particle relative to the *k* particle. Substituting (17) into (14) and gathering terms yields the equation for three body collision frequency:

$$N_{ikm} = \alpha_{ikm} N_{ik} N_{km} \tag{18}$$

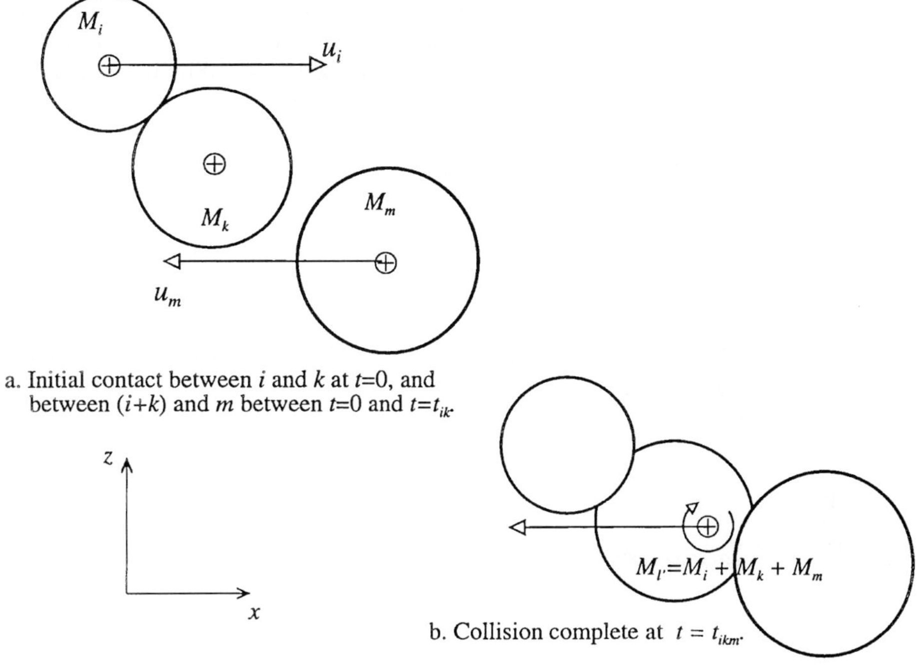

Figure 3. Schematic of three-body collision.

where α_{ikm} is the three-body collision efficiency parameter given by:

$$\alpha_{ikm} = \frac{2F_p(D_i + D_k)(D_k + D_m)}{3n_k[u_m(D_i + D_k) + u_i(D_k + D_m)]} \tag{19}$$

The number of three-body collisions in sediment suspensions is usually small in comparison with the number of two-body collisions. Nevertheless three-body collisions can significantly affect aggregation processes when the sediment particle size distribution is near equilibrium McAnally (1999).

4.3. Four-body collisions

The logic used to derive (14) can also be used to develop an expression for the frequency of four-body collisions (McAnally, 1999):

$$N_{iklm} = \frac{t_{kl}t_{ikm}}{t_{iklm}} \frac{N_{kl}N_{ikm}}{n_k} \tag{20}$$

where t_{iklm} = duration of the four-body collision. McAnally (1999) concluded that the number of four-body and higher order collisions as a percentage of three-body collisions will be very small, and they are not included in the following analysis.

5. SHEAR STRESSES ON AGGREGATES

5.1. Two-body-collision-induced stresses

Figure 2 illustrates the collision of two particles that bond during the collision. At time $t = 0$ a particle of mass M_i, moving at speed u_i relative to a particle of mass M_k, touches the k particle. At time $t = t_{ik}$ the combined particle has mass $M_{ik} = M_i + M_k$ and translational velocity u_{ik}, which by conservation of linear momentum is given by:

$$u_{ik} = \frac{u_i M_i}{M_i + M_k} \tag{21}$$

and by Newton's second law:

$$F_{ik} = \frac{u_{ik} M_k}{t_{ik}} \tag{22}$$

If either particle breaks as a result of the applied force the probable fracture plane will intersect the contact point, and run through the minimum area path of the particle as shown in Figure 2. Combining the above two equations, using (15) for t_{ik}, and dividing by the area of the fracture surface in Figure 2 yields the collisional shear stress experienced by the k aggregate:

$$\tau_{ik,k} = \frac{8u_i^2 M_i M_k}{\pi F_p D_k^2 (D_i + D_k)(M_i + M_k)} \tag{23}$$

5.2. Three-body-collision-induced stresses

The collision of three bodies has more degrees of freedom than the two-body case, in part since the angle of approach of the second particle can vary through 360 degrees; however, if we consider only the most physically probable collision sequence, which is shown in Figure 3, the shear stress experienced by the k particle is:

$$\tau_{ikm,k} = \frac{8}{\pi F_p D_k^2} \left\{ \frac{M_i M_k u_i^2}{(D_i + D_k)(M_i + M_k)} + \frac{(M_i + M_k) u_m}{(D_k + D_m)} \left[\frac{u_i M_i + u_m M_m}{M_i + M_k + M_m} - \frac{u_i M_i}{M_i + M_k} \right] \right\} \tag{24}$$

For the case of two or three same-size particles colliding, dividing (24) by (23) shows that three-body collisions produce shear stresses up to two-thirds greater than two-body shear stresses. Thus, even though the number of three-body collisions under typical conditions has been shown to be substantially smaller than the number of two-body collisions (McAnally, 1999), their net effect on disaggregation may be significant. This theoretical finding is consistent with the experimental finding of Burban et al. (1989) that three-body collisions are important for disaggregation, but not for aggregation.

5.3. Flow-induced stress

A suspended particle in a vertical velocity gradient will experience a net torque from differential drag on its top and bottom surfaces and rotate about its center until the rotational drag applied on the downstream and upstream surfaces balances the applied drag on the surfaces parallel to the flow. Assuming a linear velocity gradient across the diameter, Krone (1963) derived the maximum flow-induced shear stress in a spherical particle as:

$$\tau_u = \frac{\mu}{8}\frac{du}{dz} \tag{25}$$

where du/dz = viscous flow velocity gradient across particle. The shearing rate near the bed, where du/dz is greatest, can be adequately described by a viscous relationship such as the law of the wall.

6. COLLISION EFFICIENCY AND COLLISION DIAMETER FUNCTION

An accurate determination of collision efficiency, α_a, in (3) and collision diameter function, F_c, in (7), (10), and (11) is essential for a successful description of fine sediment aggregation. A new formulation of the collision efficiency and collision diameter function is given here, based on the following assumptions: a) data available in the literature on collision efficiency actually represent the net effect of separate collision, aggregation, and disaggregation efficiencies; b) collision efficiency, representing the fraction of near encounters that result in collisions, is a function of fluid, flow, and sediment physical properties; c) aggregation efficiency, representing the fraction of collisions that result in bonding between the colliding particles, is a function only of degree of cohesion; d) for sub-20 μm mineral grains with some organic material in suspension of at least 1 ppt salinity, aggregation efficiency is assumed to be equal to one; e) disaggregation efficiency, representing the fraction of aggregating collisions that result in disaggregation of at least one of the colliding particles, is a function of the particle shear strength and the shearing forces imposed by the flow and collisions.

The literature on aggregation contains a few data points on aggregation efficiencies in shear-dominated flows that are fit to a new efficiency term α' as given in a rearranged form of (3):

$$N_m = \alpha'\left(\frac{\beta_{S,im}}{F_c^2}\right) n_i n_m \tag{26}$$

where

$$\alpha' = \alpha_a \alpha_d F_c^2 \tag{27}$$

and where the term inside the brackets of (26) is approximately equal to the traditional flow shear collision frequency functions as employed in the literature, α_a = aggregation efficiency (the fraction of collisions that result in a bond), α_d = disaggregation efficiency (the fraction of collisions that result in disaggregation of some of the colliding particles), and F_c = the collision diameter function of (4).

Assumption (d) can be expressed as:

$$\alpha_a = \begin{cases} f_g & S \geq 1 \ ppt \\ f_g \dfrac{S}{S_0} & S < 1 \ ppt \end{cases} \tag{28}$$

where f_g = decimal weight fraction of material in the suspension that is cohesive, i.e., with elemental grain diameter less than 20 μm; S_0 = reference salinity and S = fluid salinity.

The difficulties with determining collision efficiency are demonstrated by the fact that the literature reports values of α' ranging from 1x10^{-5} to 1 (Edzwald et al., 1974; Zeichner and Schowalter, 1977; Adler, 1981; McCave, 1984; O'Melia, 1985; Han, 1989; Lick et al., 1992). However, the analyses yielding the low end of that range of efficiencies have lumped a number of assumptions and effects (e.g., hydrodynamic effects, disaggregation, heterogenous sizes, and degree of sediment cohesion) into that single parameter, making it an extremely rough calibration coefficient for which there is no intuitive estimate or consistent theoretical expression of reasonable values. The more approximate the calculations, the wider range of efficiencies needed to reproduce the data.

Inspection of the variables listed in assumptions (a)-(e) above yields several non-dimensional parameters that may contribute to the probability of an aggregating collision between two encountering particles. They are here expressed by (28) plus:

$$F_c^2 = \Pi_c\left[\left(\frac{\Delta\rho_i D_i^3}{\Delta\rho_m D_m^3}\right), \left(\frac{u_i(D_i + D_m)}{\nu}\right), \left(\frac{D_g}{D_1}\right), \left(\frac{S}{S_0}\right), \left(\frac{T_c}{T_0}\right), \left(\frac{CEC}{CEC_0}\right)\right] \tag{29}$$

and

$$\alpha_d = \Pi_d\left[\frac{\tau_m}{\tau_{im,m} + \tau_u}\right] \tag{30}$$

where Π indicates some function of the bracketed non-dimensional terms; D_g = diameter of primary grain; D_1 = reference particle size; S_0 = reference salinity; T_c = fluid temperature, deg Celsius; T_0 = reference temperature; and CEC_0 = reference cation exchange capacity.

The grain size and *CEC* of sediment samples can be measured, but often are not, so they are here estimated by the respective expressions:

$$\begin{aligned} D_g(\text{sample}) &= 0.36f(K) + 0.062f(I) + 0.011f(M) + 0.062f(Ch) \\ CEC(\text{sample}) &= 9f(K) + 25f(I) + 115f(M) + 34f(Ch) \end{aligned} \tag{31}$$

where $f(K)$, $f(I)$, $f(M)$, $f(Ch)$ = weight fractions of the sample composed of kaolinite, illite, montmorillonite, and chlorite, respectively, and the constants are representative values of equivalent grain diameter and *CEC* for the minerals indicated. These calculations are approximate since the grain size and *CEC* for a given mineral may vary considerably from the nominal values employed, but they do provide some measure of the mineral-specific characteristics of the sediment.

Data on collision efficiency for real sediments (i.e., not manufactured materials such as latex grains or pure mineral sediments such as kaolinite) were compiled from the literature by McAnally (1999) and the resulting 40 data sets were used to calculate the non-dimensional parameters listed in Table 2. The experiments of Edzwald et al. (1974), Gibbs (1985), and Tsai and Hwang (1995) measured efficiencies for the earliest stages of the aggregation process, where particle sizes would be small, but the former two did not measure particle sizes of the aggregated suspension. The median particle size for those experiments have been estimated here by using the equations of Lick et al. (1992) to calculate: a) an equilibrium particle size for the experimental conditions, and b) the time to achieve that equilibrium size; then estimating the median particle size at the end of the measurement period by interpolation over time using the time history curves of Burban et al. (1989) as representative of the growth rate.

Tsai et al. (1987) and Burban et al. (1989) measured the grain size distribution of the samples and then fit an aggregation rate equation to the measured sizes class by class to obtain two sets of collision efficiencies--one for aggregation (P_{aim}) and one for disaggregation (P_{dim})--which are functions of the grain sizes of the colliding i and m particles. For the purposes of this analysis, it has been assumed that α' is equal to the sum of P_{aim} and P_{dim}, since that is how it is expressed in the model of Burban et al. (1989). The tabulated results of Tsai et al. (1987) and Burban et al. (1989) are thus based on a particle size distribution and vary with the size of the colliding particles plus other variables. They are also representative of a near-equilibrium suspension, where particle sizes are large and relatively stable.

Two particle sizes are used in the Table 2 parameters. The larger diameter, D_m, represents a typical particle size during the period over which collision efficiency has been measured and has been specified in the table as the simple

Table 2
Non-dimensional parameters for observed collision efficiency data

Reference-Sediment	$\Delta\rho_i D_i^3 / \Delta\rho_m D_m^3$	$u_i(D_i+D_m)/\nu$	D_g/D_1	S/S_0	$\Phi/1.776$	CEC/CEC_0	$\tau_m/(\tau_{im,m}+\tau_u)$	α'
Burban-Detroit River[a]	3.43	0.0558	0.00417	17.5	0.62	3.5	334	0.216
Burban-Detroit River	3.85	0.250	0.00325	17.5	0.62	3.5	66	0.168
Burban-Detroit River	4.85	0.132	0.00168	17.5	0.62	3.5	309	0.140
Burban-Detroit River	4.63	0.199	0.00197	17.5	0.62	3.5	156	0.150
Burban-Detroit River	5.21	0.461	0.00124	17.5	0.62	3.5	86	0.122
Burban-Detroit River	5.29	0.262	0.00116	17.5	0.62	3.5	201	0.117
Burban-Detroit River	2.96	0.0732	0.00548	17.5	0.62	3.5	189	0.242
Burban-Detroit River	2.57	0.0780	0.00694	17.5	0.62	3.5	142	0.272
Burban-Detroit River	4.21	0.0608	0.00260	17.5	0.62	3.5	477	0.174
Burban-Detroit River	3.85	0.167	0.00325	17.5	0.62	3.5	113	0.194
Burban-Detroit River	3.85	0.0834	0.00325	17.5	0.62	3.5	266	0.191
Burban-Detroit River	5.73	1.34	0.000704	17.5	0.62	3.5	34	0.092
Burban-Detroit River	5.84	0.896	0.000606	17.5	0.62	3.5	79	0.085
Burban-Detroit River	5.60	1.94	0.000833	17.5	0.62	3.5	15	0.101
Edzwald-Lower Pamlico[b]	4.55	0.0544	0.00323	0.90	0.62	2.7	741	0.130
Edzwald-Lower Pamlico	4.63	0.0599	0.00306	8.75	0.62	2.7	705	0.165
Edzwald-Lower Pamlico	4.55	0.0544	0.00323	2.20	0.62	2.7	741	0.080
Edzwald-Pamlico River[b]	4.63	0.0599	0.00410	8.75	0.62	2.3	705	0.220
Edzwald-Pamlico River	4.55	0.0544	0.00433	2.20	0.62	2.3	741	0.140
Edzwald-Pamlico River	4.55	0.0544	0.00433	0.90	0.62	2.3	741	0.045
Edzwald-Upper Pamlico[b]	4.63	0.0599	0.00358	8.75	0.62	2.5	705	0.095
Edzwald-Upper Pamlico	4.55	0.0544	0.00378	2.20	0.62	2.5	741	0.180
Edzwald-Upper Pamlico	4.55	0.0544	0.00378	0.90	0.62	2.5	741	0.100
Gibbs-Amazon Delta[c]	2.57	0.444	0.000514	1.00	0.62	4.9	141	0.072
Gibbs-Amazon Delta	4.57	0.00599	0.00414	1.00	0.62	4.9	5068	0.697
Gibbs-Guiana[c]	4.96	0.0101	0.00247	0.45	0.62	6.8	4000	0.219
Gibbs-Delaware Bay[c]	4.49	0.0273	0.00474	8.75	0.62	2.1	1009	0.207
Gibbs-Delaware Bay	4.53	0.0286	0.00461	2.20	0.62	2.1	986	0.171
Gibbs-Delaware Bay	4.57	0.0299	0.00449	0.55	0.62	2.1	964	0.077
Gibbs-Yukon River[c]	4.96	0.0101	0.00134	0.30	0.62	5.1	4000	0.198
Tsai-Detroit River[d]	5.39	0.638	0.00104	0.25	0.62	3.5	66	0.053
Tsai-Detroit River	5.10	0.421	0.00130	0.25	0.62	3.5	92	0.062
Tsai-Detroit River	4.80	0.124	0.00174	0.25	0.62	3.5	320	0.071
Tsai-Detroit River	3.05	0.0790	0.00521	0.25	0.62	3.5	181	0.113
Tsai-Detroit River	3.50	0.0594	0.00401	0.25	0.62	3.5	322	0.104
Tsai-Detroit River	4.55	0.359	0.00208	0.25	0.62	3.5	66	0.078
Tsai-Detroit River	5.44	0.350	0.000992	0.25	0.62	3.5	165	0.053
Tsai-Detroit River	4.21	0.122	0.00260	0.25	0.62	3.5	213	0.088
Tsai-Tanshui Estuary[e]	4.85	0.220	0.000866	17.5	0.62	4.7	224	0.047
Tsai-Tanshui Estuary	3.59	4.69	0.000173	0.03	0.62	4.7	3	0.005

Sources: a-Burban et al. (1989), b-Edzwald et al. (1974), c-Gibbs (1985), d-Tsai et al. (1987), e-Tsai and Hwang(1995).

average of the initial and ending particle of the experiment. The second size, D_i, has been specified as the simple average of the initial particle size and D_m, representing a typical size involved in collisions with the *m* particle. Several other size combinations were tested, but within a reasonable range the choice did not have a major effect on the end result as long as the method was consistent.

Particle densities used in Table 2 were computed for the specified particle sizes using (1) and, in the absence of site-specific data, coefficients from San Francisco Bay (McAnally, 1999). Shear strength of the particles was computed by (2) with coefficients as given by Partheniades (1993), and collision and flow-imposed shear stresses were computed by (23) and (25), respectively. Temperature was not varied systematically in the experiments, so the results of Jiang (1999) were used, which expresses the effect of temperature on mean settling velocity as:

$$\Phi = 0.875\left(1 - \frac{T_c}{T_0}\right) \tag{32}$$

The reference values in (29) were set to be: T_0 = 30 deg C, S_0 = 2 ppt, CEC_0 = 9 meq/100 g, and D_1 = the ending particle size in the suspension.

A cursory graphical inspection of α' versus the non-dimensional parameters of Table 2 indicated obvious correlations for all parameters except *CEC*. The lack of an obvious *CEC* correlation agrees with observations that real sediments in estuarial waters exhibit a nearly uniform surface charge because of organic and oxide coatings (Hunter and Liss, 1982; Kranck et al., 1993; Gibbs, 1985). Nevertheless, *CEC* was retained as a variable, since it slightly improved the fit to data when used in combination with the other variables.

As an initial test, the non-dimensional terms of Table 2 were combined in a simple product form given by:

$$\Pi_\alpha = \frac{\left(\frac{D_g}{D_1}\right)\left(\frac{S}{S_0}\right)\left(1 - 0.875\,\frac{T_c}{T_0}\right)\left(\frac{CEC}{CEC_0}\right)\left(\frac{\tau_m}{\tau_{im,m} + \tau_u}\right)}{\left(\frac{\Delta\rho_i D_i^3}{\Delta\rho_m D_m^3}\right)\left(\frac{u_i(D_i + D_m)}{\nu}\right)} \tag{33}$$

and plotted against the observed α' of Table 2. Figure 4 shows the result--a surprising degree of correlation for such a simple relationship. One data point, Gibbs (1985) Amazon sample value of α' = 0.697, may either be an outlier or indicate a sharp upward inflection in the curve at higher values of Π_α.

A least squares fit of a power law function to the points of Figure 4 yields:

$$\alpha' = 0.0848\,\Pi_\alpha^{0.163} \tag{34}$$

Equation (34) fits the data with a correlation coefficient (R^2) of 0.59 and a standard error (Y_e) of α' estimate of 0.07 (0.71 and 0.04, respectively, if the Amazon point is omitted).

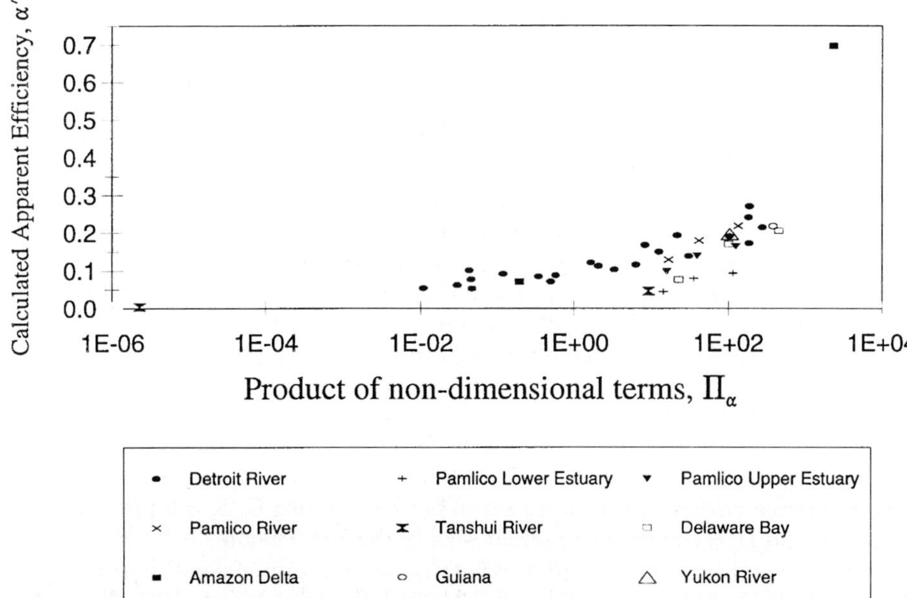

Figure 4. Comparison of observed apparent collision efficiency with product of non-dimensional terms of (33). Sources of data: Edzwald et al., 1974; Gibbs, 1983; Tsai et al., 1987; Burban et al., 1989; and Tsai and Hwang, 1995.

Numerous regression trials were performed on the sample set for α′ versus the parameters of Table 2 in various combinations. The optimum mathematical relationship, which was selected on the criteria that it: a) be a product of terms and not a sum, b) vary between 0.05 and 0.7 over the measured range, c) exhibit no local maxima or minima over that range, d) maximize correlation coefficient, and e) minimize standard error of estimate was:

$$\alpha' = \alpha_a 0.242 \ \Pi_\alpha^{0.195} e^{\left(\frac{\Pi_d}{4670}\right)^2} \tag{35}$$

where $\alpha_a = 1$ and

$$\Pi_c = \left[\frac{\dfrac{D_g}{D_1} \ \dfrac{S}{S_0} \left(1 - 0.875 \ \dfrac{T}{T_0}\right) \dfrac{CEC}{CEC_0}}{\dfrac{\Delta\rho_i D_i^3}{\Delta\rho_m D_m^3} \ \dfrac{u_i (D_i + D_m)}{\nu}} \right] \tag{36}$$

$$\Pi_d = \left[\frac{\tau_m}{\tau_{im,m} + \tau_u} \right] \tag{37}$$

Equation (37) yields an $R^2 = 0.91$ and $Y_e = 0.03$. The fit is good and the form is reasonable in that it agrees qualitatively with the relationships of (33).

Figures 5a and 5b show the agreement between observed and calculated apparent collision efficiencies for (34) and (35), respectively, along with the perfect fit 45 deg line. Equation (35) is clearly the better predictor, albeit with a more complex relationship. Equation (34) is an adequate interpolator for $0 < \alpha' < 0.25$, but fails to capture the high value of 0.697. If the 0.697 value is a valid point, and the theoretical upper limit of 1 on α' suggests that it is, then (34) is an unsatisfactory descriptor of the parameter. Visual inspection confirms that the functions are well-behaved over the range of independent parameter values given in Table 2. The goodness of fit appears to confirm at least the broad validity of assumptions (a)-(e) and the adequacy of the selected non-dimensional terms to characterize the aggregation process. The remaining scatter, yielding a standard error of estimate of about ± 4 percent of the full scale, is attributed to: a) variations in experimental method among the data sources, b) inadequacy of using only two representative particle sizes to characterize a distribution of sizes, and c) incomplete incorporation of mineral composition effects. The latter deficiency can be seen most clearly in Figure 5, where sediments from three locations in the Pamlico River estuary, apparently differing mainly in mineral composition (although organic content and grain size variability cannot be discounted) yield observed values of α' ranging from 0.095 to 0.165 for the same salinity.

Given the success of the rationally-based, but empirical, equation (35), the remaining task is to separate the F_c and α_d parts of (27). Given that F_c must fall between 0 and 1, and α_d can be assumed to lie between 0 and 1 for the experimental shear stress conditions shown in Table 2, (35) can be set equal to (27) and decomposed to yield:

$$F_c^2 = 0.805 \;\; \Pi_c^{0.195} \quad 0 < \Pi_c < 3.04 \tag{38}$$

and

$$\alpha_d = 0.308 \;\; e^{\left(\frac{\Pi_d}{4670}\right)^2} \quad 0 < \Pi_d < 5068 \tag{39}$$

and thus for flow shear-induced collisions it can be concluded that:

$$F_c = 0.897 \;\; \Pi_c^{0.0975} \tag{40}$$

with α_a given by (28).

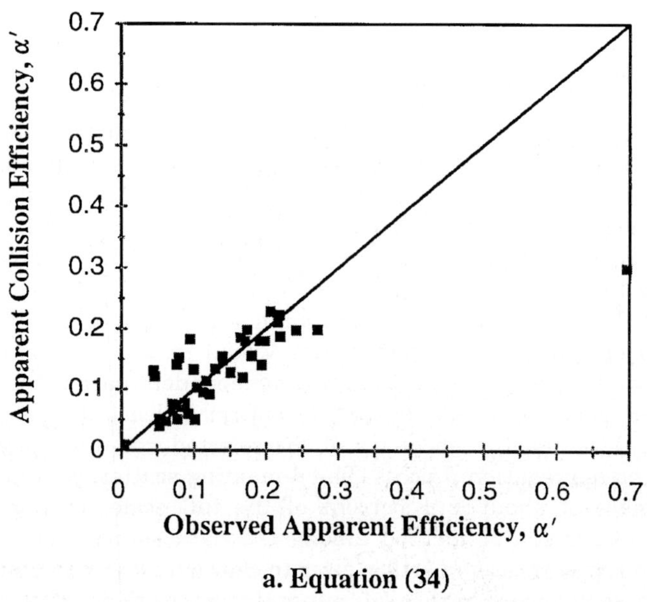

a. Equation (34)

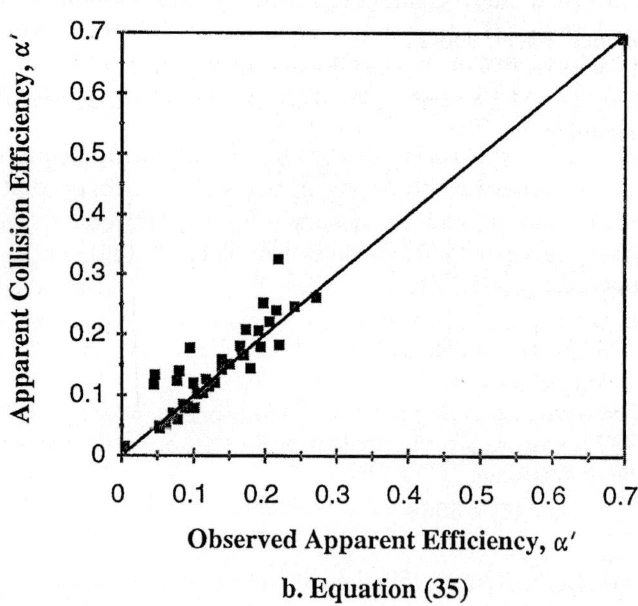

b. Equation (35)

Figure 5. Observed versus predicted apparent efficiency for two regression equations.

7. CONCLUSIONS

A collision diameter function, F_c, expressed as a function of non-dimensional numbers in (40), provides what the limited data show to be a well-behaved relationship between measurable parameters and the rate of aggregating interparticle collisions of fine estuarial sediment. It is both heuristic and empirical, but it is based on known physical dependencies and is clearly preferable to a tuned set of efficiencies covering five orders of magnitude. It can be used in a model to compute fine sediment aggregation rates, subject to suitable testing.

8. ACKNOWLEDGMENT

The work described here was supported by the U.S. Army Engineer Research and Development Center, Vicksburg, MS. The Chief of Engineers has given permission to publish this paper.

REFERENCES

Adler, P. M., 1981. Heterocoagulation in shear flow. *Journal of Colloid and Interface Science*, 83(1), 106-115.

Ariathurai, R., MacArthur, R. C., and Krone, R. B., 1977. Mathematical model of estuarial sediment transport. *TR D-77-12,* USAE Waterways Experiment Station, Vicksburg, MS.

Burban, P. Y., Lick, W., and Lick, J., 1989. The flocculation of fine-grained sediments in estuarine waters. *Journal of Geophysical Research,* 94(C6), 8323-8330.

Clarke, J. F., and McChesney, M., 1964. *The Dynamics of Real Gases*, Butterworth, Washington, DC.

CTH., 1960. Soil as a factor in shoaling processes: a literature review. *Technical Bulletin 4,* USAE Committee on Tidal Hydraulics, Vicksburg, MS.

Edzwald, J. K., Upchurch, , J. B., and O'Melia, C. R., 1974. Coagulation in estuaries. *Environmental Science and Technology*, 8, 58-63.

Gibbs, R. J., 1977. Clay mineral segregation in the marine environment. *Journal of Sedimentary Petrology,* 47(1), 237-243.

Gibbs, R. J., 1985. Estuarine flocs: their size, settling velocity and density. *Journal of Geophysical Research,* 90(C2), 3249-3251.

Grim, R. E., 1968. *Clay Mineralogy*. McGraw-Hill, New York.

Han, M., 1989. Mathematical modeling of heterogeneous flocculent sedimentation. *Ph.D. Thesis,* University of Texas, Austin.

Hunt, J. R., 1982. Self-similar particle-size distributions during coagulation: theory and experimental verification. *Journal of Fluid Mechanics*, 122, 169-185.

Hunter, K. A., and Liss, P. S., 1982. Organic matter and the surface charge of suspended particles in estuarine waters. *Limnology and Oceanography,* 27(2), 322-335.

Jiang, J., 1999. An examination of estuarine lutocline dynamics. *Ph.D. Thesis,* University of Florida, Gainesville.

Kranck, K., 1980. Sedimentation processes in the sea. *Handbook of Environmental Chemistry,* 2(A), Springer-Verlag, Berlin, 61-75.

Kranck, K., Petticrew, E., Milligan, T. G., and Droppo, I. G., 1993. *In situ* particle size distributions resulting from flocculation of suspended sediment. In: *Nearshore and Estuarine Cohesive Sediments Transport,* A.J. Mehta ed., American Geophysical Union, Washington, DC, 60-74.

Kranenburg, C., 1994. The fractal structure of cohesive sediment aggregates. *Estuarine , Coastal and Shelf Science,* 39(5), 451-460.

Krone, R. B., 1963. A study of rheologic properties of estuarial sediments. *Technical Bulletin 7,* USAE Committee on Tidal Hydraulics, Vicksburg, MS.

Krone, R. B., 1986. The significance of aggregate properties to transport processes. In: *Estuarine Cohesive Sediments Dynamics, Lecture Notes on Coastal and Estuarine Studies,* Vol. 14, A.J. Mehta ed., Springer-Verlag, New York, 66-84.

Lick, W., Lick, J., and Ziegler, C. K., 1992. Flocculation and its effect on the vertical transport of fine-grained sediments. In: *Sediment / Water Interactions, Hydrobiologia,* B.T. Hart, and P.G. Sly eds., 235/236, Kluwer, Dordrecht, The Netherlands, 1-16.

McAnally, W. H., 1999. Transport of fine sediments in estuarial waters. *Ph.D. Thesis,* University of Florida, Gainesville.

McCave, I. N., 1984. Size spectra and aggregation of suspended particles in the deep ocean. *Deep Sea Research,* 31(4), 329-352.

Mehta, A. J., and Lee, S.-C., 1994. Problems in linking the threshold condition for the transport of cohesionless and cohesive sediment grains, *Journal of Coastal Research* 10(1), 170-177.

Mehta, A. J., and Parchure, T. M., 2000. Surface erosion of fine-grained sediment revisited. In: *Muddy Coasts Dynamics and Resource Management,* B. W. Flemming, M. T. Delafontaine and G. Liebezeit eds., Elsevier, Amsterdam, The Netherlands (in press).

O'Melia, C. R., 1985. The influence of coagulation and sedimentation on the fate of particles, associated pollutants, and nutrients in lakes. In: *Chemical Processes in Lakes,* W. Strom ed.,Wiley, New York, 207-224.

Overbeek, J. T. G., 1952. Kinetics of flocculation. *Colloid Science, Volume 1,* Chapter VIII, H.R. Kruyt ed., L. C. Jackson, Translator, Elsevier, Amsterdam, 278-300.

Partheniades, E., 1993. Turbulence, flocculation, and cohesive sediment dynamics. In: *Nearshore and Estuarine Cohesive Sediments Transport,* A.J. Mehta ed., American Geophysical Union, Washington, DC, 40-59.

Saffman, P. G., and Turner, J. S., 1956. On the collision of drops in turbulent clouds. *Journal of Fluid Mechanics,* 1(1), 16-30.

Smoluchowski, M. Z. von, 1917. Versuch einer mathematischen theorie der koagulations-kinetic kolloider losungen. *Zietschrift für Physicalische Chemie,* 92-99.

Tambo, N., and Watanabe, Y., 1979. Physical characteristics of flocs I. The floc density function and aluminum floc. *Water Resources Research,* 13, 409-419.

Tsai, C. H., Iacobellis, S., and Lick, W., 1987. Flocculation of fine-grained lake sediments due to a uniform shear stress. *Journal of Great Lakes Research,* 13(2), 135-146.

Tsai, C. H., and Hwang, S. C., 1995. Flocculation of sediment from Tanshui River estuary. *Marine and Freshwater Research,* 46(1), 383-392.

Winterwerp, J. C., 1999. On the dynamics of high-concentrated mud suspensions. *Ph.D. Thesis,* Technical University of Delft, The Netherlands.

Zeichner, G. R., and Schowalter, W. R., 1977. Use of trajectory analysis to study stability of colloidal dispersions in flow fields. *American Institute of Chemical Engineering Journal,* 23 (3), 243-254.

Coastal and Estuarine Fine Sediment Processes
W.H. McAnally and A.J. Mehta (Editors)
2001 Elsevier Science B.V.

Erosion of a deposited layer of cohesive sediment

I. Piedra-Cueva[a] and M. Mory[b]

[a]Institute of Fluid Mechanics and Environmental Engineering (IMFIA), Faculty of Engineering, C.C. 30, Montevideo, Uruguay

[b]Ecole Nationale Supérieure en Génie des Technologies Industrielles, BP 576, 64012 Pau Cédex, France

Results of an experimental study on the erosion of a deposited bed of cohesive sediment by steady current are presented. The experimental procedure measured layer by layer erosion of the bed that occurred when the current velocity was increased step by step. The focus is on the asymptotic process of erosion at the end of each velocity step, when the excess shear stress vanishes. The so-called floc erosion process, which is said to occur when finite erosion continues for zero excess shear stress, was not observed. The floc erosion rate, if present, was smaller than the smallest erosion rate measured in our experiments. The erosion rate varied linearly with the excess shear stress in the range of excess shear stress considered. A dimensional analysis of erosion rate variation indicates that the gradient of the bed shear strength in the eroded layer was a significant parameter in scaling erosion.

1. INTRODUCTION

The prediction of fine cohesive sediment transport is an important challenge for coastal engineers. Navigation in channels and harbors is often limited by fine sediment transport. More recently, the role of fine sediment in capturing and transporting contaminants has become a major question that can no longer be ignored. These examples highlight the usefulness of understanding the erosion and settling of cohesive sediments.

We present the results of a laboratory experiment in which the rate of erosion of a deposited and consolidated cohesive sediment bed by steady current was measured. In this two layer system, the current in the upper layer, which carries sediment in suspension, applies a bottom shear stress $\tau_b = \rho u_*^2$ on the bed, which forms the lower layer. Here, ρ is the fluid density and u_* is the friction velocity. The bed's mechanical behavior is characterized (among other parameters) by its rigidity, defined here as the bed shear strength τ_s. Bed erosion occurs when $\tau_b >$

$\tau_s(z_b)$, where $z_b(t)$ is the vertical position of the bed surface at time t. For a deposited and consolidated bed the vertical shear strength increases within the bed, implying that under small bed shear stresses the condition for erosion $\tau_b > \tau_s(z_b)$ is met only in the surface layer of the bed. Due to the non-uniformity of the bed shear strength, the erosion rate varies with time when a steady shear stress is applied on the bed. The erosion rate per unit width of the flume ε is defined as

$$\varepsilon = -\rho_d(z_b)L_b\frac{dz_b}{dt} = hL_f\frac{d\bar{C}}{dt} \tag{1}$$

In (1), z_b denotes the vertical coordinate in the bed (directed upward) and L_f and L_b are the length of the flume and the bed, respectively. The erosion rate is defined in two ways because of mass conservation: as a function of the variation in bed thickness (ρ_d being the dry sediment concentration within the bed), and as a function of the time dependent change in suspended sediment concentration $\bar{C}(t)$ in the water column, averaged over flow depth h.

Numerous studies have been conducted on cohesive sediment erosion. Ours is a continuation of the experimental work of Parchure (1984) and Parchure and Mehta (1985). They quantified the decrease in time of the erosion rate when a constant bottom shear stress was applied on a deposited bed. Their data were analyzed in terms of an exponential decay of the erosion rate,

$$\varepsilon(t) = \varepsilon_o\exp(-\lambda t) \tag{2}$$

which implies that the erosion rate decreases to zero when τ_b-$\tau_s(z_b)$ also decreases to zero. Experiments indicated that the excess shear stress τ_b-$\tau_s(z_b)$ was the main parameter governing erosion. We note, however, that interpreting erosion rate variation in terms of excess shear stress variation is not straightforward. Erosion laws published in the literature fall in two classes. The first type of erosion law estimates erosion rate as a power function of the excess shear stress

$$\varepsilon(z_b) = M\{\tau_b - \tau_s(z_b)\}^m \tag{3}$$

Although the well known linear erosion law (with m=1) is often attributed to Partheniades (1965), Ariathurai et al. (1977) actually introduced this law by fitting a linear equation to Partheniades' data. The other type of erosion law for deposited beds is written in the form

$$\varepsilon(z_b) = \varepsilon_f\exp\left[\alpha\{\tau_b - \tau_s(z_b)\}^n\right] \tag{4}$$

Parchure and Mehta (1985) analyzed their results in terms of such an erosion law, for which they obtained n=1/2. It is important to point out that the two types of laws differ significantly at the asymptotic condition τ_b-$\tau_s(z_b)$=0. Equation (4) implies that erosion is finite when τ_b-$\tau_s(z_b)$=0. This type of erosion is called floc erosion and ε_f is the residual floc erosion rate. Although floc erosion might be quantitatively small, the consequence of including floc erosion may become important for long term modeling. Determining the floc erosion rate is not straightforward. Measurements are difficult, because the expected values for the floc erosion rate are small. Actually, Parchure and Mehta (1985) estimated the floc erosion rate by plotting ln(ε) against $(\tau_b-\tau_s)^{1/2}$, and extrapolating the resulting line to the $(\tau_b-\tau_s)^{1/2}$=0 axis.

For a deposited bed exhibiting a linear increase in the bed shear strength within the bed,

$$\tau_s(z_b) = \tau_{so} + (\tau_b - \tau_{so})\frac{z_b(t)}{H_o} \tag{5}$$

it is easy to show that (3) with m=1 is obtained. In (5), τ_{so} and τ_b are the bed shear strengths at the beginning and at the end of the velocity step, respectively, and H_o is the depth of the layer eroded during the velocity step. Integrating (1) for an exponential time variation of the erosion rate (2), the time variation of the position of the bed surface is deduced as

$$z_b(t) = H_o - \frac{\varepsilon_o}{\rho_d \lambda}\exp(-\lambda t) \tag{6}$$

Equation (6) is obtained under the assumption that the dry density $\rho_d(z_b)=\rho_d$ is uniform within the thin layer eroded during a velocity step. Using (6) and (5) the erosion rate can be rewritten as a function of the bed surface position and the excess shear stress. We obtain

$$\varepsilon(t) = \rho_d \lambda H_o\left[1 - \frac{z_b(t)}{H_o}\right] = \frac{C_\infty \lambda h L_f}{L_b}\left[\frac{\tau_b - \tau_s(z_b)}{\tau_b - \tau_{so}}\right] \tag{7}$$

Introducing the dimensionless erosion rate E = $\varepsilon / \rho_d u_*$ and the dimensionless excess shear stress $T_s = (\tau_b - \tau_s) / \tau_b$, (7) is rewritten in dimensionless form as

$$E = KT_s \quad \text{with} \quad K = \frac{\lambda C_\infty h}{\rho_d u_*}\frac{L_f}{L_b}\frac{\tau_b}{\tau_b - \tau_{so}} \tag{8}$$

We will initially focus on the asymptotic behavior of erosion when the excess shear decreases to zero. Our first goal is to examine the possibility of floc erosion based on experimental data. In the second phase, the erosion rate variations measured in our experiments will be analyzed in the dimensionless form.

2. EXPERIMENTAL PROCEDURE

Experiments were carried out in the sediment transport flume at the Laboratoire d'Hydraulique de France; see Piedra-Cueva et al. (1997) for a description of this flume. Figure 1 shows a schematic diagram of the layout. The water depth was 40 cm and the channel width was 50 cm. The current was produced by vertical rotating discs. Before starting an experiment, mud was left to deposit and consolidate in a limited section (8.2 m long) in the straight reach of the flume that did not contain the disc pump. The initial depth of the bed was 8 cm. As shown by Piedra-Cueva et al., uniform erosion was obtained over the entire bed surface within the straight reach. More severe erosion, usually in the bends and in the vicinity of the pump, was avoided.

Experiments were carried out with the procedure formerly employed by Thorn and Parsons (1979) and Parchure and Mehta (1985). The current speed was increased step by step. Each test usually had 3 to 7 velocity steps. The erosion rate decreased during each step. Depending on the consolidation time (13 to 24 hours), the step lasted 60 - 180 minutes. The step duration was set to be sufficiently long, so that the suspended sediment concentration was almost constant and the erosion rate was very small at the end of each step.

The erosion rate was determined from the suspended sediment concentration measured continuously at three levels along a vertical located 0.5 m upstream of the deposited bed. Due to the very strong vertical mixing of suspended sediment that occurred when the fluid passed through the disc pump, it was observed that the distribution of sediment was uniform over the vertical. An accurate determination of the sediment concentration averaged over depth was therefore obtained.

Natural mud from the Gironde estuary (France) was employed. The mud was chemically treated with potassium permanganate and passed through a 100 μm sieve. Sediment analysis performed by de Croutte et al. (1996) revealed that the median diameter was about D_{50}=12 μm, and the sand content (63-100 μm) was below 3%. The clay fraction consisted of illite (35%), kaolinite (27%), smectite

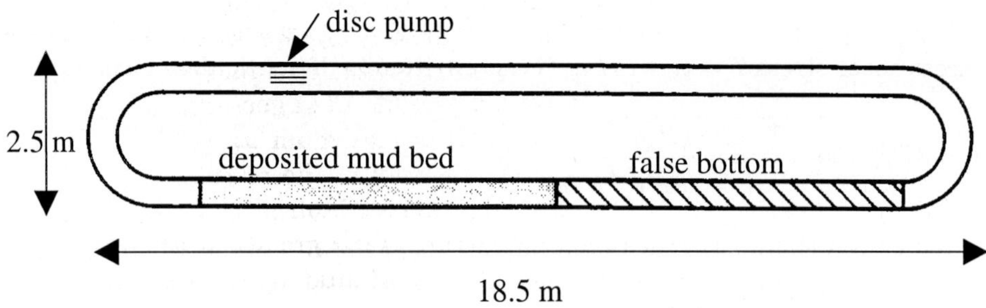

Figure 1. Schematic diagram of the experimental set-up.

(24%) and chlorite (14%). Although the yield stress of the mud was also measured for several initial conditions, this quantity is not found to be related in a simple manner to the critical bed shear strength. Migniot (1989) indicated that the bed shear strength is usually smaller than the yield stress and proposed an empirical relationship between the two quantities. However, this correlation is of no use if the yield stress is not determined *in situ*, because consolidation and thixotropic effects cause it to vary with time. The bed shear strength profile was therefore determined indirectly from the bottom shear stress applied during the different velocity steps in the experiment. The depth of erosion during each step was calculated from mass conservation using the suspended sediment concentration increase measured at the end of each step, and dry density profiles measured within the bed using an ultrasonic ECOVUS probe. Figure 2 shows an example of the vertical dry density profile in the bed and the derived bed shear strength profile for an experiment. Stratification within the bed is evident except near the bed surface, where the dry density was on the order of 100 g/l or less (Figure 2a), and significant consolidation did not occur. This upper layer was eroded during the first velocity step and was not considered for analysis.

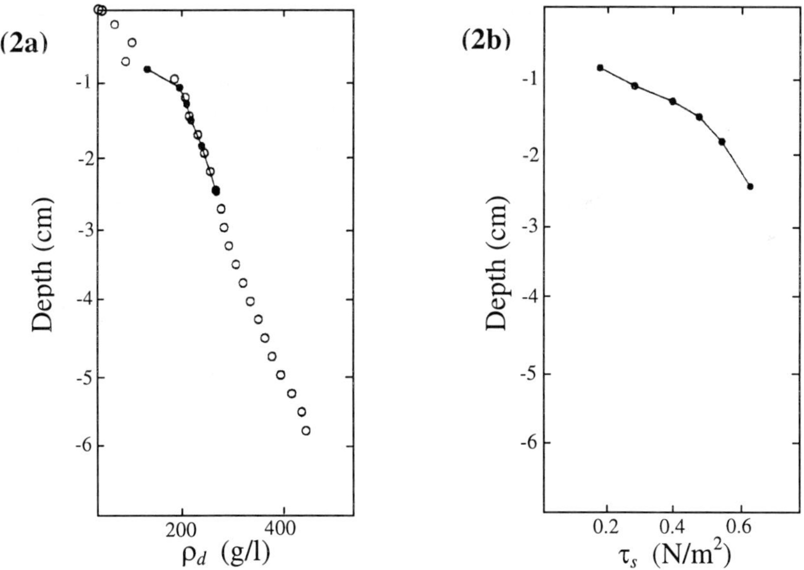

Figure 2. (2a) Vertical dry density profile within the bed; (2b) Bed shear strength profile within the bed. Open circles are measured data. Dark symbols are estimates of dry density and bed shear strength at the end of the six velocity steps of the experiment.

3. EROSION RATE DETERMINATION

Figure 3 shows a typical time dependent change in suspended sediment concentration during a velocity step. Oscillations are observed in the record. They appeared because the deposited bed did not cover the entire bottom surface of the flume. The mean change in suspended sediment concentration during a velocity step was fitted using an exponential function of the form

$$\bar{C}(t) = C_{max} - A\exp\{-\lambda t\} \tag{9}$$

which implies that the erosion rate decreased exponentially with time. The three constants C_{max}, λ and A were determined from an adjustment procedure. It was found that a set of these three constants could never be obtained to define a function that fit the data over the entire record period. This is observed in Figure 3, in which the interpolated function is superimposed on the experimental data. The initial erosion (0 to 10 minutes) is more rapid than estimated by the function. The constants C_{max}, λ and A were therefore selected to obtain the best adjustment of (9) over the longest part of the record at the end of the step. The adjustment was considered to be significant when the adjustment time interval conformed with the condition $t_m > 8t_1$; t_1 and t_m respectively being the beginning and the end of the segment of the record chosen for adjustment.

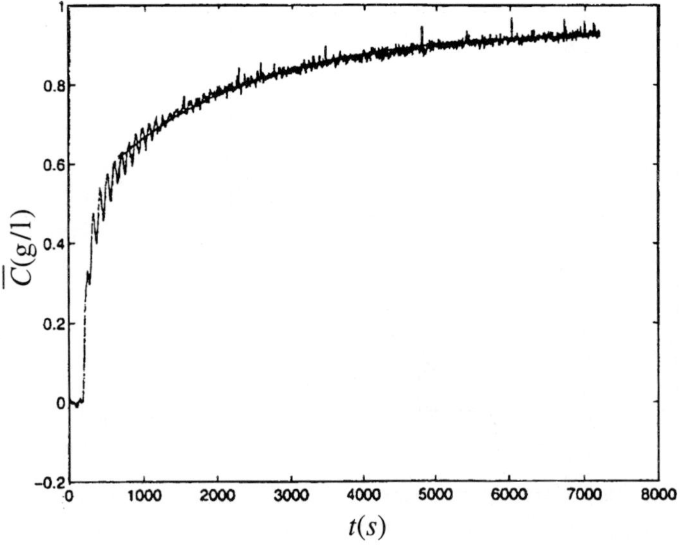

Figure 3. Time variation of suspended sediment concentration during a velocity step. Experimental data are superimposed by the adjustment function (9).

The good overlap observed in Figure 3 between the interpolated function and the experimental data (similar agreement was obtained for all plots) indicates that the erosion rate decreased exponentially with time at the end of each velocity step. For the durations of our experiments, the theoretical analysis presented earlier therefore indicates that the erosion rate varied linearly with the excess shear stress in the range considered. This implies that floc erosion, if it took place, occurred at very small rates, smaller than the smallest erosion rate measured during each velocity step. For the velocity step shown in Figure 3, the asymptotic value of the suspended sediment concentration was almost attained. The erosion rate at the end of the velocity step was 7.6×10^{-6} kg m^{-2} s^{-1}. If floc erosion had occurred when the excess shear stress decreased toward zero, the time-variation of the suspended sediment concentration would have deviated from the exponential function over a long time. This was not observed in our experiments.

For all velocity steps, the variation in the dimensionless erosion rate E was determined as a function of variation in the dimensionless excess shear stress T_s. As shown in Section 1, the two quantities are linearly related for each step. The proportionality coefficient K was determined for each step from the experimental parameters and data as

$$K = \frac{\lambda C_\infty h}{\rho_d u_*} \frac{L_f}{L_b} \frac{\tau_b}{\tau_b - \tau_{so}} \tag{10}$$

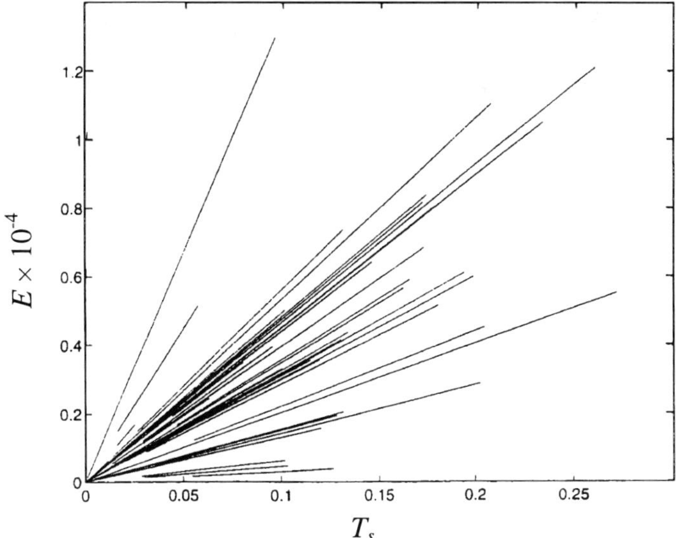

Figure 4. Variation of erosion rate with excess shear stress for all velocity steps.

The value of K varied over a wide range when all velocity steps were considered. This is shown in Figure 4. Each straight line represents the variation of E with T_s during a step, and the slope variation indicates the range of K.

4. DIMENSIONAL ANALYSIS

While the excess shear stress τ_b - τ_s is the main parameter governing erosion, it is not the only one. Erosion laws published in the literature are not expressed in dimensionless forms. We discuss below a dimensional analysis of our data by considering various mechanistic parameters. Our hypothesis is that physico-chemical and biological effects are accounted for in these parameters when erosion is considered. Erosion at any time is assumed to be local, and to depend only on the value of the different mechanistic parameters at that instant. Considering that flow turbulence and the mechanical properties within the bed enter the scaling of erosion, we express the following functional form for the erosion rate

$$\varepsilon(t) = \varepsilon\{g, \rho, \mu, h, \tau_b, \rho_d, \mu_d, H(t), \tau_s(t)\} \tag{11}$$

Flow properties in the water layer are accounted for by the density ρ, the dynamic viscosity μ, the depth of water h and the bottom shear stress τ_b. Also, g is the acceleration due to gravity. The bed shear strength τ_s, the dry concentration of sediment within the bed ρ_d and the dynamic viscosity μ_d characterize bed properties. The dynamic viscosity μ_d is arbitrarily introduced in (11) assuming a Bingham rheological model, although the rheological properties of the bed were not actually determined. The bed shear strength τ_s is taken at the surface of the bed. It varies in time as erosion proceeds and the position of the bed surface $z_b(t)$ changes. $H(t)$ denotes the depth of the layer in the bed where the bed shear strength is smaller than the bottom shear stress. This layer is continuously eroded and $H(t)$ therefore also varies with time.

The dimensional analysis leads to the following functional dependence of the dimensionless erosion rate:

$$E = \frac{\varepsilon(t)}{\rho_d u_*} = F\left\{\begin{array}{l} T_s = \dfrac{\tau_b - \tau_s}{\tau_b}, Ri = \dfrac{\rho_d g h}{\tau_b}, I = \dfrac{\tau_b H}{(\tau_b - \tau_s)h}, \\ Re_e = \dfrac{h\sqrt{\rho \tau_b}}{\mu}, R = \dfrac{\rho_d \tau_b^3 H^2}{\mu_d^2 (\tau_b - \tau_s)^2} \end{array}\right\} \tag{12}$$

where $u_* = \sqrt{\tau_b / \rho}$ is the friction velocity at the bed surface. The effect of the Richardson number Ri (comparing gravity to the gradient of stress in the water layer) on erosion was considered previously (Mehta and Srinivas, 1993). The quantity I compares the gradient of stress in the water layer and in the bed. To our knowledge, its role in the scaling of erosion has not been mentioned before. The two remaining numbers, Re_e and R, quantify dissipative effects in the water layer and in the bed, respectively. The definition of R is quite specific. It is based on the Bingham model, but its value is unknown in our experiments. For our experiments, it is important to mention that Ri, I, Re_e and R remained constant during a velocity step. The only varying dimensionless number was T_s. Comparing (8) and (12) shows that the erosion law can be analyzed through the following dependence of the proportionality coefficient K

$$E = K(I, Ri, Re_e, R)T_s \tag{13}$$

The variation of K with the different dimensionless quantities is not fully available from our experiments. Also, the dependence of K on R is not determined, which means that the dependence of K on the rheological properties of the bed is not known. On the other hand, the flow was fully turbulent (the flow velocity was in the range 0.32-0.78 m/s and the friction coefficient $\tau_b/\rho U^2$, where U is the flow velocity, was less than 0.0018). Thus, the effect of Reynolds number Re_e would not be considered to be significant when turbulent velocity and length scales are introduced in scaling erosion. The Richardson number was large because the bed was consolidated and the dry concentration of sediment was large. In most experiments Ri was in the range 1200-2000. This range is too narrow to yield a significant effect of the Richardson number in scaling erosion. Finally, the quantity I varied from one step to the other between 0.01 and 0.30.

Figure 5 shows the variation of K with I, which exhibits a high degree of scatter. The contribution to scatter attributable to the adjustment procedure for calculating the erosion rate is shown by error bars, which are narrow. This implies that the erosion rates were determined with a reasonable degree of accuracy. Dark symbols indicate the velocity steps for which the concentration variation C_∞ during a step was less than 0.35 g/l. The measurement uncertainty associated with the turbidity meters may explain the wide scatter in data for the low values of the parameter I. In spite of the scatter, an evident increase in K with increasing I is observed in Figure 5. Erosion is consequently shown to depend on the gradient of bed shear strength inside the bed. The erosion rate decreases when the gradient of the bed shear strength increases.

5. CONCLUSIONS

Experiments were carried out on the erosion of a deposited and consolidated bed by a steady current. The subject is not new but we believe that our data contribute to the understanding of several issues that are left open in the literature. Our analysis focused on the case in which the excess shear stress is low.

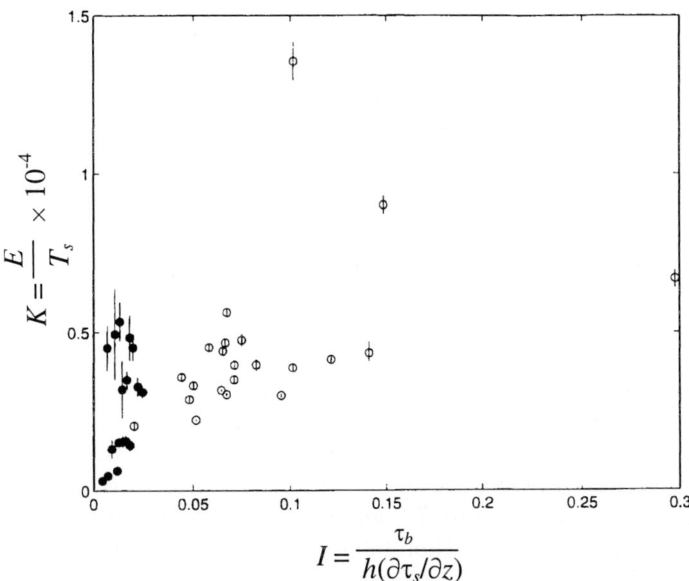

Figure 5. Variations of the proportionality coefficient K (between erosion rate and excess shear stress) with I.

Data indicate that the erosion rate decreases exponentially with time while the excess shear stress decreases toward zero. We show that this time behavior is indicative of a linear relationship between the erosion rate and the excess shear stress. Whether or not floc erosion occurs cannot be rigorously demonstrated, and requires an extrapolation of the erosion law for vanishing excess shear stress. In our experiments, floc erosion was not observed and probably would not have occurred provided the linear erosion law had held for vanishing excess shear stress.

The erosion law was analyzed in a dimensionless form, and as a result some of the functional dependences could be considered using our data. It was first shown that gravity effects estimated through the Richardson number were not significant. This was attributed to the large values of the Richardson number in our experiments because the bed was consolidated. This case is thus different from the case of the erosion of fluid mud. Our formulation confirmed that the excess shear stress is the main parameter governing erosion. It was moreover shown by our data that the gradient of the bed shear strength enters the scaling of erosion. Erosion decreases when the gradient of the bed shear strength increases.

It should be recognized that the results concern erosion at the end of the velocity steps, when the excess shear stress is small (smaller than 0.25 in dimensionless value in all cases). The most significant unknown in our experiments was the rheological property of the bed. This is a noteworthy limitation because it is known that consolidation and thixotropic effects tend to vary the mechanical properties of bed. Nevertheless, it is likely that time variations in the mechanical properties of bed are limited for the low values of the excess shear stress considered.

6. ACKNOWLEDGMENT

This work was undertaken as part of the MAST-2 G8M Coastal Morphodynamics Program. It was funded partially by the Commission of the European Communities Directorate General for Science, Research and Development under contract MAS2-CT-92-0027. The Laboratoire d'Hydraulique de France is acknowledged for providing access to its sediment flume. The scholarship granted by the Government of France to the first author and financial support from CONICYT and the Facultad de Ingenieria, Universidad de la Republica, Uruguay are sincerely acknowledged. Subsequent analysis of the data was carried out as part of the MAST-3 COSINUS program (contract MAS3-CT-97-0082).

REFERENCES

Ariathurai, R., MacArthur, R. C., and Krone, R. C., 1977. Mathematical model of estuarial sediment transport. *TR D-77-12*, U.S. Army Engineer Waterways Experiment Station, Vicksburg, MS, 79p + appendices.

de Croutte, E., Gallissaires, J. M., and Hamm, L., 1996. Flume measurements of mud processes over a flat bottom under steady and unsteady currents. *Sogréah Ingénierie Report No. R3*, Sogréah, Grenoble, France.

Mehta, A. J., and Srinivas, R., 1993. Observations on the entrainment of fluid mud by shear flow. In: *Nearshore and Estuarine Cohesive Sediment Transport,* A. J. Mehta ed., Coastal and Estuarine Studies, American Geophysical Union Washington, DC, 224-246.

Migniot, C., 1989. Tassement et rhéologie des vases: deuxième partie. *La Houille Blanche*, 2, 95-112.

Parchure, T. M., 1984. Effect of bed shear stress on the erosional characteristics of kaolinite. *Ph.D. Thesis*, University of Florida , Gainesville, FL, 339p.

Parchure, P. M., and Mehta, A.J., 1985. Erosion of soft cohesive sediments deposits. *Journal of Hydraulic Engineering*, 111(10), 1308-1326.

Partheniades, E., 1965. Erosion and deposition of cohesive soils. *Journal of the Hydraulics Division*, American Society of Civil Engineers, 91(1), 105-138.

Piedra-Cueva, I., Mory, M., and Temperville, A., 1997. A race-track recirculating flume for cohesive sediment research. *Journal of Hydraulic Research*, 35(3), 377-396.

Thorn, M. F. C., and Parsons, J. G., 1979. Properties of Balawan beds. *Technical Report EX 880*, Hydraulics Research Station, Wallingford, UK.

Coastal and Estuarine Fine Sediment Processes
W.H. McAnally and A.J. Mehta (Editors)

Critical shear stress for cohesive sediment transport

K. Taki

Department of Civil Engineering, Chiba Institute of Technology, 2-17-1, Tsudanuma, Narashino, 275-8588, Japan

The relation between the viscosity of mud and the electrochemical properties of surfaces of particles constituting the mud layer are examined based on the "theory of electrochemical anchoring force" for the prediction of the critical shear stress. This stress is calculated from the condition under which the van der Waals force and the force due to the surface charge of particles balance the shear due to water flow. The critical shear stress is shown to decrease exponentially with increasing specific gravity of mud particles and the water content of the mud layer, and is further shown to asymptotically approach the critical stress of non-cohesive particles when the water content increases. The derived equation is applicable to mud of grain size less than several tens of microns and relative water content of two or higher in dry weight percent.

1. INTRODUCTION

Mud particles deposited at the bottom of rivers, lakes and seas are resuspended mainly by flood flows, wind currents, waves and tidal flows, and are dispersed into flowing water as suspended soil. Organic and inorganic substances dissolved in the porewater are rapidly entrained into water flowing over the bed as result of resuspension of mud particles. As a result, the suspended mud directly affects water quality. Hence, elucidation of how fine particles are eroded and resuspended is highly desirable.

When the particles are smaller than about twenty microns, the cohesive force due to the surface charge acting on each particle becomes the dominant stabilizing force, rather than the weight of the particle and the frictional resistance to flow. For this reason, in the hydraulic study of mud transport, it has been common to deal with mud as if it were a single fluid having a super-concentration and a high viscosity, disregarding the discrete particles. The viscosity of the fluid has been known to be strongly dependent on the surface charge of particles (Taki, 1986). Few studies have been carried out, however, to theoretically examine the effect of this dependence on the erosion of mud.

2. MUD LAYER SURFACE PATTERNS

Geometric patterns obtained at the surface of mud layers are attributable to changes in the rheological characteristics that accompany changes in water content, as a result of the change from dominance of the frictional resistance of particles to cohesion (Allen, 1982,1969; Partheniades, 1962). Based on this consideration, the geometric forms at the surface of the mud layer can be classified into the following three patterns depending on the relative water content (W/W_L, where W is the water content and W_L is the liquid limit of mud) in dry weight percent: a) Streak line pattern: relative water content about 1.5 or less, b) Ripple-like pattern: relative water content about 2.0 and c) Internal wave pattern: relative water content 2.5 or more.

Pattern a) appears when the water content of settling mud does not exceed the liquid limit, and the bottom shear stress is larger than the critical shear stress. In this case, the flow streak lines are parallel to the surface of the mud layer, and particle resuspension is limited to the height (several millimeters) of viscous sublayer. Pattern b) appears when the bottom shear stress slightly exceeds the critical shear stress, i.e., the threshold of particle movement. In this case, particles occur individually on the mud surface. Pattern c) appears when the bottom shear stress is much larger than the critical shear stress (limit of bed destruction). It can be further considered that the particles affect one another, and the matrix behaves like a wave because of the anchoring force due to the surface charge of particles.

3. ELECTROCHEMICAL EFFECT

The surface charge on particles is mainly responsible for the high viscosity of mud. This charge (zeta potential) is found to range approximately from -20 mV to +20 mV, suggesting that the particles have weak coagulation (Overbeek, 1991). The relation between the yield stress of mud and the zeta potential has been investigated in experiments (Ito and Matsui, 1975), and found to exhibit one-to-one correspondence in the range where the relative water content (W/W_L) is higher than 2.

For mud particles of diameter d that is relatively small, approximately 20 to 30 microns, the particle surface is negatively charged at –20 mV, and these particles are deposited under weak contact or "partial non-contact condition" with one another, as in Figures 1 (a) and (b). Various ions (designated by + signs) and organic matter, that have been dissociated or dispersed in the pore water, are attracted to the surface of each particle or are combined with each other as bridging arrays A, B and C in Figure 1 (b). It can be assumed that thesc partial non-contact particles are mutually anchored by the adhesion of the surrounding organic substances such as fibrous roots, or the bridging effect of ions contained in the pore water.

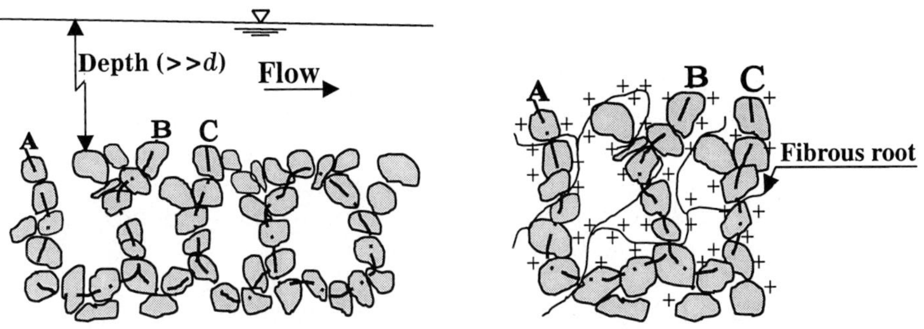

(a) Case of $d < 20$ microns. (b) Magnified A, B and C lines in (a).

Figure 1. Schematic drawings showing configurations of fine sediment particles.

A deposit structure tends to be such that the adjacent particles are homogeneously configured in the deposit as agglomerates or chains, and the voids between these chains are filled with ions of organic and inorganic substances. For this reason, and due to the bridging force between the non-touching particles, chains A, B and C generate anchoring forces. This anchoring implies that cohesion is the main factor accounting for the resistance against resuspension at high water contents.

4. FORCES ON PARTICLES

Mud particles under consideration are deposited under weak contact or partial non-contact condition with one another. The term "anchoring force" refers to the vertical component of the resulting force of attraction among the particles. Consequently, at the mud layer surface the vertical force binds the particles at the surface, as in Figure 2. Within the mud layer, however, the resultant of forces from all directions (as shown for particle B of Figure 2) is zero.

In the case of non-cohesive particles, the weight of the particles and the frictional resistance between particles are dominant forces resisting resuspension. In contrast, attraction due to the electric charge on the fine particle surface becomes the dominant force resisting fine particle resuspension.

Based on the above considerations, the following force balance equation related to a cohesive particle is obtained:

$$F_D + F_W \sin\theta_b = (F_W \cos\theta_b - F_L - F_B) \tan\phi + F_C \tan\phi \quad (1)$$

in which, F_D is the drag force due to water flow ($= 0.5\,\rho_w C_D k_1 d^2 U_d^2$), F_W is the weight of the particle ($= \rho_s k_3 d^3 g$), F_B is the buoyancy force acting on the particle ($= \rho_s k_3 d^3 g \cos\theta_b$), F_C is the anchoring force due to cohesion ($= k_1' d^2 f_c$), F_L is the lift

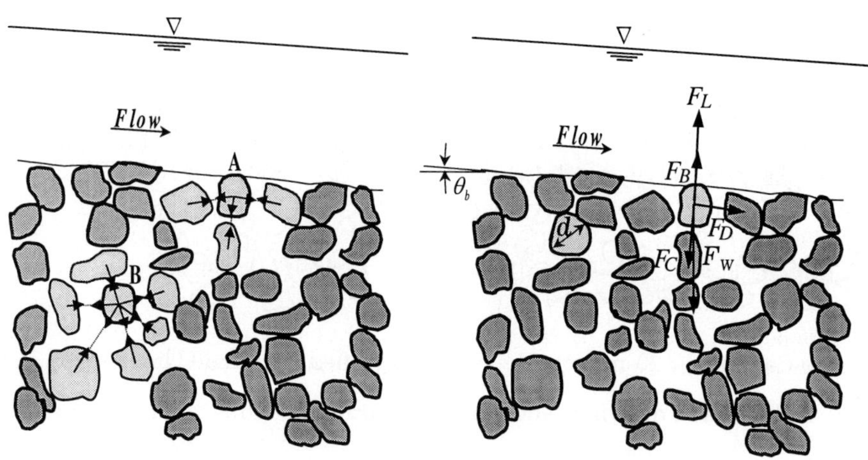

(a) Schematic of anchoring forces on particles A and B.

(b) Forces act on a mud particle.

Figure 2. Anchoring forces (a) and components on a surface particle in the mud layer (b).

force ($=0.5\rho_w C_L k_2 d^2 U_d^2$), and θ_b and ϕ are bed slope and the angle of repose, respectively. Further, g is the acceleration due to gravity, ρ_w and ρ_s respectively are the density of water and mud particles, U_d, $k_1' d^2$ and f_c respectively are the representative mean velocity close to a particle ($\propto U_*$, where U_* is the shear velocity), the particle surface area, and the cohesive force acting on a unit area of the bed. C_D is the coefficient of drag, C_L is the coefficient of lift, and k_1, k_2, k_3 are shape factors ($k_1 = k_2 = \pi/4$, $k_3 = \pi/6$ for spherical particles). Finally, k_1' is another shape factor.

5. ANCHORING FORCE

The critical shear stress can be related to the vertical component of the anchoring force. This stress is calculated from the balance between the cohesive force obtained from the sum of the van der Waals attraction between particles, the force caused by the surface charge of particles and the shear induced by water flow (Hunter, 1981; Kitahara et al., 1971).

The combined electrochemical potential E_T is commonly expressed as:

$$E_T = \frac{8k^2T^2\varepsilon\gamma^2}{z^2(d_s/2)}\exp(-\kappa x) - \frac{Ad_s}{24x} \tag{2}$$

where k is the Boltzman constant $\left(=1.38\times10^{-23}, J/^{\circ}K\right)$, T is the absolute temperature ($^{\circ}K$), ε is water permissivity, $\gamma = \{\exp(ze\varphi_0/2kT)-1\}/\{\exp(ze\varphi_0/2kT)+1\}$, z is the ionic valence, e is the electronic charge (esu), φ_0 is the surface charge of particle, $1/\kappa$ is the Debye length, A is the Hamaker constant (7.7×10^{-20} J), x is the distance from the particle surface, and d_s is the equivalent spherical diameter of the particle. The first term of (2) is the repulsive potential energy for the electric charge shown in Figure 3 as E_R, and the second is the attractive potential energy E_A. Cohesion, F_c, which acts on the particle surface, can be obtained from the condition of equilibrium between the repulsive potential energy and the attractive potential energy, i.e., $dE_T/dx = 0$. In other words, F_c in (1) is obtained as the derivative of the sum of van der Waals attraction between adjacent particles according to (2):

$$F_c = \sum_{i=1}^{n}\left[\frac{Ad_s}{24x_i^2}\left\{-\kappa\frac{96k^2T^2\varepsilon\gamma^2x_i^2}{Az^2(d_s/2)^2}\exp(-\kappa x_i)+1\right\}\right] \tag{3}$$

In (3), x_i is inter-particle spacing (Figure 4). The right hand side of (3) can be restated implicitly in terms of the mean clearance length $\bar{x}$ of adjacent particles as:

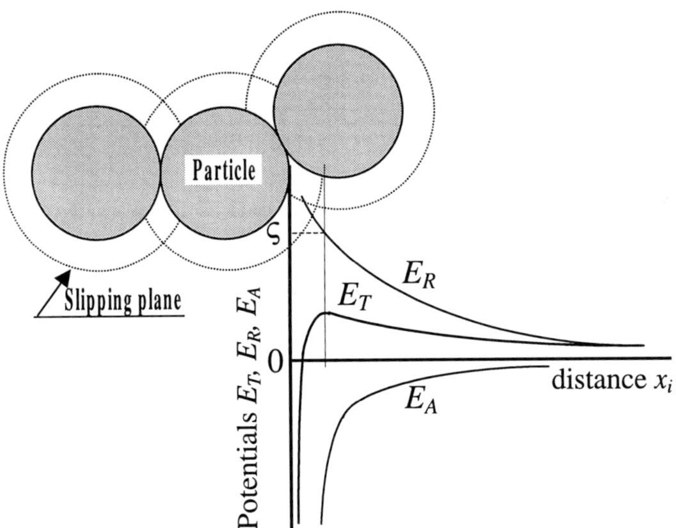

Figure 3. Schematic drawing of the slipping plane and the distributions of attractive potential (E_A), repulsive potential (E_R) and combined potential (E_T) at the particle surface, where the slipping plane is defined as the surface of water layer that moves with the particle.

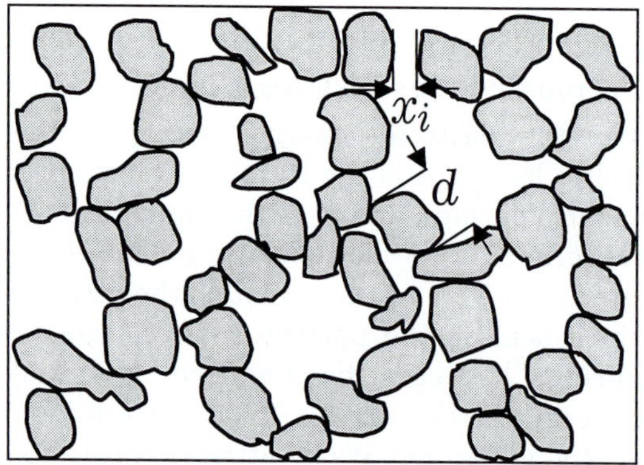

Figure 4. Schematic drawing of particle arrangement in the mud layer.

$$\alpha' = \sum_{i=1}^{n}\left[\frac{Ad_s}{24(x_i/\bar{x})^2}\left\{-\kappa\frac{96k^2T^2\varepsilon\gamma^2x_i^2}{Az^2(d_s/2)^2}\exp(-\kappa x_i)+1\right\}\cos\theta_b\right] \tag{4}$$

Therefore, (3) can be rewritten as:

$$F_c = \frac{\alpha'}{\bar{x}^2} \tag{5}$$

Now, the water content is given by,

$$W = \frac{\rho_w(V - V_s)}{\rho_s V_s} \tag{6}$$

where, V and V_s are the total mud volume and the total equivalent spherical volume, respectively. If V and V_s are replaced by the equivalent spherical diameter d_s and the mean clearance $\bar{x}$, (6) can be rewritten as:

$$W = \frac{\rho_w\left\{(d_s + \alpha''\bar{x})^3 - (\pi/6)d_s^3\right\}}{\rho_s(\pi/6)d_s^3} \tag{7}$$

Next, the equation relating W to $\bar{x}$ can be introduced as:

$$\alpha''\frac{\bar{x}}{d_s} = \left[\frac{\pi}{6}\ (1+sW)\right]^{1/3} - 1 \tag{8}$$

where α'' depends on the particle distribution in the mud layer, and s is the specific weight of particle. Then, the final expression for the cohesive force F_c can be obtained by substituting (8) for (4) as:

$$F_c = \frac{\alpha_0}{d_s^2}\left[\frac{1}{\{(\pi/6)(1+sW)\}^{1/3}-1}\right]^2 \tag{9}$$

where α_0 is an electrochemical anchoring coefficient, equal to $\alpha'\alpha''$.

6. CRITICAL SHEAR STRESS

The critical shear stress at the beginning of particle movement, i.e., the stress calculated from (1) under the equilibrium of the cohesive force and shear due to water flow, is

$$\tau_{*c} = \left(\cos\theta_b \tan\phi - \frac{\rho_w}{\rho_s-\rho_w}\sin\theta_b\right)\frac{\Phi_s}{\tan\phi} + \beta\left[\frac{1}{\{(\pi/6)(1+sW)\}^{1/3}-1}\right]^2 \tag{10}$$

where Φ_s is the Shields parameter $(=2\kappa_3 \tan\phi/\{C_D[\kappa_1+\kappa_2(C_L/C_D)\tan\phi]\Phi_0^{\,2}\})$. The quantity β is an electrochemical anchoring coefficient related to attraction caused by the surface charge of particles:

$$\beta = \frac{6\alpha_0\Phi_s}{\pi(\rho_s-\rho_w)gd_s^{\,3}} \tag{11}$$

Observe that the first term on the right hand side of (10) is the critical shear stress for non-cohesive particles, and the second term represents the viscous shear stress. Because the viscosity of mud strongly depends on the water content of the deposit, W in (10) can be estimated from the non-dimensional viscosity η_* of mud according to (Taki, 1991):

$$\frac{W}{W_L} = 32\eta_*^{\,-0.4} \tag{12}$$

in which, η_* $[= \nu_b/\{(s-1)gd_s^{\,3}\}^{1/2}]$ embodies the degree of resistance of a particle deposited on the mud surface and ν_b is the kinematic viscosity of mud.

With regard to non-cohesive fine particles, the critical shear stress was experimentally determined as 0.96 N/m^2 by Dunn (1959), 0.043 N/m^2 by Ohtsubo (1983) and 0.0473 (non-dimensional) by Sawai and Ashida (1977). If the (non-dimensional) critical shear stress can therefore be assumed to be 0.05, (10) can be rewritten as:

$$\tau_{*c} = 0.05 + \beta \left[\frac{1}{\{(\pi/6)(1 + 32sW_L\eta_*^{-0.4})\}^{1/3} - 1} \right]^2 \tag{13}$$

Observe that the critical shear stress decreases as the water content increases, and asymptotically approaches to the critical stress of non-cohesive particles at a high water content. Equation (13) is applicable to mud of grain size less than several tens of microns and for a high relative water content (W/W_L), i.e., 2 or more in dry weight percent. The coefficient β in (10) is obtained in the range of 0.1 to 2.0, with a mean value of 0.3 according to the experimental data shown in Figure 5.

7. CONCLUSIONS

This study has led to the following important results:

(1) For mud with a high water content, the electrochemical anchoring force is derived from the vertical component of the anchoring force, which in turn is calculated from the balance between the cohesive force of mud particles obtained from the sum of the van der Waals attraction between particles. The anchoring force can be represented in terms of a simplified formula by inclusion of an anchoring coefficient for the surface charge of particles.

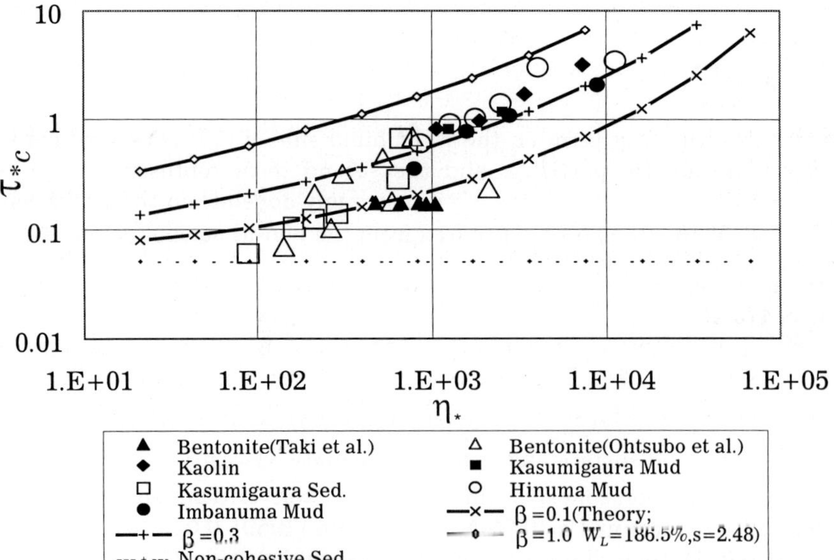

Figure 5. Non-dimensional critical shear stress variation with non-dimensional mud viscosity.

(2) The final formula, (13), for the calculation of the (non-dimensional) critical shear, is applicable to mud whose grain size is less than several tens of microns, and whose relative water content is 2 or more in dry weight percent. The best-fit value for the electrochemical anchoring coefficient β is found to be 0.3 from experimental data.

REFERENCES

Allen, J. R. L., 1969. Erosional current marks of weakly cohesive mud beds. *Journal of Sedimentary Petrology*, 39(2), 607-623.

Allen, J. R. L., 1982. Entrainment and transport of sedimentary particles. In: *Developments in Sedimentology 30A, Sedimentary Structures, their Character and Physical Basis*. Vol.1, Elsevier, Amsterdam, 69-74.

Dunn, I. S., 1959. Tractive resistance of cohesive soil. *Journal of the Soil Mechanics and Foundations Division of ASCE,* 85(3), 1-24.

Hunter, J. R., 1981. *Zeta Potential in Colloid Science, Principles and Applications.* Academic Press, New York.

Ito, I., and Matsui, T., 1975. Plastic flow mechanism of clay. *Proceedings of the Japanese Society of Civil Engineers*, Tokyo, 236, 109-123.

Kitahara, A. Fujii, T., and Katano, S., 1971. Dependence of zeta-potential upon particle size and capillary radius at streaming potential study in nonaqueous media. *Bulletin, Chemical Society of Japan*, 44, 3242-3245.

Ohtsubo, K., 1983. Experimental studies on the physical properties of mud and the characteristics of mud transportation. *Research Report No. 42*, National Institute for Environmental Studies, Tokyo, 177p.

Overbeek, J. Th. G., 1991. Stability of hydrophobic colloids and emulsions. In: *Colloid Science*, H. R. Kruyt ed., Elsevier, Amsterdam, 302-341.

Partheniades, E., 1962. A study of erosion and deposition of cohesive soils in salt water. *Ph.D. Thesis*, University of California, Berkeley, 182p.

Sawai, K., and Ashida, K., 1977. Erosion and cross section on the cohesive stream bed. *Proceedings of the Japanese Society of Civil Engineers*, Tokyo, 266, 73-86.

Taki, K., 1986. Rheological behavior on high density mud flow. *Proceedings of the 30th Japanese Conference on Hydraulics,* Japanese Society of Civil Engineers, Tokyo, 475-498.

Taki, K., 1991. Critical shear stress on cohesive mud particles in river. *Proceedings of the 3rd IAWPRC Conference Asian Waterqual'91, Development and Water Pollution Control*, F. Wang, T. Deng and X. Li eds., IAWPRC, Shanghai, China, V71-V76.

Coastal and Estuarine Fine Sediment Processes
W.H. McAnally and A.J. Mehta (Editors)

Mud scour on a slope under breaking waves

H. Yamanishi[a], O. Higashi[b], T. Kusuda[a] and R. Watanabe[c]

[a]Department of Urban and Environmental Engineering, Kyushu University, 6-10-1 Hakozaki, Higashi-ku, Fukuoka, 812-8581, Japan

[b]City and Environment Planning, Research, EX Corporation, 2-17-22 Takada, Toshima-ku, Tokyo, 171-0033, Japan

[c]Department of Civil Engineering, Fukuoka University, 8-19-1 Nanakuma, Jonan-ku, Fukuoka, 814-0180, Japan

Scour of mud by breaking waves was investigated in an experimental flume with a sloping bed and in a tidal river. It was found that: (1) the impact of breaking wave on the slope can be formulated in terms of the conservation equation of momentum considering reflection of waves; (2) the impact of breaking wave decreased with increasing wave steepness under the same wave height; (3) cohesive sediment was scoured as pieces and accumulated as clasts at the foot of the slope; and (4) the mass of sediment scoured by breaking waves, W_s, per N number of waves and per unit area can be formulated as $(W_s g/A_s)/\tau_s/N = m[(p_m/\tau_s)-(p_m/\tau_s)_c]$, where p_m and τ_s are the maximum impact of breaking wave and the bottom shear strength, respectively. From the experiments, m and $(p_m/\tau_s)_c$ were found to be 0.14 and 0.37, respectively.

1. INTRODUCTION

Waves and currents cause sediment transport, and turbidity maxima are formed and travel upstream and downstream in estuaries. Futawatari and Kusuda (1993) performed long term field observations in the Rokkaku River in Japan. This river is located in Ariake Bay, on the western side of Kyushu Island. The results showed that the minimum and maximum concentrations of suspended solids in a fortnightly cycle took place over two to three days after a neap tide and a spring tide, respectively. Suspended solids concentration in the upper water layer tended to be lower during a neap tide than usual, which means that suspended solids settled and accumulated on the bottom and banks of the river. Kusuda (1994) reported that an annual sedimentation rate at the banks was over 20 cm in height. Thus, it is difficult to maintain the cross-section of rivers with a high turbidity water body. Not only mud accumulated on banks prevents maintenance of the cross-section, but also its dredging can be costly. Therefore, it becomes essential to remove mud efficiently. A paucity of studies on natural removal of mud prompted us to investigate this problem. Breaking waves impose a high pressure on a slope, and their successive attacks may cause scour. The purpose of this paper is to examine the effects of breaking waves as a method to scour cohesive sediment accumulated on the banks of rivers.

2. IMPACT OF BREAKING WAVE ACTION

Figure 1 defines the notation for the estimation of the impact of breaking wave action on a slope. The x-axis is taken along the slope, and the z-axis normal to the slope. Here, θ is the incident angle, β is the slope angle, γ is the reflected angle, H_b is the breaking wave height, h_b is the breaking water depth, C_b is the phase velocity, and v is the free fall velocity. For estimation of the impact of breaking waves, conservation of momentum is stated for the direction normal to the slope as:

$$\int_0^{t^*} F_z dt = \int_0^{t^*} \rho Q_2 V_{2z} dt - \int_0^{t^*} \rho Q_1 V_{1z} dt$$
$$= \int_0^{t^*} \rho A_2 V_2 V_{2z} dt - \int_0^{t^*} \rho A_1 V_1 V_{1z} dt \qquad (1)$$

Here, F_z is the vertical breaking wave force onto the slope, t^* is the impulse duration, ρ is the water density, Q_1 is the incident flow rate, Q_2 is the splash flow rate, and V_{1z} and V_{2z} are the vertical components of V_1 and V_2, respectively. As seen, this equation considers splash in the breaker zone.

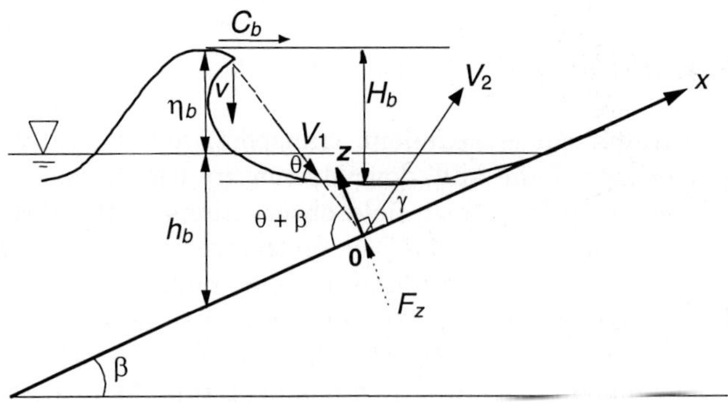

Figure 1. Notation for the formulation of the wave breaking and scour problem.

The following relations are derived from geometry:

$$\left.\begin{aligned} A_1 &= A\sin(\theta+\beta) \\ A_2 &= A\sin\gamma \end{aligned}\right\} \tag{2}$$

$$\left.\begin{aligned} V_{1z} &= -V_1\sin(\theta+\beta) \\ V_{2z} &= V_2\sin\gamma \end{aligned}\right\} \tag{3}$$

Then, substituting (2) and (3) into (1), the following equation is obtained:

$$\int_0^{t*} F_z dt = \int_0^{t*} \rho A V_1^2 \sin^2(\theta+\beta) dt + \int_0^{t*} \rho A V_2^2 \sin^2\gamma dt \tag{4}$$

Assuming that ρ, A, V_1, V_2 and θ are constant when a wave breaks on the slope, (4) is transformed as follows:

$$\int_0^{t*} p dt = \rho t^* V_1^2 \sin^2(\theta+\beta) + \rho t^* V_2^2 \sin^2\gamma \tag{5}$$

where, the impact of breaking wave pressure p is F_z/A.

Further, the impulse term on the left-hand side of (5) is approximated by (6):

$$\int_0^{t^*} p dt \approx k p_m t^* \tag{6}$$

where, k is a coefficient of the impulse term, the product of p_m and t^*, in which p_m is the maximum breaking wave pressure. Substituting (6) into (5), (7) is obtained:

$$p_m = \frac{\rho}{k}\left[V_1^2 \sin^2(\theta+\beta) + V_2^2 \sin^2\gamma\right] \tag{7}$$

Here, we assume that a portion of Q_1 becomes Q_2, i.e.,

$$Q_2 = k_1 Q_1 \tag{8}$$

Then, (9) is obtained as:

$$V_2 = K_1 V_1 \tag{9}$$

Finally, (10) is obtained by substituting (8) and (9) into (7):

$$\frac{p_m}{\rho g} = \alpha \frac{V_1^2}{2g}\left[\sin^2(\theta+\beta) + K_1^2 \sin^2\gamma\right] \tag{10}$$

Thus, the impact of breaking wave on a slope results in (10), where, g is the acceleration of gravity, and α and K_1 are the coefficients based on experimental results. Some parameters in (10) must be determined to calculate the impact of breaking wave. The incident velocity V_1 and the incident angle θ are described by (11) and (12):

$$V_1 = \sqrt{v^2 + C_b^{\,2}} \tag{11}$$

$$\theta = \tan^{-1}\left(v / C_b\right) \tag{12}$$

The free falling velocity v and the phase velocity C_b are defined by (13) and (14), respectively:

$$v = \sqrt{2gH_b} \tag{13}$$

$$C_b = \sqrt{g\left(h_b + \eta_b\right)} \tag{14}$$

where, H_b is the breaking wave height, η_b is the wave crest height (the height from mean water level to the top of breaking wave height, see Figure 1), and h_b is the breaking water depth. Here, we define the phase velocity C_b as the wave velocity of a solitary wave, because the small amplitude wave theory is not applicable to the breaking wave phenomenon. Reflected angle γ and coefficients α and K_1 must be estimated from experimental results.

3. BREAKING WAVE EXPERIMENTS

The banks of the Rokkaku River have steep slopes (1/4~1/6). Because many experiments on sediment transport have in the past been conducted on gentle slopes, we performed some experiments on breaking waves on a rigid/mud bed with a slope of 1 in 5. The wave tank used for the experiments was 0.5 m wide, 0.3 m high and 14 m long, with a sloping bed. The experimental apparatus is shown in Figure 2. Regular waves were generated under various wave conditions, and all waves were plunging breakers over the slope.

3.1. Breaking wave height H_b

Le Méhauté and Koh (1967) suggested the following empirical equation to describe the relation between the deepwater wave height and the breaking wave height,

$$\frac{H_b}{H_0'} = 0.76\left(\tan\beta\right)^{1/7}\left(\frac{H_0'}{L_0}\right)^{-1/4} \tag{15}$$

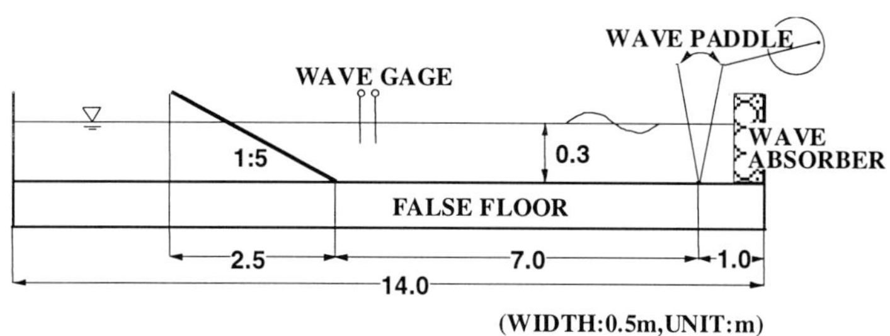

Figure 2. Elevation view of the experimental apparatus.

$$H_0' = K_r K_d K_f H_0 \tag{16}$$

where, $\tan\beta$ is the slope, H_0' is the equivalent deepwater wave height, H_0 is the deepwater wave height, L_0 is the deepwater wave length ($=gT^2/2\pi$), K_r is the refraction coefficient, K_d is the diffraction coefficient, and K_f is the friction coefficient. Here, for the sake of simplicity, we have assumed that all coefficients in (16) are unity. Therefore, we use H_0/L_0 as the characteristic wave steepness.

Figure 3 shows the relation between H_0/L_0 and H_b/H_0. In the plot, (15) does not agree with the experimental data, but the trend can be regarded as similar. Equation (15) is applicable only to a gentle slope (~1/10), and its applicability to a steep slope has not been verified. However, it is convenient to use the form of (15), in order to simulate measured data. On the basis of (15), two empirical equations were obtained for the initial and steady states of various experimental conditions. Here, the state of a few waves except for the first wave is defined as the initial state [see Figure 11(a)], and after that it is considered to be steady state. Each experimental result shows the mean value. Figure 3 relates to the initial state, and plots (17) with a dotted line:

$$\frac{H_b}{H_0} = 0.63(\tan\beta)^{1/7}\left(\frac{H_0}{L_0}\right)^{-1/4} \tag{17}$$

Figure 4 is for the steady state, and plots (18) with a dotted line:

$$\frac{H_b}{H_0} = 0.55(\tan\beta)^{1/7}\left(\frac{H_0}{L_0}\right)^{-1/4} \tag{18}$$

Comparison of Figures 3 and 4 shows that the trend gradually changes due to the effect of return flow. Equations (17) and (18) are empirical, but are applicable in this experimental region, because of the observed agreement with experimental data. In this way, it is possible to predict measured values of H_b.

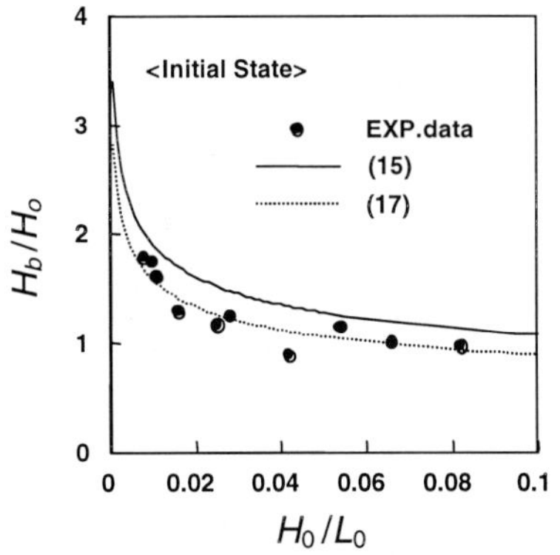

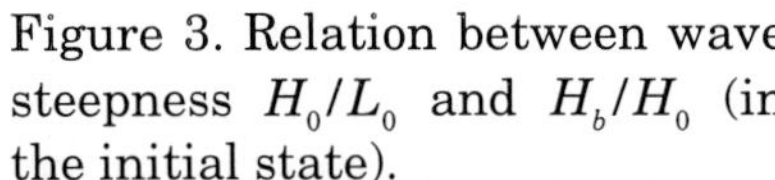
Figure 3. Relation between wave steepness H_0/L_0 and H_b/H_0 (in the initial state).

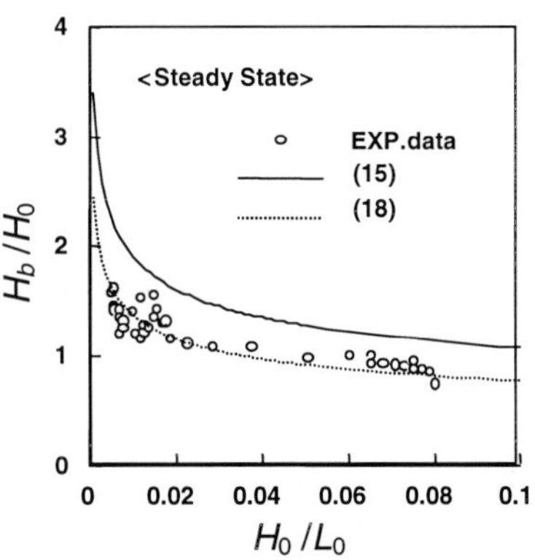

Figure 4. Relation between wave steepness H_0/L_0 and H_b/H_0 (at steady state).

3.2. Breaking water depth h_b

Sunamura (1983) derived an empirical equation describing a critical breaking wave on the basis of experimental data:

$$\frac{H_b}{h_b} = 1.09(\tan\beta)^{0.17}\left(\frac{h_b}{L_0}\right)^{-0.1} \tag{19}$$

Figures 5 and 6 show the relation between h_b/L_0 and H_b/h_b. In order to estimate the breaking wave water depth h_b, the following empirical equations were obtained from these results:

$$\frac{H_b}{h_b} = 0.97(\tan\beta)^{0.17}\left(\frac{h_b}{L_0}\right)^{-0.1} \tag{20}$$

$$\frac{H_b}{h_b} = 0.89(\tan\beta)^{0.17}\left(\frac{h_b}{L_0}\right)^{-0.1} \tag{21}$$

Equation (20) is in agreement with the initial state, and (21) with the steady state. Both are applicable in the experimental domain.

3.3. Estimation of breaking wave pressure

Figure 7 shows the relation between H_0/L_0 and the incident angle θ, and comparison of calculated results with experimental data. Each angle θ was measured by images from a digital video camera. Some lines in Figure 7 have

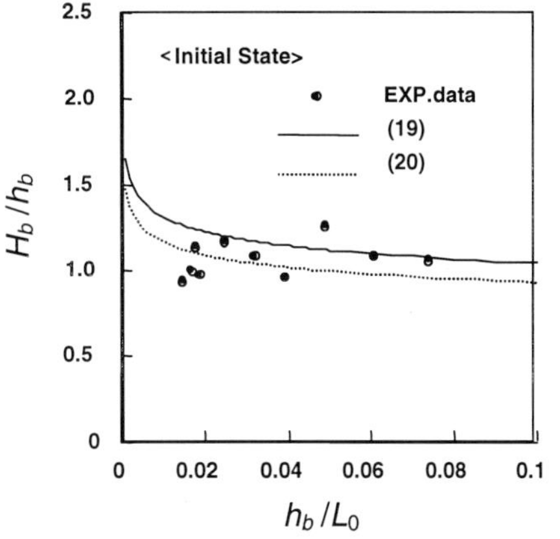

Figure 5. Relation between h_b/L_0 and H_b/h_b (in the initial state).

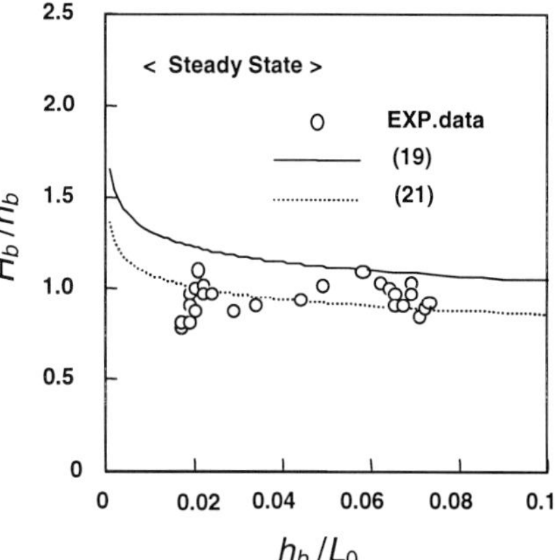

Figure 6. Relation between h_b/L_0 and H_b/h_b (at steady state).

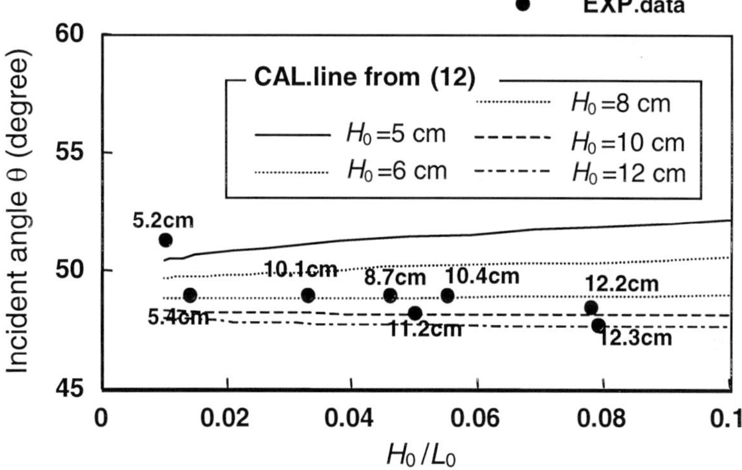

Figure 7. Comparison of calculated results with experimental data on the relationship between θ and H_0/L_0.

been obtained from (12). In order to calculate the incident angle θ from (12), it is necessary to set values of the unknown parameters included in (13) and (14). Therefore, we use (17) or (18) to obtain the breaking wave height H_b, (20) or (21) to obtain the breaking water depth h_b and (22) to estimate the wave crest height η_b (see Figure 8):

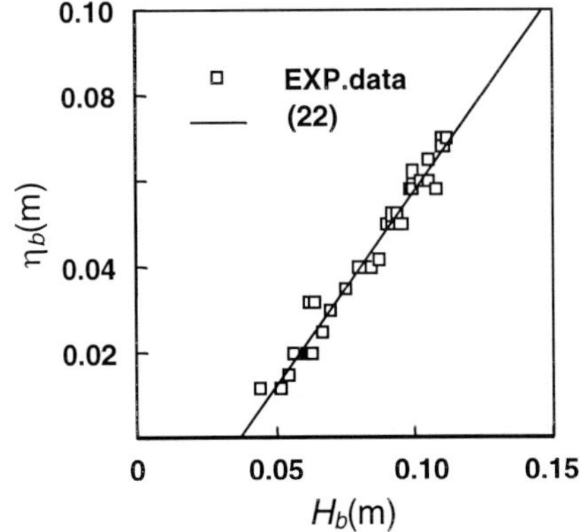

Figure 8. Relation between H_b and η_b

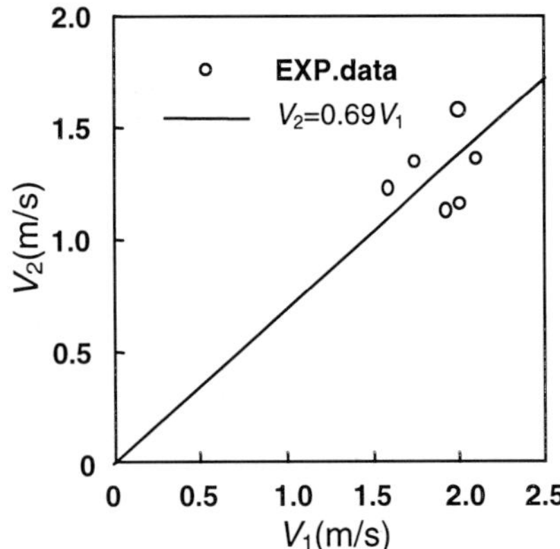

Figure 9. Relation between incident velocity V_1 and reflected velocity V_2.

$$\eta_b = 0.91(H_b - 0.036) \tag{22}$$

Extrapolated values of K_1 and γ in (10) are required to determine the impact of breaking wave from (10). Figure 9 shows the relation between the incident velocity V_1 and the reflected velocity V_2 from experimental results. If V_2 is assumed to be linear with V_1, then K_1 becomes 0.69. Figure 10 shows the relation between the deepwater wave steepness H_0/L_0 and the reflected angle γ. The angle γ was estimated from water motion monitored by the digital video camera. Equation (23) as the best-fit relation with the experimental results are shown in Figure 10.

$$\gamma = 28.6\left(\frac{H_0}{L_0}\right)^{-0.07} \tag{23}$$

Figures 11(a) and (b) indicate typical records of the impact of breaking wave on the rigid slope. The impact pressure was measured with semiconductor type transducers. The sampling speed to record the impact was 200 Hz. As shown in Figure 11, the impact pressure shows spikes during a short time period. However, none of the measured maxima of the pressure p_m indicated the same value, but were scattered. We defined each maximum of p_m under the initial and the steady state as a mean value over several waves.

The point where plunging water penetrated the water surface at first is referred to as the first plunging point (P.P.1). Further, the point where splashed water enters through water surface again is called the second plunging point

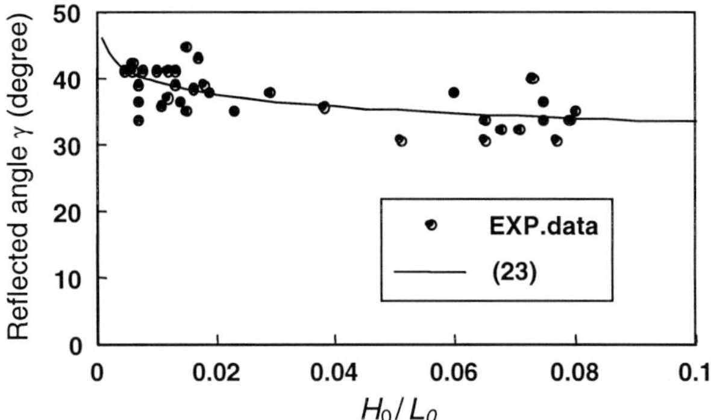

Figure 10. Relation between H_0/L_0 and γ.

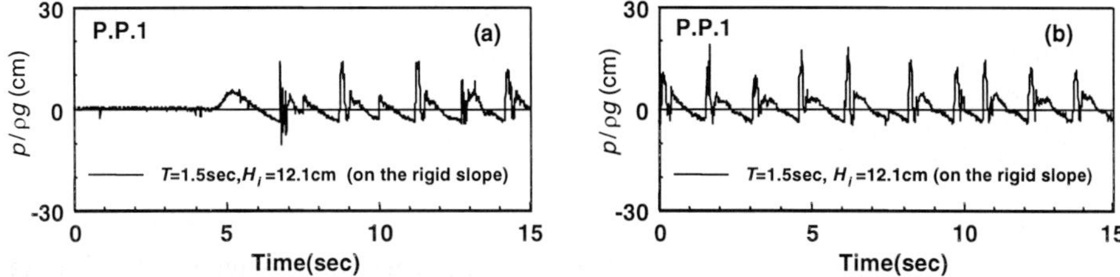

Figure 11. Temporal changes of breaking wave pressure for: (a) initial state and (b) steady state.

(P.P.2). Figure 12 shows the relation between the maximum values of the breaking wave pressure ($P_{1,m}$) at the first plunging point and values of $\rho g(V_1^2/2g)[\sin^2(\theta+\beta)+K_1^2\sin^2\gamma]$ based on (10). The best-fit line from Figure 12 yields $\alpha = 0.79$ ($k = 2.5$) in the initial state and 0.72 ($k = 2.7$) at steady state. For these results, Figure 13 can be used to estimate the impact of breaking wave. Figure 13 also shows variations of breaking wave impact at the first plunging point $P_{1,m}$ with wave steepness H_i/L_i. As the wave steepness increases, the breaking wave pressure decreases and the value of $P_{1,m}$ at steady state is seen to be smaller than in the initial state. Using this figure, breaking wave impact for a given wave steepness can be estimated.

4. RHEOLOGICAL CHARACTERISTICS

Shear strength is one of the typical rheological characteristics of mud accumulated on riverbanks. It depends on the state of accumulated mud. Kusuda et al. (1989) showed that the critical shear stress for erosion and the mass of

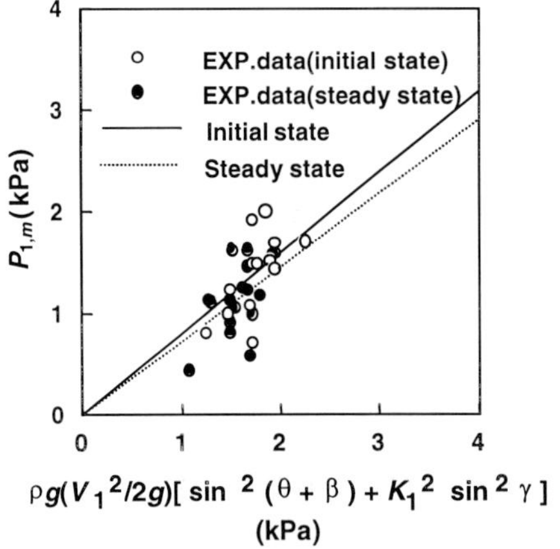

Figure 12. Relation between $P_{1,m}$ and $\rho g(V_1^2/2g)[\sin^2(\theta+\beta)+K_1^2\sin^2\gamma]$.

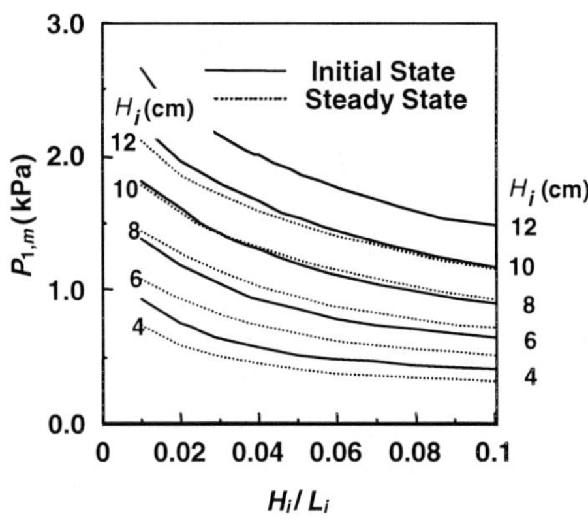

Figure 13. Variations of $P_{1,m}$ with wave steepness H_i/L_i.

erodible mud depend on the disturbance of mud. The yield stress was measured in the present study by a vane shear meter. After inserting a vane to a fixed depth, the maximum of working torque was measured by rotating the meter with a constant speed of 0.5 deg/s. In addition, in order to examine the relation between the water content and the shear strength, natural muds were collected in acrylic columns and the vertical profiles of water content were measured.

Mud used was obtained in the undisturbed state from the Rokkaku River, where the maximum tidal range is about 5 m. The mean particle diameter and density were 6.0 mm and 2,540 kg/m^3, respectively. The vertical profiles of the water content W are shown in Figure 14. The water content decreases toward the bottom due to consolidation.

Figure 15 indicates changes in shear strength τ_s with W. The shear strength of both undisturbed and disturbed mud decreases with increasing water content. According to these results, the relation between τ_s and W is expressed as:

$$\text{Undisturbed muds: } \tau_{s,1} = 1.42\times10^{10}W^{-4.47} \tag{24}$$

$$\text{Disturbed muds: } \tau_{s,2} = 4.49\times10^{9}W^{-4.47} \tag{25}$$

where, $\tau_{s,1}$ and $\tau_{s,2}$ are shear strength of undisturbed and disturbed muds, respectively.

Figure15 implies a decrease in shear strength due to remolding of mud. Here, a parameter r for gaging the remolding effect is selected to be

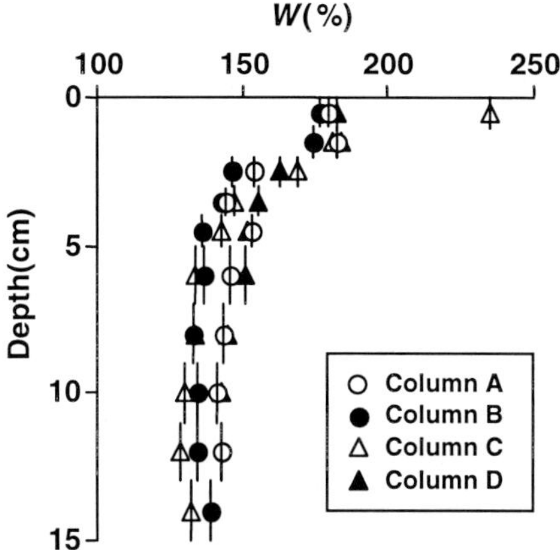

Figure 14. Water content profiles of bottom mud in the Rokkaku River, Ariake Bay.

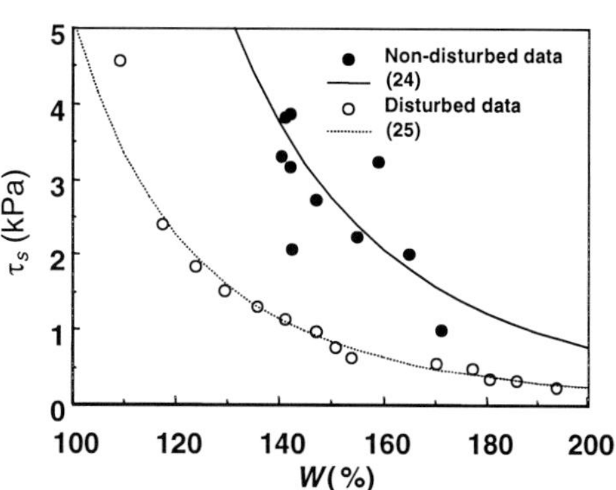

Figure 15. Changes in shear strength τ_s with water content W.

$$r = \frac{\tau_{s,1} - \tau_{s,2}}{\tau_{s,2}} \tag{26}$$

Substituting (24) and (25) into (26), r is found to be equal to 2.16. Therefore, the shear strength of undisturbed mud is about twice that of disturbed mud.

5. SCOUR EXPERIMENTS

Experiments to estimate the amount of sediment scoured by breaking waves were performed by using the described experimental flume with a slope. Along the slope, a trench was installed as the test section, in which undisturbed mud was filled to a thickness of 0.1 m. In all experiments, the water depth was 0.3 m. The experimental conditions are summarized in Table 1. The breaking wave pressure at the first plunging point ($P_{1,m}$) was obtained from Figure 13 for steady state, because of sustained wave action on the slope.

Figure 16(a) shows a profile of the maximum breaking wave pressure on the slope (for an incident wave period T=1.7 sec and height H_i=11.5 cm). The results indicate that the breaking wave pressure p_m was high around the plunging point, especially at the first plunging point (P.P.1), and was stronger than at the second point (P.P.2). Figure 16(b) shows temporal changes of scour under the same conditions as Figure 16(a). There is a correlation between the impact pressure and the scour of the inclined mud bed. Mud scoured *en masse* and accumulated as clasts at the foot of the slope. This phenomenon, which is also found in the field experiments under ship waves, depends on the mud characteristics.

Table 1
Experimental conditions

T(sec)	H_i(cm)	τ_s (kPa)	$P_{1,m}$(kPa)
1.0	9.2	2.30	1.00
1.5	9.6	1.39	1.24
2.0	6.6	2.30	1.05
1.5	12.3	1.39	1.51
1.5	11.3	1.39	1.42
1.5	9.9	2.46	1.28
1.5	8.4	2.30	1.13
1.3	11.2	4.20	1.33
1.5	11.0	1.96	1.40
1.7	11.5	2.54	1.51

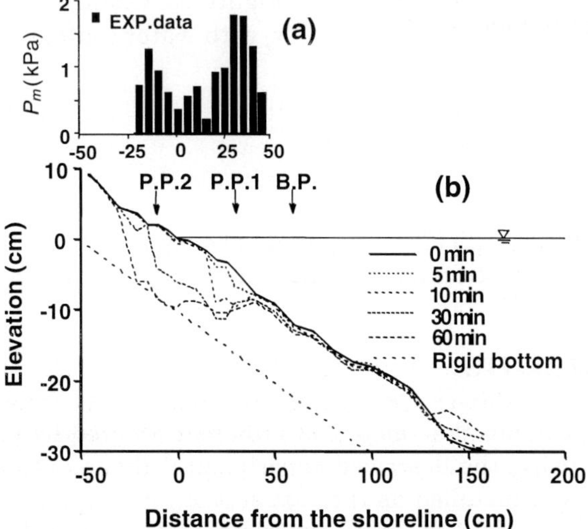

Figure 16. (a) Profiles of maximum impact pressure and (b) temporal changes of mud surface elevation.

Figure 17 shows relations between N, the number of breaking waves, and the amount of scoured mud per unit width W_s/B. Here, W_s is the mass of mud scoured by breaking waves, and B is the width of the experimental flume. W_s was evaluated by assuming that mud with a constant water content within the scoured layer was eroded uniformly along the transverse direction of the experimental flume. As seen in Figure 17, W_s/B increases with N. Furthermore, it is believed that mud was scoured step by step as the shear strength of mud decreased by successive attacks of breaking waves.

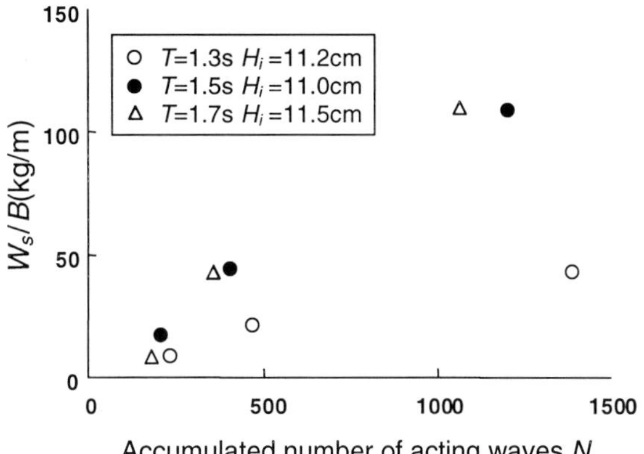

Figure 17. Relation between number N of breaking waves acting on a mud slope and mass of scoured mud per unit width (W_s/B).

When the breaking wave pressure exceeds the sum of the weight and internal shear strength of mud, scour occurs. Accordingly, W_s per wave per area A_s is formulated as

$$\frac{(W_s g/A_s)}{\tau_s N} = m\left[\left(\frac{p_m}{\tau_s}\right) - \left(\frac{p_m}{\tau_s}\right)_c\right] \tag{27}$$

where p_m and τ_s are the maximum breaking wave pressure and mud shear strength, respectively. Figure 18 shows the relation between (p_m/τ_s) and $(W_s g/A_s)/\tau_s/N$, where p_m is at steady state at the first plunging point (from Figure 16), W_s is the value for 500 waves (N=500), and A_s is the area scoured. In addition, τ_s was calculated from vane shear test results. Figure 18 suggests an increase in $(W_s g/A_s)/\tau_s/N$ with (p_m/τ_s), and the existence of a critical shear strength $(p_m/\tau_s)_c$. When (27) is applied to data obtained in the laboratory tests, the coefficient m and $(p_m/\tau_s)_c$ are found to be 1.4 and 0.37, respectively. Then, (26) is transformed as:

$$\tau_{s,2} = \frac{\tau_{s,1}}{1+r} = 0.32\tau_{s,1} \tag{28}$$

Comparison of $\tau_{s,2}$ obtained from (28) and the critical breaking wave pressure, p_{mc}(=$0.37\tau_{s,1}$), obtained from Figure 18 shows that both are almost equal. Therefore, the maximum of breaking wave pressure p_m to scour an undisturbed mud is seen not to necessarily exceed the shear strength of mud $\tau_{s,1}$, but $\tau_{s,2}$. Now, considering

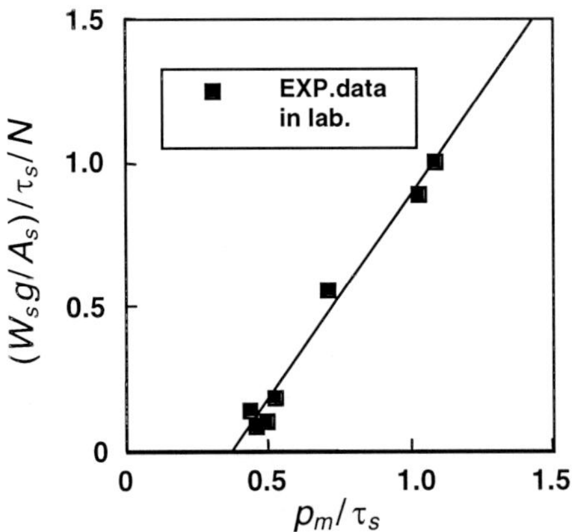

Figure 18. Relation between (p_m/τ_s) and $(W_s g/A_s)/\tau_s/N$.

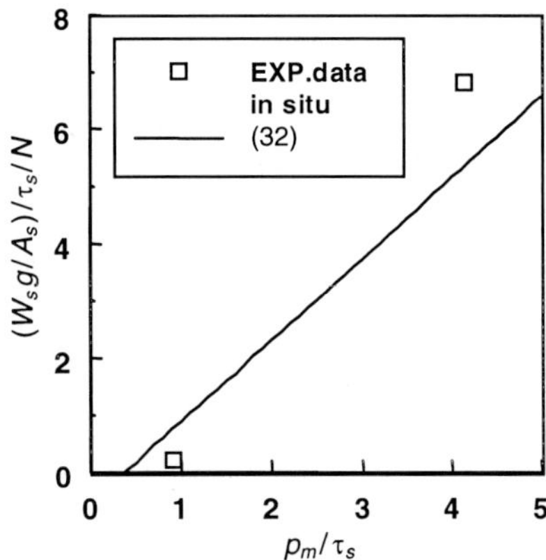

Figure 19. Relation between (p_m/τ_s) and $(W_s g/A_s)/\tau_s/N$.

$$p_{mc} = n_c \tau_c \tag{29}$$

and substitution of (24) and (25) into (29) yields

$$p_{mc,1} = \delta_1 W^n \tag{30}$$

$$p_{mc,2} = \delta_2 W^n \tag{31}$$

where, n_c is determined from experimental data, $\delta_1 = 5.11 \times 10^9$, $\delta_2 = 1.62 \times 10^9$ and $n = -4.47$. From (30) and (31), it is possible to predict the value of p_{mc} for a given state of mud and water content.

For an estimation of the mass of sediment scoured in the field, *in situ* experiments were carried out at the riverbank 6.8 km upstream from the river mouth, where the maximum tidal range was 5 m. Ship waves were produced by a fishing boat, which moved about 30 times per set of experiments. During these experiments, wave height was measured and the surf zone was recorded by a video camera. In the same manner as above, results of the field experiments were derived. Figure 19 is a comparison of (32) with results of the field experiments. Equation (32) was obtained from the laboratory experiments as follows:

$$\frac{(W_s g / A_s)}{\tau_s N} = 1.4\left[\left(\frac{p_m}{\tau_s}\right) - 0.37\right] \tag{32}$$

The field data are not sufficient; however, they are in accordance with the results from (32). Consequently, the results obtained both in the laboratory and in the field indicate the predictability of sediment mass scoured by breaking waves, especially by ship waves.

6. CONCLUSIONS

We investigated the effects of the impact of breaking waves on mud scour. The conclusions of this study are as follows:

(1) Breaking wave pressure acting on a slope can be formulated by the equation of conservation of momentum considering reflection. From pressure measurements, a relation between $P_{1,m}$ and wave steepness H_i/L_0 was derived.

(2) As wave steepness increases, the breaking wave pressure decreases, and $P_{1,m}$ at steady state is smaller than in the initial state.

(3) The shear strength of mud, measured by a vane shear meter, is a function of the water content. Due to the remolding effect, the shear strength of undisturbed mud was about twice the shear strength for disturbed mud.

(4) The mass of sediment scoured by breaking waves, W_s, per wave and per unit area is given by $(W_s g/A_s)/\tau_s/N=m[(p_m/\tau_s)-(p_m/\tau_s)_c]$. From the experiments, m=1.4 and $(p_m/\tau_s)_c$ =0.37 were obtained.

7. ACKNOWLEDGMENT

This study was initiated as a joint work by the Takeo office of the Ministry of Construction, and was supported in part by a Grant in Aid from the Fundamental Scientific Research and Scholarship of the Wesco Foundation.

REFERENCES

Futawatari, T., and Kusuda, T., 1993. Modeling of suspended sediment transport in a tidal river. In: *Nearshore and Estuarine Cohesive Sediment Transport*, A. J. Mehta ed., American Geophysical Union, Washington, DC, 504-519.

Kusuda, T., 1994. *Application of Self-Purification's Functions*, Gihodo Publications, Tokyo, Japan, 166p. (in Japanese).

Kusuda, T., Yamanishi, H., Yoshimi, H., and Futawatari, T., 1989. Experimental study on erosion of disturbed and non-disturbed mud. *Proceeding, Coastal Engineering Conference*, Japanese Society of Civil Engineers, Tokyo, Japan, 36, 314-318 (in Japanese).

Le Méhauté, B., and Koh, R.C.Y., 1967. On the breaking waves arriving at an angle to the shore. *Journal of Hydraulic Research*, 5(1), 67-88.

Sunamura, T., 1983. Determination of breaker height and depth in the field. *Annual Report No. 8*, Institute of Geoscience, University of Tsukuba, Tsukuba, Japan, 53-54.

Coastal and Estuarine Fine Sediment Processes
W.H. McAnally and A.J. Mehta (Editors)

Fluid mud in the wave-dominated environment revisited

Y. Li[a] and A. J. Mehta[b]

[a]Tetra Tech, Inc./Infrastructure Southwest Group, 17770 Cartwright Road, Suite 500, Irvine, California 92614, USA

[b]Department of Civil and Coastal Engineering, University of Florida, 345 Weil Hall, P.O. Box 116580, Gainesville, Florida 32611, USA

Generation of fluid mud by short period water waves is revisited following previously reported analytic work of the authors. Liquefaction of a soft cohesive mud bed due to wave-induced pressure loading is considered to occur when the effective submerged weight of the particle or floc is exceeded by the upward inertial force due to bed oscillation. Bed dynamics, characterized by a uniaxial extensional-Voigt viscoelastic solid response, is modeled as a spring-dashpot-mass analog for obtaining fluid mud layer thickness at equilibrium under sustained wave motion. Model results are compared with data obtained in wave flumes using weakly cohesive muds. Order of magnitude agreement between theory and data suggests that shear rheometry can be used to determine the appropriate constitutive relationship required to calculate fluid mud thickness. It is shown that the model provides reasonable estimates of fluid mud thickness in Lake Okeechobee in south-central Florida. A noteworthy feature of the method is that mud properties are described by easily measurable parameters including density and rheological constants, without the need to specify mud composition.

1. INTRODUCTION

The transportability of fine-grained sediment and the fate of associated nutrients and contaminants are strongly influenced by fluid mud generated by water wave action over mud beds. Fluid mud is a highly viscous, slurry-like mixture of fine sediment and water with practically no shear strength or elasticity (Mehta et al., 1995). Its density characteristically ranges between about 1,030 kg/m^3 and 1,300 kg/m^3. However, in the approximate range of 1,200 kg/m^3 to 1,300 kg/m^3, mud can occur either as a bed with a weakly elastic particulate matrix and a measurable shear strength, or as fluid mud, depending on sediment composition and the state of agitation. When waves act on a comparatively soft bed its particulate matrix is disturbed, and when this matrix is practically destroyed the bed looses its elasticity, the effective normal stresses vanish and fluid mud is generated (Mehta, 1996). The breakdown of bed matrix is a complex, time-dependent process that may occur by "soil piping", in which resonance is initiated within vulnerable cavities in the bed, followed by progressive failure of the matrix around such cavities (Foda et al., 1991). Ultimately, under steady and sustained wave motion a fluid mud layer of equilibrium thickness determined by wave parameters and mud properties is generated.

For an examination of many practical problems, the determination of the equilibrium thickness of cohesive fluid mud layer is a matter of first order importance. The method for obtaining this thickness rests on two requirements. The first is related to the oscillatory motion of bottom mud, and the second concerns the criterion to distinguish fluid mud from bed. Bottom mud motion is commonly modeled by solving the equations of motion of the coupled water-mud system. The use of such models in conjunction with laboratory flume data has shown that mud motion is dominantly effected by normal stress loading of the mud-water interface, and that pressure-induced shear stresses are largely responsible for the breakup of bed matrix through internal friction (Dalrymple and Liu, 1978; Isobe et al., 1992; de Wit, 1995). It follows that pressure work is ultimately responsible to a great degree for fluid mud generation. Criteria for liquefaction of mud must take into account the characteristically low permeability of the material, and associated low rates of pore water motion. Thus, for example, for weakly cohesive beds composed of a kaolin, Lindenberg et al. (1989) determined the loss of effective stress by assuming the material to be analogous to a sandy bed with a very low permeability. On the other hand, Foda et al. (1993) distinguished fluid mud from bed by liquefaction-induced changes in the rheology of mud assumed as a continuum.

In the above approaches, the essential feature of liquefaction, namely particle-particle separation as the cause of fluid mud generation, is accounted for in an implicit way. Li and Mehta (1997) treated this process directly by considering the forces on a mud particle or floc. The resultant active force was determined from the response of a mechanical analog of mud bed to wave forcing. Since this response is contingent upon mud rheology, the influence of rheological parameters on fluid mud generation was considered in an explicit way. Guidance for this development was obtained from particle-based and mechanistic approaches reported in the literature. Thus, for example, Selly (1988) considered that liquefaction of a sandy bed by vibration occurs when the upward drag exerted by pore water moving relative to particles exceeds the submerged weight of the grain. A simple mathematical model developed by Das (1983) treats the oscillating bed as a spring-mass system and yields an expression for the resonant frequency of oscillation in terms of bed thickness. Models along these lines that additionally include dissipation have been used to simulate fine-grained soil dynamics (Keedwell, 1984).

A noteworthy feature of the approach of Li and Mehta (1997) is that mud properties are specified by easily measurable parameters including density and rheological constants, without the need to specify mud composition, whose role in mud dynamics is difficult to quantify. In what follows we have recapped the main arguments of Li and Mehta (1997), then applied the method to laboratory flume data, followed by Lake Okeechobee in Florida.

2. CRITERION FOR FLUID MUD GENERATION

Consider a mud particle or a cohesive floc treated as an integral particle of mass m_p under wave motion, as shown in Figure 1(a). With reference to such a particle four types of forces are recognized [Figures 1(b) and (c)]: (i) reduced gravity or buoyancy, g', (ii) inter-particle cohesion represented as a cumulative resistive force, L_c, (iii) normally resolved forces, F_1, F_2, F_3 and F_4, and tangentially resolved forces, f_1, f_2, f_3 and f_4, both due to inter-particle contacts, and (iv) inertial force due to bed oscillation, $L_i(t)$, under wave loading. Since the actual normal and shear contact forces surrounding a particle in a cohesive bed can be considered to be randomly directed as a first order approximation, they will be assumed to have no tangible effect on bed liquefaction.

Although the bed matrix can withstand a certain degree of compression, it cannot resist significant stretching forces. Thus, when in the vertical direction the sum of the upward forces equals or just exceeds the sum of the downward forces [Figure 1(c)] the bed is critically stretched, the connections between particles are broken and incipient liquefaction occurs. We will therefore select the criterion for fluid mud generation at any elevation in the bed to be that condition at which the maximum upward force due to wave motion, L_{imax}, equals or just exceeds the sum of the effective submerged weight of the particle in water and the added effect of cohesion, i.e., $m_p\, g'+L_c$, where the reduced gravity $g'=g[\rho_2(z)-\rho_1]/\rho_1$. Here, g is the gravitational acceleration and, assigning subscripts 1 and 2 to the water and mud layers, respectively, ρ_1 is water density and $\rho_2(z)$ is bed

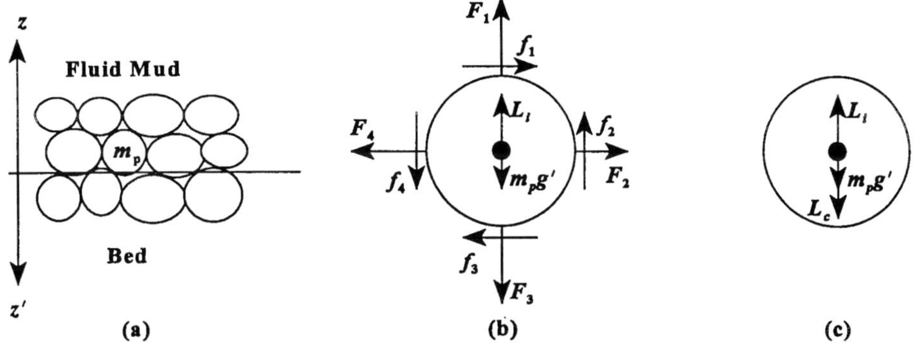

Figure 1. a) Bed and fluid mud layers; b) Forces on a particle; c) Vertical forces on a particle.

density at any elevation, z. Thus, the criterion for incipient fluid mud generation can be expressed by the equality,

$$L_{imax} = m_p g' + L_c \tag{1}$$

Assuming mud properties to vary in the vertical direction only, (1), which is based on forces on a single particle at a given elevation, will be considered to reflect the corresponding condition causing the separation of the entire fluid mud layer from the bed at the same elevation [Figure 1(a)]. Then, when the excess pore pressure is zero, this condition can be restated in the stress form as

$$\sigma_{imax} = \rho_1 g' D + C_r = \sigma' + C_r \tag{2}$$

where σ_{imax} is the maximum value of the inertial stress induced by the vertical oscillation of the bed, D is a representative particle or floc diameter, σ' is the effective normal stress and C_r is the cohesive resistance. Equation (2) has the form of the well-known Coulomb's relation for shear-induced failure (Lambe and Whitman, 1969) and C_r is analogous to the cohesion term in that relation. Equation (2) is ideally applicable to the condition of zero permeability, which precludes pore pressure changes under cyclic loading. For cohesive muds, since a gradual pore pressure buildup does occur, fluid mud generation is likely to take place more readily than implied by (2). In any event, for the simplified treatment considered here this effect of pore pressure, as well as the effect of cohesion, will be considered to modulate the reduced gravity (Partheniades, 1977). Thus, the resulting effective submerged weight of the particle will decrease due to excess pore pressure and increase due to cohesion. Then, given ζ_2 as the vertical displacement of the particle and $\ddot{\zeta}_2$ as the corresponding acceleration, (1) or (2) can be reformulated as

$$\ddot{\zeta}_{max} = \alpha g' \tag{3}$$

where $\ddot{\zeta}_{max}$ is the maximum value of $\ddot{\zeta}_2$, and α is a mud-specific coefficient which modulates gravity. Equation (3) is applicable to low density mud beds subjected to short period waves. For long waves the horizontal pressure gradient can be the dominant force for liquefaction (Madsen, 1974), whereas for fully consolidated beds with high density, cohesion may dominate and prevent liquefaction.

The acceleration, $\ddot{\zeta}_{max}$, characteristically decreases with depth and $\alpha g'$ increases. Given these trends and the threshold condition for liquefaction, i.e., $\ddot{\zeta}_{max} \geq \alpha g'$ at the surface z' (= - z) =0, the resulting fluid mud thickness, z_c, is defined in Figure 2. In general, besides wave height and period, the threshold for liquefaction and z_c depend on mud composition and density, which in turn characterize mud rheology (de Wit, 1995). To calculate $\ddot{\zeta}_{max}$, in the ensuing

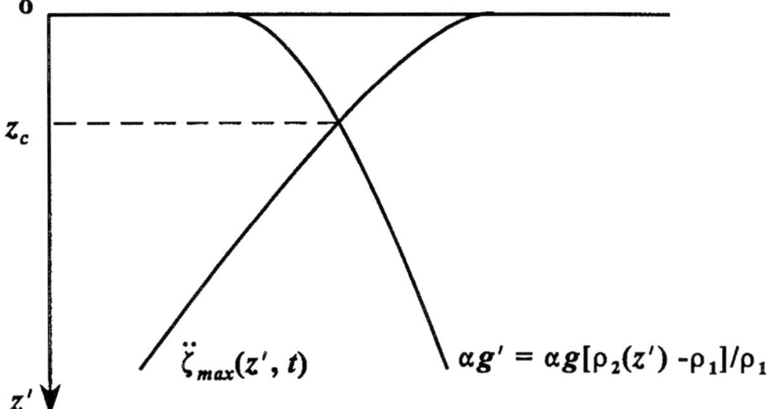

Figure 2. Definition of fluid mud thickness.

summary of the development of Li (1996) and Li and Mehta (1997) the bed is considered to be a linear viscoelastic continuum, following, among others, Chou (1989) and Maa and Mehta (1991).

3. FLUID MUD THICKNESS

For the commonly used shear-Voigt model representing bed viscoelasticity the constitutive relationship is (Malvern, 1969)

$$T'_{ij} = 2GE'_{ij} + 2\mu\dot{E}'_{ij} \tag{4}$$

where T'_{ij}= deviatoric component of stress tensor T_{ij},E_{ij}= deviatoric component of strain tensor E_{ij}, i and j denote directions and ij denotes a second order tensor, G=elastic modulus, μ=viscosity and the dot denotes derivative with respect to time. For bed liquefaction by periodic compression and stretching of mud, we are interested in the corresponding elongational or extensional stress-strain relationship applicable to forcing by normal stress (Barnes et al., 1989). Assuming the shear viscoelasticity of mud to be represented by (4), its extensional viscoelasticity can be represented by (Li, 1996)

$$T_{33} = 3GE_{33} + 3\mu\dot{E}_{33} = G_n E_{33} + \mu_n \dot{E}_{33} \tag{5}$$

where G_n=extensional elastic modulus and μ_n =extensional viscosity. Thus, bed visco-elasticity in the normal direction is described under the given assumptions by the extensional-Voigt model (5) in which G_n=3G and μ_n=3μ. When G=0, (5) satisfies the extensional relationship for a Newtonian fluid (Petrie, 1978).

Considering the bed as an impermeable continuum, it can be represented as an equivalent mechanical spring-dashpot-mass analog, with k=equivalent elastic coefficient, c=equivalent viscous coefficient and m=equivalent mass term. Equation (5) can be solved using representative mean values of the rheological parameters. Accordingly, it can be shown that given h_2=bed thickness, (5) yields

$$\zeta_2(z',t)=\zeta_2(0,t)\left(1-\frac{z'}{h_2}\right)=\zeta_2(0,t)\,\Psi \tag{6}$$

where Ψ=1-(z'/h_2) is a shape function that varies linearly with depth z', and relates the displacement at any depth to the displacement at the bed surface, z'=0. The equivalent mass term (in units of mass per area) is

$$m=\int_0^{h_2}\rho_2(z')\left(1-\frac{z'}{h_2}\right)\,dz'=\int_0^{h_2}\rho_2(z')\Psi dz' \tag{7}$$

Using the same basic procedure, it is possible to derive analogous solutions for k, c and m for a bed in which G_n and μ_n vary with depth, z', according to specified functions. In these cases, the shape function, Ψ, in general varies non-linearly with z'.

The bed is forced by wave-induced cyclic pressure, p, obtained from linear wave theory according to

$$p=p_0\sin\omega t=\frac{\rho_1 g\zeta_{01}}{\cosh k_1h_1}\sin\omega t \tag{8}$$

where p_0=pressure amplitude, ζ_{01}=wave amplitude, ω=wave frequency, k_1=wave number and h_1=water depth (Dean and Dalrymple, 1984). The dynamic equation for bed surface response is

$$m\,\ddot{\zeta}_2(0,t)+c\,\dot{\zeta}_2(0,t)+k\,\zeta_2(0,t)=p_0\sin\omega t \tag{9}$$

Denoting $\beta=\omega/\omega_o$, where $\omega_o=(k/m)^{1/2}$ is the characteristic resonance frequency and defining $\xi=c/2m\omega$, the following solution is obtained

$$\zeta_2(0,t)=\frac{p_0}{k}\frac{1}{[(1-\beta^2)^2+(2\xi\beta)^2]^{1/2}}\sin(\omega t-\phi)=\zeta_{max}\sin(\omega t-\phi) \tag{10}$$

where $\phi=\tan^{-1}[c\omega/(k-m\omega^2)]=\tan^{-1}[2\xi\beta\omega/(1-\beta^2)]$ and ζ_{max} is the amplitude of ζ_2.

From (10), which is a solution of (9), we find that $\ddot{\zeta}_2 = \omega^2 \zeta_2$, so that the fluid mud generation criterion (3) can be expressed as

$$\omega^2 \zeta_{max} \Psi(z_c) = \alpha g \frac{\rho_2(z_c) - \rho_1}{\rho_1} \tag{11}$$

where $\Psi(z_c)$ is the value of the shape function at $z = z_c$. Thus, z_c can be obtained by solving (11) through iteration.

4. FLUME DATA

Flume data on fluid mud generated by water waves over soft cohesive muds of known rheology are sparse. To illustrate the applicability of the analytic method, test conditions for data from three selected sources are summarized in Table 1. In addition, the works of Maa (1986) and Jiang (1993) were consulted for obtaining the parameters μ and G. In the tests of Table 1, a fluid mud layer developed starting from the bed surface, and the layer thickness eventually reached an approximately equilibrium value. Fluid mud was characteristically distinguished from the bed via vertical pore and total pressures profiles, with the bed having a measurable effective normal stress, σ' (equal to total pressure less pore pressure), and fluid mud defined by zero or negligible σ'.

The plot of measured versus calculated values of z_c shown in Figure 3 indicates a generally acceptable degree of overall agreement between theory and data, although at the lower end two values, one of Lindenberg et al. (1989) and the other of Feng (1992), are notably under-predicted. A likely reason is that because the mud-water interface tends to be indistinct in flume experiments, relatively small fluid mud thicknesses are not always measured accurately. It is conceivable that in the two cases, the reported thicknesses included a thin surficial layer that was actually a comparatively low density suspension above the fluid mud surface.

For the purpose of calculation, it was found that the value of 1.05 for the resistance modifying parameter α in (3) was generally applicable to all the data. The fact that this was so and that this value is close to unity is commensurate with the relatively low cohesion of all the muds listed in Table 1.

In two of the tests of Lindenberg et al. (1989) for which experimental conditions are given in Table 2, no measurable fluid mud formation could be detected. This condition can be further examined by referring to (3). Thus, it can be shown that in Test T3-3 at the surface $\ddot{\zeta}_{max}$ was equal to 2.5 m/s^2, which was smaller than $\alpha g'$=3.1 m/s^2. In Test T4-4 the two respective values were 3.4 m/s^2 and 3.1 m/s^2. Thus, in Test T3-3 the threshold condition for fluid mud generation was not reached, and in T4-4 it was barely exceeded. We will discuss this point further by considering fluid mud generation in Lake Okeechobee.

Table 1
Summary of selected fluid mud generation experiments in flumes

Source	Mud	h_1 (cm)	h_2 (cm)	ρ_1 (kg/m^3)	ρ_2 (kg/m^3)	ζ_{01} (cm)	f[a] (Hz)	μ (Pa.s)	G (Pa)
Ross (1988)	Tampa Bay	31.4-31.7	11.8-13.0	1,000	1,080	3.1-3.6	1.0-1.1	25.0	100
Lindenberg et al. (1989)	Kaolin	25	4.8-4.9	1,000	1,300	2.4-3.6	0.4-0.7	3.0	5
Feng (1992)	AK[b]	18.4-20.2	14.7-16.6	1,000	1,170	1.9-4.0	1.0	6.1	295

[a] $f=\omega/2\pi$ is the wave frequency.
[b] A 50/50 (by weight) clayey mixture of an attapulgite and a kaolin

Table 2
Flume tests of Lindenberg et al. (1989) without fluid mud generation

Test No.	Mud	h_1 (cm)	h_2 (cm)	ρ_1 (kg/m^3)	ρ_2 (kg/m^3)	ζ_{01} (cm)	f (Hz)	μ (Pa.s)	G (Pa)
T3-3	Kaolin	25	4.9	1,000	1,300	4.8	2.5	3	5
T4-4	Kaolin	25	4.8	1,000	1,300	4.7	1.5	3	5

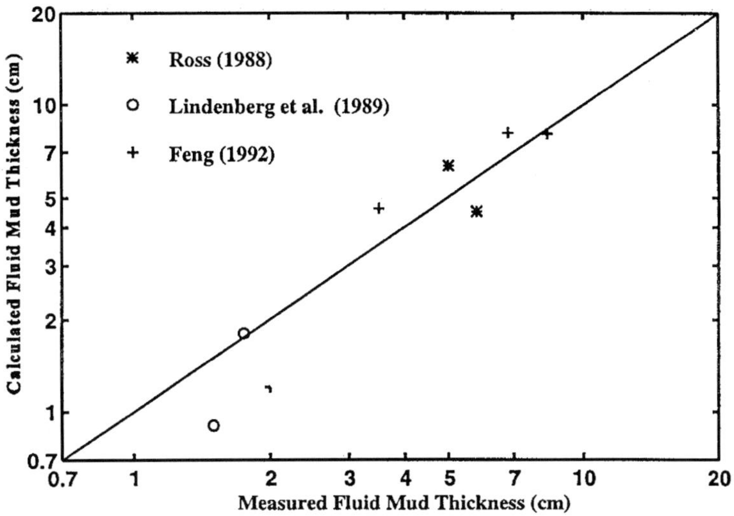

Figure 3. Comparison between calculated and measured fluid mud thicknesses.

5. LAKE OKEECHOBEE

Lake Okeechobee in south-central Florida (Figure 4) is an extensive, shallow bowl-like water body with a hard bottom that is overlain by soft, highly organic (~40% by weight) mud, especially in the deeper areas. The original source of this mud, which is as thick as 80 cm in the middle portion of the lake, is believed to be the Kissimmee River, whose waters enter the lake near its northern end. A sedimentological investigation, which included vibracoring, acoustical sub-bottom profiling and sidescan imaging of the lake bottom, was carried out to characterize mud properties and state (Kirby et al., 1994). In a complementary study, wave spectra under fair weather conditions were measured to assess the typical erodibility of mud (Jiang and Mehta, 1992). From these studies it was determined that, much of the time, typical wind-generated waves are able to maintain the top 5-20 cm thick layer of the mud in a fluid-like state.

A characteristic bottom density profile from the lake and a representative equation fit are shown in Figure 5. This profile does not distinguish between fluid mud and bed, and it is of interest to determine what portion of the profile is fluid mud. For that purpose the following characteristic parameters are selected: a representative mean mud thickness over the entire mud bottom, h_2=0.55 m; water density, ρ_1 =1,000 kg/m^3; representative wave height, ζ_{01} =0.08 m and wave frequency, f=0.4 Hz (Jiang and Mehta, 1992). The parameters, μ=21 Pa.s and G=216 Pa, were determined accurately by Jiang (1993) through applied shear

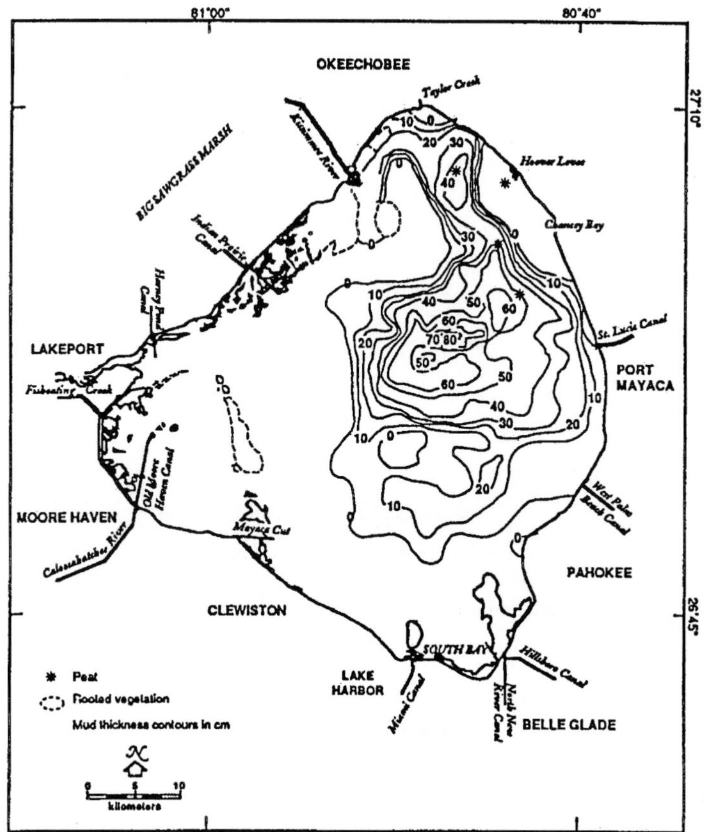

Figure 4. Lake Okeechobee in south-central Florida and isopleths (cm) of bottom mud deposit.

stress rheometry. In addition, the resistance modifying coefficient α=1.05 is selected following the earlier examples and also considering the weakly cohesive nature of lake mud, and water depth h_1 is varied from 0.5 m to 4 m.

As observed in Figure 6, fluid mud thickness calculated from (11) decreases from a little over 0.3 m in the vicinity the shallow littoral zone of the lake where the effect of waves is comparatively strong, to practically nil in the deepest part near the center, where waves are considerably less effective in liquefying mud. Despite the approximate nature of this calculation, it is illustrative of the actual range of fluid mud found in the lake. Since, however, the measured thicknesses were not resolved on a site-specific basis, it is not feasible to glean the dependence of measured fluid mud on water depth. In any event, the generation of fluid mud by wind waves, and the fact that in this lake episodic wind-driven currents on the order of 10 cm/s occur, may explain the mode by which mud from

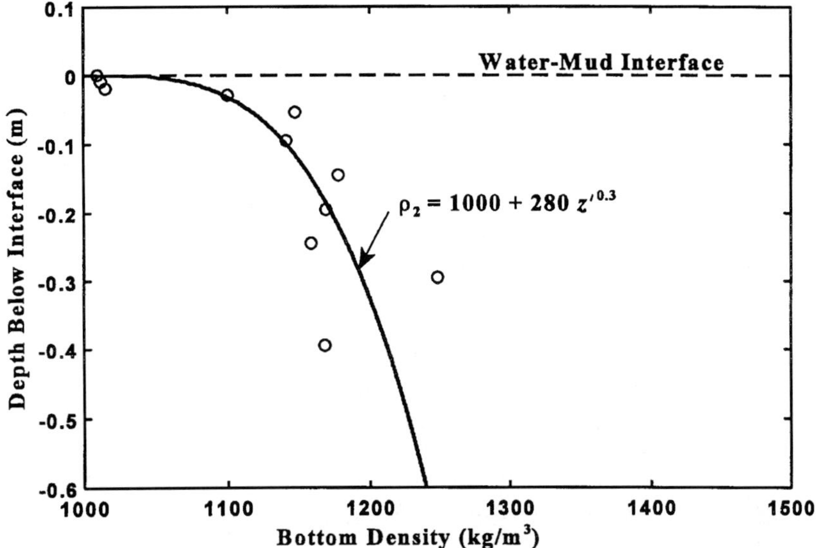

Figure 5. A representative bottom mud density profile in Lake Okeechobee.

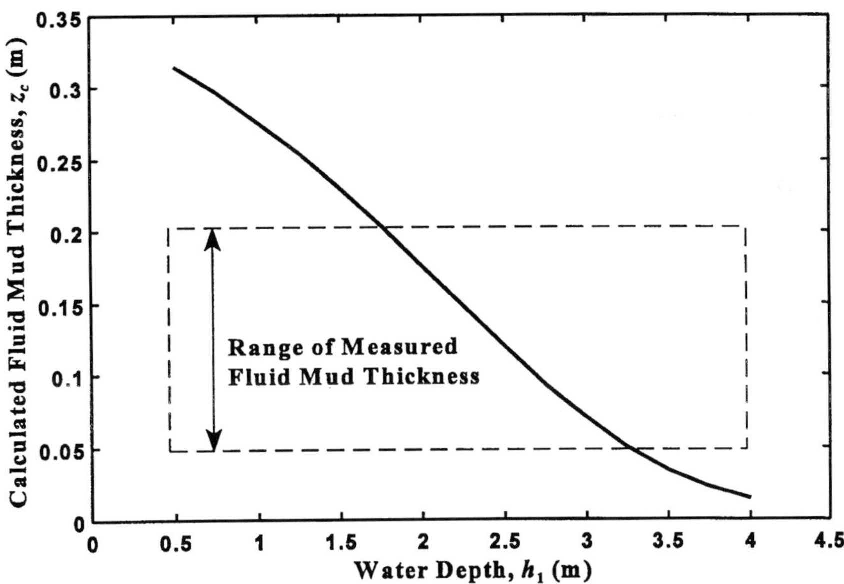

Figure 6. Calculated fluid mud thickness as a function of water depth in Lake Okeechobee, and band of measured values.

the mouth of Kissimmee River was eventually transported to the center of the lake.

It is instructive to qualitatively examine the effect of mud properties on the potential for fluid mud generation with reference to conditions during a study on mud resuspension close to the eastern bank of the lake (Jiang and Mehta, 1992). At this site all the above selected parameters are applicable except the water depth, h_1, which was 1.43 m. Equation (11) is plotted in Figure 7, in which the fluid mud thickness is seen to be about 25 cm. Observe that z_c is seemingly sensitive to the shapes of the two intersecting curves. Consideration of depth-dependent rheological parameters would tend to make z_c further sensitive to the depth-varying profile of particle acceleration, because in that case the shape of the acceleration profile would become qualitatively akin to that shown in Figure 2. Thus, it is seen that for given site conditions, fluid mud generation can be expected to be very sensitive to mud properties. The experiments of Lindenberg et al. (1989) seem to corroborate this observation based on their finding that fluid mud thickness in some of their tests was sensitive to the presence of relatively small fractions of organic matter and fine sand.

6. CONCLUDING COMMENTS

Caution is warranted with regard to the range of applicability of this method due to the numerous physical factors that can cause deviations from the underlying assumptions. Thus, for example, Lindenberg et al. (1989) found that

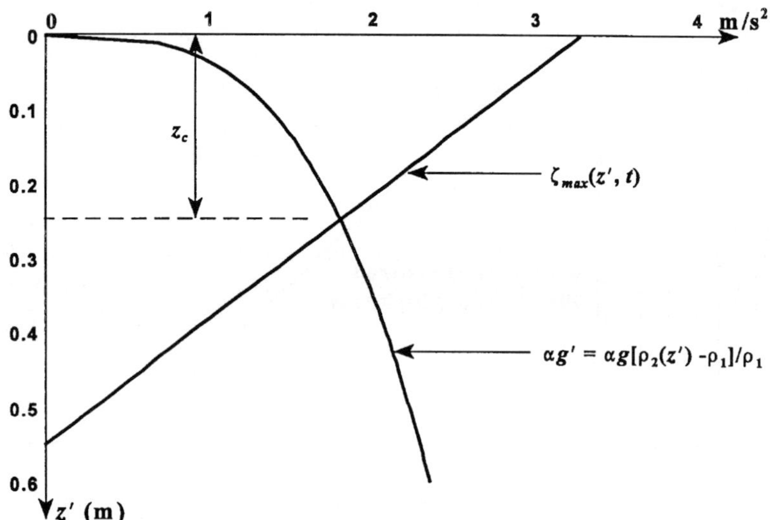

Figure 7. An example of fluid mud thickness calculation for Lake Okeechobee at a site where the water depth was 1.43 m.

gas bubbles in sediment measurably reduced fluid mud generation. Since the effect of gas on the viscoelastic parameters may not be captured easily through rheometry, such effects could render the model results qualitative. Also, if and when a further refinement of the described approach is sought, a two-dimensional wave-mud interaction model, such as that of Jiang and Mehta (1996), which accounts for both normal and shear forces, may yield better estimations of fluid mud thickness.

Jiang and Mehta (1992) noted that in Lake Okeechobee short period wind waves tend to generate forced long period, surf-beat type waves within the soft mud bottom. Although the forced wave signature in the lake was found to be weak, in those situation in which long period waves do contribute significantly, their effect on liquefaction may not be negligible. Also, at large strains the viscoelastic coefficients depend on strain amplitude, and the linear viscoelastic assumption breaks down. This in turn can lead to potentially significant scale effects associated with prototype conditions in relation to a laboratory setup (Chou, 1989).

It is conceivable that in the mild wave environment of Lake Okeechobee under fair weather conditions, scale effects are in fact not too significant. In fact, the seeming agreement between the calculated and measured values of the fluid mud thickness suggests that the physical conditions in the lake during fair weather do not violate the model assumptions drastically, and that breeze-induced waves retain the fluid mud layer in a quasi-equilibrium state.

In contrast to what is "normal" in the lake, conditions tend to become traumatic during storm events under winter cold fronts. Then, in addition to likely finite strain effects including plastic yield induced by high stresses, resuspension of sediment tends to disturb the fluid mud layer drastically (Mehta, 1996). Yet, it is interesting to note that the small strain assumption seems to yield reasonable answers in at least some applications, e.g., in predicting wave damping by bottom mud (Jiang, 1993).

It should be emphasized that in contrast to soft muds, erosion of dense and highly cohesive muds often occurs by mass failure rather than by the process considered here. In that mode of failure, bed material clasts are dislodged and the bed becomes pitted. Fluid mud is generated when the dislodged and subsequently dispersed material in suspension resettles rapidly, e.g., at the end a period of storm-induced waves. Thus, the thickness of fluid mud is governed by the rate of dewatering and consolidation of the fresh deposit, as opposed to bed failure.

7. ACKNOWLEDGMENT

Funds for this study were provided by the U.S. Army Engineer Waterways Experiment Station, contract DACW39-95-K-0022.

REFERENCES

Barnes, H. A., Hutton, J. F., and Walters, K., 1989. *An Introduction to Rheology.* Elsevier Scientific Publishing Co., Amsterdam, The Netherlands.

Chou, H. T., 1989. Rheological response of cohesive sediments to water waves. *Ph.D. Thesis*, University of California, Berkeley.

Dalrymple, R.A., and Liu, P. L-F., 1978. Waves over soft muds: a two-layer fluid model. *Journal of Physical Oceanography*, 8, 1121-1131.

Das, B.M., 1983. *Fundamentals of Soil Dynamics*. Elsevier Scientific Publishing Co., New York.

Dean, R. G., and Dalrymple, R. A., 1984. *Water Wave Mechanics for Engineers and Scientists*. Prentice-Hall, Engelwood Cliffs, New Jersey.

de Wit, P. J., 1995. Liquefaction of cohesive sediments caused by waves. *Ph.D. Thesis*, Delft University of Technology at Delft, The Netherlands.

Feng, J., 1992. Laboratory experiments on cohesive soil bed fluidization by water waves. *M.S. Thesis*, University of Florida, Gainesville.

Foda, M. A., Tzang, S. Y., and Maeno, Y., 1991. Resonant soil liquefaction by water waves. *Proceedings of Geo-Coast'91,* Port and Harbor Research Institute, Yokohama, Japan, 549-583.

Foda, M.A., Hunt J. R., and Chou, H. T., 1993. A non-linear model for the fluidization of marine mud by waves. *Journal of Geophysical Research,* 98, 7039-7047.

Isobe, M., Huyuh, T. N., and Watanabe, A., 1992. A study on mud mass transport under waves based on an empirical rheology model. *Proceedings of the 23rd International Conference on Coastal Engineering,* 3, ASCE, New York, 3093-3106.

Jiang, F., 1993. Bottom mud transport due to water waves. *Ph.D. Thesis*, University of Florida, Gainesville.

Jiang, F., and Mehta, A. J., 1992. Some observations on fluid mud response to water waves. In: *Dynamics and Exchanges in Estuaries and the Coastal Zone*, D. Prandle ed., American Geophysical Union, Washington, DC, 351-376.

Jiang, F., and Mehta, A. J., 1996. Mudbanks of the southwest coast of India IV: Wave attenuation. *Journal of Coastal Research*, 12(4), 890-897.

Keedwell, M. J., 1984. *Rheology and Soil Mechanics*. Elsevier Applied Science Publishers, Amsterdam, The Netherlands.

Kirby, R., Hobbs, C. H., and Mehta, A. J., 1994. Shallow stratigraphy of Lake Okeechobee, Florida: a preliminary reconnaissance. *Journal of Coastal Research*, 10(2), 339-350.

Lambe, T. W., and Whitman, R. V., 1969. *Soil Mechanics.* John Wiley & Sons, New York.

Li, Y., 1996. Constituent transport across the mud-water interface under water waves. *Ph.D. Thesis*, University of Florida, Gainesville.

Li, Y., and Mehta, A. J., 1997. Mud fluidization by water waves. In: *Cohesive Sediments*, N. Burt, R. Parker and J. Watts eds., John Wiley & Sons, New York, 341-351.

Lindenberg, J., van Rijn, L. C., and Winterwerp, J. C., 1989. Some experiments on wave-induced liquefaction of soft cohesive soils. *Journal of Coastal Research*, SI 5, 127-137.

Maa, P.-Y., 1986. Erosion of soft mud by waves. *Ph.D. Thesis*, University of Florida, Gainesville.

Maa, J. P.-Y., and Mehta, A. J., 1991. Soft mud response to water waves. *Journal of Waterway, Port, Coastal and Ocean Engineering*, 116(5), 634-650.

Madsen, O. S., 1974. Stability of a sand bed under breaking waves. *Proceedings of the 14th International Conference on Coastal Engineering,* 2, ASCE, New York, 777-794.

Malvern, L. E., 1969. *Introduction to Mechanics of a Continuous Medium.* Prentice-Hall, Englewood Cliffs, New Jersey.

Mehta, A. J., 1996. Interaction between fluid mud and water waves. In: *Environmental Hydraulics*, V. P. Singh and W. Hager eds., Ch. 5, Kluwer Academic Publishers, Dordrecht, The Netherlands, 153-187.

Mehta, A. J., Williams, D. J. A., Williams, P. R., and Feng, J., 1995. Tracking dynamical changes in mud bed due to waves. *Journal of Hydraulic Engineering*, 121(6), 504-506.

Partheniades, E., 1977. Unified view of wash load and bed material load. *Journal of the Hydraulics Division*, ASCE, 103(9), 1037-1057.

Petrie, C. J. S., 1978. *Elongational Flows*. Pitman Publishers, London, England.

Ross, M. A., 1988. Vertical structure of estuarine fine sediment suspensions. *Ph.D. Thesis*, University of Florida, Gainesville.

Selly, R. C., 1988. *Applied Sedimentology*. Academic Press, London.

Coastal and Estuarine Fine Sediment Processes
W.H. McAnally and A.J. Mehta (Editors)

Response of stratified muddy beds to water waves

R. Silva Jacinto and P. Le Hir

Institut Français pour la Recherche et l'Exploitation de la Mer (IFREMER), Direction de l'Environnement Littoral, DEL-EC-TP, B.P. 70, 29280 Plouzané, France

A modeling approach for assessing rheological changes of muddy beds under wave action is proposed. This approach is based on an analytical model for wave-mud interaction and an empirical characterization of the rheological properties of mud as functions of oscillatory forcing. Results show that rheological changes must be accounted for in the prediction of liquefaction, and that non-linear modeling is needed.

1. INTRODUCTION

This study was motivated by the observed erosion under wave action in the Seine estuary in France. As is known, shear stresses induced by waves and currents cause erosion of cohesive sediment, or entrainment of a fluid mud layer, from the bed to the water column. Even in sheltered areas like estuaries, cohesive sediment resuspension is strongly influenced by the presence of short-crested, locally generated waves. A second, but not secondary, effect of water waves on a deformable bed can be described by a pressure wave propagating along its surface (e.g., Maa and Mehta, 1990; Foda et al., 1993). Such a wave induces normal and shear stresses within the bed that modify the constitutive parameters and the structural strength of the sediment. These rheological changes alter the dispersive characteristics of the medium, and modify sediment erodability under waves and currents. The main consequences of such structural changes are possible mud liquefaction on the one hand, and wave and turbulence damping on the other. Rheological changes are related to the inherently non-linear behavior of the mud-as induced stresses and deformation change, rheological parameters change as well.

The Seine estuary (Figure 1) is located on the northwestern coast of France (English channel). Because the estuary is open westward, which is the direction of the most frequent winds, waves are very often able to penetrate the estuary and erode sediments deposited on tidal flats. In the Seine estuary, tides induce a residual depositional flux of fine sediment from the turbidity maximum towards the tidal flats in calm periods, while waves tend to erode these tidal flats and

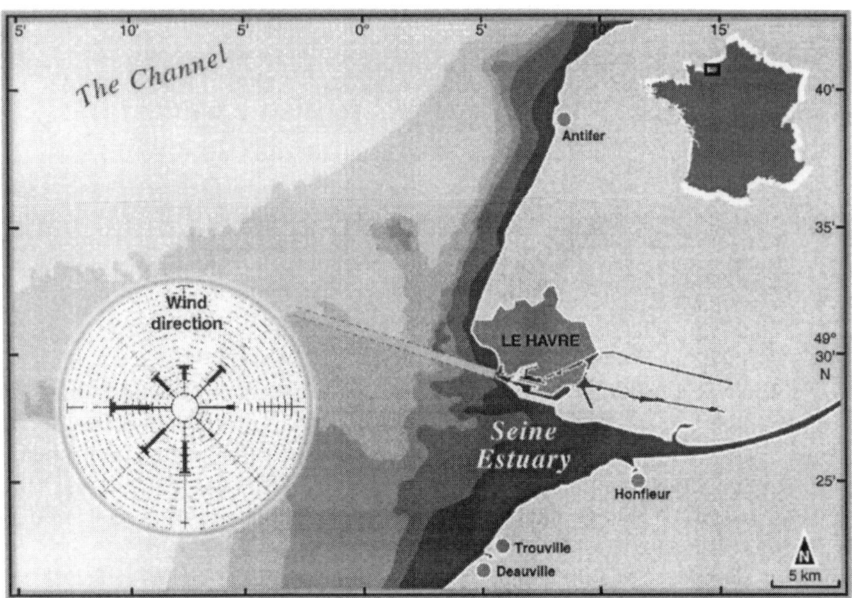

Figure 1. The Seine estuary and dominant wind directions.

supply suspended sediment mass to the estuary during storm events. Also, other sediment sources exchanging sediment with the turbidity maximum are present in the system.

Monthly surveys of a Seine estuary tidal flat were carried out from April 1996 to January 1999 (Silva Jacinto et al., 1998). Depending on the period, one can observe (Lesourd et al., 1998) three main situations corresponding to three main sediment structures in the tidal flat area (Figure 2). Sediment cores show different layered structures corresponding to different meteorological conditions. After long periods of calm, soft mud covers a consolidated layer over a stratified structure of sand layers and very stiff overconsolidated mud. When significant waves occur, the soft mud cover is eroded, and only the consolidated layers remain. When waves are strong during storm events, complete and sudden (24-48 h) erosion of the consolidated black (≈20 cm) layer is observed. The erosion of the consolidated layer takes the form of mass or bulk erosion. Structural failure occurs at the interface with the stiff bed, and "muddy pebbles" are detached and dispersed over the tidal flat.

The aim of the present study was to analyze the effects of wave-bed interactions, to examine the importance of the non-linear rheological properties on the induced stresses, and to propose explanations for the observed sudden erosion (does liquefaction occur under wave action?) and failure of the consolidated mud layer over the stiff/rigid bottom.

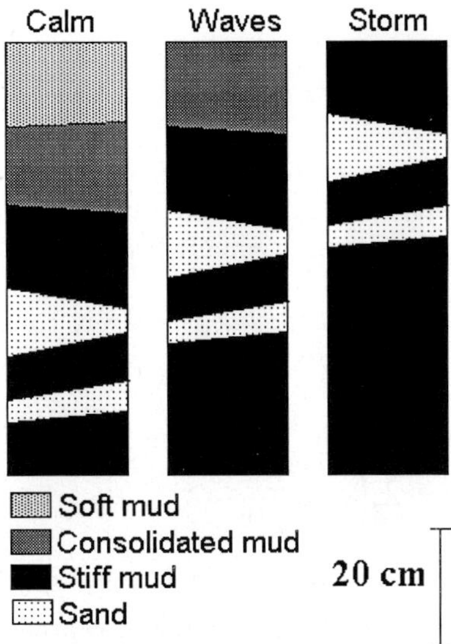

Figure 2. Sediment structure in the northern tidal flat of the Seine estuary: after a long period of calm conditions, under wave action and after a storm event.

2. ANALYTIC MODELING

Several analytical multi-layer models have been presented in the literature to account for wave-bed interactions. Most of them consider as hypothesis *a priori* rheological properties and associated parameters. Mainly, elastic (Mallard and Dalrymple, 1977), viscous (Dalrymple and Liu, 1978) and viscoelastic (e.g., Maa and Mehta, 1990) behaviors have been considered.

In the present study, an analytical model has been developed to account for wave-mud interaction. The model accounts for the dispersive and damping characteristics of monochromatic waves. It also provides velocity and stress profiles in both the water column and the mud. The model considers linear viscoelastic behavior within each layer, ranging from pure elastic near a rigid bed to a viscous fluid in the water column. The bed is split into any number of homogeneous layers in order to account for stratification.

The governing equations for each layer are the linearized continuity and momentum equations for an incompressible material, expressed as:

$$\nabla \vec{u}_i = 0$$
$$\frac{\partial \vec{u}_i}{\partial t} = -\frac{1}{\rho_i}\vec{\nabla} p_i + \upsilon_{e,i}\nabla^2 \vec{u}_i \qquad (1)$$

where subscript i represents each layer, $\vec{u}$ is the velocity vector, p is the total pressure, ρ is the density and υ_e is an equivalent complex kinematic viscosity that depends on the constitutive, or rheological, characteristics of the material in each layer.

The constitutive relationship for each layer is considered to be of the linear type, i.e., the strain and stress tensors are related to a linear differential equation:

$$a_0 T'_{ij} + \sum_{n=1}^{\infty} a_n \frac{\partial^n T'_{ij}}{\partial t^n} = b_0 E'_{ij} + \sum_{n=1}^{\infty} b_n \frac{\partial^n E'_{ij}}{\partial t^n} \qquad (2)$$

where T'_{ij} and E'_{ij} are, respectively, the stress and strain tensors. Prime indicates the deviatoric part of these two tensors and n represents the order of derivative with respect to time; a_n and b_n are characteristic rheological parameters of the material.

A basic viscoelastic model is the Voigt (or Kelvin) model. This model is based on the analogy with a parallel spring-dashpot system, i.e., two modes of behavior are present, an elastic mode and a viscous mode. The constitutive relation is:

$$T'_{ij} = 2GE'_{ij} + 2\mu \frac{\partial E'_{ij}}{\partial t} \qquad (3)$$

where G and μ characterize the shear modulus of the elastic mode and the dynamic viscosity of the viscous mode, respectively.

The Voigt model is of practical interest. It can be shown (e.g., Barnes et al., 1989) that under periodic forcing at a frequency ω (as with water waves), any linear rheological model such as (2) can be reduced to an equivalent Voigt model where:

$$G = G(\omega, a_0 \ldots a_\infty, b_0 \ldots b_\infty)$$
$$\mu = \mu(\omega, a_0 \ldots a_\infty, b_0 \ldots b_\infty) \qquad (4)$$

Hence, for sinusoidal waves, and using a complex notation, the Voigt model can be reduced to :

$$T'_{ij} = 2\left(\mu + j\frac{G}{\omega}\right)\frac{\partial E'_{ij}}{\partial t} = 2\mu_e \frac{\partial E'_{ij}}{\partial t} \qquad (5)$$

where $j = \sqrt{-1}$ and μ_e ($= \rho \nu_e$) represents an equivalent complex dynamic viscosity.

The solutions for (1) are assumed separable and periodic (i.e., harmonic) in time and the direction of propagation. For any parameter θ, these solutions are expressed as :

$$\theta(x,z;t) = \hat{\theta}(z).\mathrm{e}^{j(mx-\omega t)} \tag{6}$$

where x is the spatial coordinate in the direction of wave propagation, z the vertical coordinate, t is the time and $\hat{\theta}(z)$ represents the vertical dependence of the amplitude of the parameter θ. The quantity ω represents the wave frequency and $m = K + jD$ is the complex wave number, whose real part K represents the real wave number and the imaginary part D is a decay or damping scale for the wave amplitude.

The introduction of appropriate boundary conditions, following for instance Dalrymple and Liu (1978), leads to an eigenvalue problem solvable by a "complex" Newton method for any number of layers. The dynamic boundary conditions at each interface between two layers are the continuity of vertical normal stress and shear stress. The imposed kinematic conditions are the continuity of vertical velocity, which also means continuity of the time derivative of the vertical displacement of the interface, and the continuity of the horizontal velocity. At the rigid bottom the velocities must be zero.

As an option, the water layer may be considered inviscid. In this case, the flow in the water layer is irrotational and the pressure at the water-mud interface represents the forcing of the mud layer. Further, the continuity of shear stress and horizontal velocity must be excluded from the boundary conditions at the interface.

The present model has been validated by comparison with other analytical models presented in the literature, for which different rheological behaviors have been considered, e.g., Mallard and Dalrymple (1977), Dalrymple and Liu (1978) and Maa and Mehta (1990).

Figure 3 shows the dispersive and damping characteristics of a wave propagating over a viscous mud layer; these characteristics are represented by the dimensionless real and imaginary parts of the complex wave number as functions of the dimensionless mud layer thickness. The wave number is non-dimensionalized by ω^2/g and the mud layer thickness by $(\upsilon/2\omega)^{1/2}$. Solid lines are model results, open circles are provided by the Dalrymple and Liu (1978) model, and dark circles are experimental data of Gade (1957).

An important feature of these results, besides the observed agreement between the data and the two models, is the significant damping when the mud layer thickness approaches the wave boundary layer thickness $(\upsilon/2\omega)^{1/2}$. When an elastic mode is present (e.g., Voigt behavior), resonance can occur because the frequency of oscillatory loading approaches the natural frequency of the system, depending on the mud layer thickness and the rheological parameters (see for instance, Maa and Mehta, 1990). Near resonant configurations, computed stress

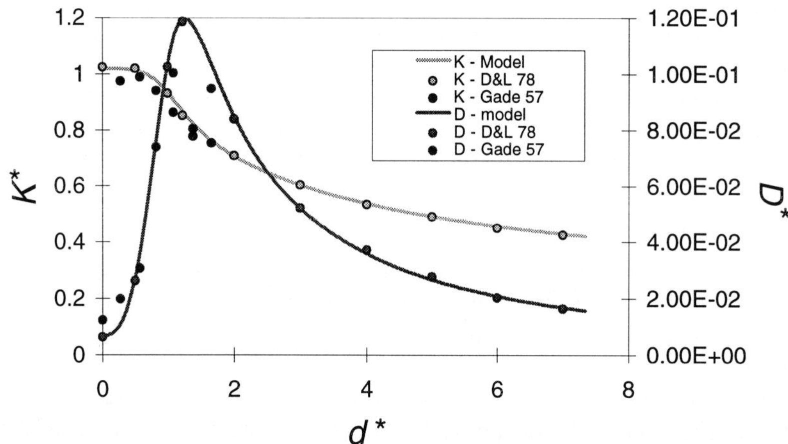

Figure 3. Dispersive and damping characteristics of a wave propagating over a viscous mud layer. Comparison with the Mallard and Dalrymple (1978) model and data from Gade (1957). Represented are the dimensionless wave number (real part K^* and imaginary part D^*) versus the dimensionless mud layer thickness (d^*).

and deformations can be unrealistically high if the rheological behavior of mud is not allowed to change. Hence, together with the depth-varying mud properties, non-linear response of mud must be considered

3. NON-LINEAR RHEOLOGICAL BEHAVIOR

In the described analytical model, solutions are obtained for unchanging rheological properties in each layer. However, the rheological behavior of natural mud is highly non-linear, i.e., as induced stress and deformation change, rheological properties change as well.

Together with non-linearity, mud exhibits thixotropic behavior, i.e., (isothermal) variation of rheological parameters as the micro-structure changes with time due to variations in deformation (Toorman, 1995). Experiments show that rheological changes occur over time scales that are on the order of minutes. In practice, rheological properties can be considered to be in equilibrium with respect to the stress and deformation amplitudes, because rheological adjustment is typically "faster" than the gradual variations of the forcing parameters (wave climate). Also, rheological parameters can be considered constant over the wave period, because the adjustment is "slow" over that time scale.

In order to characterize the rheological response of the Seine estuary mud to oscillatory forcing, rheometric tests were carried out using a Physica UDS 200 rheometer from TUDelft to obtain shear stress-strain relationships. Similar characterization for a kaolinite has been presented by Chou (1989). For different densities and oscillatory frequencies, the mud samples were subjected to a uniform shear strain γ that was sinusoidal in time t with a frequency ω :

$$\gamma = \gamma_0 \, e^{-i\omega t} \tag{7}$$

The resultant shear stress $\tau(t)$ was nearly sinusoidal but with a phase shift δ:

$$\tau = \tau_0 \, e^{-i(\omega t + \delta)} \tag{8}$$

This phase shift meant that the material had an equivalent Voigt behavior, or a complex shear modulus G_υ, inasmuch as

$$G_\upsilon = \tau/\gamma = G_0 \, e^{-i\delta} \;\; ; G_0 = |\tau_0/\gamma_0|$$

or (9)

$$G_\upsilon = G' - iG'' = G_0(\cos\delta - i\sin\delta)$$

The real part of G_υ, G', represents the storage modulus, while the imaginary part, G'', represents the loss modulus. For a zero phase shift the shear loss modulus is zero and the response is purely elastic, while a phase shift of $\pi/2$ implies a purely viscous response. Hence the phase shift δ, also called loss angle, can characterize the state of liquefaction of mud, provided one considers liquefaction as a rheological evolution of mud (due to structural break down) that allows the material to approach viscous behavior. Sometimes, the damping factor, $\tan(\delta)$, i.e., the ratio between the two modulus, is chosen instead of δ.

Using (5), mud rheology in (1) can also be represented by a complex kinematic viscosity υ_e defined as:

$$\upsilon_e = \frac{\tau}{\rho\dot{\gamma}} = \frac{G'' + iG'}{\rho\omega} \tag{10}$$

Here, real υ_e indicates viscous response, purely imaginary υ_e means an elastic response and, in general, complex υ_e characterizes a viscoelastic Voigt response.

An example of the rheological results is presented in Figure 4, where it is observed how the rheological behavior of Seine mud changes with the amplitude of forcing. For small strain amplitudes the mud has a viscoelastic behavior that is elasticity dominant, with the storage modulus G' an order of magnitude larger than the loss modulus G''. An important feature is that for small strains the rheological parameters remain almost constant; this is the linear range within which mud behavior is usually characterized.

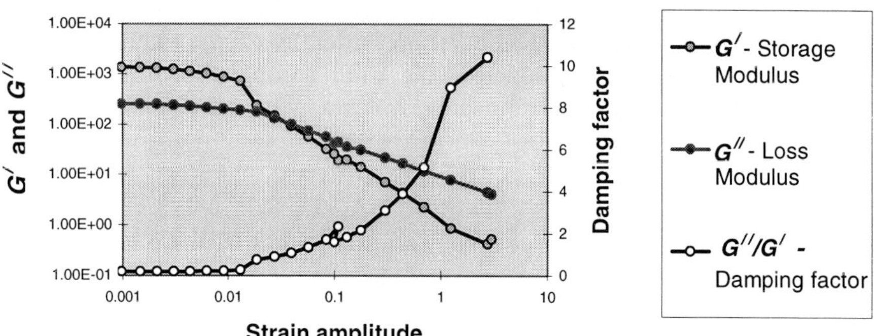

Figure 4. Non-linear rheological behavior of Seine estuary mud under oscillatory forcing (period 5 s). Storage and loss moduli are in Pa.

If the strain amplitude becomes larger, both parameters decay, but the loss modulus becomes dominant, the mud approaches viscous behavior and becomes liquefied. This feature can also be seen from the damping factor [tan(δ)], which is near zero in the linear range and becomes on the order of ten or more when the mud approaches a liquefied state.

The experimental results shown in Figure 5 were obtained for a mud bulk density of 1,260 kg/m^3 and a frequency of oscillation of 5 s. Similar tests were carried out for five other densities and frequencies. Results yielded similar features for all densities and did not show any relevant dependence on frequency in the tested range (1 to 10 s).

4. APPLICATION OF THE ANALYTICAL MODEL

Rheological changes can be taken into account using the analytical model for wave-mud interaction presented above. At noted, because in nature wave parameters change very slowly when compared to the time scale of rheological adjustment, we can assume equilibrium between forcing (e.g., strain amplitude) and the rheological state of the bed. In that way, by using the experimental stress-strain relationships, an iterative procedure can be developed. Initially, *a priori* rheological parameters are imposed on each mud layer; those from the linear range for instance. Then, stress and deformation profiles are computed using these parameters. Corresponding to the resulting stress and deformation in each layer, new rheological parameters are deduced from the rheological curves and then used for new calculations of stress and deformation. The calculation can be run in that manner until convergence occurs.

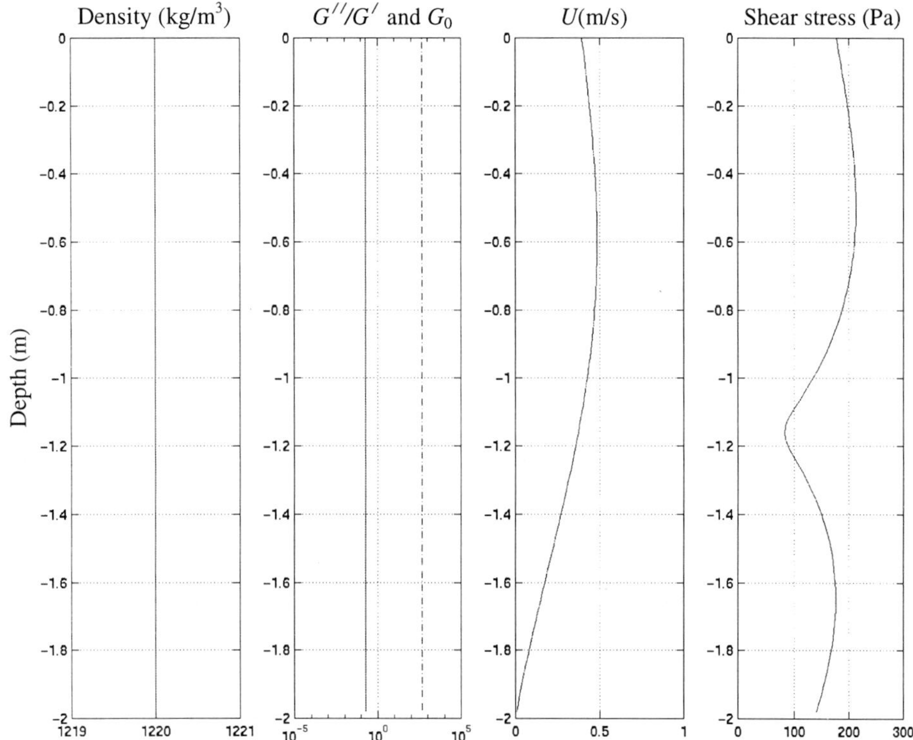

Figure 5a. Linearized calculation for a homogeneous mud layer of bulk density of 1,220 kg/m^3. Mud depth: 2 m. Wave: 0.3 m amplitude, period 5 s. The damping factor is represented by a solid line; G_0 is in Pa and shown by a dash-dot line.

It is important to note that this procedure may only be used when it is started from the linear range. Even if one can obtain convergence starting the procedure from a viscous state the result may be physically meaningless, because the rheological characterization was obtained for increasing strain amplitudes only. On the other hand, rheological adjustment for structure recovery situations may occur over long time scales (hours or longer) that were not characterized.

5. RESULTS

The main objective of the calculations to be presented is to show how non-linear rheological behavior and strong sediment stratification are important to qualitatively describe wave-mud interaction. In order to obviate the need to model turbulence closure, the water layer is considered inviscid and the sediment layer (split into 100 layers in the model) is forced by pressure at the sediment/water interface.

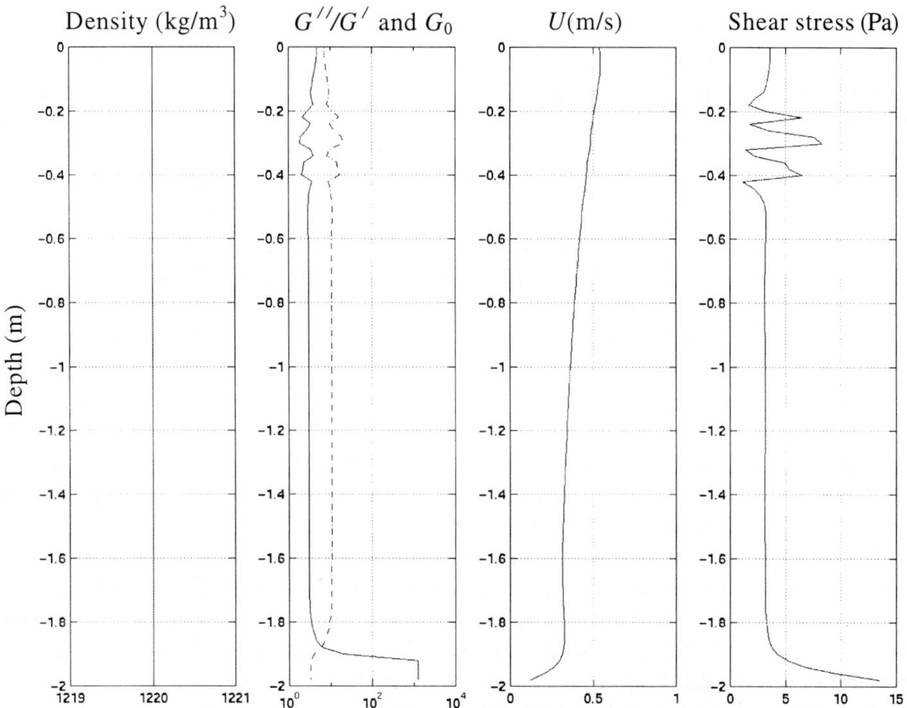

Figure 5b. Non-linear calculation for a homogeneous mud layer of bulk density of 1,220 kg/m^3. Mud depth: 2 m. Wave: 0.3 m amplitude, period 5 s. The damping factor is represented by a solid line; G_0 is in Pa and shown by a dash-dot line.

The first set of calculations (Figures 5a and 5b) corresponds to a homogeneous, 2 m thick mud layer with a bulk density of 1,220 kg/m^3. This configuration does not relate directly to the Seine estuary; however, the test provides an understanding of the processes that can occur when the mud thickness is larger than the wave boundary layer within the bed.

The total shear stress τ_{max} at a given depth in the bed can be calculated according to Mohr's stress analysis:

$$\tau_{max} = \sqrt{\tau_{xz}^2 + \tfrac{1}{4}(\sigma_x - \sigma_z)^2} \tag{11}$$

where τ_{xz} is the shear stress and σ_x and σ_z are the horizontal and the vertical normal stresses, respectively. Total shear stresses result from the "pumping" effect due to cyclic (induced) normal stresses, or from cyclic (induced) "shaking" due to shear stress.

For linear calculations (Figure 5a), the total shear stress shows a vertical structure which exhibits two maxima. The first maximum is due to "pumping" (i.e., due to normal stresses) and appears at a depth of 0.5 m, the wave boundary layer thickness within the bed. The other maximum appears further inside the bed and is due to the shear stress. It corresponds to the maximum gradient in the velocity amplitude (U) profile. Using linear calculations one would predict that, if forcing were sufficient to cause liquefaction, it would start at the point of maximum stress and deformation. A 0.5 m deep fluid layer would then be predicted. Such a critical value criterion for liquefaction has been proposed for instance by Li (1996), who used a linear uni-axial extensional-Voigt model.

Non-linear calculations (Figure 5b) show that rheological adjustment leads to qualitatively different features. The horizontal velocity amplitude has a smaller maximum and becomes relatively homogeneous; gradients (and shear stresses) are present only in the boundary layer near the rigid bottom. The boundary layer is thinner than in the linear calculation because the rheological parameters have diminished; being smaller they indicate localized failure or liquefaction of mud near the rigid bottom. This feature is also characterized by an increasing damping factor in the boundary layer.

Rheological adjustment of the type near the rigid bottom is also observed near a strongly stratified interface. The second set of calculations (Figures 6a and 6b) corresponds to an expected configuration for the bottom structure in the Seine estuary tidal flat after a long period of calm weather, when a relatively homogeneous soft mud layer occurs above a consolidated layer.

For linear calculations, the vertical structure of the velocity amplitude shows that maximum total shear stresses are associated with velocity gradients, i.e., shear stress is dominant when compared to the "pumping" effect. In this case, shear stress is dominant because the layer is thinner than the theoretical boundary layer of 0.5 m.

Inside the consolidated layer, velocities and strain rates are smaller. However, because the storage and loss moduli are greater at a greater density, increasing shear stresses occur up to the interface with the rigid bottom.

Non-linear calculations show a damping factor profile where the structural state of the upper layer is weakened; a well developed liquefied layer is present near the interface with the consolidated layer. This liquefied layer is in fact the boundary layer that has become thinner because the rheological parameters have decreased. If continuous stratification were considered, the bed would only be weakened near the rigid bottom, as in the homogeneous case.

6. CONCLUSIONS

A modeling approach for analyzing rheological changes of muddy beds under wave action is proposed. This approach is based on an analytical model for wave-mud interaction, and an empirical characterization of the rheological properties of mud as functions of oscillatory forcing. The results show that rheological

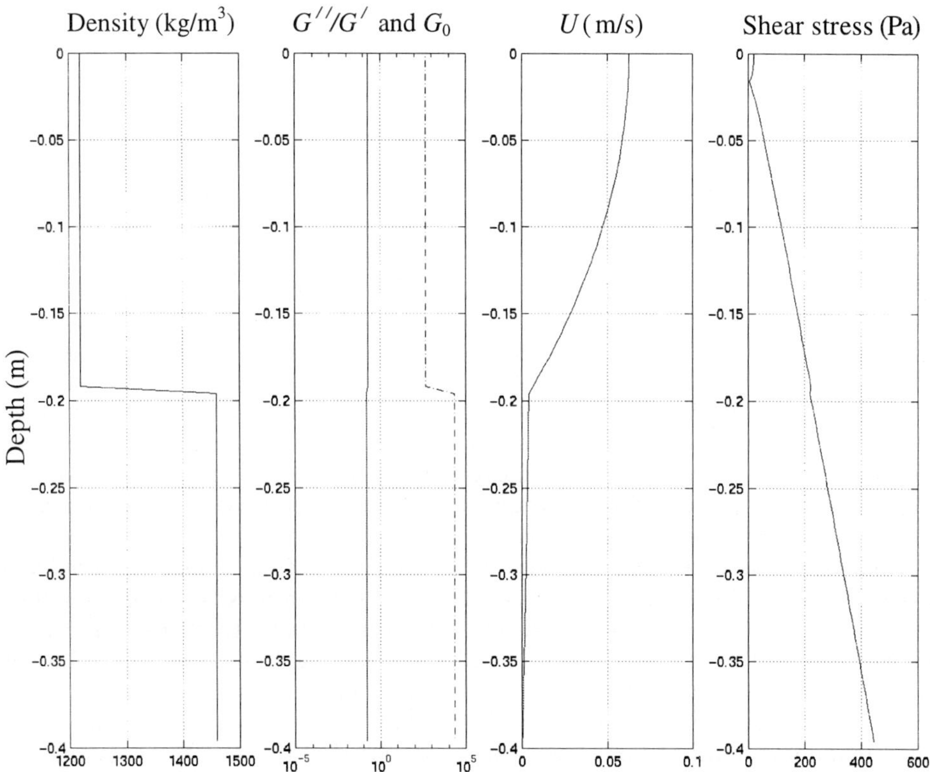

Figure 6a. Linearized calculation for a stratified mud layer. Mud depth: 0.4 m. Wave: 0.3 m amplitude, period 5 s. The damping factor is represented by a solid line; G_0 is in Pa and shown by a dash-dot line.

changes must be accounted for in order to predict liquefaction and/or structural failure of mud bed.

Further use of the model will provide a deeper insight into non-linear wave-mud interaction mechanisms. The procedure must be used for different materials and for different forcing conditions in order to fully describe non-linear wave mud interactions, including dispersion and damping of waves, onset of liquefaction and critical scales of the problem (time, depth, wave amplitude etc.).

The model has only been used for assessing the rheological behavior of Seine estuary mud. Calculations relate only to wave conditions observed under storm events. Nevertheless, the model allows us to conceptualize the observed erosion in the Seine estuary tidal flat, namely, liquefaction of the soft mud layer and failure of the consolidated mud near the stiff bottom.

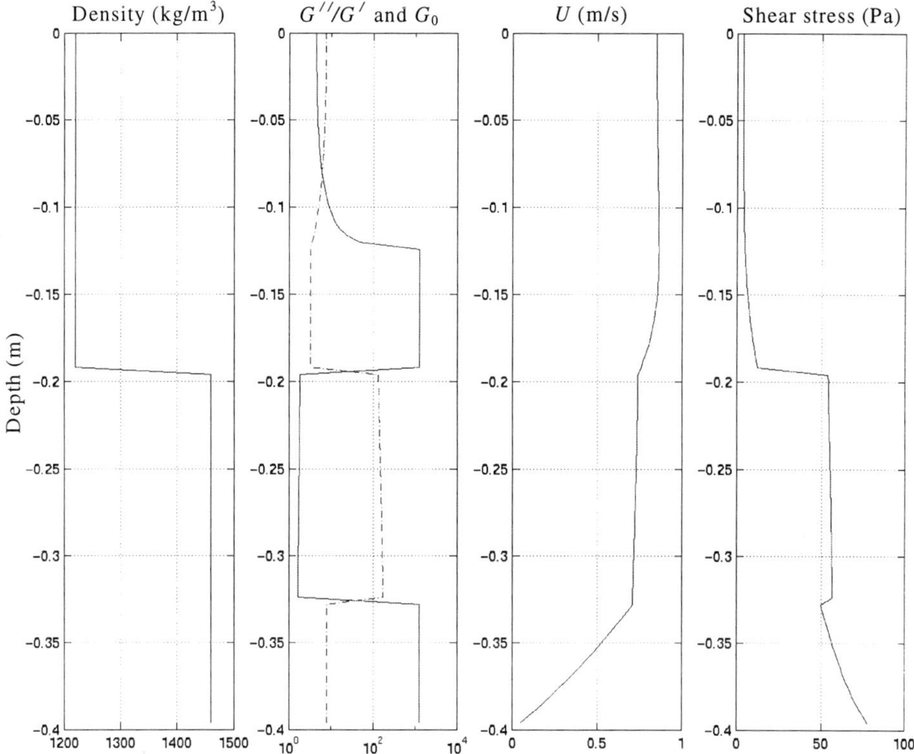

Figure 6b. Non-linear calculation for a stratified mud layer. Mud depth: 0.4 m. Wave: 0.3 m amplitude, period 5 s. The damping factor is represented by a solid line; G_0 is in Pa and a dash-dot line.

When the mud layer is thicker than the wave boundary layer inside the bed, "pumping" (due to normal stresses) seems to be the main source of the total shear stresses induced in the bed and the on-set of liquefaction. Maximum shear stresses occur at the limit of the wave boundary layer inside the bed.

When the mud layer is thinner than the theoretical wave boundary layer, total shear stresses correspond to velocity gradients and induced shear. Maximum shear stresses occur in this case near the interface with the stiffer layer underneath, where a thin boundary layer of liquefied mud occurs.

This last mechanism is surely important in order to detach the observed clasts of "muddy" agglomerates. However, it may be not the only one; other mechanisms my be present. For instance, dynamic normal stresses can generate significant compression and stretching forces. While the bed can sustain great compression, it cannot sustain stretching forces. Vertical stretching can be balanced by hydrostatic pressure. However, horizontal stretching may be critical and generate vertical failure surfaces and, hence, together with the failure surface underneath, create "muddy pebbles". If sand exists underneath the mud layer, one can also hypothesize that pore pressures would play a role.

Further research is needed in this topic. Nevertheless, the proposed approach constitutes a practical way to investigate rheological adjustment of mud to water wave forcing, and to understand the onset of liquefaction processes *in situ*.

REFERENCES

Barnes, H. A., Button, J. F., and Walters, K., 1989. *An Introduction to Rheology.* Elsevier, Amsterdam, 208p.

Chou, H. T., 1989. Rheological response of cohesive sediments to water waves. *Ph.D. Thesis*, University of California at Berkeley, 150p.

Dalrymple, R. A., and Liu, P. L., 1978. Waves over soft muds: a two-layer fluid model. *Journal of Physical Oceanography*, 8(6), 1121-1131.

Foda, M. A., Hunt, J. R., and Chou, H.-T., 1993. A nonlinear model for the fluidization of marine mud by waves. *Journal of Geophysical Research*, 98(C4), 61-82.

Gade, H. G., 1957. Effects of a non-rigid, impermeable bottom on plane surface waves in shallow water. *Ph.D. Thesis,* Texas A & M University, College Station, 35p.

Lesourd, S., Lesueur, P., and Avoine, J., 1998. Les fluctuations de lénvasement dans lémbouchure de la Seine. *Rapport du Programme Scientifique Seine Aval*, Thème Hydrodynamique et Transport Sédimentaire, 173p (in French).

Li, Y., 1996. Sediment-associated constituent release at the mud-water interface due to monochromatic waves. *Ph.D. Thesis*, University of Florida, Gainesville, 313p.

Maa, P.-Y., and Mehta, A. J., 1990. Soft mud response to water waves. *Journal of Waterway, Port, Coastal and Ocean Engineering*, 116(5), 634-650.

Mallard, W. W., and Dalrymple, R. A., 1977. Water waves propagating over a deformable bottom. *Proceedings of the 9th Annual Offshore Technology Conference*, 3(Paper OTC2895), 141-146.

Silva Jacinto, R., Bessineton, C., and Lesourd, S., 1998. Réponse de la vasière Nord aux forçages météo-océaniques. *Rapport du Programme Scientifique Seine Aval*, Thème Hydrodynamique et Transport Sédimentaire, 173p (in French).

Toorman, E. A., 1995. The thixotropic behaviour of dense cohesive sediment suspensions. *Report HYD149*, Hydraulics Laboratory, Civil Engineering Department, Katholieke Universiteit Leuven, 69p.

Coastal and Estuarine Fine Sediment Processes
W.H. McAnally and A.J. Mehta (Editors)

Assessment of the erodibility of fine/coarse sediment mixtures

H. Torfs[a], J. Jiang[b] and A. J. Mehta[c]

[a]Provincie Vlaams-Brabant, Directie Infrastructuur, Dienst Waterlopen, 3010 Leuven, Belgium

[b]ASL Environmental Sciences Inc., 1986 Mills Road, Sidney, V8L 5Y3, BC Canada

[c]Civil and Coastal Engineering Department, University of Florida, Gainesville, 345 Weil Hall, P.O. Box 116580, Florida 32611, USA

Erosion of mixtures of fine-grained sediments with sand is examined in a heuristic treatment relying on laboratory flume experiments under steady flows. Formulas for the critical stress for mixture erosion, τ_{cm}, and the rate of erosion of the fine fraction are diagnostically evaluated using data on the erosion of selected clay/sand and mud/sand mixtures. The stress, τ_{cm}, is found to vary non-monotonically with the fine sediment weight fraction, ψ. Starting with a sandy bed, as ψ increases, at first τ_{cm} seemingly decreases from its value for sand and reaches a minimum at a weight fraction ψ_{sp}, then increases. It appears that ψ_{sp} may be a measure of the space-filling concentration of fine particles within the pores of the sandy bed matrix. The logarithm of the rate of erosion is shown to vary linearly with the logarithm of the excess bed shear stress, and this variation depends on ψ. Further understanding of modeling mixture erosion must rely on additional data on the variation of τ_{cm} with ψ.

1. INTRODUCTION

Fine sediment aggregates are often found in association with sand in submerged areas. Because the fine fraction typically varies both areally and with depth, its rate of release depends on the location, and at a given site varies with time even under constant fluid forcing. This spatial and temporal variability is presently not characterized easily without extensive measurements of bed erosion. It is therefore evident that if a physical framework could be developed to explain mixture erosion, assessments of the erosion potential of heterogeneous substrates would be facilitated. Here we will deal with a simplified treatment of a bed characterized by two grain populations, one fine and the other coarse. In order to determine the erosion threshold of this mixture, we will extend the force balance based development of Mehta and Lee (1994), and use the laboratory flume data of Torfs (1995) on steady-flow erosion of selected fine-sediment/sand mixtures to make an assessment of the usefulness of the resulting formula. We will then briefly examine the ensuing rate of erosion of the fine fraction.

2. THRESHOLD FOR SINGLE GRAIN SIZE

Consider the forces acting on a grain at the surface of a bed subject to steady flow, as shown in Figure 1. When cohesion is important, i.e., for clayey sediment, a hydrodynamic lift force, L, opposing the cohesive force is required to rupture the inter-granular bonds. Likewise, shearing due to fluid drag, D, is opposed by resistance due to cohesion, friction and interlocking. The resistive effect of friction and interlocking, also present in lifting the grain, can be incorporated by way of a multiplicative coefficient as a modifier of cohesion. The resulting cumulative force, F, will be additive to the buoyant weight W of the grain. The quantity F_R is a reactive shear force, equal in magnitude and opposite in direction to D. As to the bed surface, the spatial-mean bed plane passes through point a, which characteristically lies above the center of mass of the grain at b. Consequently, while D, and hence F_R, pass through a, L and W are centered on b. Ignoring the small moment arm, ab, the condition for incipient or threshold motion in terms of the active forces is shown in Figure 2. This condition implies that, in this physical state of the grain, the resultant of the drag force, D, and the net downward force, $W + F - L$, must subtend an angle ϕ, equal to the angle of repose (Graf, 1971). By way of the definition ϕ, the resultant force therefore passes through c, the point of inter-granular contact. The angle of repose has a clear physical meaning for cohesionless sediment, and is mainly dependent on grain size, but for cohesive sediment it becomes a coefficient embodying shear resistance.

From Figure 2 the condition for incipient motion is:

$$\tan\,\phi = \frac{D}{W + F - L} \tag{1}$$

where D and L both relate to a characteristic flow velocity at the top of the grain (Graf, 1971). To simplify further treatment, we will assume the flow to be non-viscous, ranging from weakly to fully turbulent. Accordingly, given τ_c = threshold or critical bed shear stress, d = a representative grain size, α_1= an area shape factor and $\alpha_2 = \alpha_1 C_L / C_D$, where C_L and C_D are the lift and drag coefficients, respectively, we have $D = \alpha_1 \tau_c d^2$ and $L = \alpha_2 \tau_c d^2$. The buoyant weight, $W = \alpha_3 g(\rho_{ag} - \rho) d^3$, where α^3 = a volume shape factor, g = acceleration due to gravity, ρ_{ag} = aggregate or granular density depending on sediment composition and size, and ρ = fluid (water) density. Substituting these quantities in (1), the following relationship is obtained (Mehta and Lee, 1994):

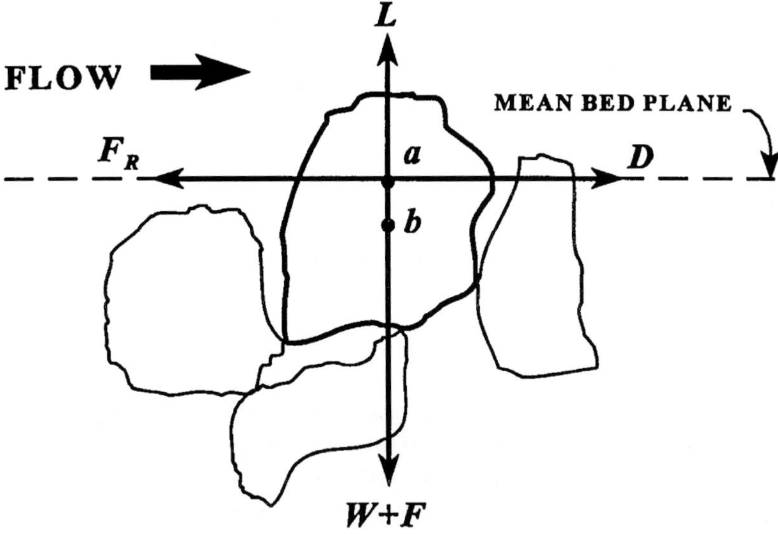

Figure 1. Forces on a grain at the bed surface subjected to a steady flow.

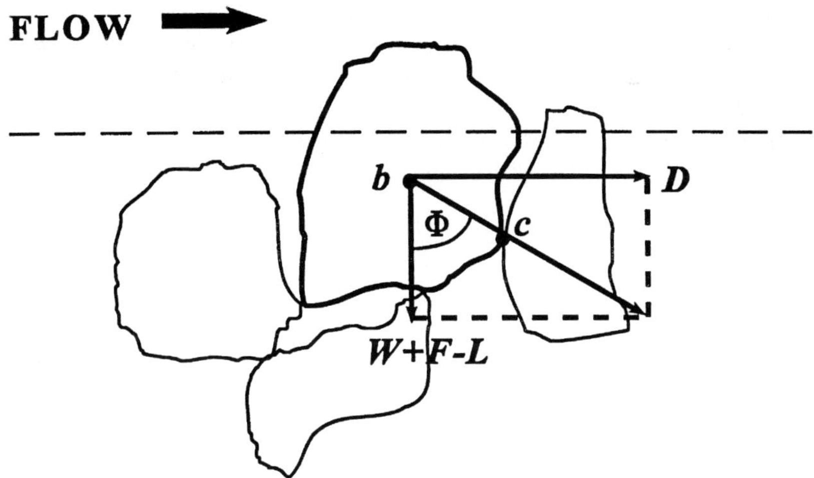

Figure 2. Force diagram at incipient or threshold grain motion.

$$\frac{\tau_c}{g(\rho_{ag}-\rho)d}=\frac{\alpha_3\tan\,\phi}{(\alpha_1+\alpha_2\tan\,\phi)}+\frac{F\tan\,\phi/(\alpha_1+\alpha_2\tan\,\phi)}{g(\rho_{ag}-\rho)d^3} \tag{2}$$

where the term on the left hand side is the Shields' entrainment parameter representing the ratio of drag force to buoyancy. The first term on the right hand side is a sediment-dependent parameter, and the second term is essentially the ratio of cohesive force to buoyancy. When cohesion is absent, i.e., F=0, the second term on the right hand side vanishes, and (2) becomes a representation of Shields' entrainment relationship for coarse sediment.

3. THRESHOLD FOR FINE/COARSE GRAIN MIXTURES

Equation (2) is strictly applicable to erosion of a bed of single-sized grains, even as it allows this size to vary from fine to coarse. When fine grain aggregates occur in a mixture with coarse grains, the interpretation of (2) becomes complex, especially because the two particle populations, having different sizes, densities and cohesion (equal to zero for coarse grains), do not necessarily entrain at the same rate. Therefore, to proceed further, it becomes useful to make a heuristic interpretation of (2) based on the mechanics of grain entrainment. To begin with, we will assume that we are dealing with the entrainment of a fine/coarse mixture which can be characterized by size, d_m, and granular density, ρ_{sm}, the latter being an indirect measure and approximate substitute for fine sediment aggregate density. For the composite (mixture) bed, the corresponding angle of repose will be ϕ_m, and the critical stress, τ_{cm}. Further, for fine grains, F must be replaced by a composite force, F_m, which depends not only on cohesion, but also on the influence of the interbedded coarse grains. Such an influence can either enhance the effect of cohesion, e.g., by a "hiding", or sheltering, effect (Graf, 1971), or reduce cohesion by decreasing inter-particle bonding. Thus, we may simply consider $F_m = KF$, where K depends on bed composition. Accordingly, (2) becomes

$$\frac{\tau_{cm}}{g(\rho_{sm}-\rho)d_m}=\frac{\alpha_3\tan\phi_m}{(\alpha_1+\alpha_2\tan\phi_m)}+\frac{KF\tan\phi_m/(\alpha_1+\alpha_2\tan\phi_m)}{g(\rho_{sm}-\rho)d_m^3} \tag{3}$$

in which α_1, α_2 and α_3 now relate to d_m. For a uniform, cohesive bed K=1, and the first term on the right hand side loses meaning because, as noted, ϕ_m is not definable for such a bed. Accordingly, equating τ_{cm} with the bed shear strength against erosion, τ_s (Parchure and Mehta, 1985), from (3) we obtain

$$\tau_s=\frac{K_1F}{d_{ag}^2} \tag{4}$$

where K_1 is a lumped, sediment-specific parameter and d_{ag} is the aggregate diameter. An analysis of cohesive bed erosion experiments has yielded the following relation (Mehta and Parchure, 2000):

$$\tau_s = \zeta(\phi_v - \phi_{vc})^{\xi} \tag{5}$$

in which ϕ_{vc} is the fine solids weight fraction obtained by dividing the dry density of solids, i.e., sediment dry weight per unit volume of fine sediment-water mixture, by the granular density of the fine particles. The quantity ϕ_v is the "threshold" value of ϕ_v below which the bed matrix is not fully particle-supported and is therefore fluid-like, and ζ and ξ are coefficients which depend on bed composition and the degree of consolidation. From (4) and (5) we obtain

$$F = \frac{\zeta d_{ag}^2}{K_1}(\phi_v - \phi_{vc})^{\xi} \tag{6}$$

Hence, (3) can be restated as

$$\frac{\tau_{cm}}{g(\rho_{sm} - \rho)d_m} = \frac{\alpha_3 \tan\phi_m}{(\alpha_1 + \alpha_2 \tan\phi_m)} + \frac{K \tan\phi_m/(\alpha_1 + \alpha_2 \tan\phi_m)}{K_1} \frac{d_{ag}^2}{g(\rho_{sm} - \rho)d_m^3}\zeta(\phi_v - \phi_{vc})^{\xi} \tag{7}$$

Equation (7) can be simplified by recognizing that the first term on the right hand side is coarse grain dominated, while the second term is dominated by the fine constituent. Therefore, in the first term we may conveniently assign the subscript, *cg*, to imply coarse grain dominance. As for the second term we will consider the ratio, $K \tan\phi_m/(\alpha_1 + \alpha_2 \tan\phi_m)K_1$, to be a bed dependent characteristic coefficient, K'. Then, (7) can be restated as

$$\tau_{cm} = \left[\frac{\alpha_{3cg} \tan\phi_{cg}}{(\alpha_{1cg} + \alpha_{2cg} \tan\phi_{cg})} + \frac{K'\zeta(\phi_v - \phi_{vc})^{\xi}}{g(\rho_{sm} - \rho)d_m}\right] g(\rho_{sm} - \rho)d_m \tag{8}$$

where we have also invoked the assumption, $d_{ag} \approx d_m$. In (8), $K'=0$ implies the absence of fine material, and increasing K' signifies increasing modulation of the critical stress for erosion of coarse material by fines.

In order to render (8) useful for application, we will introduce the fine sediment weight fraction in the mixture, ψ, such that 1-ψ is the corresponding weight fraction of the coarse material. Now, provided ρ_{sm}, d_m, α_{1cg}, α_{2cg}, α_{3cg}, ϕ_{cg}, ζ, ξ, ϕ_{vc} and K' are known, the dependence of the critical stress, τ_{cm}, on ϕ_v, and therefore indirectly on ψ, can be evaluated from (8). Alternatively, since τ_{cm} can be considered to be a surrogate for K', measurements of τ_{cm} as a function of ψ can be used to assess the behavior of K'. Towards that aim, the data of Torfs (1995) will be used as follows.

4. FLUME DATA

4.1. Experimental conditions

In a 9 m long and 40 cm wide flow recirculating flume, Torfs (1995; 1997) eroded consolidated mixtures of clayey sediments and sand in different proportions (by weight). The fine materials included a kaolinite (dispersed median size 2 μm), a natural, clay-silt mud (25 μm) and a montmorillonite (8 μm). The median diameter of sand was 0.23 mm. Relevant bed conditions are summarized in Table 1.

The critical shear stress, τ_{cm}, was determined by increasing the flow velocity starting from nil in small steps, and measuring the concentration change both in suspension and in a bottom trap. When between two consecutive steps a measurable increase occurred in the total concentration, the critical shear stress was calculated as the average bed shear stress for those two steps. The value of τ_{cm} as a function of ψ obtained in this way is plotted in Figures 3 and 4 for the three mixtures from Table 1. Starting with ψ=0 corresponding to pure sand at a nominal bulk density of $1{,}850\,\mathrm{kg/m^3}$, the fine fraction was increased to a maximum value of 0.379 (without changing the mixture bulk density). The variation of τ_{cm} with ψ is observed to be non-monotonic and seemingly passes through a minimum, yielding values of τ_{cm} that may at first be lower than those for sand (0.35 Pa), then rise to values larger than for sand, and further increase with ψ. Notwithstanding the fact that this description is constrained by the paucity of data and possible albeit unquantified uncertainties in the estimates of the bed shear stress (Torfs, 1995), it can be elaborated upon as follows.

Referring to Figure 5a, when a small quantity of fines is added to sand, a reduction in the inter-granular friction between sand particles due to partial filling of pore spaces by fines causes sand grains to erode with greater ease, thus lowering the critical stress for erosion below that for pure sand. This effect increases with increasing fine fraction until a space-filling network of fines is established, when the threshold for incipient motion becomes minimum. Given this condition, a further increase in the fine fraction is increasingly influenced by the erodibility of fines, because as clayey particles increasingly surround sand grains, sand-sand contacts decrease (Figure 5b), and the number of sand grains per unit surface area of bed also decreases (Panagiotopoulos et al., 1997).

Using (8) and the data from Figures 3 and 4, we will now determine the behavior of K' as a function of ψ. To do so, the various parameters in (8) are selected as follows.

4.2. Selection of ϕ_{cg}, α_{1cg}, α_{2cg}, α_{3cg}, d_{ag}, ζ, ξ and ϕ_{vc}

ϕ_{cg}: We will conveniently use a formulation derived from the work of Wiberg and Smith (1987):

$$\phi_{cg} = \cos^{-1}\left(\frac{d_m/d_{sand} + z_*}{d_m/d_{sand} + 1}\right) \tag{9}$$

where z_* is defined as the average level at the bottom of the almost-moving grain.

Table 1
Bed parameters in erosion experiments

Bed mixture[a]	Mixture granular density, ρ_{sm}, range[b] (kg/m^3)	Mixture grain size, d_m, range[c] (mm)	Fine grained sediment seight fraction, ψ, range[d]
Kaolinite/Sand	2,385 - 2,650	0.155 - 0.230	0 - 0.203
Natural Mud/Sand	2,408 - 2,650	0.168 - 0.230	0 - 0.183
Montmorillonite/Sand	2,215 - 2,650	0.110 - 0.230	0 - 0.379

[a] All beds had a nominal bulk density of 1,850 kg/m^3.
[b] This density is obtained by mass balance from:

$$\rho_{sm} = \frac{\omega_{fpw}\rho + \omega_{fine}\rho_{fine} + \omega_{sand}\rho_{sand}}{\omega_{fpw} + \omega_{fine} + \omega_{sand}}$$

[c] This diameter is defined as:

$$d_m = \frac{\omega_{sand} d_{sand} + \omega_{ag} d_{ag}}{\omega_{sand} + \omega_{ag}}$$

[d] By definition, this fraction is:

$$\psi = \frac{\omega_{fine}\rho_{fine}}{\omega_{fine}\rho_{fine} + \omega_{sand}\rho_{sand}}$$

In the above, ω denotes volume fraction, and the following quantities are used (where V denotes volume, and subscripts "*fpw*" and "*t*" refer to "fine sediment aggregate pore water" and "total", respectively.):

$$\omega_{fpw} = \frac{V_{fpw}}{V_t};\ \omega_{fine} = \frac{V_{fine}}{V_t};\ \omega_{sand} = \frac{V_{sand}}{V_t};\ \omega_{ag} = \frac{V_{fpw} + V_{fine}}{V_t};$$

$$V_t = V_{water} + V_{fine} + V_{sand}$$

Subscripts "*fine*", "*sand*", "*ag*" and "*water*" refer to fine (clayey) sediment, sand, fine sediment aggregate and water, respectively; aggregate being an agglomeration of fine sediment particles and water.

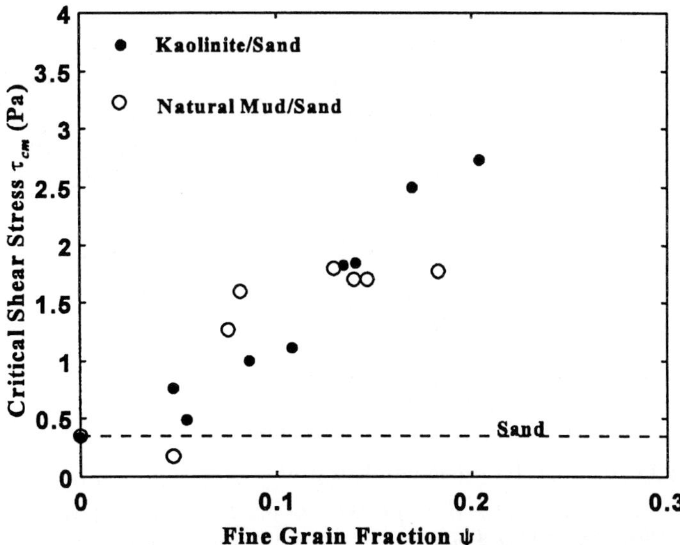

Figure 3. Measured critical shear stress as a function of fine grained weight fraction for mixtures of kaolinite/sand and natural mud/sand.

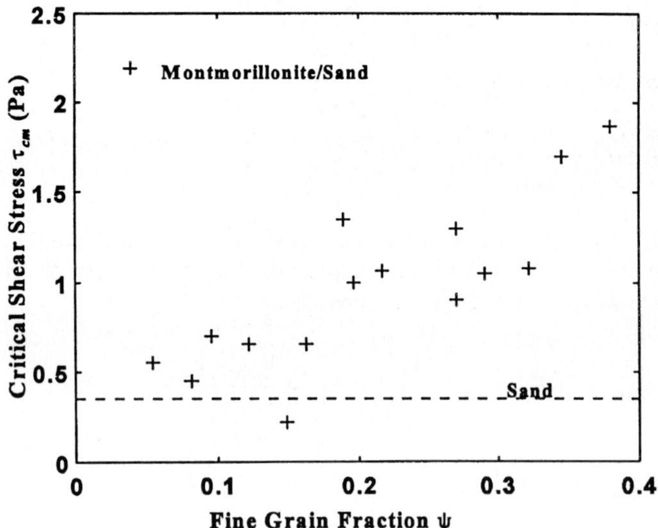

Figure 4. Measured critical shear stress as a function of fine grained weight fraction for mixtures of montmorillonite/sand.

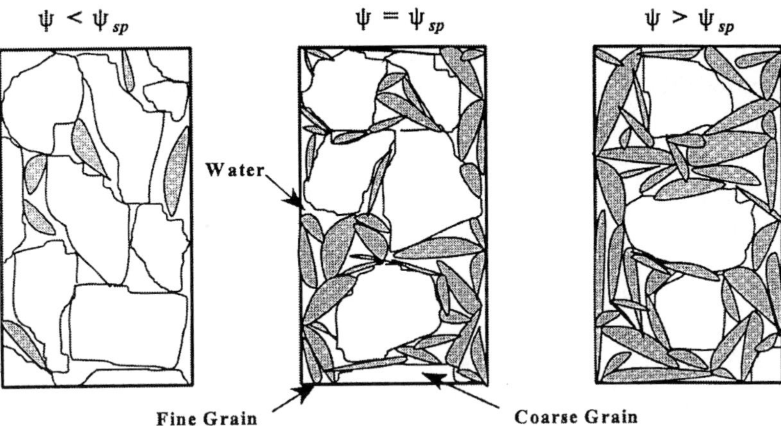

Figure 5a. Saturated bed composed of large and small grain populations. Left: small grain fraction is less than space-filling; Center: small grain fraction is space-filling; Right: small grain fraction exceeds space-filling value.

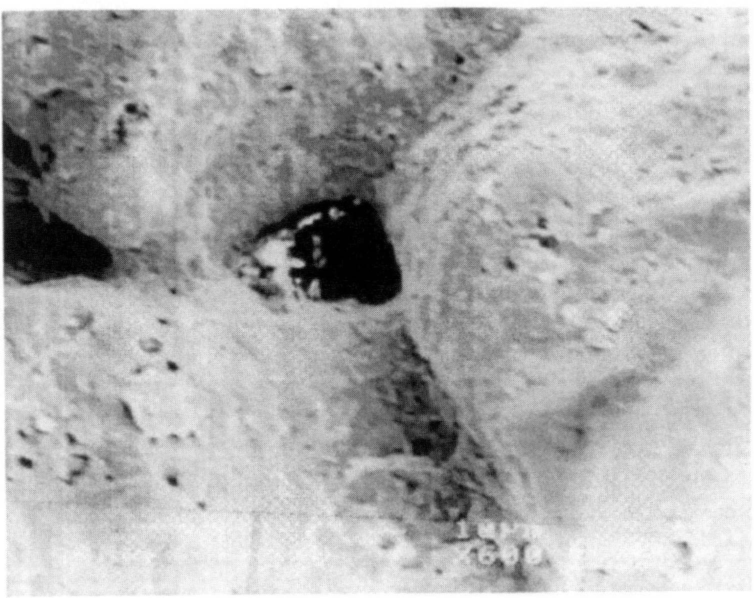

Figure 5b. Scanning electron microscope photograph of a mixture of sand and 7.1% (by weight) kaolinite. Observe the clay particle "bridges" between sand grains. Bridged distances are on the order of 10 μm (after Torfs, 1995).

α_{1cg}: We will consider the sand grains to be slightly angular, with an assumed sphericity of 0.7. For such a sand, Wiberg and Smith (1987) give $z_* = -0.02$ in (9). Then, since for this case $d_m/d_{sand} = 1$, we obtain $\phi_{cg} = 60.7°$. Now, in (8) with $K' = 0$, $\tau_{cm} = \tau_{cs} = 0.35$ Pa (measured), and given a granular density of sand equal to 2,650 kg/m^3, we obtain $\alpha_{1cg} = 3.72$.

α_{2cg}: Considering a "smooth" sandy bed, we will represent the grain lift force, L, following Dade et al. (1992):

$$L = 0.81\rho\nu^2 Re_w^3 = 0.81 Re_w \tau_{cs} d_m^2 \tag{10}$$

where ν = kinematic viscosity of water and $Re_w = (\tau_{cs}/\rho)^{1/2} d_m/\nu$ is the wall Reynolds number. Then, since $L = \alpha_{2cg}\tau_{cs}d_m^2$, we obtain

$$\alpha_{2cg} = 0.81 Re_w \tag{11}$$

α_{3cg}: Considering d_{sand} to be the nominal diameter (i.e., the diameter of a sphere having the same volume as the actual grain), by definition $\alpha_{3cg} = \pi/6$.

d_{ag}: For the aggregate diameter required to calculate the composite diameter, d_m (see c, Table 1), we will select the following relation between d_{ag} (μm) and the aggregate density, ρ_{ag} (kg/m^3), based on data compiled by Dyer (1989) and assuming its applicability in the present context:

$$d_{ag} = \exp\left[1.32\ln\left(\frac{1250}{\rho_{ag} - \rho}\right)\right] \tag{12}$$

ζ, ξ and ϕ_{vc}: For (5), a review of previous experimental works (Mehta and Parchure, 2000) suggests the following "best values" of the coefficients: ζ=12, ξ=1 and $\phi_{vc} = 0.05$. Note that these values of ζ and ξ are commensurate with shear stress measured in Pascal (Pa), and will be considered to be representative of the fine components of all three mixtures.

4.3. Determination of *K'*

From an inspection of Figures 3 and 4 in conjunction with (8), it is readily concluded that the variation of K' with ψ must be qualitatively akin to the trend of variation of τ_{cm} with ψ. Accordingly, we will select the following forms for $K'(\psi)$:

$$K' = A\left|\psi - \psi_{sp}\right|^n - K'_{min}\,;\ \tau_{cm} < \tau_{cs} \tag{13}$$

$$K' = B - (1-\psi)^m;\ \tau_{cm} \geq \tau_{cs} \tag{14}$$

Here ψ_{sp} is the space-filling value of ψ (Figures 5), K'_{min} is the minimum value of K' at $\psi = \psi_{sp}$, and *A*, *B*, *n* and *m* are mixture-specific constants. Note that although space-filling is dependent on the *volume* fraction of the aggregates, we have conveniently chosen the fine sediment *weight* fraction as a representative, albeit approximate, surrogate. The chosen values of coefficients for (13) and (14) are indicated in Figures 6 and 7 (Mehta et al., 1997). While kaolinite and the natural mud seemingly exhibit an affine behavior with respect to the dependence of K′ on ψ, the constants differ for montmorillonite, reflecting its uniqueness in this regard (Torfs, 1995).

Knowing K' for each ψ, (8) can be used to obtain the corresponding τ_{cm}. Plots of τ_{cm} against ψ thus obtained are shown in Figures 8 and 9. Note that the degree of agreement between (8) and data points is an artifact of the best-fit agreement between the curves and data points for K' variation with ψ in Figures 6 and 7.

5. RATE OF EROSION

Torfs (1995) eroded mixtures of Table 1 and measured the time-variation of resuspended fine sediment as a function of the bed shear stress, τ_b. The rate of erosion, ε, was found to agree with the relation

$$\varepsilon = \varepsilon_0(\tau_b - \tau_{cm}) \tag{15}$$

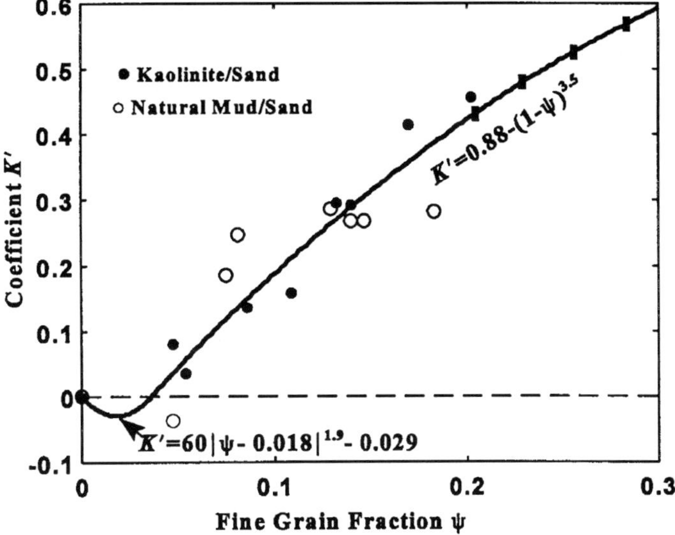

Figure 6. K' as a function of ψ for kaolinite/sand and natural mud/sand mixtures.

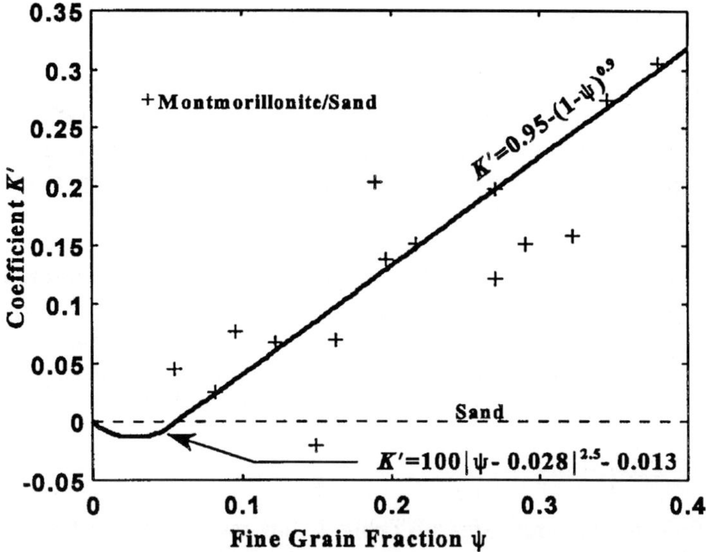

Figure 7. K' as a function of ψ for montmorillonite/sand mixtures.

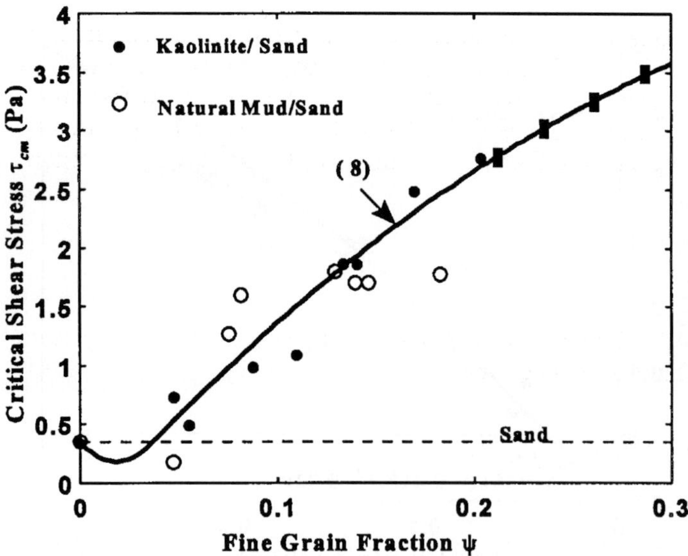

Figure 8. Critical shear stress for erosion of kaolinite/sand and natural mud/sand mixtures versus fine grained weight fraction: data and (8).

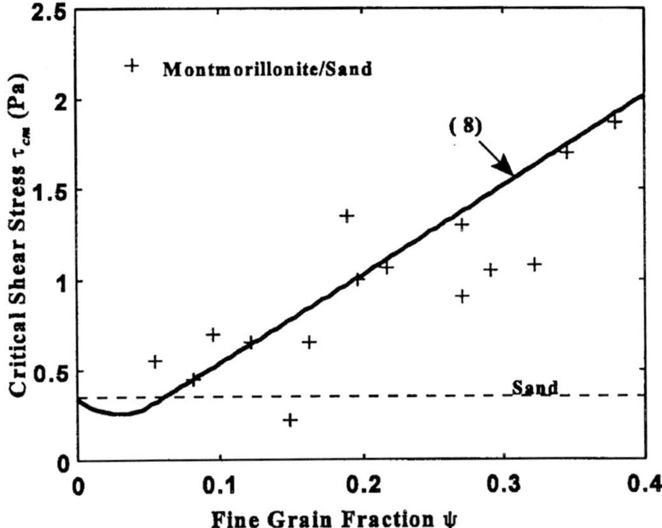

Figure 9. Critical shear stress for erosion of montmorillonite/sand mixture versus fine grained weight fraction: data and (8).

From an analysis of previous laboratory data on fine sediment erosion, Mehta and Parchure (2000) showed that the rate constant, ε_0, tends to decrease exponentially with increasing shear strength with respect to erosion. If we assume the validity of this observation even when the critical shear stress incorporates the combined effects of fine and sand fractions in the bed, the following analogous relation is obtained:

$$\varepsilon_0 = \varepsilon_{00} e^{-\kappa \tau_{cm}^{\chi}} \tag{16}$$

where ε_{00} is the limiting value of ε_0 (at τ_{cm} =0), and κ and χ are bed-specific coefficients. Hence, (15) becomes

$$\log \varepsilon = \log\left(\varepsilon_{00} e^{-\kappa \tau_{cm}^{\chi}}\right) + \log(\tau_b - \tau_{cm}) \tag{17}$$

which implies that log ε varies linearly with log (τ_b-τ_{cm}). Now (17) along with (8), (13) and (14) can be used to calculate ε as a function of the excess shear stress, τ_b-τ_{cm}, for different values of ψ. For illustrative purposes we will consider beds of kaolinite/sand. For that case, we will select ε_{00}=0.25 kg N^{-1} s^{-1}, κ=8 and χ=0.5 by calibration. The ensuing calculations are given in Figure 10 in terms of the

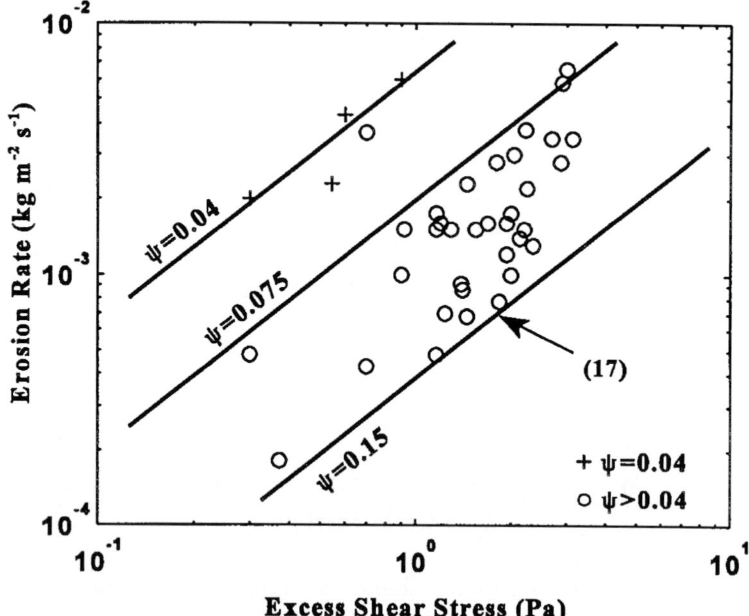

Figure 10. Erosion rate of fine fraction of kaolinite/sand mixtures plotted against excess shear stress. Plusses denote data points for ψ=0.04; circles are for ψ>0.04. Lines are based on (17).

boundary lines defined by selected ψ values. By virtue of (17), the intercept of each line varies with ψ. The agreement between the line for ψ=0.04 and the data points obtained at the same value of ψ is seen to be reasonable. Likewise, observe that the data points corresponding to values of ψ>0.04 (the actual range of ψ was not reported) seem to be largely bracketed (with one noteworthy exception) by the calculated lines using ψ= 0.075 and 0.15.

6. CONCLUSIONS

Equations (8) and (17) are limited in their application by the numerous assumptions concerning the flow field and sediment behavior in flow. As for the empirical relations used to obtain the sediment-specific parameters for (8) and (17), these relations are also restricted in their usage for two reasons: 1) their validity is confined to the original data on which they are based, and 2) the correlations assumed in some of these relations are not inherently unique, e.g., between the bed shear strength and bed solids weight fraction as expressed by (5). Nevertheless, especially given such caveats, the observation that (8) and (17) can be "matched" with flume measurements points to the empirical validity of these formulas, at least within the bounds of the selected experiments.

The highly non-linear dependence of the critical shear stress on the fine-grained weight fraction is indicative of the interactive role of the mixture components in controlling erosion, and suggests a potential route for improvement in modeling erosion by accounting for these interactions. In that context, additional experimental observations on the erosion of mixtures in the vicinity of space-filling weight (or volume) fraction are essential. It would also be helpful to observe the trend in critical stress variation as the fine fraction nears unity, i.e., as the behavior of the mixture approaches that of the fine component.

REFERENCES

Dade, W. B., Nowell, A. R. M., and Jumars, P. A., 1992. Predicting erosion resistance of muds. *Marine Geology*, 105(1-4), 285-297.

Dyer, K. R., 1989. Sediment processes in estuaries: future research requirements. *Journal of Geophysical Research*, 94(C10), 14327-14339.

Graf, W. H., 1971. *Hydraulics of Sediment Transport*. McGraw-Hill, New York, 524p.

Mehta, A. J., and Lee, S.-C., 1994. Problems in linking the threshold condition for the transport of cohesionless and cohesive sediment grain. *Journal of Coastal Research*, 10(1), 170-177.

Mehta, A. J., Kirby, R., Stuck, J. D., Jiang, J., and Parchure, T. M., 1997. Erodibility of organic-rich sediments: a Florida perspective. *Report No. UFL/COEL/MP-97/01*, Coastal and Oceanographic Engineering Department, University of Florida, Gainesville, variously paginated.

Mehta, A. J., and Parchure, T. M., 2000. Surface erosion of fine-grained sediment revisited. In: *Muddy Coast Dynamics and Resource Management*, B. W. Flemming, M. T. Delafontaine and G. Liebezeit eds., Elsevier, Amsterdam, The Netherlands (in press).

Panagiotopoulos, I. Voulgaris, G., and Collins, M. B., 1997. The influence of clay on the threshold of movement of fine sandy beds. *Coastal Engineering*, 32, 19-43.

Parchure, T. M., and Mehta, A. J., 1985. Erosion of soft cohesive sediment deposits. *Journal of Hydraulic Engineering*, 111(10), 1308-1326.

Torfs, H., 1995. Erosion of mud/sand mixtures. *Ph.D. Thesis*, Catholic University of Leuven, Leuven, Belgium, 261p.

Torfs, H., 1997. Erosion of mixed cohesive/non-cohesive sediments in uniform flow. In: *Cohesive Sediments*, N. Burt, R. Parker and J. Watts eds., John Wiley, Chichester, UK, 245-252.

Wiberg, P. L., and Smith, J. D., 1987. Calculations of the critical shear stress for motion of uniform and heterogeneous sediments. *Water Resources Research*, 23(8), 1471-1480.

Coastal and Estuarine Fine Sediment Processes
W.H. McAnally and A.J. Mehta (Editors)

Rapid siltation from saturated mud suspensions

J. C. Winterwerp[a,c], R. E. Uittenbogaard[a] and J. M. de Kok[b]

[a]WL | Delft Hydraulics, P.O. Box 177, 2600 MH Delft, The Netherlands

[b]Rijkswaterstaat, RIKZ, The Hague, The Netherlands

[c]Delft University of Technology, Faculty of Civil Engineering and Geosciences, Delft, The Netherlands

Conditions are analyzed at which rapid siltation in navigation channels and harbor basins occurs. From an analogy with salt-fresh water stratified flows, a saturation concentration is defined as the maximum sediment load that can be carried by turbulent flow. At concentrations beyond the saturation value, the turbulent field and the vertical concentration profile collapse and a layer of fluid mud is formed. Simulations are carried out with the 1DV POINT MODEL to establish the dependency of the saturation concentration on the bulk flow parameters. Next, this model is applied to simulate a series of measurements performed in the Maasmond area, the entrance to the Port of Rotterdam. From these simulations it is concluded that proper results can only be obtained if all physical processes in the vertical are included in the model, i.e., hindered settling, salt-fresh-water- and sediment-induced buoyancy effects and wave-induced mixing. Within the access channel "de Maasgeul", conditions become super-saturated and a fluid mud layer is formed.

1. INTRODUCTION

To safeguard navigation, port authorities maintain their access channels and harbor basins by frequent dredging. For the Port of Rotterdam, for instance, this is particularly required during and after rough weather periods, when rapid siltation is observed and large volumes of cohesive sediment are deposited over short time intervals. A recent series of measurements at four semi-permanent anchor stations on either side of the Maasgeul, the access channel to the port (e.g., Figure 1), shows very high suspended sediment concentrations with near-bed values beyond 10 g/l during stormy periods.

This paper focuses on establishing the conditions favoring rapid siltation and analyzing the governing physical processes on the basis of measurements, with

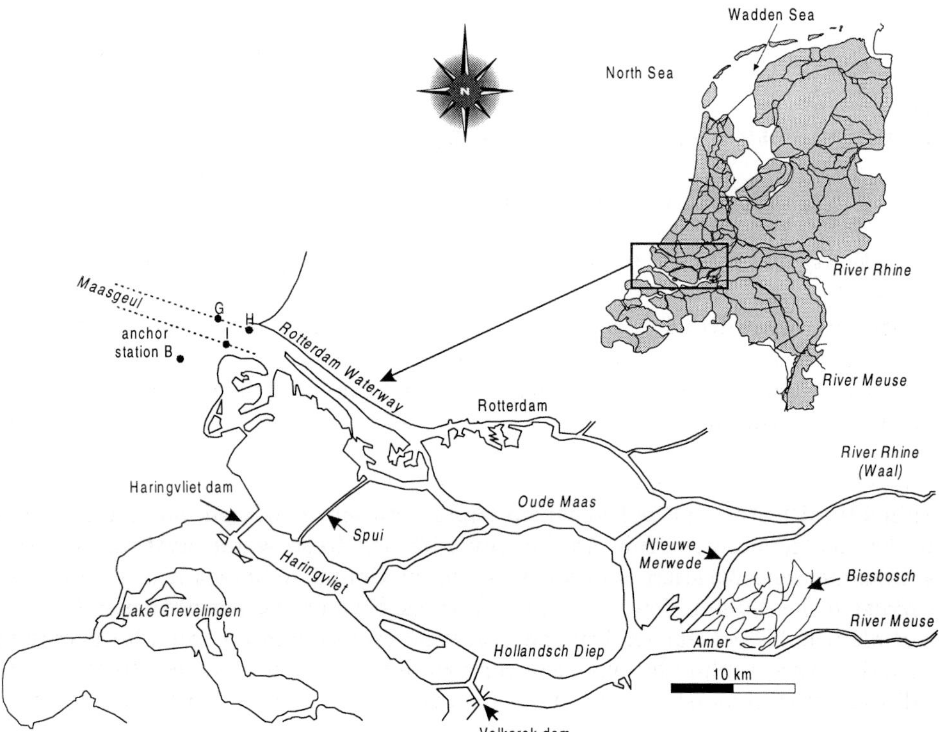

Figure 1. Maasmond area and location of Maasgeul and measuring stations B, G, H and I.

the ultimate goal to minimize maintenance costs by optimizing the design of access channels and harbor basins and/or the dredging operations themselves.

The physical processes involved, and their mutual interactions, are complex. An analysis is therefore possible only if restricting assumptions are made. The water-sediment mixture is treated as a single-phase fluid in which all particles follow turbulent movements, but for their settling velocity. Uittenbogaard (1994) argues that this is an appropriate assumption if $W_s << w'$, where W_s is the settling velocity of the sediment and w' a measure of the vertical turbulent velocity fluctuations (rms-values). As W_s is about 0.1 - 1 mm/s and $w' \approx u_*$ in open channel flow (Nezu and Nakagawa, 1993), where u_* is the shear velocity, this condition is generally met for cohesive sediments in estuarine and coastal waters. Uittenbogaard showed theoretically that sand particles with a diameter of up to 200 μm can follow turbulent movements sufficiently to allow for a single-phase description. His argument was confirmed by Muste and Patel (1997), who measured the fluctuating velocity components of suspended sand particles of 250 μm median diameter in a turbulent flow and found that their rms-value was only about 15 to 20% smaller than the rms-fluctuating fluid velocity.

Furthermore, it is assumed that all sediment particles will remain part of the water-sediment mixture, i.e., they will remain within the water column, unless stated otherwise. A third important restriction is that the suspension is treated as a Newtonian fluid, i.e., fluid mud non-Newtonian dynamics is not considered.

These assumptions allow the sediment suspension to be treated via the advection-diffusion equation, the advective and diffusive processes being governed by the turbulent water movement. It is noted that a similar approach is followed by Teisson et al. (1992) and Le Hir (1997).

The transport and accumulation of the sediment is governed by the horizontal transport processes. Their physical-mathematical description is fairly straightforward, except for interactions with processes in the vertical, such as settling, deposition, entrainment and mixing. Therefore, the analysis is focused on the processes in the vertical direction. Accordingly, the relevant mathematical formulations are implemented in a one-dimensional vertical numerical code called the 1DV POINT MODEL. The next section provides a description of this model, as used to analyze field measurements. Prior to a description of these measurements, the concept of saturated suspension is treated, which results from sediment-flow interactions. A simulation of the measurements with the 1DV POINT MODEL, and a sensitivity analysis to assess the relevant physical processes, are then presented. The consequences of analysis with respect to the rapid siltation processes in the Maasmond area are discussed in the last two sections.

2. THE 1DV POINT MODEL

The 1DV POINT MODEL is based on Delft Hydraulics' full three-dimensional hydrostatic code DELFT3D, but in which all horizontal gradients have been stripped, except for the longitudinal pressure gradient. Various versions are operational; here the focus is on the version for cohesive sediment. The model includes the momentum equation, the advection-diffusion equation for cohesive sediment including the effects of hindered settling, and a description to account for the effects of surface waves on the bed shear stress and vertical mixing. Uittenbogaard (1995) has shown that the k-ε turbulence closure model is applicable to fairly stratified conditions. This model is therefore used with a salinity- and sediment-induced buoyancy term. Finally it is noted that all (except one) simulations described in this paper are carried out without any water-bed exchange, i.e., all particles are considered to remain within the computational domain.

The horizontal momentum equation in the 1DV POINT MODEL reads:

$$\frac{\partial u}{\partial t}+\frac{1}{\rho}\frac{\partial p}{\partial x}=\frac{\partial}{\partial z}\left\{\left(\nu+\nu_T\right)\frac{\partial u}{\partial z}\right\} \tag{1}$$

in which p is the pressure, $u(z,t)$ is the horizontal flow velocity, x and z are the horizontal and vertical co-ordinates, t is time, ρ is the fluid bulk density, ν is the kinematic viscosity and $\nu_T(z,t)$ is the eddy viscosity, including the effects of wind and/or waves. The pressure term in (1) is adjusted to maintain a given time-varying depth-averaged flow velocity:

$$\frac{1}{\rho}\frac{\partial p}{\partial x} = \frac{\tau_s - \tau_b}{\rho h} + \frac{U(t) - U_0(t)}{T_{rel}} \quad ; \quad U(t) = \frac{1}{h}\int_{z_{bc}}^{\zeta} u(z,t)dz \tag{2}$$

where h is the water depth, U is the actual computed depth-averaged flow velocity, U_0 is the desired depth-averaged flow velocity, T_{rel} is a relaxation time, z_{bc} is the apparent roughness height, τ_b is the bed shear stress, τ_s is the surface shear stress, and ζ is the surface elevation. A quadratic friction satisfying the log-law is used, and the boundary conditions to (1) read:

$$\tau_b = \left\{\rho(\nu + \nu_T)\frac{\partial u}{\partial z}\right\}\bigg|_{z=z_{bc}} \quad ; \quad \tau_s = \left\{\rho(\nu + \nu_T)\frac{\partial u}{\partial z}\right\}\bigg|_{z=\zeta} \tag{3}$$

The effect of waves on the bed shear stress and vertical mixing is modeled through the approach of Grant and Madsen (1979) in the form of an additional bed boundary condition to the flow model, applying linear wave theory to relate wave length, period, orbital excursion and velocity. This approach gives good results for large waves, but for smaller waves the wave-effect is overestimated by about 20% (e.g., Soulsby et al., 1993). The rms-value of the wave-induced bed shear stress is defined by:

$$\tau_w = \left\langle \tilde{\tau}_w^2 \right\rangle^{1/2} = \rho u_{*w}^2 \quad ; \quad u_{*w}^2 = \frac{1}{2} f_w \hat{u}_{orb}^2 \tag{4}$$

in which the near-bed horizontal orbital velocity amplitude $\hat{u}_{orb}$ is determined from the rms wave height H_{rms}, and the friction coefficient f_w from the Swart's formula (e.g., Soulsby et al., 1993). The wave-affected boundary layer thickness δ_w is given by (Grant and Madsen, 1979):

$$\delta_w = \frac{2\kappa}{\omega}\left|u_{*fw}\right| = \frac{2\kappa}{\omega}\sqrt{u_{*f}^2 + u_{*w}^2} \tag{5}$$

where the subscript w reflects wave-related parameters, f flow-related parameters, fw the effects of current-wave interaction and κ is the Von Kármàn constant. Within the turbulent wave-boundary layer the (mean) eddy viscosity ν_T reads:

$$z_0 \le z \le \delta_w \ : \quad \nu_T = \kappa\left|u_{*fw}\right|z = \kappa\sqrt{u_{*f}^2 + u_{*w}^2}\,z \tag{6}$$

Note that the near-bed shear stress $\tau_b = \rho_w u_{*f}^2$ is constant, but yet unknown, throughout the wave-boundary layer, and yields a logarithmic velocity profile based on $|u_{*fw}|$. Above the wave-boundary layer, wave-induced turbulence is not notable and Grant and Madsen assume:

$$z > \delta_w \;:\; \nu_T = \kappa |u_{*f}| z \tag{7}$$

which yields the usual logarithmic velocity profile with $|u_{*f}|$. The two velocity profiles are matched at $z = \delta_w$. Wave-induced turbulence contributes to the mean flow above the wave-boundary through an increase in the effective roughness z_{bc}:

$$\delta_w \geq z_0 \;:\; \frac{z_{bc}}{z_0} = \left(\frac{\delta_w}{z_0}\right)^{\beta} \;;\; \beta = 1 - \left(1 + \left(\frac{u_{*w}}{u_{*f}}\right)^2\right)^{-½} \tag{8}$$

The k-ε turbulence model, implemented in the 1DV POINT MODEL, consists of transport equations for the turbulent kinetic energy k and the turbulent dissipation rate per unit mass ε, neglecting horizontal transport components:

$$\frac{\partial k}{\partial t} = \frac{\partial}{\partial z}\left\{\left(\nu + \Gamma_T^{(k)}\right)\frac{\partial k}{\partial z}\right\} - \overline{u'w'}\frac{\partial u}{\partial z} - \frac{g}{\rho}\overline{\rho' w'} - \varepsilon \tag{9a}$$

$$\frac{\partial \varepsilon}{\partial t} = \frac{\partial}{\partial z}\left\{\left(\nu + \Gamma_T^{(\varepsilon)}\right)\frac{\partial \varepsilon}{\partial z}\right\} - c_{1\varepsilon}\frac{\varepsilon}{k}\overline{u'w'}\frac{\partial u}{\partial z} - \left(1 - c_{3\varepsilon}\right)\frac{\varepsilon}{k}\frac{g}{\rho}\overline{\rho' w'} - c_{2\varepsilon}\frac{\varepsilon^2}{k} \tag{9b}$$

in which a prime denotes turbulent fluctuations and an overbar averaging over the turbulent time scale. The turbulent transport terms are modeled as a diffusion process, and the eddy viscosity ν_T and eddy diffusivity $\Gamma_T^{(\varphi)}$ are given by:

$$\nu_T = c_\mu \frac{k^2}{\varepsilon} \;;\; \Gamma_T^{(\varphi)} = \frac{\nu_T}{\sigma_T^{(\varphi)}} \tag{10}$$

in which $\sigma_T^{(\varphi)}$ is the turbulent Prandtl-Schmidt number for any quantity φ. The various coefficients in the k-ε turbulence model are the result of calibration against grid-generated turbulence and a log-law velocity profile from homogeneous flow experiments, and from a series of stratified flow experiments (e.g., Uittenbogaard, 1995). The selected values of these coefficients are: $c_\mu = 0.09$, $c_{1\varepsilon} = 1.44$, $c_{2\varepsilon} = 1.92$, $\sigma_T^{(k)} = 1.0$, $\sigma_T^{(\varepsilon)} = 1.3$, $\sigma_T^{(\rho)} = 0.7$, $\kappa = 0.41$, and $c_{3\varepsilon} = 1$ for stable stratification.

The model is closed with the following boundary conditions (no wind stresses):

$$k\big|_{z=z_{bc}} = \frac{u_{*b}^2}{\sqrt{c_\mu}} \;,\; \varepsilon\big|_{z=z_{bc}} = \frac{u_{*b}^3}{\kappa z_{bc}} \;,\; k\big|_{z=\zeta} = 0 \;,\; \varepsilon\big|_{z=\zeta} = 0 \tag{11}$$

The transport of sediment is modeled with the advection-diffusion equation for various fractions identified by the superscript (ℓ):

$$\frac{\partial c^{(\ell)}}{\partial t} - \frac{\partial}{\partial z}\left\{W_{s,ef}^{(\ell)} c^{(\ell)}\right\} - \frac{\partial}{\partial z}\left\{\left(D^{(\ell)} + \Gamma_T^{(\ell)}\right)\frac{\partial c^{(\ell)}}{\partial z}\right\} = 0 \ , \quad \text{with}$$
$$W_{s,ef}^{(\ell)} = W_0^{(\ell)}\left(1 - \sum_{\ell}\phi^{(\ell)}\right)^{\beta} \quad \text{to account for hindered settling} \tag{12}$$

in which the exponent β generally has the value 5 for fine grained sediment. The volume concentration ϕ in (12) is related to the mass concentration c through $\phi = c/c_{gel}$, where c_{gel} is the gelling concentration, i.e., the sediment concentration at which a space-filling network is formed as a result of flocculation.

The fluxes at the bed and at the water surface are set to zero in the present simulations; the 1DV POINT MODEL however does provide a facility to prescribe the water-bed exchange processes explicitly, either by the classical formulas for cohesive sediment, or by the formula of van Rijn (1987) for non-cohesives. The buoyancy term in (9a) and (9b) accounts for the effect of vertical sediment and salinity gradients:

$$\rho\left(S, c^{(\ell)}\right) = \rho_w(S) + \sum_{\ell}\left\{\left(1 - \frac{\rho_w(S)}{\rho_s^{(\ell)}}\right)c^{(\ell)}\right\} \tag{13}$$

with $\rho_w(S)$ the density of water due to salinity only. These equations are solved with reference to the so-called σ-coordinate system. Time discretization is based on the θ-method; for $\theta = 1$ the Euler-implicit time integration method is obtained. Convection is discretized by a first-order upwind scheme in conjunction with a three-point scheme for the diffusion operator.

The 1DV POINT MODEL is validated against analytical solutions of the vertical sediment concentration profile provided by Malcherek et al. (1993), and against measured vertical velocity and concentration profiles for sediment-laden flow in a straight flume as published by Coleman (1981). Results of this validation are not presented herein.

3. THE CONCEPT OF SATURATION

An important item is the concept of the local "saturation concentration" $c_s(z)$ and its depth averaged value C_s. It will be shown that this concept is one of the key issues in understanding the process of rapid siltation. For loose granular sediments this concept is well known and is related to equilibrium concentration profiles and equilibrium transport (e.g., Vanoni, 1975). Its importance for the behavior of high-concentrated mud suspensions was discussed by Winterwerp (1996), and the terminology was introduced by Uittenbogaard et al. (1996). The first ideas on the existence of such a saturation concentration for fine grained cohesive sediment were presented by Teisson et al. (1992); however, at that time no explicit physical meaning was attributed to this concentration.

The concept of the saturation concentration for cohesive sediment is based on the empirical evidence (e.g., Turner, 1973) that a turbulent flow field collapses when the flux Richardson number Ri_f exceeds a critical value of about $Ri_{f,cr} \approx 0.15$. Ri_f is defined as the ratio of buoyancy to the production term in the turbulent kinetic energy equation, neglecting the diffusive contribution :

$$Ri_f = -\frac{g\overline{w'\rho'}}{\rho\overline{u'w'}\,\partial u/\partial z} = -\frac{\alpha g\overline{w'c'}}{\rho\overline{u'w'}\,\partial u/\partial z} \tag{14}$$

where α is the relative sediment density $[\alpha = (\rho_s - \rho_w)/\rho_s]$.

Evaluating sediment dynamics starting from low concentrations, a zero-order approximation is justified. By assuming a logarithmic velocity profile (with a parabolic viscosity profile) and local equilibrium between settling and mixing, the suspended sediment concentration profile at which saturation will occur is obtained as:

$$c_s(z) = \frac{Ri_{f,cr}\rho}{\alpha g \kappa}\frac{u_*^3}{hw_s}\left(\frac{h}{z} - 1\right) \tag{15}$$

where u_* is the shear velocity, h the water depth, and w_s the local settling velocity. At concentrations beyond c_s (super-saturated) turbulence collapses ($Ri_f > Ri_{f,cr}$) and the flow is no longer able to carry the sediment in suspension. As c_s represents a vertical profile, a more convenient parameter is the depth-averaged saturation concentration C_s, which can be regarded as a scaling parameter for saturated suspensions (e.g., Galland et al., 1997):

$$C_s = \frac{1}{h}\int_0^h c_s dz = K_s \frac{\rho}{\alpha g}\frac{u_*^3}{hW_s} \tag{16}$$

where K_s is a proportionality constant. It is noted that for tidal flow a somewhat different scaling law can be derived; however, this is not elaborated upon here.

The saturation concept is well demonstrated through two simulations with the 1DV POINT MODEL for a hypothetical open channel flow of 16 m depth, a constant, depth-averaged flow velocity of $U = 0.2$ m/s at a roughness value $z_0 = 1$ mm and a uniform initial sediment concentration C_0 with a settling velocity $W_s = 0.5$ mm/s. The water depth is discretized into 100 equidistant grid points, the time step is set at $\Delta t = 1$ min, and the relaxation time $T_{rel} = 2\Delta t$. Figure 2 shows the computed isolutals for $C_0 = 0.023$ g/l, indicating a Rousean vertical concentration profile for a nearly parabolic eddy diffusivity (not shown here).

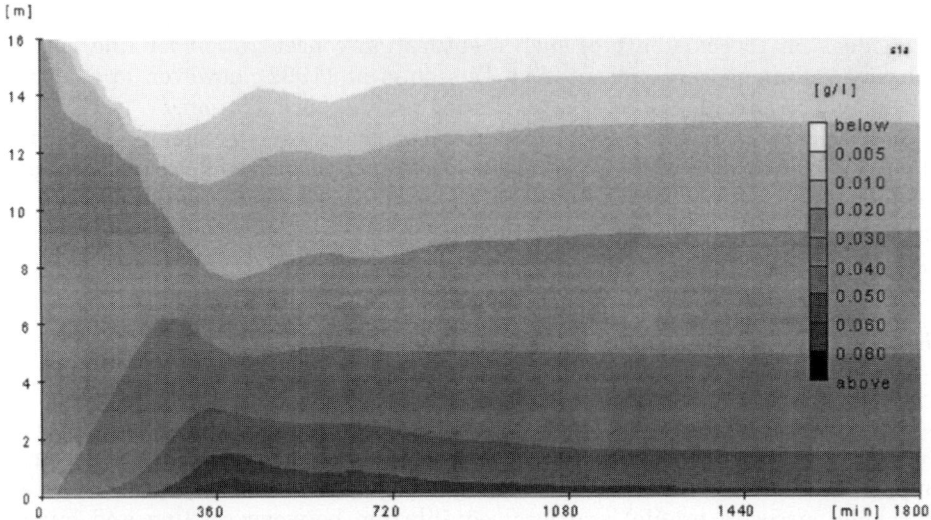

Figure 2. Isolutals for (sub-)saturated suspension in open channel flow; C_0 = 0.023 g/l.

Figure 3 presents the computed isolutals for a slightly larger initial concentration, i.e., C_0 = 0.024 g/l, showing a complete collapse of the concentration profile and of the turbulence (not shown here), forming a high-concentrated near-bed (fluid mud) layer.

It is inferred that, contrary to the case of non-cohesive sediment, the collapse of turbulence is irreversible as long the fluid mud layer remains soft, i.e., as long as no turbulence can be generated at the water-mud interface. This implies that, in addition to the scaling law (16), time scales for gelling and consolidation influence the problem.

From the behavior shown in Figures 2 and 3 it is concluded that for these particular conditions the saturation concentration C_s = 0.023 g/l, and further that, for uniform flow conditions the proportionality constant K_s in (16) amounts to about 0.7.

Next, a series of simulations with the 1DV POINT MODEL is carried out to establish the variation of C_s with U and W_s. The results are shown in Figure 4, where the vertical flux W_sC_s is plotted versus U. It is observed that the numerical results follow (16). The influence of numerical parameters, e.g., the time step and the number of layers, is also studied, the results of which are also shown in Figure 4; note that the latter effects are so small that they are indistinguishable from the reference data points.

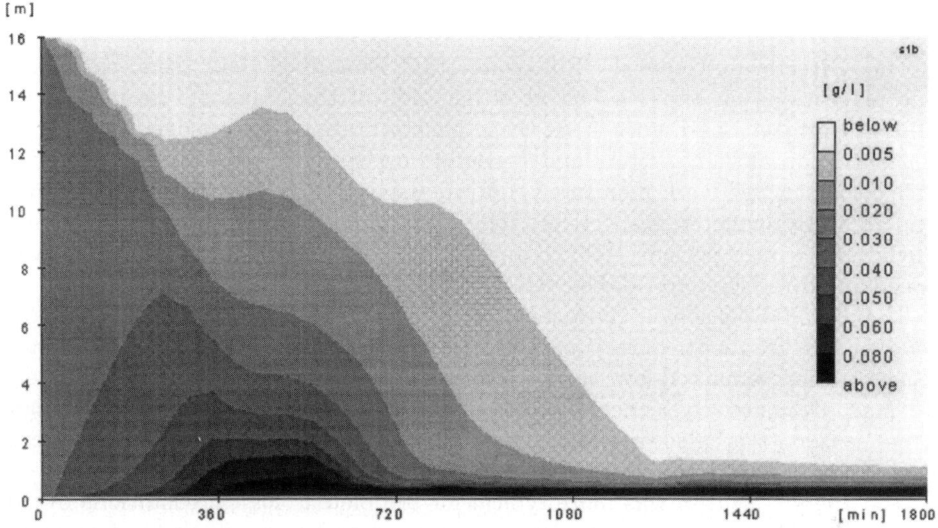

Figure 3. Isolutals for super-saturated suspension in open channel flow; C_0 = 0.024 g/l.

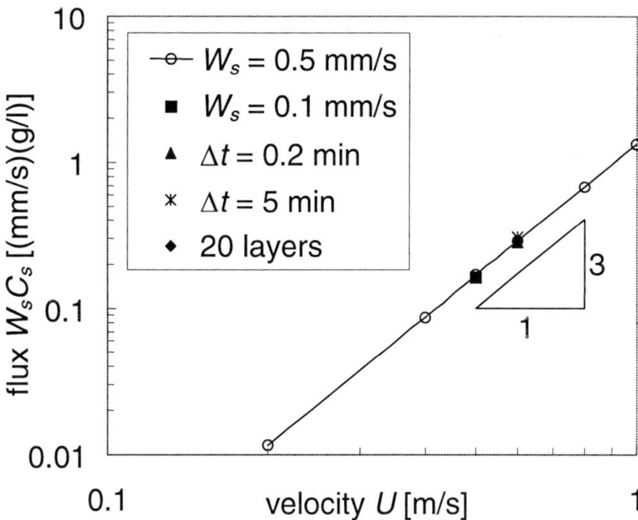

Figure 4. Saturation concentration flux as a function of flow velocity (results 1DV POINT MODEL).

4. FIELD MEASUREMENTS

The Directorate-General "Rijkswaterstaat" installed four semi-permanent anchor stations on the sea bed at either side of the Maasgeul as part of the SILTMAN-project; see Figure 1. At these anchor stations the suspended sediment concentration at 0.15, 0.55, 2 and 7 m and flow velocity and direction at 0.35 m above the sea bed were monitored continuously with an optical turbidity sensor and an EMS-flowmeter, respectively. The instruments were operated continuously during the winter periods 1995/96 and 1996/97. Also, data on wave height, wind speed and direction and tidal elevation at nearby stations were available.

The 1995/96 season is characterized by exceptionally calm weather with no long and/or frequent stormy periods. During the entire season the sediment concentration remained low, rarely exceeding the mean summer values of several 10 mg/l. Occasionally, concentrations of up to a few 100 mg/l were measured. This season was also marked by an exceptionally low siltation rate and, consequently, low dredging needs.

The 1996/97 season was more typical for the meteorological conditions in that part of the North Sea, and dredging operations had to be carried out as usual. During this period very high suspended sediment concentrations were measured frequently, exceeding 10 g/l at the lower measuring stations. As an example, Figure 5 shows the data at anchor station B for November 13 and 14, 1996. Wind speed and wave height were considerable, but not extreme. It is probably the sequence of rough weather conditions that determine the availability of sediment; once available, the local hydro-meteorological conditions govern sediment dynamics.

Finally, it is noted that measurements were also carried out simultaneously a few km within the breakwaters. Vertical profiles of salinity, velocity and suspended sediment concentration revealed a sharp lutocline low in the water column. The sediment concentration below this lutocline amounted to more than several 10 g/l, whereas above this lutocline virtually no suspended sediment was found, e.g., concentrations of a few 10 mg/l at most (van Woudenberg, 1998).

The next sections discuss the analysis of these measurements. The data from anchor station B were selected for this purpose, as the flow at this location was only slightly distorted by the converging flow effects near the head of the breakwaters.

5. NUMERICAL SIMULATIONS OF FIELD MEASUREMENTS

The 1DV POINT MODEL was run to simulate the measurements on November 13/14, 1996. Flow velocity measured 0.35 m above the bed was used to obtain a depth-averaged value assuming a logarithmic velocity profile. The time-varying significant wave height was schematized into a few time functions (blocks with wave height either constant or linearly varying with time).

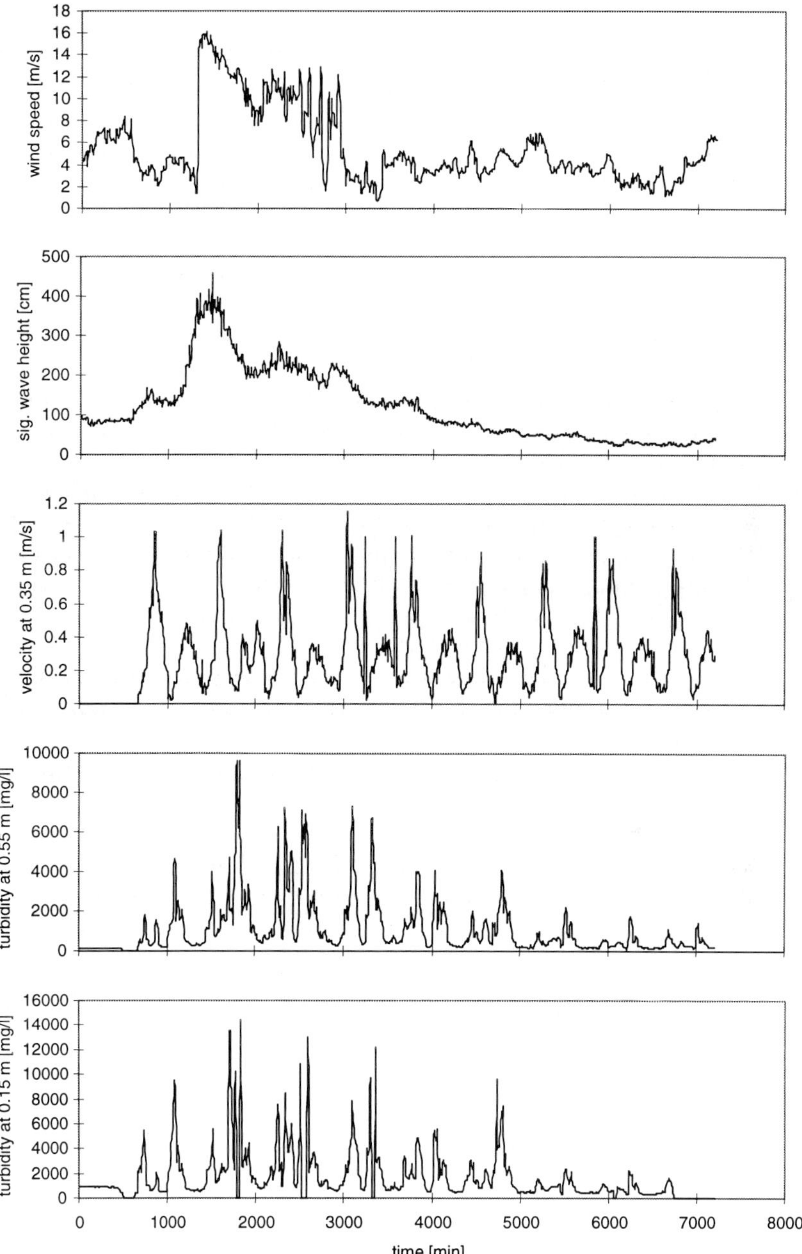

Figure 5. Measurements of wind speed, wave height, flow velocity and turbidity (measured as suspended sediment concentration) at Station B, November 13 and 14, 1996.

rms-values for the wave height were obtained from a procedure outlined in van der Velden (1990), and the relevant wave period was deduced from Roskam's (1995) data base on the wave climate in the Dutch coastal zone.

The coastal zone in this area exhibits a strong vertical fresh-saline water induced stratification due to the outflow of the Rhine. As no salinity profiles were measured during the measurement period, the vertical salinity profile measured on April 8, 1997 was applied, which is believed to be typical for conditions throughout the year, and characterized by a constant salinity S of about 30 ppt in the lower 8 to 9 m of the water column, down to 10 to 15 ppt at the water surface. With respect to the vertical salinity distribution, the 1DV POINT MODEL was run in the diagnostic mode. This implies that the initial salinity profile remained constant during the simulations and only affected the vertical density profile, hence the vertical mixing process.

All simulations were carried out for a single sediment fraction, and they start with an initially homogeneous concentration profile C_0 and a spin-up time of 24 h. After trial and error, a fair agreement with the measurements was obtained for the parameters given in Table 1.

Figure 6a compares the computed and measured suspended sediment concentration at 0.55 m and 0.15 m above the bed. It is observed that the numerical results follow the measured trend and variations, although some of the detailed patterns are missing, including a series of individual peaks in the concentration measurements. The latter are most probably related to episodic events (bursts?) that are not accounted for in the model. However, these deviations can also be attributed to short-term variations in wave activity or fresh/saline water stratification (note that the simulations are not run with the actual salinity profile), or to advection and/or patchiness of the suspension. It

Table 1. Parameters used in 1DV POINT MODEL simulations

parameter	value	parameter	value
h	16 m	c_{gel}	80 & ∞ kg/m^3
H_{rms}	from data	W_s	0.6 mm/s
U	from data	C_0	0.5 kg/m^3
z_0	0.001 m	n	109 grid points[a]
ρ_w	1,020 kg/m^3	Δt	1 min
ρ_s	2,650 kg/m^3	T_{rel}	2 min
Buoyancy effects		yes	no; see Figure 8
surface waves		yes	no; see Figure 9
salinity profile		yes	no, see Figure 10
water-bed exchange		no	
wind stress		no	

[a]n = number of grid points; near the bed a logarithmic distribution is used

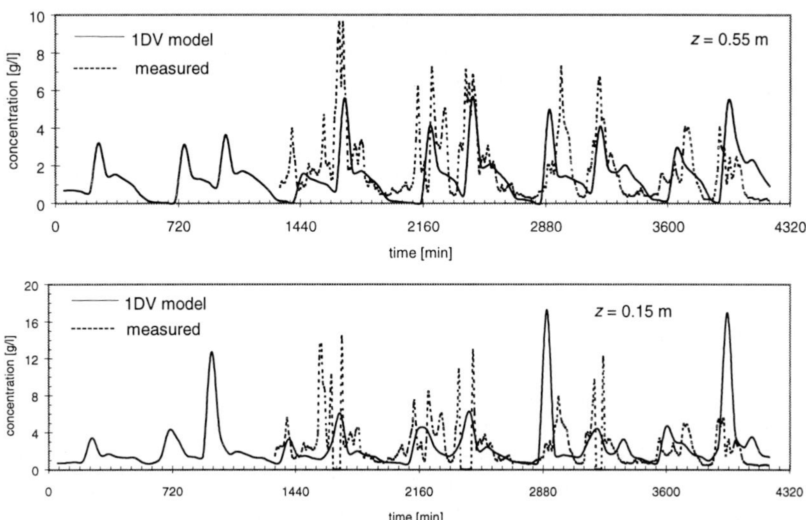

Figure 6a. Computed and measured concentration at Station B, November 13 and 14, 1996.

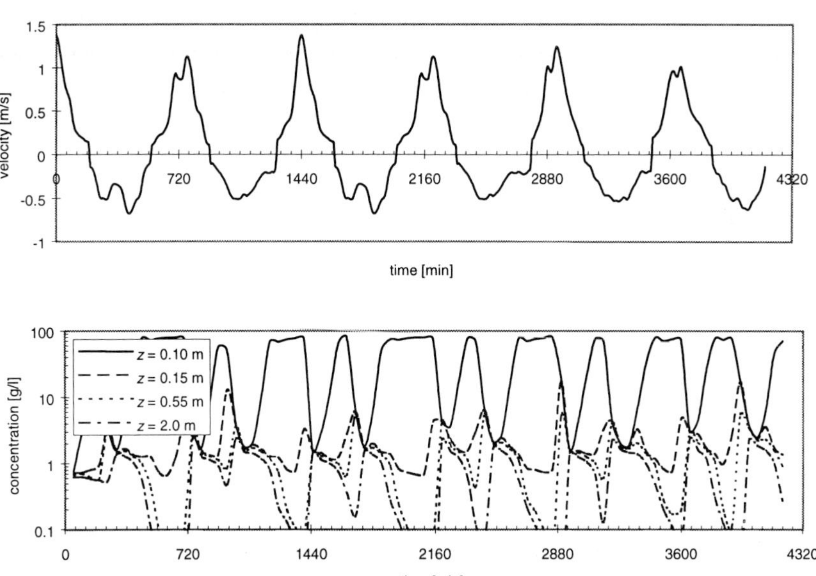

Figure 6b. Measured velocity profile and computed sediment concentration at four elevations at Station B, November 13 and 14, 1996.

is concluded, however, that the agreement between the simulations and measurements is sufficient to assess the dominant physical processes governing sediment dynamics around the Maasgeul.

Figure 6b shows the variation in concentration at four levels (z) for the same simulation, together with the flow velocity. At z = 2 m, c varies around 1 g/l. At this height no data were available from Station B, but the measurements at Station H gave concentrations within the same range. Close to the bed, a temporal, about 10 cm thick, layer of fluid mud appears to have formed around slack water with $c_{max} = c_{gel}$. This layer is entrained rapidly with increasing flow velocity, causing peaks in $c_{0.15}$ and $c_{0.55}$. The numerical results are also plotted in Figure 7, showing that no sediment could penetrate the upper water column due to low vertical mixing caused by salinity-induced stratification.

Simulations were also carried out with the Partheniades-Krone bed-boundary conditions, the results of which are presented elsewhere (Delft Hydraulics, 1998). However, in this case it was not possible to reproduce the measurements in any reasonable way for any realistic or unrealistic parametric setting. The erosion formula causes the water column to load up with sediment until it becomes super-saturated and leads to a collapse of the concentration profile; the deposition formula causes the water column to unload, as a result of which the buoyancy effects vanish. It is noted that this assessment is consistent with the sediment composition of the sea-bed, which contains only about 3% fine-grained sediment.

It is stressed however, that it is not concluded that erosion-deposition processes are not important with respect to the dynamics of high concentrated mud suspensions in general, as ultimately they are responsible for the amount of sediment available.

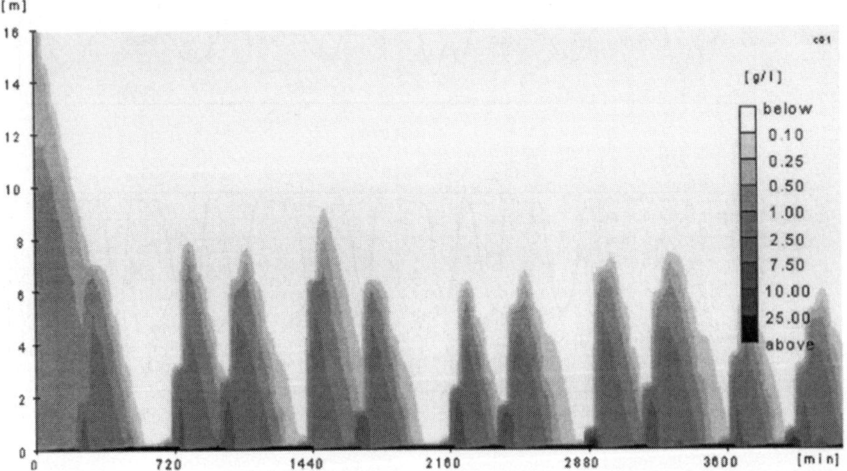

Figure 7. Computed evolution of concentration profile at Station B, November 13 and 14, 1996; reference conditions: with buoyancy, waves, salinity and hindered settling.

The results, as presented in Figure 7, are the basis of sensitivity analyses in which the various processes in the 1DV POINT MODEL are switched off one by one. Figure 8 shows the results of a simulation without sediment-induced buoyancy effects. Quantitatively this picture is considerably different from Figure 7, and it is not possible to reproduce the measurements. However, from a qualitative point of view Figures 8 and 7 are comparable. This agreement is lost, however, if the effect of waves is excluded (Figure 9), or the vertical salinity profile (Figure 10), or hindered settling ($c_{gel} \rightarrow \infty$); see Figure 11. Inspection of Figure 9 shows a much smaller agreement between computational and experimental results in comparison with the reference situation, i.e., Figure 6a. Figure 11 even shows a complete collapse of the concentration profile. It is estimated that the contribution of waves to the vertical mixing is about 20%, implying that the suspension is close to its saturation value. These analyses further show that hindered settling tends to maximize the vertical density gradients, hence buoyancy effects, and also appears to be crucial for a proper reproduction of the sediment vertical structure.

It is concluded that the vertical salinity profile prevents the sediment from mixing over the entire water column. This effect appears to be important in the Maasmond area and agrees with remote-sensing observations which have never revealed highly turbid waters in this area, because such observations are limited to the near-surface water layers. It is noted, however, that no data are available on the suspended sediment concentration in the upper part of the water column to verify this conclusion.

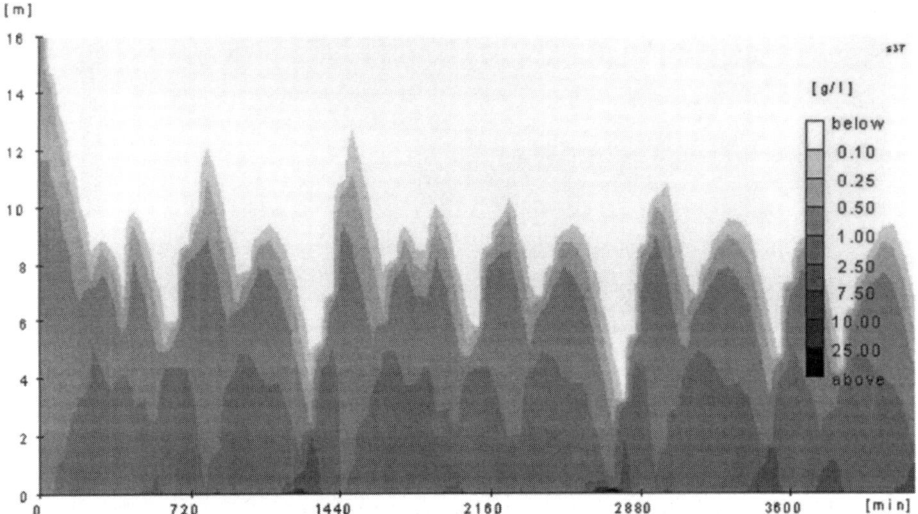

Figure 8. Computed evolution of concentration profile at Station B, November 13 and 14, 1996; sensitivity analysis: with waves, salinity and hindered settling, but no buoyancy.

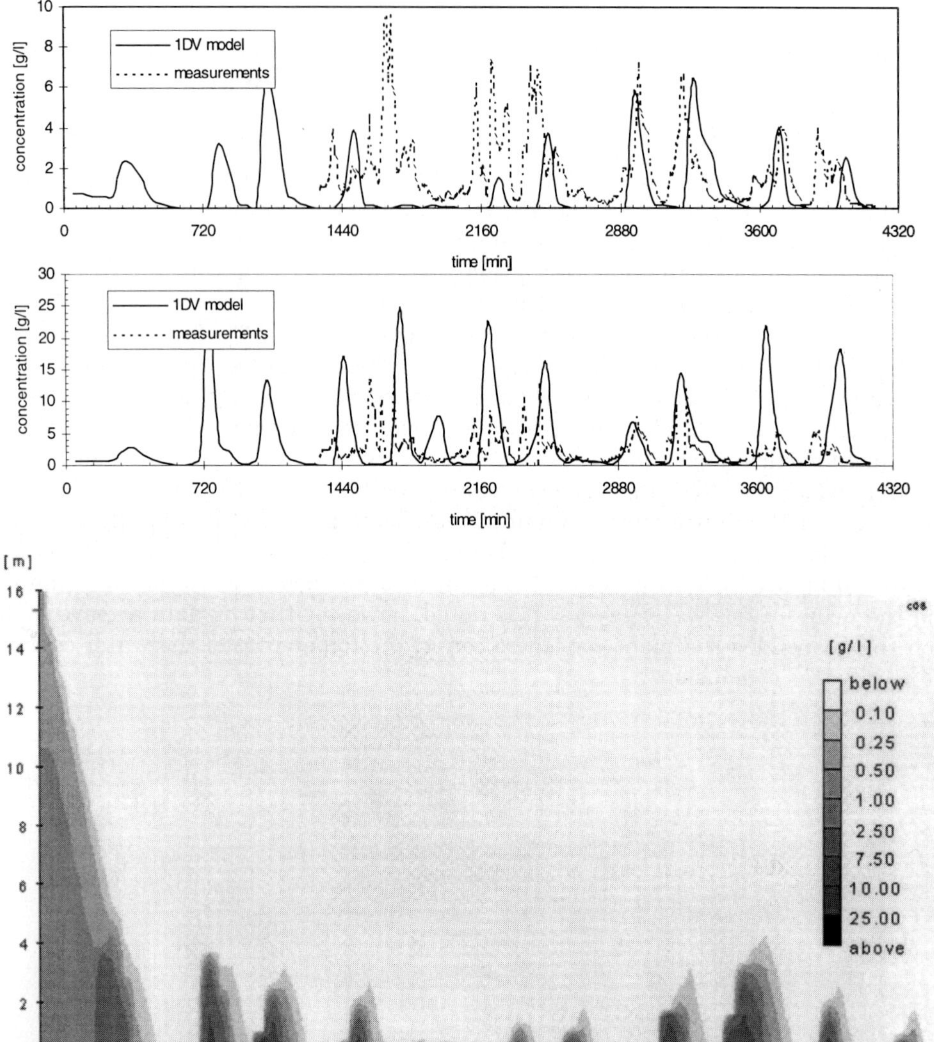

Fig. 9. Computed evolution of concentration profile at Station B, November 13 and 14, 1996; sensitivity analysis: with buoyancy, salinity and hindered settling, but no waves - upper graph: time series at z = 0.55 m; middle graph: time series at z = 0.55 m.

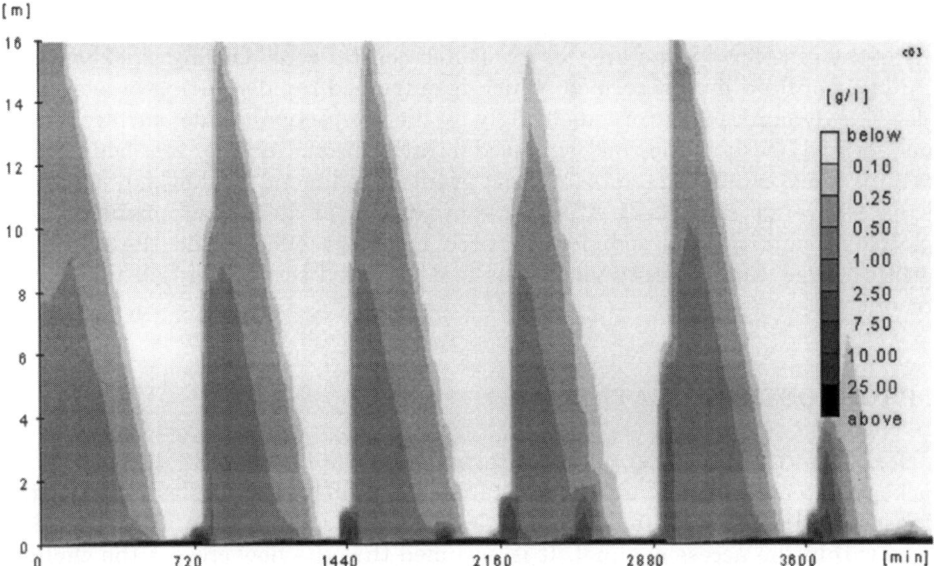

Figure 10. Computed evolution of concentration profile at Station B, November 13 and 14, 1996; sensitivity analysis: with buoyancy, waves and hindered settling, but no salinity.

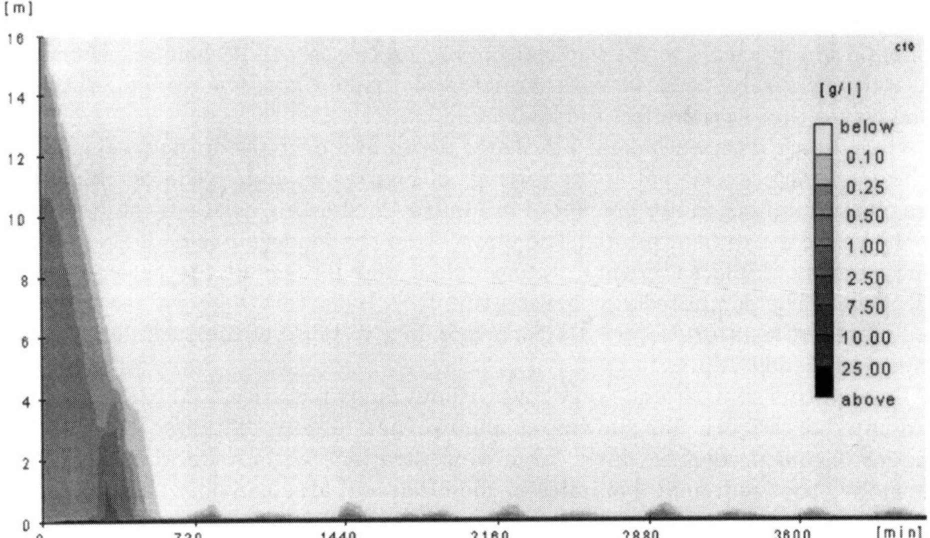

Figure 11. Computed evolution of concentration profile at Station B, November 13 and 14, 1996; sensitivity analysis: with buoyancy, waves and salinity, but no hindered settling.

From the analyses it is concluded that the sediment concentration, measured at anchor station B, is close to its saturation value, and it is further inferred that this is the case for a large area in the Dutch coastal zone. During slack water a thin layer of fluid mud is formed, which is entrained rapidly during accelerating tide. This dynamics is governed strongly by the interaction of tide, surface waves and vertical fresh-saline-water-induced stratification. As the sea bed locally contains only small amounts of fine grained sediment, water-bed exchange processes are not important. A proper reproduction of the measurements is only possible if salinity- and sediment-induced buoyancy effects, and the effects of surface waves, tidal velocity variations and hindered settling are all taken into account.

6. PROGNOSTIC SIMUATIONS

Next, the 1DV POINT MODEL is interpreted as a Lagrangean model, i.e., a package of water with sediment at Station B is followed on its course into the Maasgeul. Along this route, the water depth increases from 16 m at Station B to 24 m within the access channel. It is assumed that the flow crosses the channel perpendicularly, as a result of which the flow velocity decreases by 1/3. In reality, a considerable along-channel component is generated due to the inflow and outflow through the Rotterdam Waterway and the various harbor basins. However, for the present analysis the above assumption is sufficient, as the suspension is almost saturated, and a small decrease in flow velocity will already be sufficient to cause a complete collapse of the vertical profile. Moreover, the contribution of waves to the vertical mixing processes will become small or even disappear entirely, because the channel is too narrow to allow (re-)generation of long waves that would affect the bed.

Thus, apart form an increase in water depth and decrease in flow velocity, all other parameters are kept at the settings of Figures 6a and 7. The results of this prognostic simulation are presented in Figure 12, showing a catastrophic collapse of the vertical sediment profile. The flow within the Maasgeul is no longer able to carry the available sediment in suspension, and the conditions become super-saturated. The computed Richardson numbers have very high values near the sediment-water interface, with a corresponding collapse of the turbulent kinetic energy and eddy diffusivity.

Finally, this simulation is repeated without buoyancy effects. These results are presented in Figure 13, showing a picture not largely different from that of Figure 8, and predicting fairly large concentrations within the Maasgeul. It is concluded that sediment dynamics in the Maasgeul area can only be represented properly by a numerical model that includes sediment-induced buoyancy effects, the effects of the local salt-stratification and the contribution of surface waves to vertical mixing.

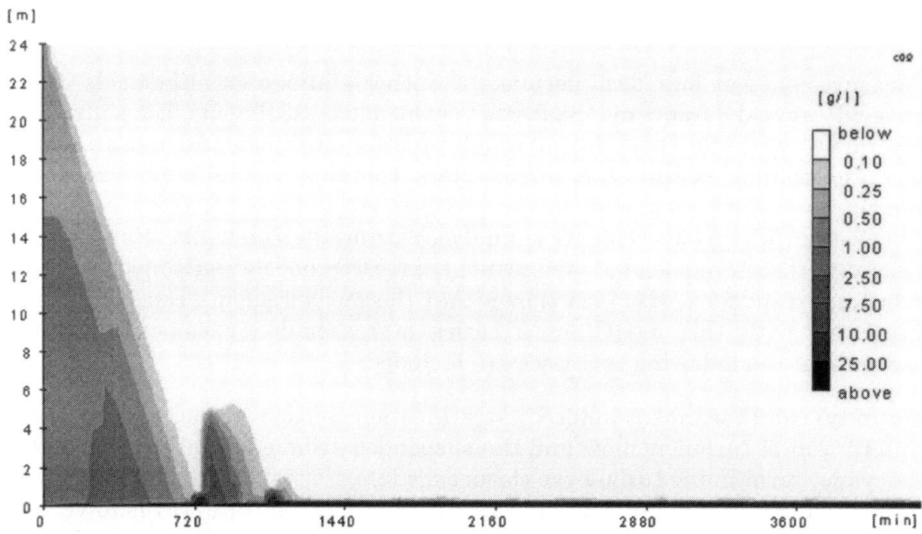

Figure 12. Prognostic simulation of concentration profile in Maasgeul, November 13 and 14, 1996; reference conditions: with buoyancy, waves, salinity and hindered settling.

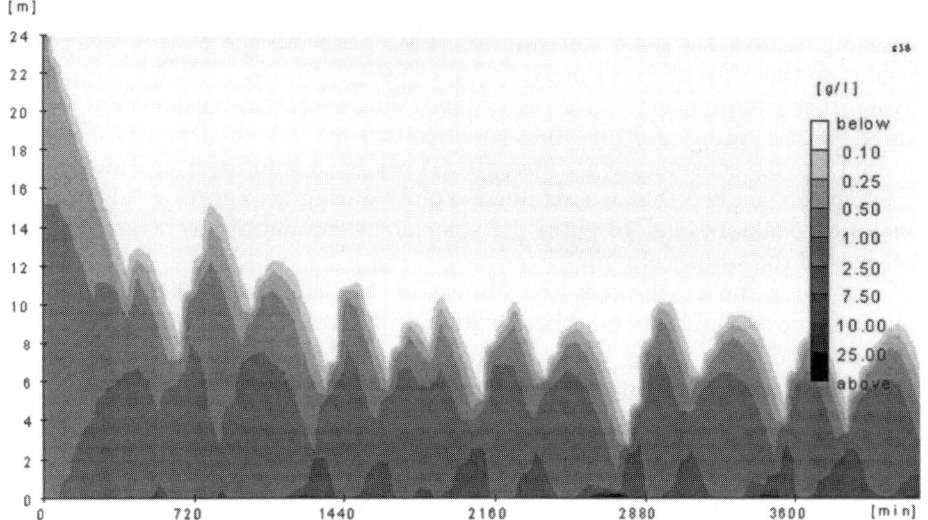

Figure 13. Prognostic simulation of concentration profile in Maasgeul, November 13 and 14, 1996; sensitivity analysis: with waves, salinity and hindered settling, but no buoyancy.

7. DISCUSSION, SUMMARY AND CONCLUSIONS

Measurements at four semi-permanent anchor stations on either side of the Maasgeul reveal frequent near-bed suspended sediment concentrations temporarily exceeding 10 g/l during rough weather conditions. At the same time, the water column within the breakwaters contains virtually no suspended sediment.

One set of these measurements is analyzed using the 1DV POINT MODEL, which is a fully three-dimensional sediment transport model omitting horizontal gradients, except for $\partial p/\partial x$. It is shown that the measurements can be properly represented with this model for a series of reasonable parameter settings, provided that the following processes are included:

1. Salinity- and sediment-induced stratification effects to account for the interaction of turbulent flow and the suspended sediment, modeled through a buoyancy term in the turbulence closure module,
2. Wave-induced mixing to simulate the contribution of surface waves to vertical mixing of suspended sediment through an increase in effective roughness height, and
3. Hindered settling for its tendency to maximize the vertical sediment-induced density gradients, hence buoyancy effects.

Water-bed exchange processes should not be modeled in the present case; they are unrealistic because these tend to load or unload the water column beyond reasonable values. Moreover, the mud content of the sea bed at this location is very low, on the order of 3% at most.

From these simulations it is concluded that sediment suspension in the vicinity of the Maasgeul is almost saturated for the hydro-meteorological conditions under consideration. During slack water, a thin layer of fluid mud is formed on the sea bed that is entrained rapidly during accelerating tide causing pronounced peaks beyond 10 g/l in the suspended sediment concentration in the lower part of the water column.

When water moves through the Maasgeul, the suspension becomes super-saturated, and turbulence and concentration profile collapse, forming a fluid mud layer. This collapse takes place at the time scale of settling $T_s = h/W_s = 12/(0.6\times10^{-3})(3600) \approx 6$ hrs ($h_{eff} \approx h/2$ because of the salt stratification). Hence, this collapse cannot explain the observed absence of suspended sediment in the water column in-between the breakwaters, as the travel time of this package of water-sediment mixture between the two locations is only one to two hours. However, from soundings and dredging it is known that sediment enters the channels and basins within the breakwaters frequently.

It is inferred therefore that the collapse of the concentration profile is a necessary, but not sufficient condition to form such a high-concentrated near-bed suspension. It is hypothesized that this collapsed suspension triggers the generation of a sediment-driven density current, which cleanses the water column rapidly, causing the rapid siltation through the transport and accumulation of fluid mud into the channels and harbor basins.

8. ACKNOWLEDGMENT

This study was commissioned by the Netherlands Ministry of Transport, Public Works and Water Management and the Port of Rotterdam under the SILTMAN project. We would like to thank these institutes for financing this study and making the data available for publication. This work was co-financed by the European Commission, Directorate General XII for Science, Research & Development through the COSINUS project under the MAST-3 program, contract MAS3-CT97-0082. We would like to thank Mr. Jan van Kester and Mrs. Heelen Leepel for their skillful assistance in the numerical implementations and Mr. R. Bruinsma for his help in post-processing. We would also like to acknowledge the fruitful discussions with Dr. C. Kranenburg and Prof. Dr. J. A. Battjes of the Delft University of Technology.

REFERENCES

Coleman, N. L., 1981. Velocity profiles with suspended sediment. *Journal of Hydraulic Research*, 19(3), 211-229.

Delft Hydraulics, 1998. SILTMAN, analysis of field measurements. *Report Z2263*, Delft, The Netherlands.

Galland, J.-C., Laurence, D., and Teisson, C., 1997. Simulating turbulent vertical exchange of mud with a Reynolds stress model. In: *Cohesive Sediments.* T. N. Burt, W. R. Parker and J. Watts eds., John Wiley, Chichester, UK, 439-448.

Grant, W. D., and Madsen, O. S., 1979. Combined wave and current interaction with a rough bottom. *Journal of Geophysical Research*, 84(C4), 1797-1808.

Le Hir, P., 1997. Fluid and sediment "integrated" modelling application to fluid mud flows in estuaries. In: *Cohesive Sediments.* T. N. Burt, W. R. Parker and J. Watts eds., John Wiley, Chichester, UK, 417-428.

Malcherek, A., Markofsky, M., Zielke, W., Le Normant, C., Lepeintre, F., and Teisson, C., 1993. Three-dimensional numerical modelling of cohesive sediment transport in estuarine environments. *Final Report MAST-0034-C*, EDF-DER and University of Hannover, Hannover, Germany.

Muste, M., and Patel, V. C., 1997. Velocity profiles for particles and liquid in open-channel flow with suspended sediment. *Journal of Hydraulic Engineering*, ASCE, 123(9), 742-751.

Nezu, I., and Nakagawa, H., 1993. Turbulence in open-channel flows. *International Association for Hydraulics Research, Monograph Series*, Balkema, Rotterdam.

van Rijn, L. C., 1987. Mathematical modelling of morphological processes in the case of suspended sediment transport. *Ph.D. Thesis*, Delft University of

Technology, also Delft Hydraulics Communications No 382, Delft, The Netherlands.

Roskam, B., 1995. Wave climate at EUR- and LEG- observation platform for Maasvlakte-studies. *EIA-working document, Report RIKZ/OS-95.111x*, Rijkswaterstaat/RIKZ, The Netherlands (in Dutch).

Soulsby, R. L., Hamm, L., Klopman, G., Myrhaug, D., Simons, R. R., and Thomas, G. P., 1993. Wave-current interaction within and outside the bottom boundary layer. *Coastal Engineering*, 21, 41-69.

Teisson, C., Simonin, O., Galland, J. C., and Laurence, D., 1992. Turbulence modelling and mud sedimentation: a Reynolds stress model and a two-phase flow model. *Proceedings of the 23rd International Conference on Coastal Engineering*, American Society of Civil Engineers, New York, 2853-2866.

Turner, J. S., 1973. *Buoyancy effects in fluids.* University Press, Cambridge, UK.

Uittenbogaard, R. E., 1994. Physics of turbulence: technical report on sub-task 5.2. *Report Z649*, MAST Veriparse Project, Delft Hydraulics, Delft, The Netherlands.

Uittenbogaard, R. E., 1995. The importance of internal waves for mixing in a stratified estuarine tidal flow. *Ph.D. Thesis*, Delft University of Technology, Delft.

Uittenbogaard, R. E., Winterwerp, J. C., van Kester, J.A.Th.M., and Leepel, H. H., 1996. 3D cohesive sediment transport - a preparatory study about implementation in DELFT3D. *Report Z1022*, Delft Hydraulics, Delft, The Netherlands.

Vanoni, V. A., 1975. *Sedimentation Engineering*. ASCE Manuals and Reports on Engineering Practice, No 54, American Society of Civil Engineers, New York.

van der Velden, E.T.J.M., 1990. Coastal engineering. volume II., *Lecture Series f7*, Delft University of Technology, Department of Civil Engineering, Delft.

Winterwerp, J. C., 1996. HCBS - high concentrated benthic suspensions - SILTMAN desk study. *Report Z1013,* Delft Hydraulics, Delft, The Netherlands (in Dutch).

van Woudenberg, C., 1998. First results of preliminary measurements of suspended sediment in the Calandkanaal/Beerkanaal. *Report N98.02*, Rijkswaterstaat, Directorate North Sea (in Dutch).

Coastal and Estuarine Fine Sediment Processes
W.H. McAnally and A.J. Mehta (Editors)
2001 Elsevier Science B.V.

Density development during erosion of cohesive sediment

C. Johansen[a] and T. Larsen[b]

[a]NIRAS, Vestre Havnepromenade 9, DK-9100 Aalborg, Denmark

[b]Hydraulics & Coastal Engineering Laboratory, Aalborg University, Sohngaardsholmsvej 57, DK-9000 Aalborg, Denmark

Density development of cohesive sediment beds during erosion and consolidation experiments is described. The determination of the bed shear strength using the density profile obtained from the consolidation experiment, and the need for *in situ* determination of density in the erosion apparatus, are examined. Due to differences in the distribution of particle size within the deposit formed in the two setups, the density profiles are found to differ. It is therefore essential to measure the density profile directly in the erosion device, especially because the bed shear strength depends on density.

1. INTRODUCTION

Bed density development during the erosion experiments was investigated. The calculation of the erosion rate requires knowledge of the bed density profile with respect to the consolidation time. At present, a common assumption made in these calculations is that the density profile can be obtained from consolidation experiments conducted with the same consolidation period as in the erosion measuring apparatus. Furthermore, the density is assumed constant throughout the erosion experiment. Experiments were conducted in order to investigate the validity of assumptions. Kaolinite was used to obtain sediment properties during the experiments.

2. EXPERIMENTAL SETUP

The annular flume used in the erosion tests was stationary, and a rotating lid placed at the water surface was used to induce flow. The flume was 1.90 m in outside diameter, 0.20 m wide and 0.26 m deep. The lid surface fitted inside the flume with a tolerance of 3 mm on both sides, and could be rotated with variable speed between 0 and 3 m/s. Further details can be found in Johansen (1998). The bed shear stress was kept constant during each erosion step. The erosion experiments were conducted in tap water with temperature ranging from 20 to 23 °C. Bed density profiles were measured for each step in the experiments starting with the density profile at the end of the consolidation phase. Thereafter, measurements were made at the end of each increase in the applied shear stress.

Consolidation experiments were conducted in a 0.54 m high and 0.20 m wide column. In this column, the possible influence on the settling behavior due to wall friction was minimized by the relatively large diameter of the column. A minimal diameter of 0.1 m is recommended (MAST, 1993). The density profiles were measured at the same time intervals as in the erosion experiment.

The density distributions in both the flume and the consolidation column were measured using a penetrating acoustic probe, the Ultra High Concentration Meter (UHCM) (Delft, 1994). The probe consisted of two sensors of 0.5 cm diameter and placed 4 cm apart, and was able to measure the dry density in the interval of 0 - 1000 kg/m^3 (Delft, 1994; MAST, 1993). The measuring method was based on the attenuation of the acoustic wave signal. Therefore, it was necessary to calibrate the instrument for the sediment used. In order to be able to reproduce the experiments, commercially available kaolinite was used, with characteristics given in Table 1.

Density measurements in the flume were conducted through a hole at the center of the lid. In both experimental setups, vertical traversing of the acoustic probe was done manually, and the duration of each measurement was approximately 5 minutes.

Preliminary experiments were conducted to determine both the reproducibility and possible disturbing influences from the UHCM. These tests were carried out with suspended sediment concentrations of 25 kg/m^3 and 50 kg/m^3 and a water depth of 0.23 m. Initially, the density was measured continuously by recording the density at a constant elevation above the bed. Thereafter, the experiment was repeated and the density profile was measured at different time intervals by lowering the UHCM into the sediment bed to different depths. The results confirmed that the probe did not disturb the sediment noticeably, and that the experiments were reproducible (Johansen, 1998).

Additional, experiments were conducted to investigate the significance of the initial mixing effect of the suspension. In general, the difference in mixing between the erosion experiments and the consolidation experiments can be expected to be significant. The particle size distribution of the bed formed from settling suspension in the circular flume was not uniform, mainly due to the

Table 1. Properties of kaolinite

Properties of kaolinite	
Median diameter (d_{50})	2.2 microns
Cation exchange capacity (CEC)	3 meq/100g dry weight of sediment
Liquid limit	47 %
Plasticity limit	35 %
Plasticity index	12 %

effect of the secondary currents, as also confirmed by visual inspection. In contrast, the size distribution in the consolidation column could have been considered to be uniform across the column. Therefore, experiments were conducted to achieve the same particle sorting mechanism as found in the circular flume. The mixing in the consolidation experiment was performed using a metal stick rotating at a high speed close to the outer wall. In this case, large particles would be transported and deposited at the center of the column when the settling velocity of the particles is larger than the upward fluid velocity at the center of the column. A preliminary consolidation experiment was conducted with an initial suspended sediment concentration of 25 kg/m³ and a consolidation period of 24 hours. The result of this experiment is presented in Figure 1.

The measurements show uniform density profiles across the column width except at the center of the column, where larger particles were able to deposit during the mixing process. This observation implies that sorting of particles during mixing was significant. Therefore, when using density profiles obtained from the conventional setup in consolidation experiments, the mixing of the suspension should be "tea cup like" in order to achieve approximately the particle size distribution as in the annular flume. However, it should be emphasized that this conclusion is only valid for stationary annular flumes, i.e., ones in which only the top lid is moving.

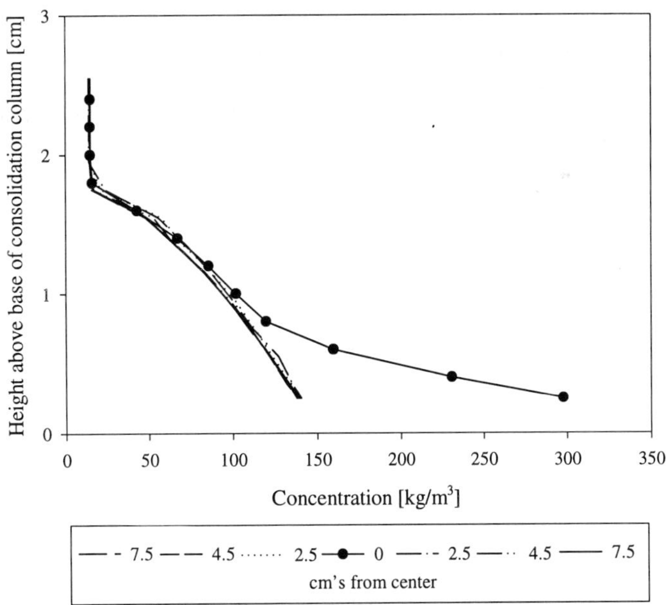

Figure 1. Concentration profiles across the consolidation column.

Experiments to determine density development during erosion were conducted with initial suspended sediment concentrations of 25 kg/m^3 and 50 kg/m^3 and consolidation periods of 12 hours and 24 hours. In both experimental setups a water depth of 0.23 m was used. Table 2 summarizes the conditions in the erosion experiments.

3. RESULTS

The density measurements conducted in the annular flume are presented in Figures 2 through 5. The profile at the upper left corner was measured after 12 or 24 hours, and subsequent profiles are read from left to right at 4 hour intervals. The measurements indicate that the increase in density due to consolidation was minor at the bed surface.

The measurements performed in the consolidation column are presented in Figure 6 and Figure 7. These measurements show that the height of the bed surface was practically constant during the consolidation experiments. The measurements further indicate, within the resolution limit, that the density profile remained constant during consolidation, implying that primary consolidation had practically ended by the time the first profile was measured.

In Figures 8 and 9, the measured densities from the erosion and the consolidation experiments are presented together with a continuous recording of the density level measured in the consolidation column. These densities were measured 1 cm above the flume/column bottom. The results indicate that the difference between the measured densities in the erosion experiment and the consolidation experiment was significant. This difference is believed to have been due to difficulties in achieving the same particle size distribution in the two experimental setups, especially considering the complex hydrodynamics of the annular flume.

Table 2. Conditions for the erosion experiments

Mixing phase	Duration	[hours]	2
	Bed shear stress	[N/m^2]	1.8
Settling/consolidation phase	Duration	[hours]	12 and 24
	Bed shear stress	[N/m^2]	0
Erosion phase	Duration	[hours]	4
	Bed shear stress	[N/m^2]	0.05, 0.10, 0.15, 0.20, 0.25, 0.30 and 0.36

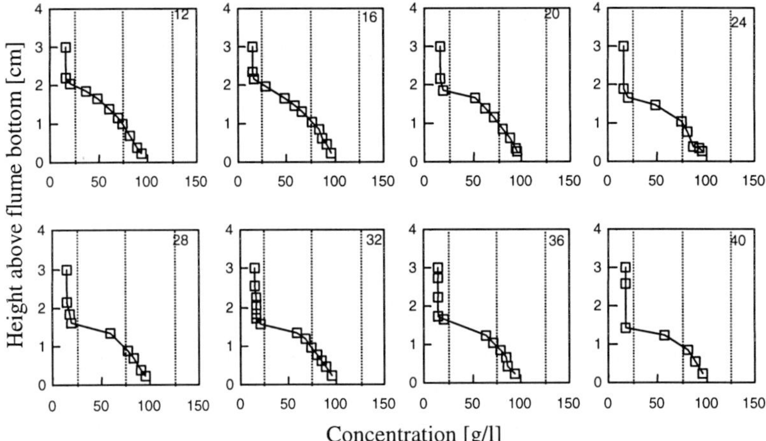

Figure 2. Density profiles during an erosion experiment conducted with a 12 hour consolidation period for the initial bed and an initial suspension concentration of 25 kg/m^3.

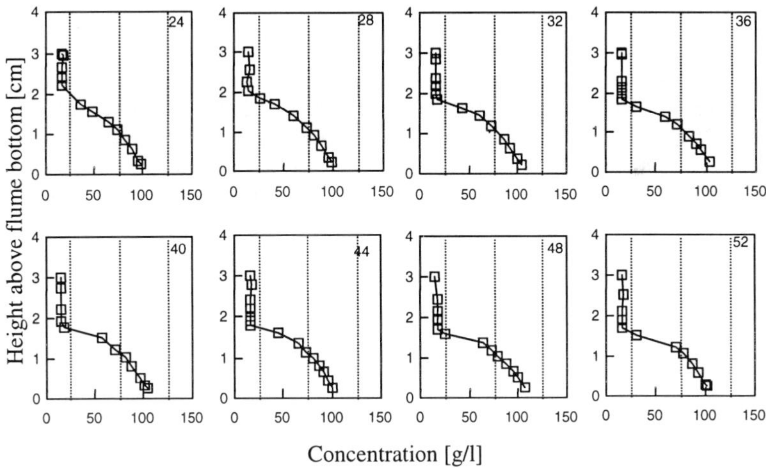

Figure 3. Density profiles during an erosion experiment conducted with a 24 hour consolidation period for the initial bed and an initial suspension concentration of 25 kg/m^3.

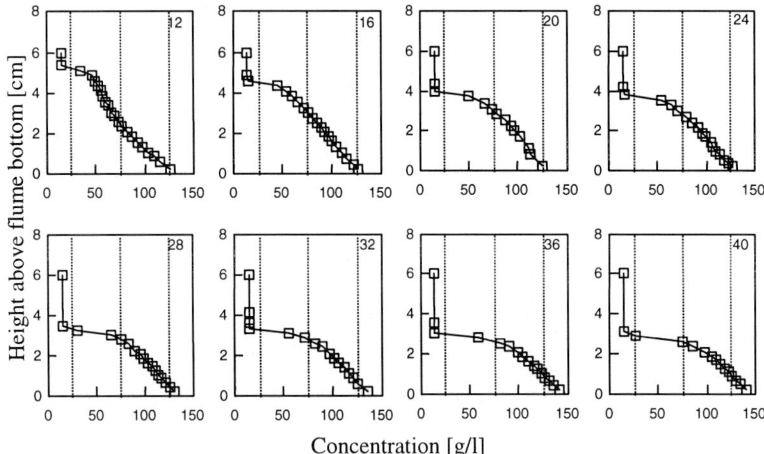

Figure 4. Density profiles during an erosion experiment conducted with a 12 hour consolidation period for the initial bed and an initial suspension concentration of 50 kg/m^3.

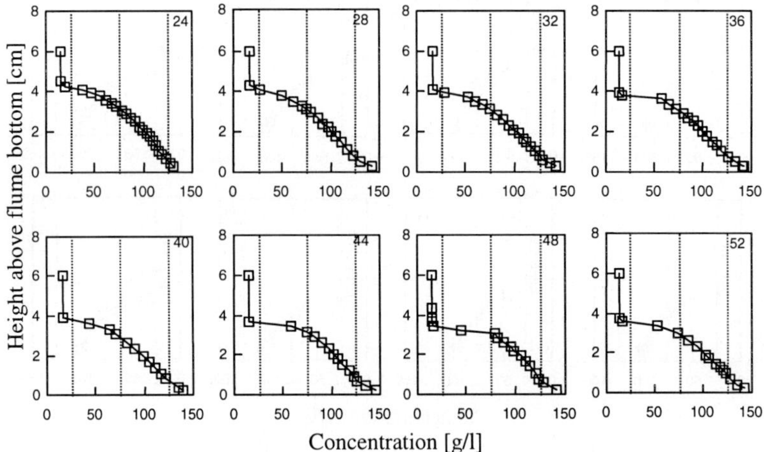

Figure 5. Density profiles during an erosion experiment conducted with a 24 hour consolidation period for the initial bed and an initial suspension concentration of 50 kg/m^3.

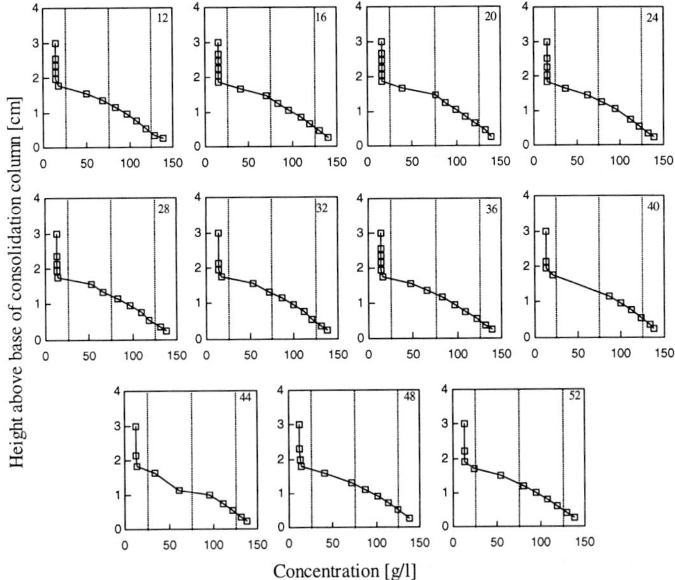

Figure 6. Density profiles from the consolidation column. Initial concentration 25 kg/m^3.

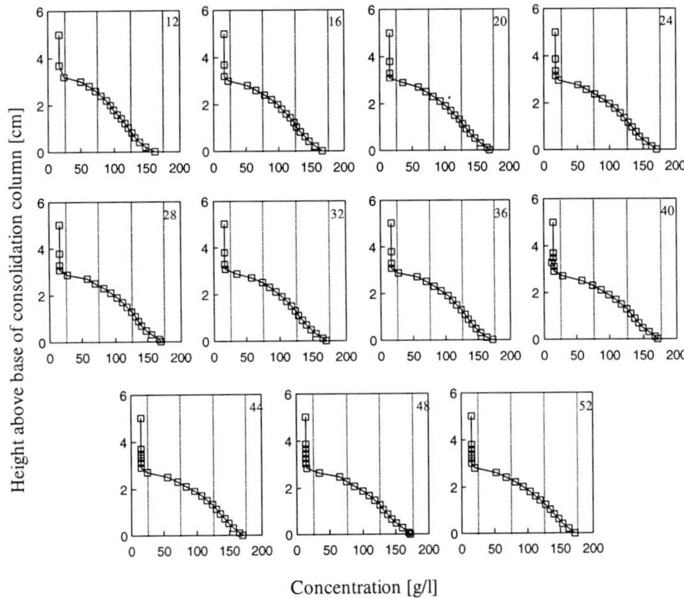

Figure 7. Density profiles from the consolidation column. Initial concentration 50 kg/m^3.

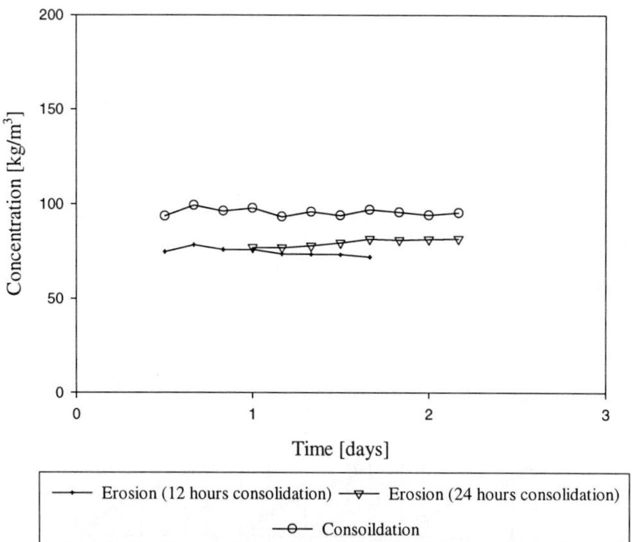

Figure 8. Comparison between measured densities corresponding to an initial concentration of 25 kg/m^3. The density was measured 1 cm above the flume/column bottom.

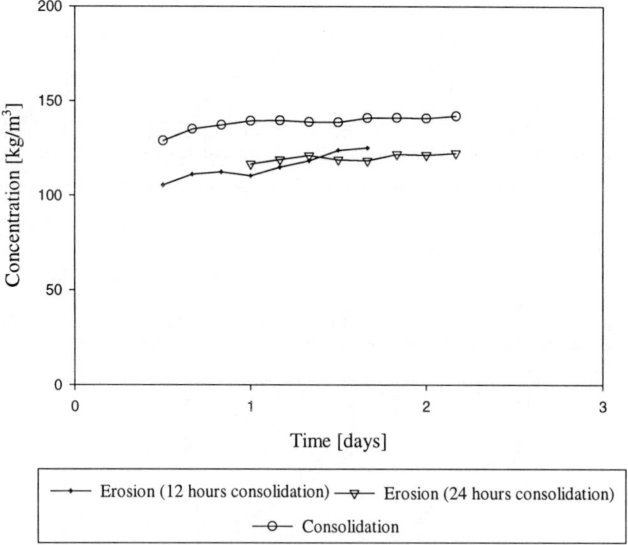

Figure 9. Comparison between measured densities corresponding to an initial concentration of 50 kg/m^3. The density was measured 1 cm above the flume/column bottom.

4. CONCLUSIONS

When using density profiles obtained from the conventional setup in the consolidation column experiments, mixing of the initial suspension should be performed in such a way that approximately the same particle size distribution as in the flume for erosion tests is achieved.

The investigations of the density development during the erosion experiments indicate that the commonly made assumption regarding a time-invariant density profile throughout the erosion experiment appears to be acceptable. However, when the consolidation period is on the order of a few hours only, the outcome may be questionable, because in that case the density changes within the same time scale as in the erosion experiment.

It was found that the bed density was different in the annular flume and in the consolidation column. This difference must be considered in the calculation of the bed shear strength, which depends on density. In general, the investigation demonstrates the need for *in situ* determination of the density in the erosion apparatus.

REFERENCES

Delft Hydraulics, 1994. *Manual for UHCM.* Ultra High Concentration Meter, Delft, The Netherlands, 19p.

Johansen, C., 1998. Dynamics of cohesive sediments. *Ph.D. Thesis.* Department of Civil Engineering. Series paper no. 16. University of Aalborg, Denmark, 127p.

MAST, 1993. On the methodology and accuracy of measuring physical-chemical properties to characterize cohesive sediments. *G6M report*, EU Marine Science and Technology Programme Project, Commission of the European Communities Directorate General XII, Brussels.

Coastal and Estuarine Fine Sediment Processes
W.H. McAnally and A.J. Mehta (Editors)
2001 Elsevier Science B.V.

Clay-silt sediment modeling using multiple grain classes. Part I: Settling and deposition

A. M. Teeter

Coastal and Hydraulics Laboratory, U.S. Army Engineer Waterways Experiment Station, 3909 Halls Ferry Road, Vicksburg, Mississippi 39180, USA

Settling and deposition process descriptors are developed for a multiple grain class numerical sediment transport model. Grain class settling rates are calculated to span floc settling rate distributions. Depositional fluxes are coupled from the coarsest to the finest grain class in proportion to class concentrations, consistent with the analytic model of Kranck and Milligan (1992) and other previously observed grain distributions. Numerical deposition experiments display characteristic features of observed grain spectra, as well as trends in overall distribution moments, and stress-dependent steady state concentrations. Floc settling experiments were performed to examine the combined effects of suspension concentration and fluid-shear. A new settling function is proposed and compared to experimental results.

1. INTRODUCTION

A challenge in understanding and modeling fine-grained sediment settling and deposition is to describe the relationships between the sediment grains and flocs. Since individual flocs contain appreciable sub-populations of grain sizes (Mehta and Lott, 1987; Gibbs, 1977), coupling or interaction between grain classes occurs during settling (Kranck, 1980). Flocculation is an important consideration for any fine-grained sediment model. Flocs greatly affect cohesive sediment transport properties, and form both in fresh waters and to a greater extent in salt water. Electrostatic cohesion and biochemical adhesion bind together sediment grains, small flocs, and organic matter. Flocculation has been found to produce aggregates of 0.1-0.5 mm modal diameter D (Kranck and Milligan, 1992) that settle and deposit at orders of magnitude greater rates than their constituent grains.

For modeling multiple sediment classes, sediment grains are better model state variables than flocs since they are conservative constituents in both suspension and bed, affect both erosion and deposition, and easily measured in the environment. Sediment grains strongly affect the flocculation process (Kranck and Milligan, 1992) along with a host of other conditions such as temperature, salinity, ionic content, pH, clay mineralogy, organic constituents, etc. Grain classes are coupled by cohesion, and this coupling must be accounted for in multiple grain class models. In this paper, a previous a grain distribution model is reviewed and extended to a numerical model framework for multiple fine-grain clay-silt classes.

Settling rates for cohesive suspensions depend on concentration and fluid shear rate conditions rather than dispersed particle size. A previously-proposed settling function (Malcherek and Zielke, 1996; Teisson, 1997) assumes concentration and shear effects to be multiplicative. Settling experiments were performed using flocculators and columns, and a new function relating to settling was developed and compared to laboratory results.

The issues addressed here involve important details of the multiple grain class implementation, specifically how various particle classes interact in a flocculent suspension and during deposition. In Part II (this volume), single and multiple fine grained numerical sediment transport models are compared. The multiple grain class formulation improved the deposition and, to a lesser extent, the erosion process description. Both algorithms used in that study assumed mutually-exclusive erosion and deposition, consistent with previous laboratory investigations. The multiple grain class formulation reproduced the equilibrium suspensions observed in laboratory tests, and resuspension in a large, shallow lake.

2. SIZE-SPECTRA RESPONSE TO DEPOSITION

Kranck and Milligan (1992) measured floc and particle grain size spectra at an anchor station over flood and ebb tidal phases in San Francisco Bay. This work extended and generalized previous work performed in the laboratory (Kranck, 1980). Floc and grain size spectra were parameterized in terms of three variables Q, m, and K. Q depends on the total particle concentration and the shape of the distribution, and is defined by the concentration C_o at 1 μm diameter. The variable m defines the slope of the fine end (small size) of the distribution when plotted log-log. Kranck and Milligan (1992) found m to be constant for both floc and grain distributions. Together with Q,

$$C_o = Q\, D^m \tag{1}$$

defines the fine end of the particle diameter (D) spectra for flocs or grains. The analytic time-dependent solution for the concentration C of a well-mixed suspension with settling is:

$$C = C_o \exp\left(-\frac{W_s t}{H} \right) \tag{2}$$

where W_s is settling velocity, H is the suspension depth, and t is time (Krone, 1962). Kranck and Milligan (1992) took W_s proportional to D^2 as in Stokes Law, and t/H to represent a settling decay term K. The variable K is related to the fall off at the coarse end of the distribution. By combining (1) and (2), and substituting for $W_s t/H$ the equation describing the distributions becomes:

$$C(D) = QD^m \exp(-K\, AD^2) \tag{3}$$

where AD^2 defines Stokes' settling rate, $A = g\,(\rho_f - \rho_l)/18\,\nu\,\rho_l$, g is the acceleration of gravity, ν is the kinematic viscosity of the fluid, ρ_f is the floc density, and ρ_l is the fluid density. Their distribution model (3) was fit to observed spectra taken during both decreasing and increasing suspension concentrations, and reflects how changing suspension concentrations effect grain spectra. Figure 1 shows grain and floc spectra covering the range of distribution parameters found for San Francisco Bay suspended sediments. Flocs formed well-sorted size spectra with $m = 2$, $Q_f = 1.1\exp(-3)$, and $K_f = 2.3\exp(2\ C^{-0.92})$ where the subscript f refers to floc components. Dispersed particles, on the other hand, were poorly sorted with $m = 0$, $Q_g = 7.8\ \exp(-3\ K_g^{-0.54})$, and $K_g = 3.3\ \exp(4\ C^{-1.86})$ where the subscript g refers to grains. In Figure 1, mean D's are plotted at mean C's (each averaged over 65 discrete D channels) for grains (crosses) and flocs (dots).

The values of the Q_g, K_g, and K_f distribution parameters changed with total suspension concentrations, while m values remained about constant. Coupling

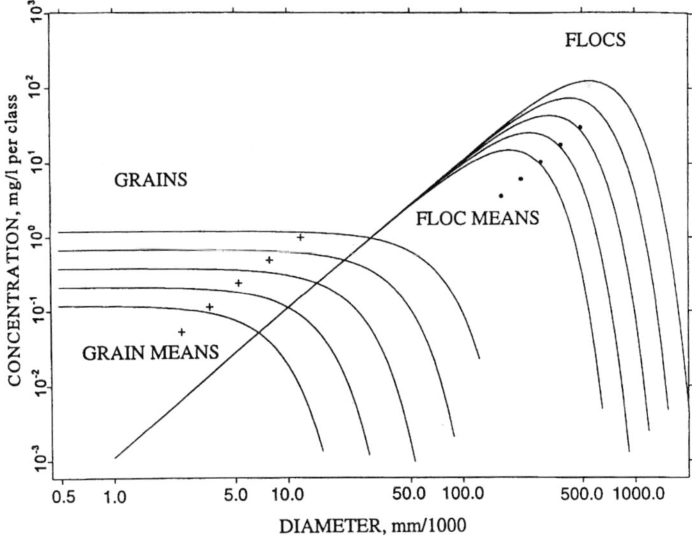

Figure 1. Floc (right) and grain (left) spectra based on tidal measurements of Kranck and Milligan (1992) for varying suspended sediment concentration.

between grain classes caused fine sediment to deposit when a well-sorted sediment of the same size would have otherwise remained in suspension. Links between floc and grain settling were also demonstrated by Kranck and Milligan (1992) as both maximum grain and floc size varied with total concentration and with bottom shear stress.

Though their distribution model (3) was developed based on particle removal by settling, spectra observed by Kranck and Milligan (1992) during resuspension followed the same pattern in reverse. As material was resuspended, increments of material were added first to the fine and then to progressively coarser regions of the spectra. This observation is consistent with erosion of all grain sizes up to a limit of erodible grain size (shear stress limited erosion). Features which produce this effect were incorporated into the multiple grain class erosion algorithm described in Part II (this volume).

3. NUMERICAL METHODS

The distribution model (3) is most useful in describing measured size spectra, and explaining the effects of various settling modes on spectral shapes. Numerical fine-grained sediment models, however, normally use source and sink fluxes at the bed/suspension interface to affect concentration changes. Numerical algorithms were developed which produce results similar to observed and distribution model-grain spectra, and will be presented next. Two numerical algorithms will be described; a settling velocity algorithm to relate floc and grain settling, and a depositional algorithm that couples grain classes during deposition.

3.1. Effects of concentration on settling velocity

Floc settling velocity is defined as the sinking speed in quiescent fluid. It affects vertical transport and distribution in the water column and maximum rate of deposition. Settling velocity of cohesive sediments varies with concentration and with fluid shear rate (Camp, 1946; Krone, 1962; van Leussen, 1989; Kranck and Milligan, 1992). The effect of differential settling is ignored here as the suspensions under consideration are assumed to be relatively deep and turbulent. The effect of salinity is also ignored here, though if a model were to be applied through the fresh-to-brackish zone, salinity effects should be considered.

Suspension concentration affects cohesive sediment aggregate collision frequency, floc size, and settling rate. For example, an empirical relations between median settling velocity and concentration can be developed from the results of Kranck and Milligan (1992): $W_s = 30.9\ C^{0.99}$ (mm/sec) where C is the total concentration in kg/m^3. Previous laboratory quiescent settling tests indicated $W_s \sim C^{4/3}$, and the difference in the exponent were attributed to differences in turbulence conditions.

Enhanced settling occurs over a concentration range from a lower concentration limit C_u to an upper concentration limit C_{ul}. Below C_u, particle collisions are to infrequent to promote aggregation. Figure 2 shows a schematic of concentration limits for enhanced settling. C_{ll} is typically 50-300 mg/l depending on sediment characteristics. At C_{ul}, collisions are so numerous that particles interact completely, causing all floc settling rates to converge to one value. At concentrations greater than C_{ul}, particle interactions begin to hinder settling, and dense suspensions settle as masses. Camp (1946) found the onset of concentration-hindered settling to be 1-5 kg/m^3 for turbid river water. The author has found C_{ul} to be 1-10 kg/m^3 for estuarine sediments.

For the multiple grain size model the general form for grain class settling velocity $W_s(gs)$ is:

$$W_s(gs) = a_1 \left(\frac{C}{C_{ul}} \right)^{n(gs)} \quad ; C_{ll} \leq C \leq C_{ul} \tag{4}$$

where a_1 is a grain class mass-weighted average maximum floc settling velocity, C is the total concentration for all grain classes, gs is the grain class index

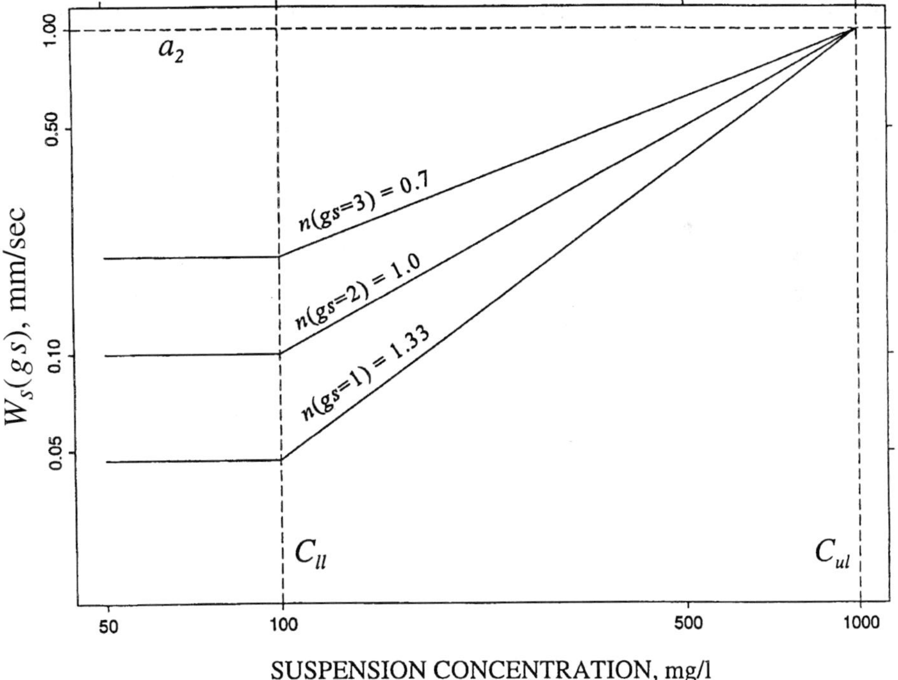

Figure 2. Definition of upper and lower limiting concentrations for enhanced settling, and example grain class exponents.

ranging from 1 to the number of grain classes, $n(gs)$ is an exponent, and C_u and C_{ul} are mass-weighted average lower and upper reference concentrations, respectively, over which concentration-enhanced settling occurs. The mass-weighted averages are taken over grain classes. For example:

$$a_1 = \frac{1}{C}\sum_{gs=1}^{NS} a_1(gs)C(gs) \tag{5}$$

where *NS* is the number of grain classes. Unlike the power law expressions given earlier, the normalized concentration C/C_{ul} is used in (4) to make the dimensions of a_1 (gs) and a_1 mutually consistent with W_s, and not affected by the magnitude of $n(gs)$. At $C \geq C_{ul}$, $W_s(gs)$ equals a_1 for all gs. At $C \leq C_{ll}$, settling rates are independent of concentration, and equal to $W_s(gs)$ evaluated at C_u.

The range of $n(gs)$ defines the span of the floc settling distribution as in the hypothetical case shown in Figure 2. Though the smallest grains are not always associated with the smallest flocs and the largest grains with the largest flocs, (4) produces distributions of $W_s(gs)$ which reflect the effect of grain composition on floc settling spectra. Additional coupling between grain classes is imposed during deposition, as described later, such that deposition of a given grain class is not necessarily related to its settling velocity.

The exponent $n(gs)$ can be determined empirically using information from settling tests conducted over a range of concentrations. Then, by selecting appropriate settling velocity percentiles to represent the grain classes, fits to settling data are made. The exponent n has been determined to range below a value of about 1.33 using this method. Teeter and Pankow (1989a) found that n's for the 50 and 75 percentile values were progressively less than for the 25 percent slowest settling fraction, similar to Figure 2. Alternately, settling experiments can be performed with particle size analyses performed concurrently, and settling velocities estimated for specific size classes. For example, settling velocities were determined by size class using sediments from New Bedford Harbor estuary by Teeter (1993). The median settling velocity $W_{s50} = 1.13\ C^{4/3}$ (mm/sec) where C is the total concentration in kg/m^3.

3.2. Effects of fluid shear and concentrations on settling velocity

Fluid-shear promotes particle collisions, and can, up to a point, promote floc growth. Since aggregate collisions and turbulence at the microscale can also break flocs apart (Camp, 1946; van Leussen, 1989), only relatively low shear rates are effective at optimizing floc size. Specific data on the effect of shear rate on floc growth and settling is relatively rare for natural sediments. The measurements of Kranck and Milligan (1992) captured fluid shear effects on floc characteristics in the field. Much work has been done on shear coagulation in waste water treatment (Camp, 1946). Fluid shear rate G has been termed root-mean-square velocity gradient, related to work input per unit volume per unit time and viscosity, and can be related to overall mean velocity and to overall turbulence intensity (McConnachie, 1991).

A settling velocity function including the effects of both concentration and fluid shear rate has been proposed by Malcherek and Zielke (1996), and Teisson (1997). In their models:

$$W_s = W_{so} \frac{(1 + a_2 G)}{(1 + a_3 G^2)} \tag{6}$$

where a_2 and a_3 are constants, and W_{so} is a concentration-dependent settling velocity function like those presented earlier. According to (6), increasing concentration will increase W_s, for any given G. However, data from Lick et al. (1993) suggest that, as concentration increased from 5 to 200 mg/l, maximum floc size and floc settling rate steadily decreased, as shown in Figure 3. Tests used natural river sediments in salt water, and rotating disk flocculators. They found similar trends in fresh water, and for chemically treated sediments. Winterwerp (1998) presented data from a series of settling tests on the Elms estuary sediment in a flocculator column. Three tests at a constant G = 0.9 Hz indicated that as concentration increased (from 150, to 790, and 970 mg/l), floc settling velocity first decreased then increased, contrary to (6). In both data sets, reductions in settling velocity occurred at concentrations well below those associated with concentration-hindered settling. Since little information is available to check the applicability of (6), laboratory tests were performed, and will be described in the remainder of this section.

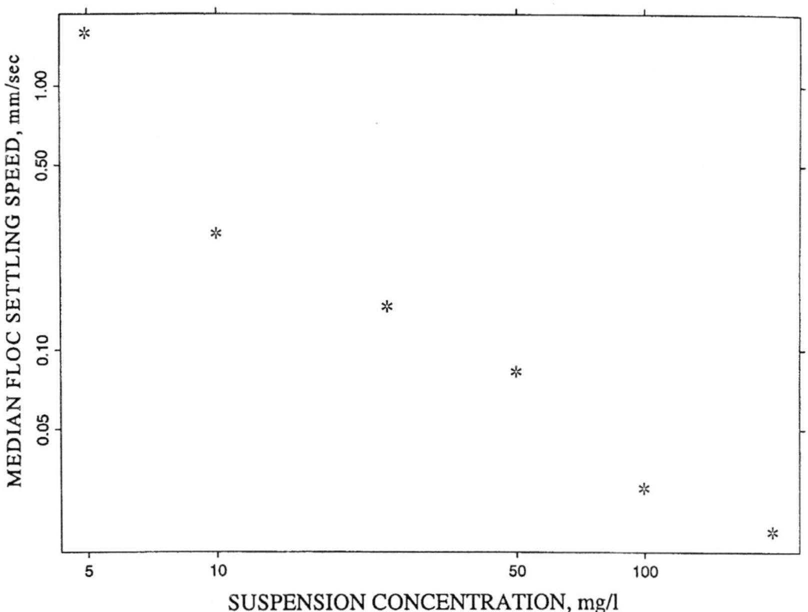

Figure 3. Effect of concentration on floc settling rate (data from Lick et al., 1993).

Experiments on the effects of both concentration and shear on settling velocity were performed. Tests were carried out on a model clay and on resuspended estuarine-channel sediments. A model clay mix was prepared from pure, dry clays. Dispersed particles greater than 10 μm Stokes' settling diameter were removed, and a stock mix with 40 percent illite and kaolinite, and 20 percent montmorillinite by dry weight was prepared. All three minerals were found by Whitehouse et al. (1960) and Edzwald et al. (1974) to flocculate at ionic concentrations lower than those used here. The stock mix was allowed to stand a week after 15 ppt Instant Ocean® salts were added. Clay stock mix was admixed with 15 ppt artificial seawater to make test suspensions. A natural estuarine sediment from the Gulf Intracoastal Waterway channel about 11.2 km north of Port Isabel, Texas, was also tested. This sediment had mean and median dispersed grain diameters of 26 and 7 μm, 7 percent sand (> 74 μm) and 64 percent less than 16 μm. The material had an *in situ* bulk wet density of 1,408 kg/m^3, and 8 percent organic content by weight.

Rotating flocculators were used to generate large flocs. Cylindrical flocculators 12.5 cm diameter by 21 cm long/deep were used to allow for settling velocity determinations after floc generation. Cylinders of 11.75 cm diameter by 28 cm deep were used for two higher shear rate and quiescent settling experiments. For higher shear rate model clay experiments, these cylinders were filled to 18-cm depth in a Particle Entrainment Simulator (PES). The PES is a standard device developed by Tsai and Lick (1986) that has a 2.5 cm stroke oscillating grid.

The flocculators were filled, rotated such that the outside edge or wall moved at about 1 cm/sec for 3.5-7 hours until floc sizes stabilized, then removed from the rotator, up-ended slowly to bring the axis to the vertical, and sampled over time with a pipette to determine floc settling velocities. In the first series of model-clay experiments, laminar shear rates were varied by partially filling the cylinders, and spanned those which produced maximum or optimum floc sizes. Shear rates were estimated visually by observing small particle movement at 5 to 10 locations in the disk flocculator. Photo-micrographs and grids were used to visually estimate floc dimensions which were found to vary between < 0.2 mm and 0.8 mm diameter depending on the test. In the second series of model-clay experiments, only the optimum shear rate was used, and compared to quiescent settling tests.

The quiescent cylinder experiments were conducted by mixing and introducing sediments, and withdrawing samples from 20 cm depth for 6 hours. Care was taken to keep cylinders under constant temperature conditions. PES settling experiments were performed while grid oscillation was underway and producing mild turbulence. Oscillation rates were 70 and 140 per second producing turbulent shear rates of about 20 and 40 per second estimated using procedures presented by McConnachie (1991). The PES was operated at a rate below previous calibration range to produce mild turbulence and a shear stress of less than 0.05 Pa on the bed. Samples were withdrawn from 12.5 cm depth for 6 hours.

Two series of tests were performed on the natural sediment: one using rotating cylindrical flocculators and the other in a 10 cm diameter by 1.9 m high column. The disk flocculator was operated at the shear rate found to be optimum for the model clay and produced large flocs. Samples were analyzed for total suspended material, and median settling velocities estimated.

The effects of shear rate G alone on settling are shown in Figure 4 for model clay suspensions with initial concentrations of 46 mg/l. Also shown is a fit to (6) with W_{so} set to a constant equal to the median W_s value observed in the corresponding quiescent test. The median W_s for the quiescent test was about the same with a G value of about 30 Hz. The maximum floc settling rates were more than 100 times greater than with G = 0, occurred at a very low shear rate (about 0.5 Hz) and dropped off sharply with increasing G.

Results on the effects of concentration and shear rate on W_s for model clay and natural sediments are shown in Figures 5 and 6. Test data were median W_s's for disrupted flocs (quiescent tests which began with floc disruption), and for optimum flocs (G = 0.5 Hz). The model clay floc settling velocity decreased as initial concentration increased from 57 to 196 mg/l (Figure 5). Eleven floccculator

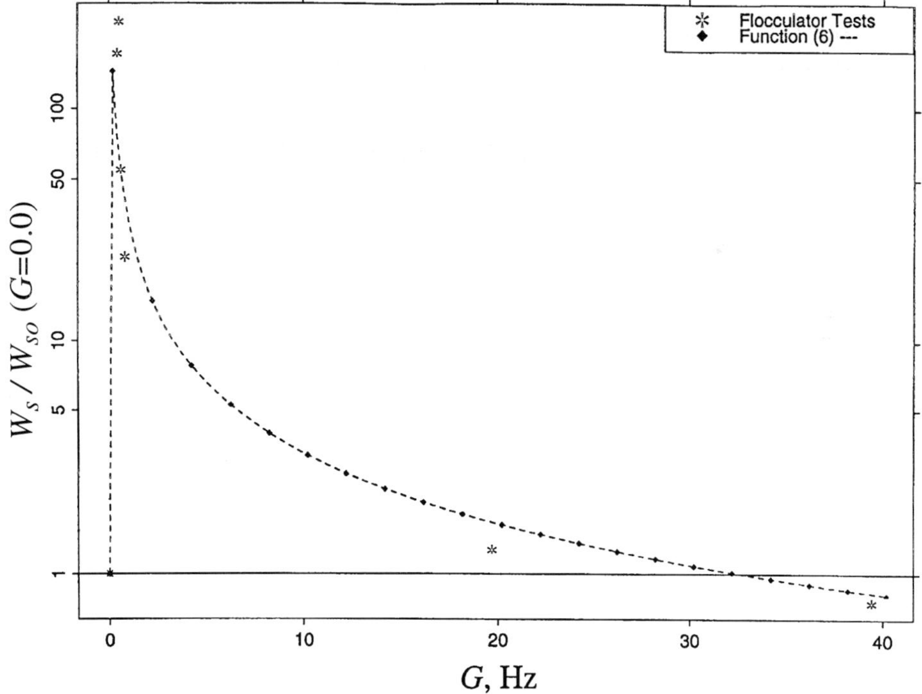

Figure 4. Experimental results for the effect of shear on settling velocity, and a fit of (6) to the data.

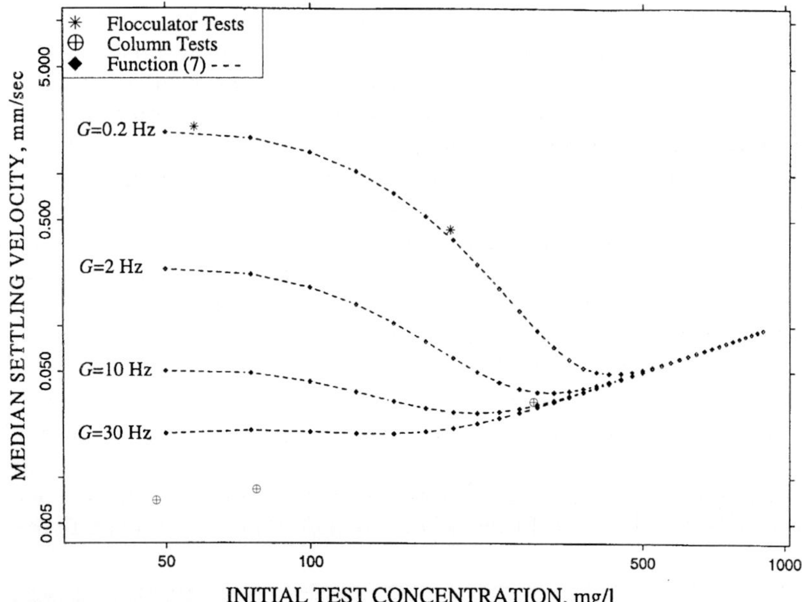

Figure 5. Model clay settling velocity results for varying shear and concentration, and a fit of (7) to the data.

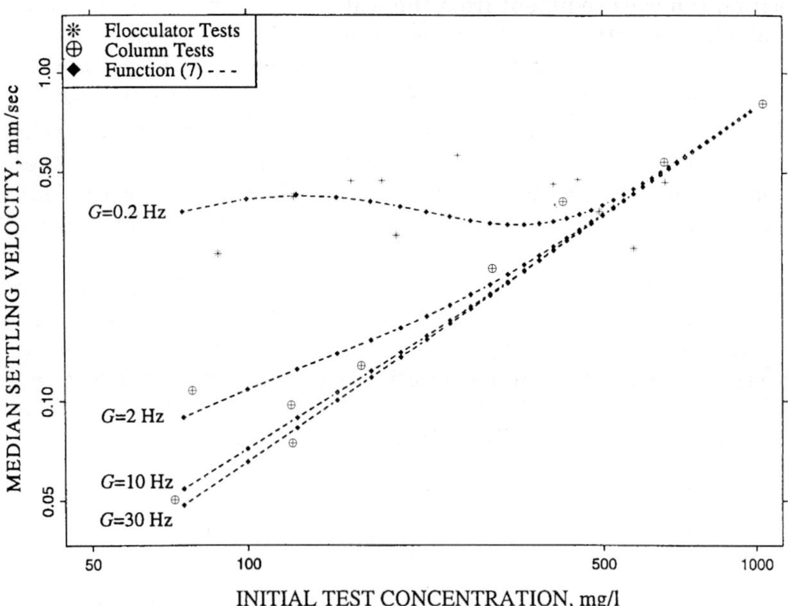

Figure 6. Natural sediment settling velocity results for varying shear and concentration, and a fit of (7) to the data.

tests were performed at G = 0.5 Hz with the natural sediment over a concentration range of 60 to 660 mg/l (Figure 6). There was a slight upward trend in the linear regression line between initial concentration and median optimum-floc W_s, but the slope was not statistically different from zero. Nine quiescent column tests were performed between initial concentrations of 72 and 1,012 mg/l (Figure 6). Above 122 mg/l median floc W_s's increased steadily, indicating that suspension concentrations were not hindering settling, and were generally in the range of $C_{ll} \leq C \leq C_{ul}$. Optimum and disrupted-floc W_s's matched at about 500 mg/l (Figure 6). Therefore, contrary to (6), floc settling rates did not increase with concentration in either model clay or natural sediment tests.

Based on experimental results, the effects of concentration and shear rate on settling velocity are not taken to be multiplicative as in (6), and the following functional relationship is proposed:

$$W_s(gs) = a_1 \left(\frac{C}{C_{ul}} \right)^{n(gs)} \left[\left(\frac{1 + a_2\, G}{1 + a_3\, G^2} \right) \exp\left(-a_4 \frac{C}{C_{ll}} \right) + 1 \right] \tag{7}$$

for $C_{ll} \leq C \leq C_{ul}$. Equation (7) is shown plotted in Figures 5 and 6 as approximate fits to the data.

3.3. Deposition rate

Deposition removes sediment from the water column and is a rate equal to the product of effective settling and concentration. To deposit, sediment must transit the zone just above the bed which can have very high shear rates. Previous laboratory experiments (Krone, 1962; Teeter and Pankow, 1989b) have observed that effective W_s based on deposition are lower than those measured in the water column. The calculation procedures presented in this section first assess the deposition process for individual grain classes, then couple grain classes such that final result depends on deposition of the coarsest active class and the grain size spectra.

Potential deposition of each grain class is first assessed. Deposition is assessed differently for the cohesive fraction than for silts. The cohesive fraction is taken to follow Krone's deposition law (Krone, 1962) using the concept of a critical shear stress for deposition and depositional probability. The effective settling velocity is the settling velocity times depositional probability P (Krone, 1962) defined for the finest cohesive fraction as:

$$P = \left(1 - \frac{\tau}{\tau_{cd}} \right); \ \tau < \tau_{cd} \tag{8}$$

where τ is the bed shear stress and τ_{cd} is the critical threshold shear stress for deposition. According to (8), all sediment eventually deposit at shear stresses less than the critical value.

For each silt class, upper and lower shear stress threshold values are defined slightly differently than those used for the cohesive fraction. The lower or critical depositional shear stress is defined as that value below which all material is free to deposit. Below this threshold, silt deposition depends only on concentration and settling velocity (P = 1.0). At shear stresses between the upper and lower critical values, silt sediment fractions erode or deposit at rates linearly related to the bed and threshold shear stresses.

The upper or critical erosional shear stress is defined as that value above which all material of this class will remain in suspension (P = 0), but is not linked explicitly to erosion. Erosion of a silt fraction is first dependent on the erodibility of the cohesive fraction. That is, silt is held in a cohesive matrix and is not free to erode unless the bed shear stress exceeds the erosional threshold for the cohesive fraction, as described in Part II (this volume). Grain class shear stress thresholds are specified to be contiguous. That is, the upper shear stress threshold τ_{ce} for one class is the same value as the lower threshold τ_{cd} for the next coarser class.

After the potential deposition from each grain class has been assessed, grain classes are coupled such that spectral shapes follow the distribution model of (3). As noted earlier, coupling between grain classes causes some sediment to deposit when a well-sorted sediment of the same class would remain in suspension. For the deposition of grain classes to be proportional as in (3), settling velocities over those grain classes depositing must be equal. Other controls are introduced to ensure that the algorithm is capable of forming steady state suspensions at a given bed shear stress level. Thus, if F is the depositional flux, gs is a grain class with P ~ 0, and gs+1 is a larger-sized depositional grain class:

$$F(gs) = \frac{d_1 C(gs)\, F(gs+1)}{C(gs+1)+d_2}, \quad P(gs) < 0.05 \qquad (9)$$

where d_1 controls the proportion between the fluxes, and d_2 limits flux of smaller grains as $C(gs + 1)$ tends toward zero. The deposition algorithm introduces two parameters to control the proportional deposition. Data indicate that d_1 ~ 1 or slightly below.

4. RESULTS OF NUMERICAL DEPOSITION EXPERIMENTS

The algorithms (4), (8), and (9) were used in a numerical scheme that included mass conservation equations for each grain class in a layered sediment bed and in a suspension at a point (1-dimension vertical). Addition details of the numerical scheme and descriptions of other algorithms for erosion and sediment bed processes are presented in Part II (this volume). The intention was to demonstrate the algorithms, realizing that application to most real problems would require multi-dimensional numerical modeling.

Model simulations were performed with seven grain classes arbitrarily assigned at nominal sizes of 4, 18, 22, 28, 37, 44, and 56 μm. Normally, grain classes would be spaced logarithmically by size, but in this case the finest fraction was spaced to accentuate the fine tail of the spectrum. A shear stress was initially imposed (0.16 Pa) greater than τ_{cd} for the clay fraction and less than the τ_{ce} for the next larger class. Numerical deposition experiments were performed by initializing suspensions with size distributions of m = -1, 0, and 1.

Time-series grain spectra from the three experiments shown in Figure 7. The model spectra maintained slopes on their fine ends about the same as initial m's while the coarse ends deposited. This pattern is consistent with field observations described earlier and with (3). Portions of the clay fraction of the distribution can be seen in Figure 7 to deposit, even though the imposed bed shear stress exceeds this class's threshold for deposition. Statistical measures of the mean size, sorting (standard deviation), and skewness were calculated in phi units with time for the three experiments. The mean grain sizes became progressively finer and skewness became more negative in all three cases. Distribution means correlated to their deposition rate. The distribution with M=-1 was the finest and settled the slowest. The distribution with M=1 settled fastest, especially at the coarse end.

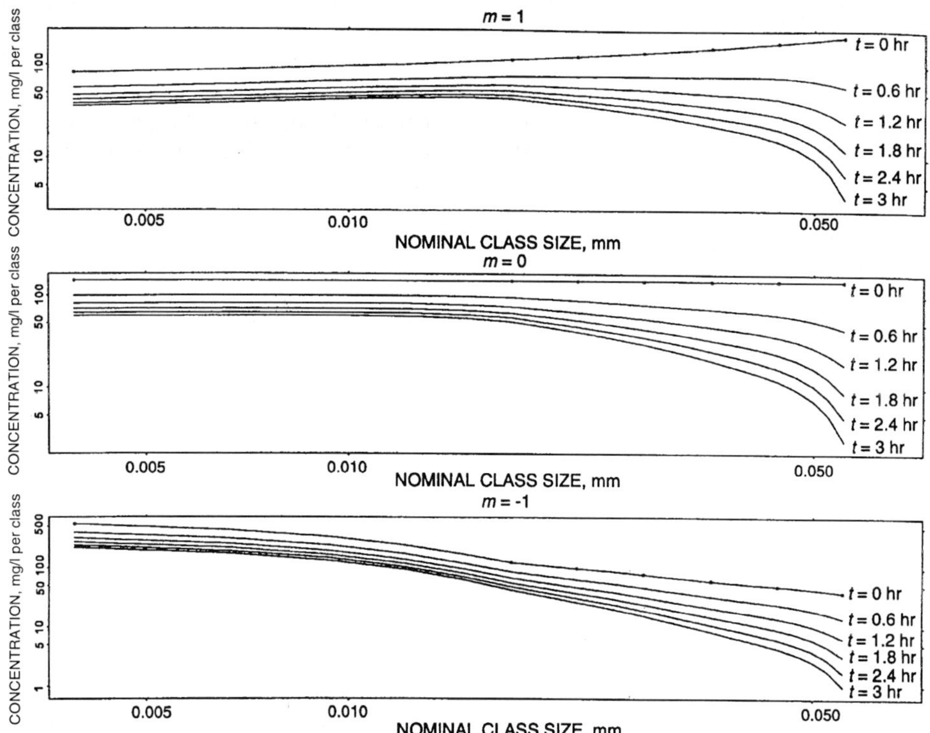

Figure 7. Time-series numerical experiment results for m = -1, 0, 1 for deposition under a shear stress of 0.16 Pa.

The sorting for the case of m = 1 became larger, for the case of m = 0 remained about constant, and for the case of m = -1 became appreciably smaller. Trends similar to the latter were reported by Teeter (1993) who found that sorting decreased slightly during laboratory settling tests on a well-graded sediment (m about equal to 0). Trends similar to the former were reported by Stow and Bowen (1980) who found that clayey-silts with positive m became more poorly sorted in the direction of down-slope transport on the Scotian continental margin.

5. CONCLUSIONS

Cohesive suspensions have grain spectra which vary with depositional conditions and settling. A limited number of coupled grain classes can be used to mimic spectral changes during quiescent or stress-limited deposition. Coupling of numerical grain classes was based on a previous analytic description of grain spectra change during settling. Settling experiments using flocculators and columns indicated that concentration and shear rate are not independent multiplicative factors. The concentration effect on settling was found to depend on fluid shear rate, and a new functional relationship proposed.

REFERENCES

Camp, T. R., 1946. Sedimentation and the design of settling tanks. *Transactions,* American Society of Civil Engineers, 111, 895-958.

Edzwald, J. K., Upchurch, J. B., and O'Melia. C. R., 1974. Coagulation in estuaries. *Environmental Science and Technology*, 8(1), 58-63.

Gibbs, R. J., 1977. Clay mineral segregation in the marine environment. *Journal of Sediment Petrology*, 47(1), 237-243.

Kranck, K., 1980. Experiments on the significance of flocculation in the settling of fine-grained sediment in still water. *Canadian Journal of Earth Science*, 17, 1517-1526.

Kranck, K., and Milligan, T. G., 1992. Characteristics of suspended particles at an 11-hour anchor station in San Francisco Bay, California. *Journal of Geophysical Research*, 97(C7), 11,373-11,382.

Krone, R. B., 1962. Flume studies of the transport of sediment in estuarial shoaling processes. University of California, Hydraulic Engineering Laboratory and Sanitary Engineering Research Laboratory, Berkeley, CA.

Lick, W., Huang, H., and Jepsen, R., 1993. Flocculation of fine-grained sediments due to differential settling. *Journal of Geophysical Research*, 98(C6), 10,279-10,288.

Malcherek, A., and Zielke, W., 1996. The role of macroflocs in estuarine sediment dynamics and its numerical modeling. *Proceedings, Estuarine and Coastal Modeling Conference*, American Society of Civil Engineers, San Diego, CA, 695-706.

McConnachie, G. L., 1991. Turbulence intensity of mixing in relation to flocculation. *Journal of Environmental Engineering*, 117(6), 731-750.

Mehta, A. J., and Lott, J. W., 1987. Sorting of fine sediment during deposition. *Proceedings, Coastal Sediments'86*, American Society of Civil Engineers, 348-362.

Stow, D.A.V., and Bowen, A. J., 1980. A physical model for the transport and sorting of fine-grained sediment by turbidity currents. *Sedimentology*, 27, 25-40.

Teeter, A. M., 1993. Suspended transport and sediment-size transport effects in well-mixed, meso-tidal estuary. In: *Nearshore and Estuarine Cohesive Sediment Transport*, A. J. Mehta ed., American Geophysical Union, Washington, DC.

Teeter, A. M., 2000. Clay-silt sediment modeling using multiple grain classes. Part II: Application to shallow water resuspension and deposition. (this volume).

Teeter, A. M., and Pankow, W., 1989a. The Atchafalaya River Delta; report 2, field data; settling characteristics of bay sediments. *Technical Report HL-82-15,* U.S. Army Engineer Waterways Experiment Station, Vicksburg, MS.

Teeter, A. M., and Pankow, W., 1989b. Deposition and erosion testing on the composite dredged material sediment sample from New Bedford. *Technical Report HL-89-11,* U.S. Army Engineer Waterways Experiment Station, Vicksburg, MS.

Teisson, C., 1997. A review of cohesive sediment transport models. In: *Cohesive Sediments*, N. Burt, R. Parker and J. Watts eds., John Wiley, Chichester, UK, 367-381.

Tsai, C.-H., and Lick, W., 1986. A portable device for measuring sediment resuspension. *Journal of Great Lakes Research*, 12(4), 314-321.

van Leussen, W., 1989. Aggregation of particles, settling velocity of mud flocs - a review. In: *Physical Processes in Estuaries*, J. Dronkers and W. van Leussen eds., Springer-Verlag, N.Y.

Whitehouse, U. G., Jeffrey, L. M., and Debrecht, J. D., 1960. Differential settling tendencies of clay minerals in saline waters. *Clays and Clay Minerals. Seventh National Conference*, Pergamon Press, New York, NY, 1-79.

Winterwerp, J. C., 1998. A simple model for turbulence induced flocculation of cohesive sediment. *Journal of Hydraulic Research*, 36(3), 309-326.

Coastal and Estuarine Fine Sediment Processes
W.H. McAnally and A.J. Mehta (Editors)
2001 Elsevier Science B.V.

Clay-silt sediment modeling using multiple grain classes. Part II: Application to shallow-water resuspension and deposition

A. M. Teeter

Coastal and Hydraulics Laboratory, U.S. Army Engineer Waterways Experiment Station, 3909 Halls Ferry Road, Vicksburg, Mississippi 39180, USA

Single and multiple grain class model formulations for the erosion and deposition of clay and silt sized particles are compared to each other for a series of laboratory experiments and field data from a shallow lake where resuspension occurs from wind-waves. The approach for coupling between grain classes during settling and deposition is presented in Part I (this volume). Threshold shear stresses for mutually-exclusive erosion and deposition are used in the model formulation to be consistent with previous laboratory investigations. Model deposition and erosion laws treat silt and clay fractions differently, yet couple them during certain modes of vertical transport. Models with up to 7 grain classes are compared to laboratory flume tests which formed steady state suspension concentrations during deposition. Models with 1 and 4 grain classes are compared to field data. An automated, objective method is used to adjust coefficients for both model simulations. Comparisons indicate that the multiple grain class formulation improves erosion and deposition process descriptions.

1. INTRODUCTION

Resuspension of fine estuarine and lacustrine sediments by wind-waves has been studied using empirical methods (Pejrup, 1986; Petticrew and Kalff, 1991; Arfi et al., 1993; Hamilton and Mitchell, 1996), and numerical models (Luettich et al., 1990; Hawley and Lesht, 1992; Lick et al., 1994). Fine-grained sediment transport models are used in hydro-environmental studies of water quality and ecological conditions such as turbidity, dissolved oxygen, and contaminants, often involving relatively low concentrations of fine suspended material. Model process descriptions are not derived from first principles, and much empirical information is needed to successfully describe fine sediment transport at a given site. To qualify as a "good" model, sufficient physical processes must be represented to capture important observed behavior and non-observable system feedbacks that might become important after a physical change. The purpose here is to evaluate two model formulations: single and multiple grain class representations.

Single grain sediment models which allow simultaneous erosion and deposition (Luettich et al., 1990; Hawley and Lesht, 1992; Lick et al., 1994) have been previously found to predict resuspension more accurately than single grain models with mutually exclusive erosion and deposition, despite considerable laboratory experimental evidence supporting the latter assumption (Sanford and Halka, 1993). Numerous laboratory studies have determined that steady state suspensions, which can occur during either deposition or erosion, do not reflect a balance between simultaneous operation of these processes. Sanford and Halka (1993) list many references to experimental studies, and Lau and Krishnappan (1994) is an important recent study. One explanation for the failure of single grain models with mutually exclusive erosion and deposition is that natural sediments with silt and clay have multiple thresholds (Sanford and Halka, 1993), while one reason for the success of the models that allow simultaneous erosion and deposition is that concentration-dependent depositional flux acts as a penalty to limit erosion, similar to a decrease in sediment bed erodibility with erosion depth observed in laboratory experiments (Dixit, 1982). While the issue of simultaneous erosion and deposition remains to be resolved, the modeling approach used here is to assume no simultaneous erosion and deposition, consistent with most experimental results.

Grain size coupling or amalgamation occurs due to cohesion, causing individual flocs or aggregates to be a mix of particle sizes (Kranck, 1980; Lau and Krishnappan, 1994) and bed sediments to be poorly sorted. The paradox is that, even though these sediments occur as mixed-grain aggregates, they exhibit particle-size dependent transport behaviors. Spatial variation of dispersed grain size distributions correlate to field transport processes (McLaren and Bowles, 1985; Teeter, 1993). During deposition, suspension particle-size spectra first decrease at their coarser ends, and modal sizes become progressively finer as deposition proceeds (Kranck, 1980). Dispersed particle size of fine-grained suspensions increases with shear stress during erosion (Teeter et al., 1997). At a given bed shear stress level, the particles remaining in suspension are generally finer than those which deposit (Mehta and Lott, 1987). Both settling velocities and critical shear stresses for deposition vary between clay and silt fractions (Teeter and Pankow, 1989). While a well-sorted cohesive suspension will steadily deposit in a flow below a critical shear stress (Krone, 1962), a suspension of silts and clays will partially deposit to a steady state constant suspension concentration level (Partheniades et al., 1968). A similar paradox is that clay minerals segregate during transport in a manner that is similar to their settling rates based on dispersed particle size (Gibbs, 1977). Thus, even though grain classes are coupled by cohesion, dispersed particle size affects transport properties, and size distribution imprints form clearly detectable patterns in estuarine and lake sediments.

2. METHODS

This section describes multiple grain class transport modeling at and near the sediment/water interface, the data sets used to compare model results, and the method used to adjust model coefficients. Emphasis will be on model features related to multiple grain class erosion laws. Deposition laws are described in Part I (this volume). A numerical fine-grained sediment transport model was formulated with optionally one or more conservative grain classes to test the effectiveness or applicability of multiple grain class model for resuspension and deposition modeling. For modeling purposes, the fine-grained sediment is assumed to be cohesive in nature, not to liquefy under waves, and to follow separate and exclusive process descriptions for particle erosion and deposition.

Since fine-grained cohesive transport processes are not dependent on individual particle size intrinsically, size designations of grain classes are nominal. The finest class consists of the most cohesive clay and fine-silt sized particles, and is referred to here as the cohesive fraction. Evidence suggest that this class behaves as a unit, and dominates several transport processes and the character of the sediments as a whole (Stevens, 1991a,b; Teeter and Pankow, 1989). Silts are divided into a variable number of classes, though evidence suggests that they act as a continuous distribution.

2.1. Model description

The basic model is configured as a vertical suspension and a unit area of sediment bed at a point, similar to the models developed by Luettich et al. (1990), and Hawley and Lesht (1992). Horizontal gradients are ignored in favor of vertical ones. The focus of the model is the sediment/water interface where deposition and erosion algorithms compute flux boundary conditions for the suspension and the bed. Mass conservation equations are solved for each grain class in the suspension and in the bed. A bed model component describes layers of sediment properties, performs bookkeeping with respect to sediment mass, and adjusts layer properties for consolidation and silt content.

For a well-mixed suspension at a point, model sediment mass conservation equations are:

$$H\frac{dC(gs)}{dt} = E(gs) - F(gs) \tag{1}$$

where H is the water depth, C is depth-mean sediment concentration (mass per unit volume), t is time, E is the erosion flux, F is the depositional flux, and the index gs refers to the grain class. Dimensions of the erosion and deposition fluxes are mass per unit area per unit time. Deposition flux for multiple grain classes was described in Part I (this volume). For the point model, an analytical near-bed concentration C_b (Teeter, 1986) is used in the calculation of F:

$$C_b = \left[1 + \frac{Pe}{1.25 + 4.75P^{2.5}}\right]C \qquad (2)$$

where a particle Peclet number $Pe = HW_s/K_z$, W_s is a concentration-dependent settling velocity described in Part I (this volume), $K_z = 0.067U_*H$ is the vertical diffusivity (Fischer, 1973), U_* is the friction velocity, and P is the depositional probability. The Pe value was constrained in the model such that $Pe < 5$.

The erosion flux depends first on the erosion threshold of the cohesive fraction, and then the thresholds for silt fractions. The form of the cohesive-fraction erosion model is similar to the single class equations of Kandiah (1974) and Ariathurai et al. (1977), and is:

$$E(gs=1) = M(gs=1, bl=a)\left[\frac{\tau}{\tau_{ce}(gs=1, bl=a)} - 1\right] \quad , \tau > \tau_{ce}(gs=1, bl=a) \qquad (3)$$

where the cohesive fraction is designated $gs = 1$, the layer bl exposed at the bed surface is designated a, M is the erosion rate constant for the cohesive grain class and bed layer, τ is the bed shear stress, and τ_{ce} is the erosion threshold. If $\tau < \tau_{ce}$ for the cohesive fraction, no sediment is eroded even if the bed shear stress exceeds the critical threshold for some silts. The critical shear stress for erosion of the cohesive fraction is estimated by a power law depending on the concentration of the cohesive fraction in the bed layer exposed to the flow (Teeter, 1987), and generally increases vertically downward in the bed. The erosion rate parameter M is functionally related to the τ_{ce} value based on Lee and Mehta (1994). The forms of these auxiliary equations are not important to the results to be presented here.

For each silt class, upper and lower shear stress threshold values are defined slightly differently than those used for the cohesive fraction. The lower or critical depositional shear stress is defined as the value below which a silt fraction is free to deposit (depositional probability $P = 1$, see Part I). At shear stresses between the upper and lower critical values, a silt sediment fraction will erode or deposit ($0 < P < 1$) at rates linearly related to the bed and threshold shear stresses, and depending on whether or not the cohesive fraction is eroding. Erosion is given precedence in the model such that, for a particular grain class, bed erosion precludes deposition.

The upper or critical erosion threshold shear stress is defined as the value above which a silt class will remain in suspension. This definition recognizes that erosion of a silt fraction depends first on the condition that the critical shear stress for erosion of the cohesive fraction has been exceeded. Silt-fractions erode in proportion to clay fraction masses to maintain similarity in the shape of bed and suspended grain size distributions as discussed in Part I (this volume), and

$$E(gs>1) = E(gs=1)\left.\frac{S(gs>1, bl=a)}{S(gs=1, bl=a)}\right|_{gs>1} \quad , E(gs=1) > 0 \text{ and } \tau > \tau_{ce}(gs>1) \qquad (4)$$

where $S(gs,bl)$ is the grain class sediment mass per unit area in a bed layer. Erosion thresholds for silt fractions are taken to be independent of their bed layer location.

Erodibility is also linked to the structure of the bed. A layered bed algorithm was developed with variable silt concentrations by layers, depending on initial conditions, and on erosional and depositional history. A fully-settled near-surface concentration distribution with respect to the cohesive fraction is assumed. After deposition occurs, hindered settling rate is calculated by bed layer, and material is transported vertically downward in the bed using class-aggregated transport parameters, until the specified density distribution is achieved. The mass conservation equation for bed layer consolidation is:

$$\frac{dS(gs,bl)}{dt} = -\frac{W_h(bl)S(gs,bl)}{H_s(bl)} + \frac{W_h(bl-1)S(gs,bl-1)}{H_s(bl-1)} \quad ,H_s(bl) > H_{so}(bl) \tag{5}$$

where $H_s\ (bl)$ is the bed layer thickness, $H_{so}\ (bl)$ is the specified fully-settled thickness, and the nominal hindered settling rate is:

$$W_h(bl) = W_{ho}\left[1 - b_1 \sum_{gs=1}^{ns} \frac{S(gs,bl)}{H_s(bl)}\right]^{b_2} \quad , \sum_{gs=1}^{ns} \frac{S(gs,bl)}{H_s(bl)} < \frac{1}{b_s} \tag{6}$$

where ns is the number of grain classes, W_{ho} is a reference settling rate, and b_1 and b_2 are grain class mass averaged coefficients. Hindered settling/consolidation is inhibited by deposition or erosion greater than 0.01 g/m^2/sec. In the bed, volumes of grain classes are taken into account when converting between mass and concentration. In general, it is assumed that:

$$H_s(bl) = \sum_{gs=1}^{ns} \frac{S(gs,bl)}{\rho_s} + \frac{O_c S(gs=1,bl)}{\rho_l} + \sum_{gs=2}^{ns} \frac{O_s S(gs,bl)}{\rho_l} \tag{7}$$

where O_c and O_s are the ratios of clay and silt masses to water masses associated with these fractions, respectively, and ρ_s and ρ_l are the particle and fluid densities. As mass is transported vertically downward as a result of consolidation, the layer concentration of the cohesive fraction is maintained constant over time, and the condition:

$$H_s(bl) = \frac{S(gs=1,bl)}{\rho_s} + \frac{O_c S(gs=1,bl)}{\rho_l} \tag{8}$$

is imposed. Bed concentration (mass per unit volume) is $S(gs,bl)/H_s(bl)$.

Bed layers are numbered vertically downward. If a layer is withered away by erosion, it disappears at least temporarily. The erosion surface thus descends through the bed as the surface layer thins, then the surface descends step-wise through progressively deeper layers. The effects of erosion on bed mass is evaluated as:

$$\left.\frac{dS(gs,bl=a)}{dt}\right|_e = -E(gs) \tag{9}$$

where a is the exposed bed layer index. Deposition, on the other hand, always occurs into the first layer (bl =1), and the effect of deposition on bed mass is evaluated as:

$$\left.\frac{dS(gs,bl=1)}{dt}\right|_e = F(gs) \tag{10}$$

In this way, the bed structure is formed by consolidation from the top layer down. After appreciable deposition has occurred, the bed (in the absence of erosion or further deposition) will return to the specified fully-settled structure with respect to the cohesive fraction.

Coupling of grain classes during erosion and deposition is apparent in grain size measurements taken under these conditions, as previously described. The grain coupling scheme used in this model assumes that fine-grained, cohesive sediments are deposited along with silts even though shear stresses are too high for them to deposit on their own. Serial coupling between grain classes is used to promote log-normal trends in the resulting size distributions, as discussed earlier. As a barrier to excessive winnowing of the cohesive fraction and sorting of grain classes under moderate shear stress conditions, the model shields a small portion of the cohesive fraction, about equal in magnitude to that proportion deposited with silts, from erosion until coarser grain classes are involved in erosion.

Shear stresses generated by a current are calculated using the Manning's equation where:

$$U_* = \frac{g^{1/2}n}{H^{1/6}}U \quad \text{and} \quad \tau = \rho_l U_*^2 \tag{11a,b}$$

where g is the acceleration of gravity, n is Manning's bed resistance coefficient, and U is the depth average flow velocity. Wave shear stresses are calculated using linear wave theory, and smooth, laminar wave friction formulations as described by Luettich et al. (1990). Wave and current components are added to estimate the total bed shear stress.

The point model used a forth-order Runge-Kutta scheme to integrate conservation equations in time using time steps of 90 to 120 seconds for the applications described next.

2.2. Model comparison data sets

Mehta and Partheniades (1975) performed annular flume deposition experiments starting at high shear stresses. Initially suspended fine-grained cohesive sediments deposited when shear stresses were reduced, forming constant, steady state concentrations that depended on the initial suspension concentrations and the bed shear stresses. Figure 1 shows typical results for one series of experiments. Each experiment had 1 g/l initial concentration of kaolinite. The kaolinite contained about 35 percent coarser than 2 μm, and a maximum particle size of about 45 μm. Similar results were obtained for coarser fine-grained sediments from San Francisco Bay, and Maracaibo Bay, Venezuela. The fractional amount remaining in suspension, C_f/C_o, for the kaolinite experiments (Mehta and Partheniades, 1975) are summarized in Figure 2a. The degree of deposition ($1 - C_f/C_o$) was found not to depend on initial concentration. This result, plus other experiments on kaolinite suspensions by Partheniades et al. (1968) and Lau and Krishnappan (1994), confirms that these steady state concentrations were not caused by a balance between erosion and deposition.

Results follow Krone's (1962) deposition law for bed shear stresses less than 0.16 Pa, when all sediment eventually deposited. At higher bed shear stresses, however, they do not follow Krone's deposition law as only a certain fraction of material, depending on shear stress, deposited. The times required for deposition to occur and for suspensions to reach steady state were not greatly affected by the bed shear stress, as can be seen in Figure 1. Material either deposited or

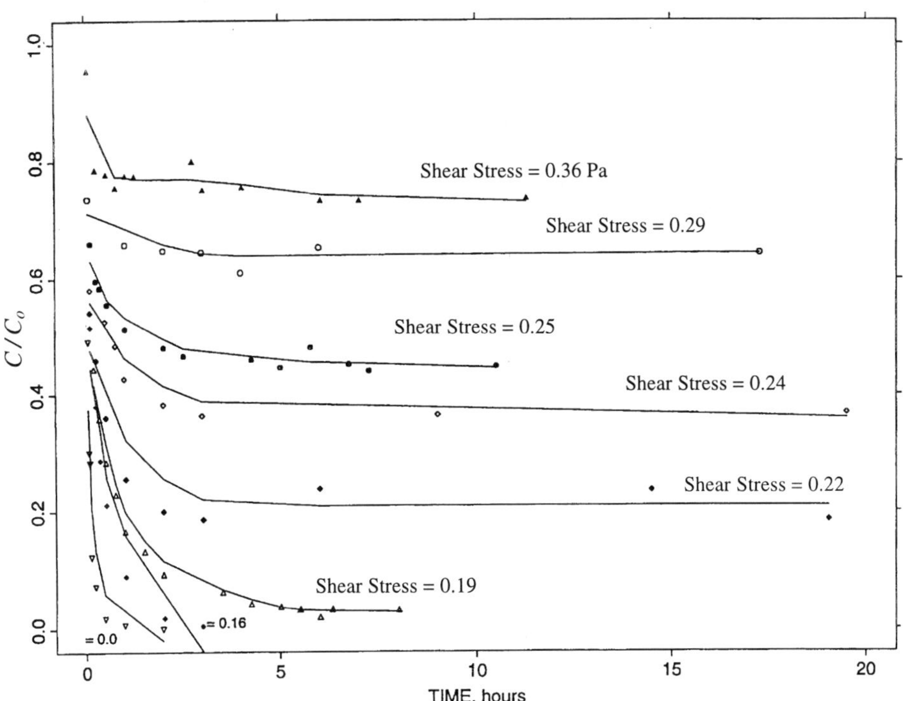

Figure 1. Relative suspension concentration during kaolinite deposition experiments (Mehta and Partheniades, 1975). Initial shear stress was 1.05 Pa for all tests.

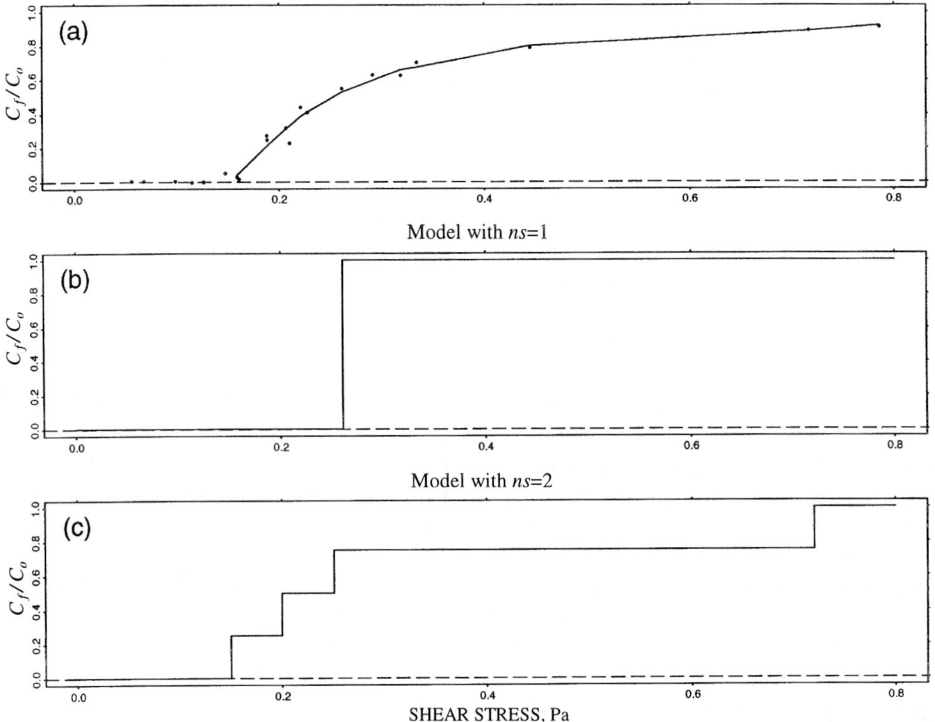

Figure 2. (a) Observed C_f/C_o corresponding to Figure 1, (b) example model result with 1 grain class, (c) example model result for 4 grain classes.

remained in suspension, with the transition time consistent with typical settling velocities.

Numerical deposition experiments were performed much the same way as the original experiments. Total suspended material concentrations were initialized at 1 kg/m^3, and shear stresses at 1.05 Pa. A series of 25 model simulations were performed in which shear stresses were reduced to allow deposition.

Wind-wave resuspension in a large shallow lake was studied by Luettich et al. (1990). Data were collected for 15 days in August 1985 in the 600 km^3, 3.2 m deep Lake Balaton, Hungary. At the time of the field experiment, the lake had Secchi depths of less than 0.2 m, was eutrophic, and had a fine-grained sediment bed. A tripod station was established in 2 m water depth to carry BASS velocity meters at two depths and a wind instrument 2 m above the water surface. Water samples were collected at mid-depth from an anchored boat. Wave information was extracted from the 2 Hz velocity data.

Two large storms occurred during the study, but wave information was only collected during the second storm. Wind speeds reached 7-9 m/sec. For the present study, current speed, wave height, and wave period data were digitized and interpolated to 0.1 hours for use as boundary conditions for the model. The model simulations covered 25.5 hours with time steps of 2 minutes. Total suspended material (TSM) data were digitized and interpolated for model comparison at 0.5 hour intervals to make the data coverage uniform in time.

2.3. Coefficient adjustment for model comparison

The comparison of models for simulating lake resuspension was complicated by the requirement to adjust model coefficients, since optimum model coefficient values could vary between model formulations. An automated model coefficient adjustment method was developed to expedite model adjustment, and to systematize this adjustment, so that alternate model formulations could be tested objectively.

The adjustment process began with model coefficients and parameters manually set to rough model-to-prototype TSM agreement using threshold values interpreted from field data, or typical values found for other systems. Important coefficients were identified, and adjustment for modeling lake resuspension involved 18 select coefficients for the single grain class model and 27 corresponding coefficients for the multiple grain class model. Each model coefficient in this set was varied by a range of factors (typically 0.6 to 1.4) in 10 model simulations, holding all other parameters constant. The variance in the difference between the model results and the field TSM data was used as a criterion of the goodness of model simulation. This criterion allowed a constant model-to-prototype offset to occur in case that a washload or background concentration existed. A background concentration was used in the previous modeling of this system (Luettich et al., 1990). If variation of a given coefficient produced a range of variances such that the ratio of the maximum variance to the minimum exceeded a threshold (typically 1.1 to 1.2), then the optimum value of the coefficient was determined for this particular set of other coefficients. After all unknown coefficients were tested in this manner, a new set of model coefficients was developed by applying a weighting to the difference between the original and optimum values and adding those weighted differences to the old coefficient values. Typical weighting values used were 0.33 to 0.5. This procedure was repeated 20 times for each model.

3. RESULTS AND DISCUSSION

Plots of example numerical C_f/C_o curves are shown in Figure 2b-c for a single grain model (*ns*=1) and a multiple grain model (*ns*=4), respectively. Multiple grain classes allow representation of the C_f/C_o curves in a step-wise fashion. The more classes, the better the representation. At times shorter than that required to reach steady state, multiple grain class model results have smoother transitions between different shear stress levels. Figure 3a and 3b shows model results with *ns* =7 after 0.1 and 1 hours for various shear stress levels. These

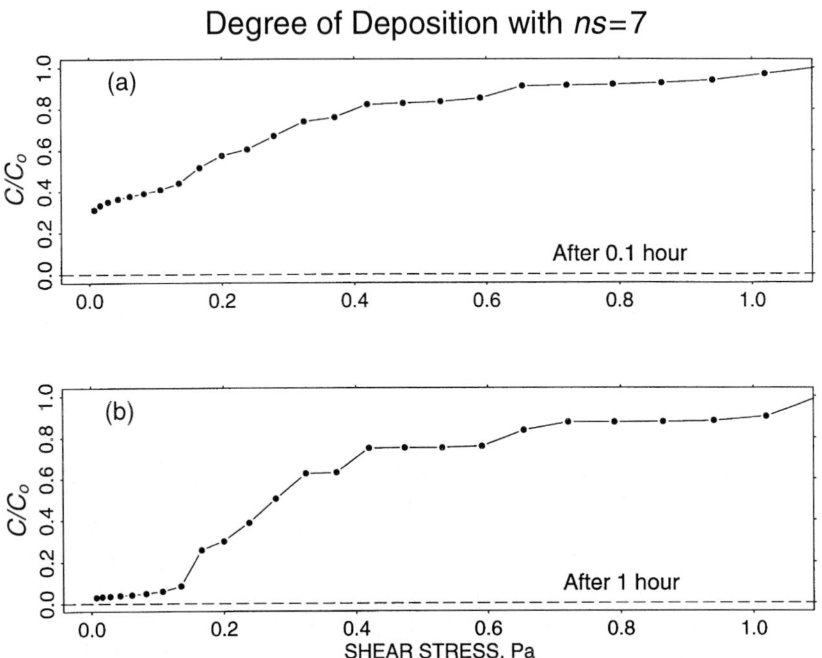

Figure 3. Model relative TSM concentration curves for *ns* = 1 for (a) 0.1 hour, and (b) 1 hour of deposition.

results indicate that, as suspensions approach steady state values, C/C_o curves take on the stair-step nature as shown in Figure 2, with the finer, slower settling grain classes requiring more time to come to steady state. Figure 3b can be compared to the observed C_f/C_o curve shown in Figure 2a. Similarly good results were obtained by a multiple grain class analytic deposition model used to simulate the same experimental data set (Mehta and Lott, 1987). In the present model, grain class coupling is expected to improve the ability to simulate sediment sorting.

Resuspension in Lake Balaton was modeled with 1 and 4 grain classes. Wave and current boundary data are shown in Figure 4. Table 1 summarizes the model to interpolated prototype TSM differences after 10 and 20 iterations of coefficient adjustment. Coefficients converged quickly during the first few iterations of the adjustment process to give results similar to those above at 10 iterations, and improved only slowly thereafter. During coefficient adjustment, the multiple grain model converged more uniformly, while the single grain model oscillated to a greater extent about the minimum variance.

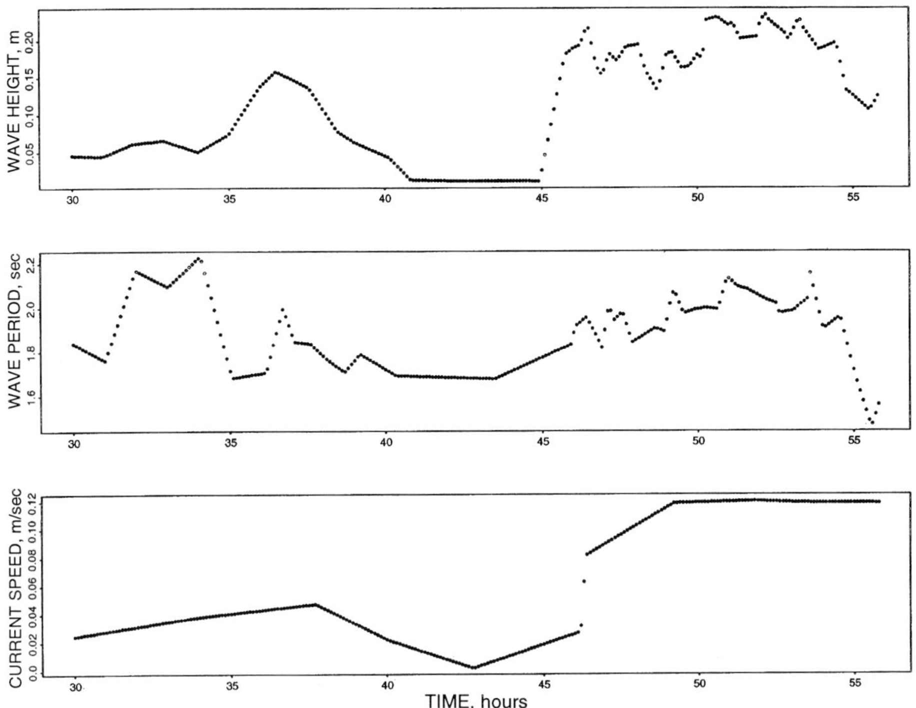

Figure 4. Model boundary conditions for wave height, wave period, and current speed. Data interpolated from Luettich et al. (1990).

Table 1
Summary of model to interpolated prototype TSM differences

		Prototype-Model TSM, g/m^3		
Iterations	*ns*	Mean	95% CI	
10	1	-7.97	-10.33	-5.61
10	4	1.36	-0.50	3.21
20	1	-9.37	-11.40	-7.29
20	4	0.29	-1.20	1.77

The coefficient adjustment process resulted in the model coefficients after 20 iterations as shown in Table 2. Model coefficients a_1, $n(gs)$, C_{ll}, and C_{ul} are defined in Part I (this volume). As indicated in the table, model coefficients were similar for the two models. While 8 bed layers were available in each model, only the top 3 layers were active in the simulations. Bed layer thicknesses were only 2-3 mm for both models indicating that erodibility decreased rapidly with depth in the bed, as has been observed in the laboratory (Dixit, 1982).

Table 2
Model coefficients

Coefficient	Model with ns = 1	Model with ns = 4
Initial $C(gs)$, g/m^3	29.8	28.2, 3.5, 1.0, 0.5
Manning's n	0.024	0.022
a_1, m/sec	0.00189	0.00175
$n(gs)$	1.0896	1.138,0.885,0.7,0.5
C_{ll}, g/m^3	24	68
C_{ul}, g/m^3	923	875
τ_{cd} , Pa	0.025	0.0375
$\tau_{ce}(gs{=}1, bl{=}1)$, Pa	0.0381	0.0375
$\tau_{ce}(gs{=}1, bl{=}2)$, Pa	0.0915	0.0873
$\tau_{ce}(gs{=}1, bl{=}3)$, Pa	0.104	0.1564
$M(gs{=}1, bl{=}1)$, g/m^2/sec	0.1421	0.1278
$M(gs{=}1, bl{=}2)$, g/m^2/sec	0.1055	0.0939
$M(gs{=}1, bl{=}3)$, g/m^2/sec	0.0984	0.0631
$\tau_{ce}(gs{>}1)$, Pa		0.121, 0.226, 0.268
$S(1 \leq gs \leq ns, bl{=}1)$, k/m^2	0.0697	0.1444
$S(1 \leq gs \leq ns, bl{=}2)$, k/m^2	0.0871	0.1818
$S(1 \leq gs \leq ns, bl{=}3)$, k/m^2	0.0879	0.1891

Plots of model predictions using final coefficients for ns=1 and ns=4 are shown in Figure 5 along with observed TSM. The single grain model had difficulty with the depositional phases of the time series, especially after hour 54, when shear stresses were higher than the threshold for deposition. If the prototype data set had extended through the end of the storm, the single grain model comparisons would have probably been poorer than those presented. The resuspension phase of the data was also better represented in the model with ns=4.

The results of the multiple grain model with ns=4 appear to be better than the single grain model results reported by Luettich et al. (1990). His model adjustment resulted in a root-mean-square error of 8.7 g/m^3 over the last 10.2 hours of the data set used here.

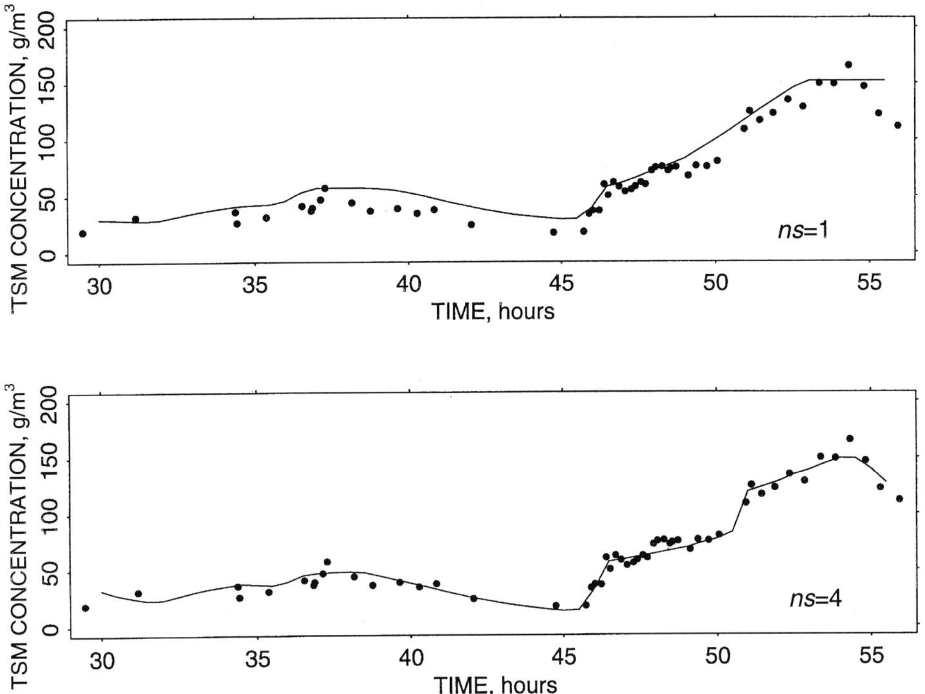

Figure 5. Model results compared to field TSM data for a single grain class (top), and for 4 grain classes. Data from Luettich et al. (1990).

4. CONCLUSIONS

The multiple grain class formulation improved model prediction of resuspension and deposition over the single grain formulation. Some of the deficiencies of the single grain formulation can be overcome by assuming simultaneous erosion and deposition (Sanford and Halka, 1993; Luettich et al., 1990). However, as Lau and Krishnappan (1994) conclude, fine-grained sediment models that assume simultaneous erosion and deposition may not have the correct physical representation. The conclusion of this study is that the multiple grain class formulation can accurately simulate suspension concentrations during resuspension and deposition, although nothing new can be added to the debate over simultaneous erosion and deposition. The multiple grain class formulation appears to be a good alternative to a single grain formulation that assumes simultaneous erosion and deposition. The added physical realism of the multiple grain class formulation allows sediment texture to adjust to hydraulic conditions, but comes at the price of increased computational burden for operating the model and increased numbers of sediment parameters requiring estimation. The suitability of the use of such a formulation is probably site or case specific, depending on the study objectives, availability of laboratory and field data. For sites with high shear stresses and/or significant quantities of silt and for which system changes in sedimentation patterns are to be evaluated, a multiple grain model formulation may be warranted.

REFERENCES

Arfi, R., Guiral, D., and Bouvy, M., 1993. Wind induced resuspension in a shallow tropical lagoon. *Estuarine, Coastal and Shelf Science*, 36, 587-604.

Ariathurai, R., MacArthur, R. C., and Krone, R. B., 1977. Mathematical model of estuarine sediment transport. *Technical Report D-77-12*, U.S. Army Engineer Waterways Experiment Station, Vicksburg, MS, 79p + appendices.

Dixit, J. G., 1982. Resuspension potential of deposited kaolinite beds. *M.S. Thesis*, University of Florida, Gainesville, FL, 167p.

Fischer, H. B., 1973. Longitudinal dispersion and turbulent mixing in open-channel flow. *Annual Review in Fluid Mechanics*, 5, 59-78.

Gibbs, R. J., 1977. Clay mineral segregation in the marine environment. *Journal of Sedimentary Petrology*, 47(1), 237-243.

Hamilton, D. P., and Mitchell, S. F., 1996. An empirical model for sediment resuspension in shallow lakes. *Hydrobiologia,* 317, 209-220.

Hawley, N., and Lesht, B. M., 1992. Sediment resuspension in Lake St. Clair. *Limnology and Oceanography*, 37(8), 1720-1737.

Kandiah, A., 1974. Fundamental aspects of surface erosion of cohesive soils. *Ph.D. Thesis*, University of California, Davis, CA, 261p.

Kranck, K., 1980. Experiments on the significance of flocculation in the settling of fine-grained sediment in still water. *Canadian Journal of Earth Science*, 17, 1517-1526.

Krone, R. B., 1962. Flume studies of the transport of sediment in estuarial shoaling processes. *Final Report*, Hydraulic Engineering Laboratory and Sanitary Engineering Research Laboratory, University of California, Berkeley, CA, 118p.

Lau, Y. L., and Krishnappan, B. G., 1994. Does reentrainment occur during cohesive sediment settling? *Journal of Hydraulic Engineering,* 120(2), 236-244.

Lee, S.-C., and Mehta, A. J., 1994. Cohesive sediment erosion. *Contract Report DRP-94-6*, U.S. Army Engineer Waterways Experiment Station, Vicksburg, MS, 82p.

Lick, W., Lick, J., and Ziegler, C. K., 1994. The resuspension and transport of fine-grained sediments in Lake Erie. *Journal of Great Lakes Research*, 20(4), 599-612.

Luettich, R. A., Harleman, D. R. F., and Somlyody, L., 1990. Dynamic behavior of suspended sediment concentrations in a shallow lake perturbed by episodic wind events. *Limnology and Oceanography*, 35(5), 1050-1067.

McLaren, P., and Bowles, D., 1985. The effects of sediment transport on grain-size distributions. *Journal of Sedimentary Petrology*, 55(4), 457-470.

Mehta, A. J., and Partheniades, E., 1975. An investigation of the depositional properties of flocculated fine sediments. *Journal of Hydraulic Research*, 13(4), 1037-1057.

Mehta, A. J., and Lott, J. W., 1987. Sorting of fine sediment during deposition. *Proceedings of Coastal Sediments '87*, N. C. Kraus ed., ASCE, New York, 348-362.

Partheniades, E., Cross, R. H., and Ayora, A., 1968. Further results on the deposition of cohesive sediments. *Proceeding of the 11th Conference on Coastal Engineering*, ASCE, New York, 2, 723-742.

Pejrup, M., 1986. Parameters affecting fine-grained suspended sediment concentrations in a shallow micro-tidal estuary, Ho Burt, Denmark. *Estuarine, Coastal and Shelf Science*, 22, 241-254.

Petticrew, E. L, and Kalff, J., 1991. Predictions of surficial sediment composition in littoral zone of lakes. *Limnology and Oceanography*, 36(2), 384-392.

Sanford, L. P., and Halka, J. P., 1993. Assessing the paradigm of mutually exclusive erosion and deposition of mud, with examples from upper Chesapeake Bay. *Marine Geology*, 114(1-2), 37-57.

Stevens, R. L., 1991a. Grain-size distribution of quartz and feldspar extracts and implications for flocculation processes. *Geo-Marine Letters*, 11, 162-165.

Stevens, R. L., 1991b. Triangle plots and textural nomenclature for muddy sediments. *Geo-Marine Letters*, 11, 166-169.

Teeter, A. M., 1986. Vertical transport in fine-grained suspension and newly-deposited sediment. In: *Estuarine Cohesive Sediment Dynamics*, A. J. Mehta ed., Springer-Verlag, New York, 170-191.

Teeter, A.M., 1987. Alcatraz disposal site investigation; Report 3: San Francisco Bay-Alcatraz disposal site erodibility. *Miscellaneous Paper HL-86-1*, U.S. Army Engineer Waterways Experiment Station, Vicksburg, MS, 120p.

Teeter, A. M., 1993. Suspended transport and sediment-size transport effects in a well-mixed, meso-tidal estuary. In: *Nearshore and Estuarine Cohesive Sediment Transport*, A. J Mehta ed., American Geophysical Union, Washington, DC, 411-429.

Teeter, A. M., and Pankow, W., 1989. Deposition and erosion testing on the composite dredged material sediment sample from New Bedford Harbor, Massachusetts. *Technical Report HL-89-11*, U.S. Army Engineer Waterways Experiment Station, Vicksburg, MS, 61p.

Teeter, A. M., Parchure, T. M., and McAnally, W. H. Jr., 1997. Size-dependent erosion of two silty-clay sediment mixtures. In: *Cohesive Sediments*, N. Burt, R. Parker, and J. Watts eds., John Wiley, Chichester, UK, 253-262.

Coastal and Estuarine Fine Sediment Processes
W.H. McAnally and A.J. Mehta (Editors)

Analysis of nearshore cohesive sediment depositional process using fractals

LI Yan and XIA Xiaoming

Second Institute of Oceanography, SOA, P.O. Box 1207, Hangzhou 310012, Peoples Republic of China

Records of the depositional process and the depositional sequence from the Jiaojiang estuary are analyzed in terms of the fractal theory. The time scale, ranging from several days to about 100 years, correlates with sediment cores of a few centimeters to about two meters. It is shown that scale invariance appears both in the time series and in the depositional sequence linked with similar fractal dimension and initiator. Accordingly, the total thickness $T(N)$ for a certain number N of interlayered sand/mud bedding in any fraction of the depositional sequence is linked to its initiator $T(1)$ related to the square mean thickness for every individual interlayered sand/mud bedding by the power function: $T(N) = T(1)\ N^{2-D}$, with its exponent expressed in terms of the fractal dimension D. This dimension, derived from the muddy layer sequence, ranged from 1.2 to 1.5. Following the power law, the square mean fluctuation range of the sedimentary series was calculated for different time scales assuming that the individual interlayered sand/mud bedding is generated primarily by the spring/neap tidal cycle in tide- and turbidity-dominated estuaries. As shown by the fractal dimension and the initiator derived from the muddy layer sequence of the Jiaojiang, the square mean fluctuation range was 8 to 28 cm for a one-year return event, 29 to 167 cm for a ten-year return event and 72 to 575 cm for a fifty-year return event. The fractal property is a useful tool to assess the potential fluctuation range of erosion and deposition events characterized by different return cycles.

1. INTRODUCTION

The cohesive sediment depositional process in turbid nearshore waters is characterized by alternate erosion and accumulation events over a range of temporal and spectral scales, thus contaminating the depositional layering sequence (Schwarzachar, 1975). As with many geologic variables, the fluctuating process series consists of at least two components, namely, the cyclic and the random. The cyclic process can be predicted by simple models within a cycle or by harmonic analyses for a series of cycles, while the random component is very difficult to predict. Traditional sedimentology is built on simple models and cyclic models. Sediment dynamics is essential to describe the depositional processes in such models. However, it is difficult to predict the long-term variables in the active environment when the processes are dominated by the random components.

The theory of fractals, presented by Mandelbrot in 1970s, recognizes the self-similarity of some figures, structures and processes without characteristic feature lengths. The exponent of the fractal dimension D reflects, in some cases, the underlying physical characteristics. Based on the analysis of long-term stratigraphic data from the Aswan dam (Egypt) in 1950s, the theory of fractals has been applied to a variety of geophysical processes, and it has been generally found that $D \leq 1.5$ (Mandelbrot and Wallis, 1995). Moreover, it is shown that riverine, oceanic and atmosphere dynamical processes which govern modern sedimentation and sea floor topography tend to be self-similar (Feder, 1988; Mandelbrot and Wallis, 1995; Malinverno, 1995; Shabalova and Konnen, 1995). In view of these observations, the depositional process series and the resulting depositional sequence series may also be expected to be self-similar. By identifying the interrelation between the two series in fractal terms, one may be able to make use of a valuable tool to describe the random component of the depositional process.

The Jiaojiang estuary, the experimental site (Figure 1), is one of the muddy estuaries to the south of the Changjiang estuary, and is well-known for a significant variation of runoff, strong tidal currents, high turbidity and frequent influences of storm tides (Li et al., 1993). The inner reach of this macrotidal estuary starts from the confluence of its tributaries, and extends about 14 km as a wide and shallow course to the rocky Niutou Inlet. Its offshore end, a funnel-shaped estuarine mouth called Taizhou Bay, extends at least 10 km seaward from the inlet. Together with estuarine fronts and tidal asymmetry, a zone of turbidity maximum has developed over the main reach of the inner estuary to the head of Taizhou Bay. Suspended sediment concentrations of ~ 20 kg/m^3 are reported frequently in the inner estuary. Large amount of cohesive sediment from the sea is accumulated here and a depositional center has formed, while the coarser bed-load sediment carried by fresh water is mostly distributed over the central sandy bars and troughs upstream (Figure 1).

2. METHOD

2.1. Field survey

Beginning in 1991, the authors spent about five years studying the turbid water dynamics of the Jiaojiang by way of a series of field surveys using a time series monitoring system. Two depositional process series within the turbidity maximum were obtained (Table 1). First, the fluctuation of the lutocline depth, with a time-scale of 1 to 1,000 s, was measured by an acoustic back-scatter sensor (ABS) at a 0.6 s sampling interval and a vertical resolution of 5 cm. Second, suspended sediment concentration (SSC) at 17-cm height above the seabed, with

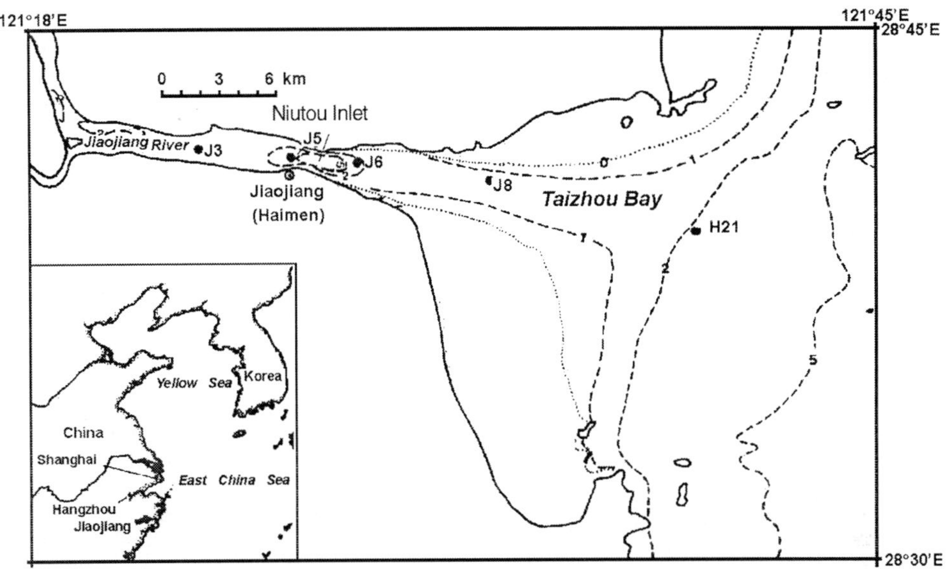

Figure 1. Location map of Jiaojiang estuary with core stations. Water depths are in meters relative to low water.

Table 1
Parameters for the time-dependent processes in the Jiaojiang estuary

Source	Instrument	Time scale t	Fractal dimension D
Lutocline (Figure 3)	ABS	1-1,000 s	1.4 - 1.5
SSC (Figure 4)	OBS	0.1-100 h	1.4 - 1.5

a time scale of 0.1 to 100 h, was recorded by an optic back-scatter sensor (OBS) at a 5 min sampling interval.

The depositional sequence series was derived from core samples J3 to J8 collected by a gravity corer on April 13, 1991. All coring stations were located within the main course of the estuary (Figure 1). Among them J3, in 2 m of water, and J5, in 3 m of water, were situated in the inner reach. J6, with a water depth of 2.5 m, was located near Niutou Inlet; and J8 was situated at the shoal head of Taizhou Bay in a water depth of 1.5 m. In addition, sediment sample H21 was cored and analyzed previously in 1982. In the laboratory, Pb^{210} activity was determined in totally digested sediment samples by measuring the activity of its derivative, Po^{210}, in an alpha spectrometer. The undisturbed cores were photographed when the water content was suitable, to visually distinguish between the silt and mud layers. Using a scanner, the sequences for the thicknesses of the mud, silt and sand layers were retrieved from core bedsets.

2.2. Fractal principle

In the context of the fractal theory, it was found that scale invariance occurred in both time series, which can be linked by similar fractal dimension and initiator. Hence, the fluctuating range $T(t)$ for a certain period t in any fraction of the sequence can be shown to be linked to its initiator $T(1)$ by a power function with an exponent expressed in terms of the fractal dimension D (Figure 2):

$$T(t) = T(1)\, t^{2-D} \tag{1}$$

This means that a complex depositional process can be represented in the frequency domain by only two parameters, the fractal dimension D and the initiator $T(1)$. Hence, if the depositional process conforms to the fractal theory, the entire series can be reconstructed from two periods of measurement, corresponding to different time scales.

The initiator $T(1)$ reflects the potential range of fluctuation corresponding to a certain period. The fractal dimension D reflects the contribution of the components mentioned. A simple model dominates the process when D is just above 1.0, while the random model dominates the process when D is around 1.5. However, the dominant fractal dimension D is larger than 1.5 when many cyclic components are involved within the time scale considered (Feder, 1988).

The Scale invariance appears both in the time series and the depositional sequence linked by a similar fractal dimension and initiator. Therefore, the total thickness $T(N)$ for a certain number N of inter-layered sand/mud bedding layers in any fraction of the depositional sequence is linked with its initiator [$T(1)$,

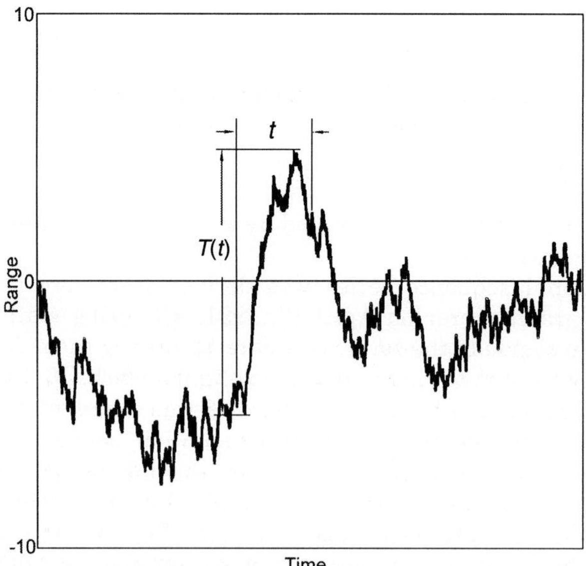

Figure 2. Depositional process from the fractal perspective.

related to the square mean thickness for every individual, interlayered sand/mud bedding] by D:

$$T(N) = T(1)\, N^{2-D} \tag{2}$$

2.3. Fractal parameters

The fractal parameters can be calculated by the power spectrum method when the series samples exceed 512 (Feder, 1988) using the formula: $D=(5-\beta)/2$, in which β is the power of the regression function of the power spectrum distribution index.

When the series contains fewer than 512 samples, the fractal parameters can be calculated by the structure function method (Feder, 1988; Shabalova and Konnen, 1995). The procedure is as follows:

(1) determine the minimum time scale;
(2) calculate successively the corresponding $T(t)$ of the entire series for the successive time scales (t=1,2,3.....);
(3) calculate successively the root mean square of $T(t)$, i.e., the structure function

$$S(t) = \overline{\left\{\left[T(t)^2\right]\right\}}^{1/2} \tag{3}$$

(4) obtain best-fit structure function values corresponding to the entire t using the formula:

$$S(t) = T(1)\, t^{2-D} \tag{4}$$

Similarly, the fractal parameters of the depositional sequence can be calculated from:

$$S(N) = T(1)\, N^{2-D} \tag{5}$$

3. RESULTS

3.1. Depositional process and fractal dimension

Figure 3 shows lutocline fluctuation driven by interfacial waves, as recorded by the ABS in November, 1995. Below the lutocline a mobile fluid mud occurred with SSC of 10 to 40 kg/m^3. The lutocline can be regarded as a depositional interface with the lowest wet density of the deposit. The fractal dimension D of its time series was 1.4 to 1.5, as calculated by the power spectrum method.

Figure 4 shows the near bottom SSC measured by the OBS. The maximum and the minimum SSC appeared alternately along with the lutocline rise and fall, which indirectly reflects depositional interface fluctuation. The D value of this process time series was also found to be 1.4 to 1.5.

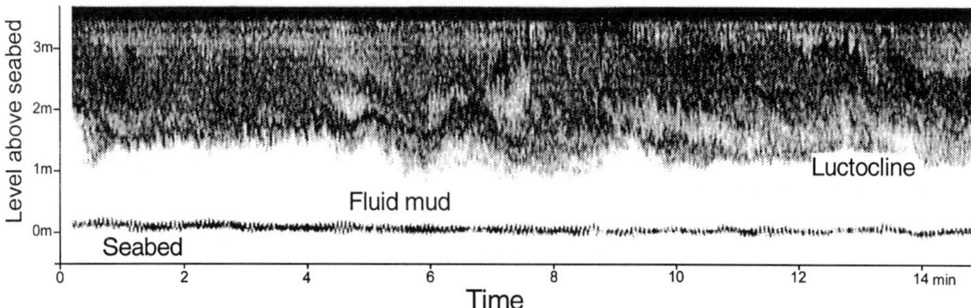

Figure 3. Lutocline fluctuation in the turbidity maximum zone of the Jiaojiang.

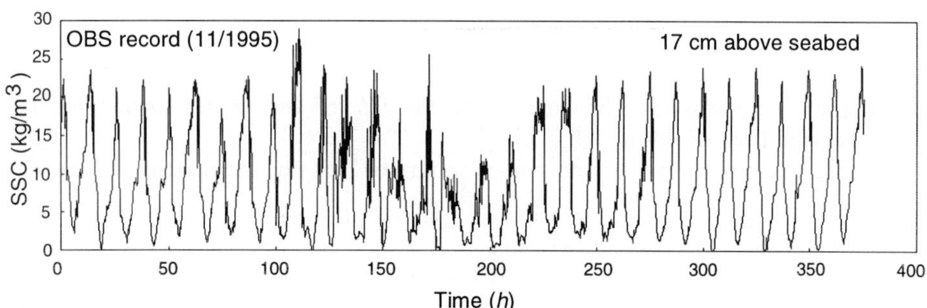

Figure 4. Near-bottom fluctuations of the SSC in the turbidity maximum zone of the Jiaojiang.

Table 1 shows the fractal dimensions for the above fluctuating processes in the Jiaojiang. As the fractal dimension D indicates, random components are dominant along the fluctuating lutocline. In addition, it was found that the sea bed processes on the tidal flat along coast north of the Jiaojiang were also characterized by a combination of simple and random components, or of cyclic and random components (Li et al., 1997). Beds with $D < 1.4$ were located in low tidal flats, and beds dominated by $D > 1.5$ were found over high tidal flats. This means that the random component decreased seaward in relation to the sedimentary processes of this tidal flat, while the simple component increased.

3.2. Depositional sequence and fractal dimension

Figure 5 shows the vertical 2D profile of excess Pb^{210} in the turbidity maximum. In addition to the lower activity events, the profile is also characterized by several obvious stages. There is an important change in the gradient of the Pb^{210} profile at depths of 25 to 40 cm in every core. The sedimentary rates in the upper and lower sections are differentiated in terms of this behavior. The data suggest that most of the recent sedimentary sequences in the turbidity maximum are made up of a fast depositional stage, followed by a

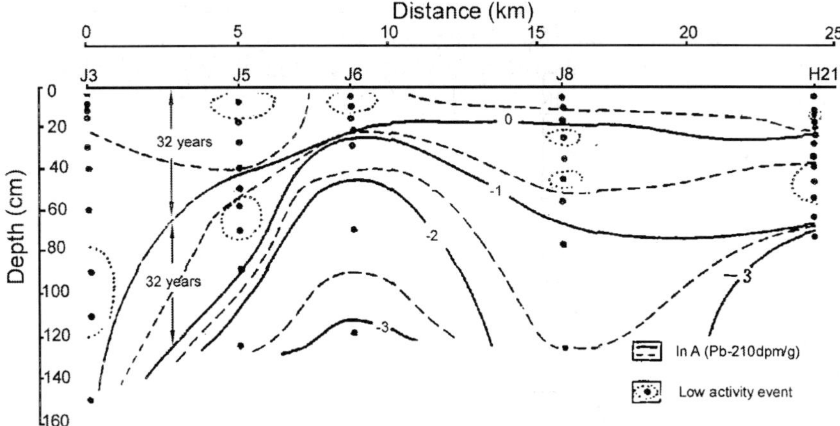

Figure 5. Pb^{210} distribution in sediment in the turbidity maximum zone of the Jiaojiang.

slow stage. Mean sedimentary rates for both stages range from 0.9 to 6.3 cm annually, with two long-term deposition-dominated regions around stations J3 and J8, and a tendency to dramatically decrease seaward (Table 2). However, at every station, the deposition process was not synchronous. Variable, short-term depositional stages were found at station J6 at the beginning of the century, J8 in the 1930s and 1940s, J5 in the 1950s, and J3 in the 1960s and 1970s. This variability is an indicator of the complexity of the depositional process within the turbidity maximum (Li et al., 1992).

Table 2
Parameters related to the depositional sequence in the cores

	Station No.				
	J3	J5	J6	J8	H21
Water depth (m)	5	5	8	2	3
Coring depth (cm)	150	121	113	122	71
Total number of layers	137	136	122	73	50
Depositional rate (cm/a)	6.3	1.3	1.2	1.9	0.9
Duration (a)	23	93	93	75	88
Initiator (cm, for a period of 15d)	0.91	0.85	0.96	1.64	1.62
Fractal dimension	1.33	1.43	1.38	1.23	1.29
Predicted potential fluctuation range (cm)					
1 year return event	11	8	10	28	23
10 years return event	53	29	43	167	118
50 years return event	157	72	116	575	371

Sediment in the turbidity maximum of the Jiaojiang is dominated by clayey silt with a median diameter of 6.5 to 8.0 phi, and with a phi sorting coefficient of 1.5 to 1.8. The beds are characterized by interlayering of clayey silt, silt and fine sand. Based on the thicknesses and the ordering of the layers, and taking the eroded beds or coarser sand layers as their surfaces, four types of bedsets could be identified (Figure 6): thinly interlayed silt/mud bedding (Bedset A), interlayered silt/mud bedding with a thick mud layer (Bedset B), coarsely interlayered sand/silt bedding (Bedset C) and coarsely interlayered sand/mud bedding (Bedset D). The four bedsets represent slow deposition, rapid deposition, abnormally high turbid deposition, and alternate suspended load and bed-load deposition. The recent depositional process, as reflected by the sequence structure, was almost identical with that inferred from excess Pb^{210} (Li et al., 1992).

Figure 7 shows the sequence for mud layering in cores J3, J5, J6, J8 and H21. Their initiators and fractal dimensions D, which are calculated using (5) and related sedimentation parameters, are listed in Table 2. The initiator is equal to the average layer thickness and can be regarded as the basic depositional unit. The dimension D reflects the variance of maximum range of different layer scales. When D is equal to 1.5, the mud layer sequence corresponds to a random distribution. When D is between 1.5 and 2, the mud layer sequence includes more cyclic components. The sequence may include more linear components when D is between 1.0 and 1.5.

The data indicate that the fractal dimension derived from the muddy layer sequence were in a range of 1.2 to 1.5, with a maximum located at J5 and decreasing both eastward and westward from that station.

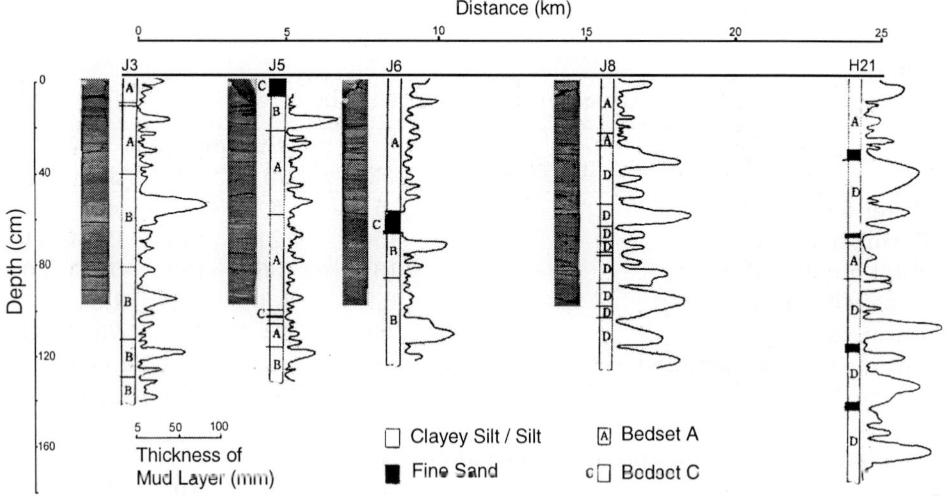

Figure 6. Sequence of sedimentary structures and mud laminae size profiles from cores in the turbidity maximum zone of the Jiaojiang.

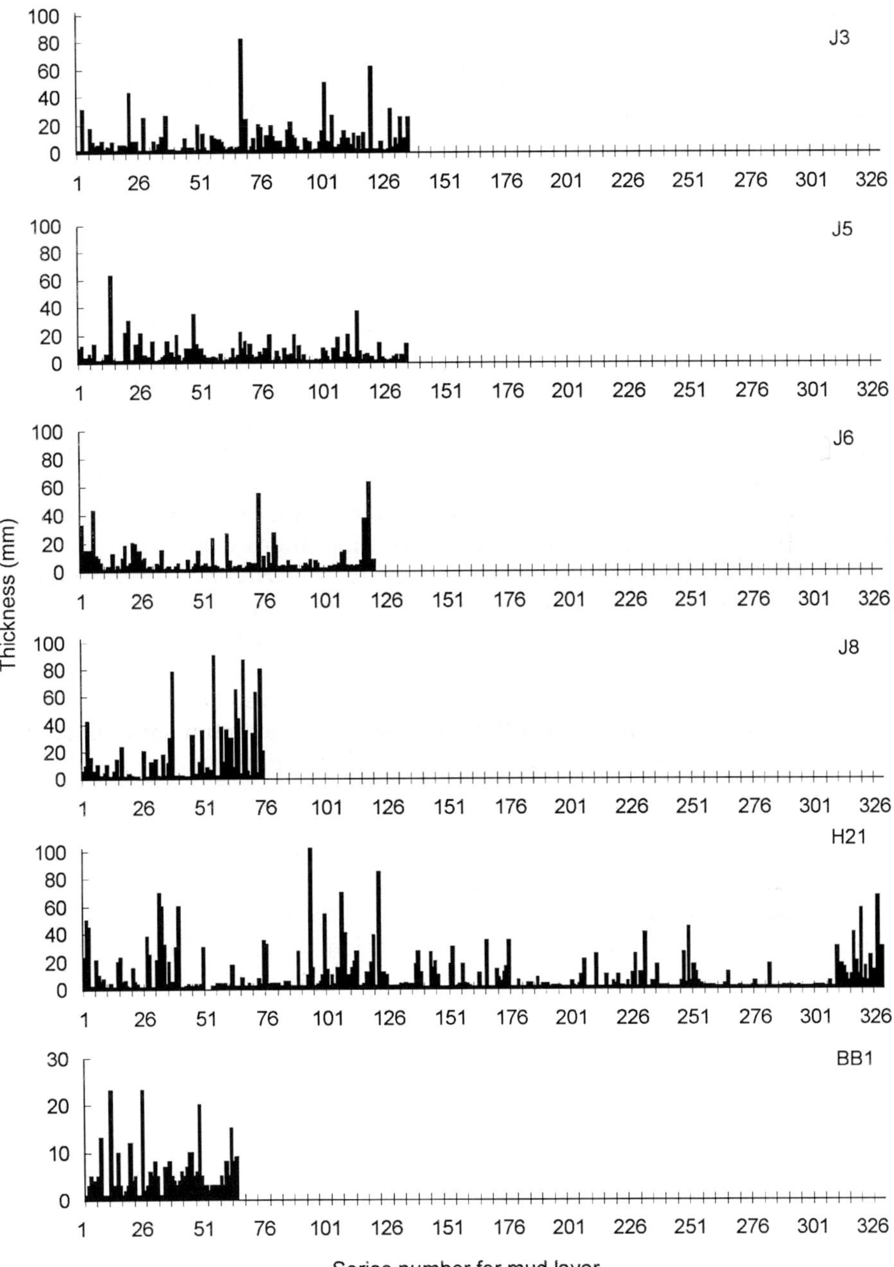

Figure 7. Vertical distribution series of the mud layer thickness of cores collected in the Jiaojiang.

4. DISCUSSION

The cyclic- and simple-component model, based on short-term depositional data, cannot predict the long-term depositional process, especially when the random component dominates. It is usually difficult to obtain sufficiently long-term measurements of deposition in the field. Traditional sedimentary analyses such as qualitative sequence descriptions or Pb^{210} sedimentation rates from seabed cores also cannot do so. A simple but quantitative model is needed which considers not only simple or cyclic depositional process but also fractal effects, including deposition and erosion events. The depositional sequence record overshadows some information which can be revealed by fractal analysis.

4.1. Depositional process and depositional sequence

In the Jiaojiang estuary, the basic depositional units are mud (silty clay or clayey silt) layers and silt layers. Rhythmic interlayering, or tidal bedding, can be caused by the cyclic effect of flood and ebb or spring and neap (Reineck and Wunderlich, 1967; Li et al., 1987). In Figure 4, the SSC behavior seems to support the role of spring-neap tide variation in governing interlaying.

The fractal dimensions characterizing the depositional process and the depositional sequence are almost identical. Moreover, in depositional sequences conforming to self-similarity, there is complete transfer of the fractal dimension from the whole to its parts. Undoubtedly, in the estuary the dynamic processes and the seabed topography should also have the same fractal features as the depositional process and sequence. Hence, one can understand the complexity of the depositional environment and the seabed topography only by analyzing the sequence records of some cores from the fractal viewpoint.

4.2 Potential fluctuation range

The ratio of the thickness of the mud lamina to the silt lamina is about 2:1, as determined from statistical analysis of the core samples (Figure 6), and the basic units forms over ~15 days. Under the condition of linear transfer of the basic depositional units and the fractal dimension, the potential fluctuation ranges are estimated by (1) (Table 2).

At every station, the potential fluctuation range was 8 to 28 cm for a one-year return event, 29 to 167 cm for a ten-year return event and 72 to 575 cm for a fifty-year return event. The estimated yearly mean fluctuation range at stations J3, J5 and J6 approach the sedimentation rates from Pb^{210} separately. However, at J8 and H21, the calculated fluctuation range seems to be too high to be acceptable. The lower fractal dimension may explain this deviation. The two stations are situated over an open and shallow shoal where strong flow and sediment dynamics resulted in an abnormal depositional process being recorded in the silt and sand layers. Thus, the fractal dimension derived from the mud layers is on the low side. Considering mud, silt and sand records collectively may reduce the deviation.

5. CONCLUSIONS

In a muddy estuary, the depositional process series and the depositional sequence appear to be self-similar. Between them, transfer of the initiator and the fractal dimension exist with scale invariance. In every core sample, the fractal dimension from the mud layer series varied between 1.2 to 1.5, which reflects different depositional environments. The higher values show a more complicated depositional process and sequence, which corresponds with the results from the excess Pb^{210} profile and qualitative analysis of the sequence bedsets. Moreover, from the fractal principle, we have estimated the potential fluctuation range at every core station to be 8 to 28 cm for a one-year return event, 29 to 167 cm for a ten-year return event and 72 to 575 cm for a fifty-year return event. The fractal properties lend themselves as a valuable tool to differentiate erosion/deposition events having different return periods.

6. ACKNOWLEGMENT

This study was carried out with support from the Zhejiang Province Natural Science Foundation (grant no. 494007), the Foundation of the State Oceanic Administration and the Croucher Foundation. We also thank Dr. Eric Wolanski, Prof. Xie Qinchun and Prof. Su Jilan for their help in organizing the joint studies in the Jiaojiang estuary. Most raw data used in this paper were collected from the cruises made under those studies.

REFERENCES

Feder, J., 1988. *Fractals*. Plenum Press, New York, 283p.

Li, Y., Xie, Q., and Zhang, L., 1987. Cycle of the tidal flat at Damutu, Zhejiang,China. *Acta Oceanologica Sinica*, 9(6), 725-734.

Li, Y., Pan, S., Shi, X., and Li, B., 1992. Recent sedimentary rates for the zone of the turbidity maximum in the Jiaojiang Estuary. *Journal of Nanjing University (Natural Sciences Edition)*, 28(4), 623-632.

Li, Y., Wolanski, E., and Xie, Q, 1993. Coagulation and settling of suspended sediment in the Jiaojiang river estuary, China. *Journal of Coastal Research*, 9(2), 390-402.

Li, Y., Chen, X., and Li., B., 1997. Analysis on fluctuation range of coastal sedimentary process, In: *Proceedings of the Symposium on Coastal Ocean Resources and Environment (SCORE'97)*, HKUST School of Science & CAAR, Hong Kong, 206-216.

Malinverno, A., 1995. Fractals and ocean floor topography: a review and a model. In: *Fractals in the Earth Sciences*, C. C. Barton and P. R. LaPointe eds., Plenum Press, New York, 107-130.

Mandelbrot, B. B., and Wallis, J. R., 1995, Some-long-run properties of geophysical records. In: *Fractals in the Earth Sciences*, C. C. Barton and P. R. LaPointe eds., Plenum Press, New York and London, 41-64.

Reineck, H. E., and Wunderlich, F., 1967 Aeitmessungen an gezeitenschichten. *Natur und Museum*, 97(6), 193-197.

Shabalova, M., and Konnen, G. P., 1995. Scale invariance in long-term time series. In: *Fractal Reviews in the Natural and Applied Sciences*, M. M. Novak ed., Chapman & Hall, London, 309-410.

Schwarzacher, W., 1975. *Sedimentation Models and Quantitative Stratigraphy.* Elsevier Scientific Publishing, Amsterdam-Oxford-New York. 232p.

Coastal and Estuarine Fine Sediment Processes
W.H. McAnally and A.J. Mehta (Editors)

Laboratory experiments on consolidation and strength of bottom mud

L. M. Merckelbach[a], G. C. Sills[b] and C. Kranenburg[a]

[a]Delft University of Technology, Faculty of Civil Engineering, P.O. Box 5048, 2600 GA Delft, The Netherlands

[b]University of Oxford, Department of Engineering Science, Parks Road, Oxford OX1 3PJ, United Kingdom

Three settling column experiments were carried out on the Caland-Beer Channel mud to identify constitutive relationships for self-weight consolidation modeling. We measured densities using an X-ray apparatus and pore water pressures using a single transducer for multiple ports, and we performed shear vane tests on the three columns at three different times. Effective stress and permeability were calculated from the density profiles and pore water pressure profiles. Peak shear stresses were obtained from the torque data. We interpreted both effective stress and permeability data using the concept of fractal structure, which implies that the aggregates that build the bed are self-similar. According to this concept, effective stress and permeability relate to the volume fraction of solids through the fractal dimension. We found a fractal dimension of 2.71 ± 0.05 for all experiments. Peak shear stresses were evaluated in terms of effective stress, yielding a remarkably linear relationship, with positive shear strength at zero effective stress (true cohesion). This true cohesion seemed to increase with time, indicating thixotropic effect.

1. INTRODUCTION

Many ports suffer from high siltation rates in their basins and entrance channels. To guarantee safe shipping, port authorities are required to maintain the navigable depth by having large amounts of mud dredged at substantial costs. A better understanding of relevant processes such as: 1) transport towards harbor basins and entrance channels of fluid mud layers formed by deposition from sediment suspensions, and 2) consolidation of these fluid mud layers, may eventually lead to more economic dredging strategies.

Essential for all these processes is the behavior of (fluid) mud layers with respect to self-weight consolidation and the development of strength, as the strength influences navigability and resistance against erosion. The behavior of a fluid mud layer *in situ* may be predicted by a mathematical model that uses constitutive relationships obtained from laboratory experiments. Settling column experiments are very useful in this respect and have been reported by a number of authors (Berlamont et al., 1992; Migniot and Hamm, 1990; Sills, 1997; Been and Sills, 1981).

In engineering practice, experimental effective stress data are usually interpreted using Terzaghi's effective stress principle (Terzaghi, 1936). This principle is twofold: 1) the effective stress is defined as the difference between the total stress and the pore water pressure, and 2) strain in soil is uniquely related to change in effective stress. For very soft soils, however, as discussed by Sills (1995), there is experimental evidence that time or rate-dependent processes exist that are not associated with changes in effective stress.

Among the reported settling column experiments that were aimed at determining bed strength, McDermott (1992) measured shear stiffness with bender elements, while van Kessel (1997) measured undrained shear strength with a small-scale sounding test. Others have measured undrained shear strength by using a sensitive vane (Elder, 1985; Bowden, 1988). These tests indicate that a relationship exists between shear strength and effective stress, and possibly time. In the present work we describe laboratory experiments on consolidation and strength evolution of a natural mud. First, the experimental set-up and measurement techniques used are discussed. Then, effective stress and permeability data are analyzed using the concept of fractal structure applied to cohesive sediment aggregates by Kranenburg (1994). Finally, peak shear stresses are evaluated in terms of effective stress.

2. EXPERIMENTAL SET-UP

The sediment used for our settling column experiments was Caland-Beer Channel mud dredged from the port of Rotterdam, The Netherlands. This silty clay had a particle size grading of 40% smaller than 2 μm and 90% smaller than 63 μm. The density of the solids was 2,528 kg/m^3.

We measured the shear stress using a shear vane. An inherent drawback of this measurement technique is that the structure of the sample is destroyed, so that the consolidation experiment must be terminated after the vane test. Therefore, monitoring the evolution of strength with time requires multiple settling columns. To that end three acrylic settling columns each with a height of 40 cm and an inner diameter of 10.2 cm were filled with a slurry of uniform density of 1,133 kg/m^3 and a pore water density of 1,017 kg/m^3, which corresponds to a salinity of 2.4%. The initial height of the mud layer was 35.4 cm for all three columns. These three columns are identified as T8, T13 and T24, where the numbers refer to times (in days) after which the vane tests were carried out. The three vane tests were initiated at the same time. The experiments were carried out at the Soil Mechanics Laboratory of the University of Oxford.

2.1. X-ray density measurement

The X-ray density profiler, available in Oxford, provides a method to measure the density distribution in a column in a non-destructive manner. The main features of this technique are discussed briefly. For further information the reader is referred to Been (1980).

For X-rays of sufficient energy, the absorption of X-rays passing through soil can be uniquely related to the soil density. A collimated beam of X-rays is directed horizontally at the settlement column and, after passing through the soil, reaches a detector. The X-ray tube and detector can be traversed up and down to produce a count rate that is linked to the soil density by $N = N_0 \exp(-\mu \rho_b g d)$, where N is the count rate, N_0 a reference value, μ the absorption coefficient, ρ_b the bulk density of the mud, g the acceleration due to gravity and d the diameter of the column. The coefficients N_0 and μ can be determined by using the overlying water as a calibration sample, and by satisfying the condition that the total stress at the bottom, obtained by a vertically integration of the density profile, remains constant. The accuracy of the system is claimed to be $\pm$ 2 kg/m^3 (Sills, 1997). The energy level of the X-rays was set at 160 keV.

2.2. Pore water pressure measurements

Pore water pressures ports were tapped into the column wall at different heights, and each port was provided with a *vyon* plastic filter. The ports were connected via small tubes to a pressure measuring unit designed by Bowden (1988). The pressure measuring unit housed one pressure transducer that could be connected to any of the pressure ports, one at a time. The transducer was connected to a variable water head for calibration before the pore water pressure readings were taken. The accuracy of the pore water pressure values from calibration was 1 mm water head or 10 Pa, but this may have been reduced somewhat during the course of a long experiment if and when filters became clogged.

2.3. Shear stress measurements

The shear vane tester was driven by a stepper motor that drove the vane at a constant speed of 0.4 degree per second. A torque transducer was mounted between the vane rod and the stepper motor. The transducer produced an output in mV, which was recorded by a data logger on a PC. The torque transducer was calibrated by applying a number of known torques on the transducer (see also, Bowden, 1988).

The vane, which consisted of four blades, was 4 cm in height and 2 cm in diameter. This 2:1 ratio is the standard for vane measurements. Strength profiles were obtained by taking measurements at a spatial interval of 4 cm. To avoid interference from consecutive measurements, the measurement positions were staggered.

In order to relate shear stress to torque, we assumed that the shear stress was uniformly distributed over the height of the vane, and that both end surfaces of the cylinder contributed to the total torque. In a formula this reads

$$T = 2\pi l r^2 \left(1 + \frac{2r}{3l}\right)\tau \tag{1}$$

where T is the measured torque, τ the shear stress, l the height of the vane and r the radius of the vane.

Peak and residual shear stresses can both be used to represent the soil strength. In our case, the residual shear stresses were too inaccurate to use with confidence, so only the peak values are presented. The accuracy of the calculated peak shear stress was about ± 7 Pa. The torque resulting from friction along the shaft was estimated to be within the accuracy range, and was therefore neglected.

2.4. Calculation of effective stress

The total stress at any level in the soil was calculated by integrating the density multiplied by g from the water level to the level of interest. Subsequently, the effective stress was calculated by subtracting the linearly interpolated pore water pressure from the total stress. Given the accuracies of the X-ray density measurements and the pore water pressure readings, the accuracy of effective stress was on the order of ± 20 Pa (Sills, 1997).

2.5. Determination of permeability

Been (1980) showed that Darcy's law can be written as

$$v_s = \frac{k}{\rho_w g} \frac{\partial p}{\partial z} + k \tag{2}$$

where v_s is the average vertical particle velocity, k is the permeability, ρ_w is the density of the pore water, g is the acceleration due to gravity and $\partial p/\partial z$ the vertical pressure gradient.

The average particle velocity at a certain level, i.e., the settle rate, was estimated as follows. The column was divided in N imaginary levels. Each level corresponds to a certain percentage of the total mass in-between the bottom and that level. This percentage must be constant with time. Then, from two consecutive density profiles, the settlement rate of each level was calculated. The pressure gradient was calculated from the corresponding pore water pressure profiles for the averaged height of the two levels. By substituting the values into (2), the local permeability was evaluated. Due to averaging over both time and space, the accuracy was estimated to be not better than $\pm 1.5 \times 10^{-7}$ m/s.

3. EXPERIMENTAL RESULTS

3.1. Density profiles

Figure 1 shows that the interface levels for experiments T13 and T24 were practically the same, but the levels for experiment T8 deviated slightly. It may be that, had T8 lasted longer, the differences would have reduced.

The initial density distribution was quite uniform (Figure 2). As expected, the density increased with the passage of time, most significantly at the base of the columns. The maximum density after 24 days was 1.22×10^3 kg/m^3. The peaks in density in the upper few cm of the bed for T8 and T13 were most probably due to

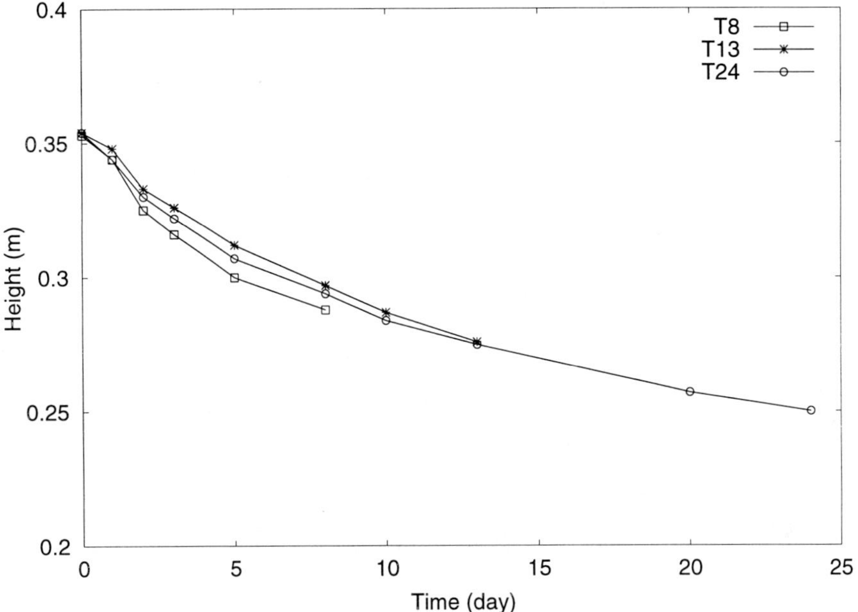

Figure 1. Interface height with time.

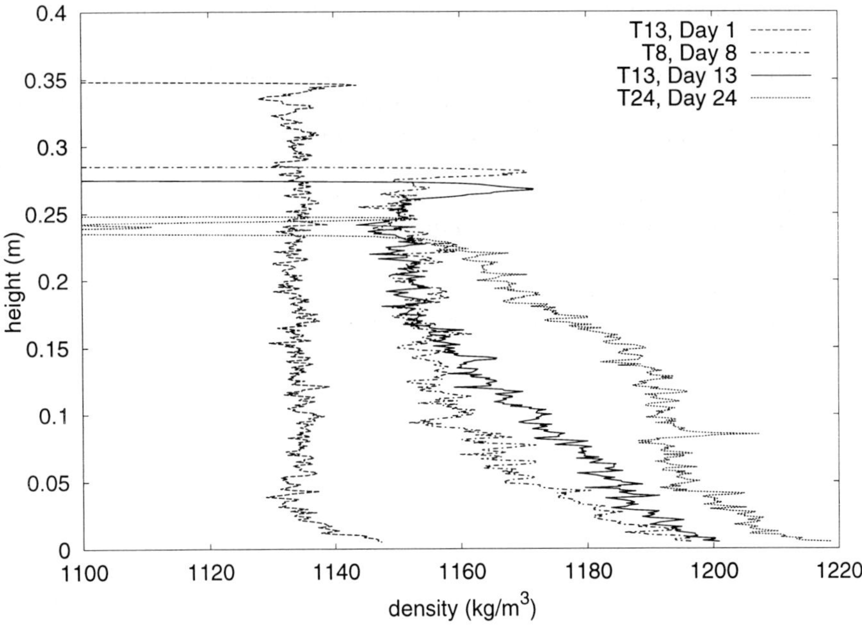

Figure 2. Density profiles after 1, 8, 13 and 24 days of consolidation.

chemical reactions. This was also observed for T24 during 20 days of consolidation. After this, as a result of the chemical changes, gas bubbles developed in this thin layer, reducing the density significantly.

3.2. Pore water pressure profiles

The excess pore water pressure, p_e, is the difference between the pore water pressure and the hydrostatic pore water pressure. Initially, the total load of the bed is carried by the pore water, which gives a more or less linear excess pore water pressure profile (Figure 3). Subsequently, the excess pore water pressure dissipates, first at the bottom, so that consolidation proceeds upwards from the base, consistent with the density profiles shown in Figure 2.

3.3. Peak shear stress profiles

As expected, the peak shear stress increased with time and depth (Figure 4). To some extent the behavior of the peak shear stress resembles that of the density, but in addition this stress increases with time. The densities at h = 0.15-0.25 m for days 8 and 13, for example, are nearly the same, whereas in this interval the peak shear stresses at day 13 are larger than those at day 8.

4. EFFECTIVE STRESS AND PERMEABILITY

The conventional way of presenting effective stress data in soil mechanics is as a relationship between effective stress and void ratio e, which for saturated soils is defined as the ratio of the pore water volume fraction and the solids volume fraction, i.e., $\phi_p = 1/(1+e)$. For the sake of convenience later, we here follow a different approach and relate the effective stress to the solids volume fraction, ϕ_p.

In the suspension phase cohesive sediment particles tend to form complex structures, called aggregates or flocs. These aggregates tend to be self-similar (Jullien and Botet, 1987; Family and Landau, 1984). The concept of fractal structure is a tool to model self-similar structures and introduced by Mandelbrot (1982). Basically, a fractal structure is a structure of which certain geometrical properties are scale-invariant irrespective of the complexity of the structure. An important parameter is the fractal dimension D, which relates the number of primary particles in the aggregate, N, to the ratio of the size of the aggregate R_a and the size of the primary particles, R_p, according to

$$N \sim \left(\frac{R_a}{R_p}\right)^D \tag{3}$$

The dimension of a fractal structure in three dimensions may be any number in the interval [1,3]. Very tenuous or string-like aggregates have a D close to 1. In the case of pure coalescence of particles, D would be equal to 3.

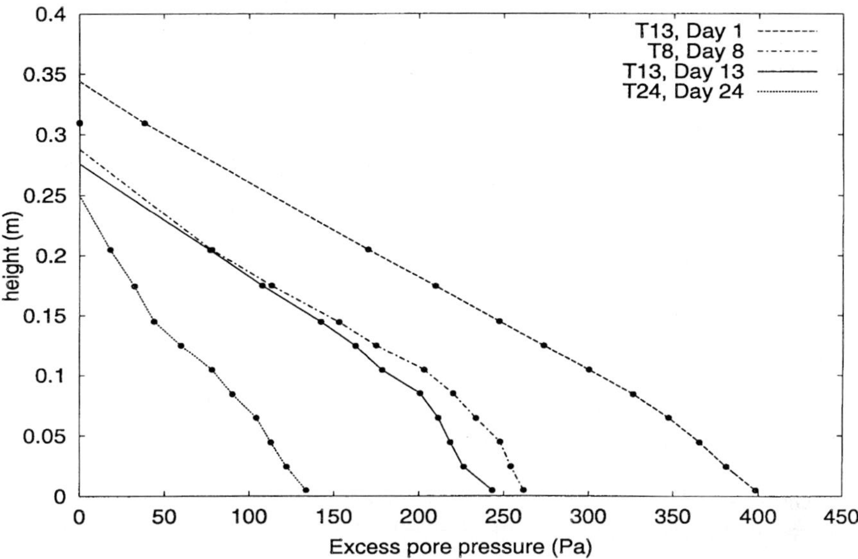

Figure 3. Excess pore water pressure profiles after 1, 8, 13 and 24 days of consolidation.

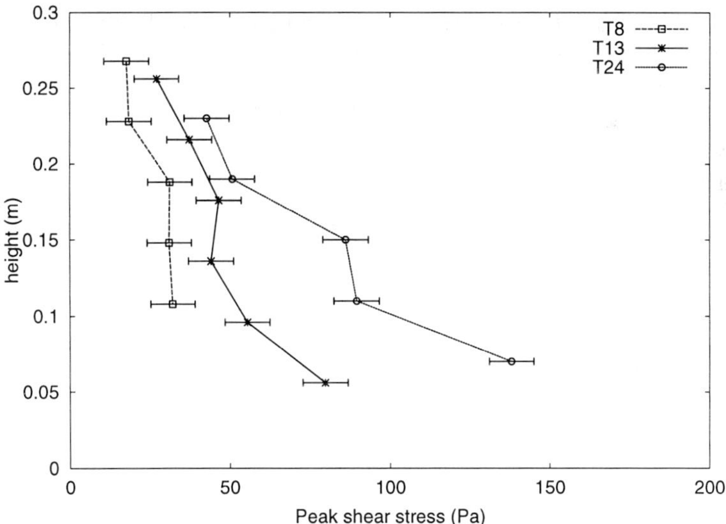

Figure 4. Peak shear stress profiles after 8, 13 and 24 days of consolidation.

In dilute suspensions of cohesive sediment the fractal dimension is 2.0 ± 0.3 (Winterwerp, 1998). However, at concentrations above the gelling or structure point the aggregates form a volume filling network so that the fractal dimension of the aggregates making up the network is larger, that is, 2.5 ± 0.3. It is noted that the structure of real aggregates is random rather than deterministic. This means that fractal properties are recovered only after averaging over sufficient aggregates.

With respect to cohesive sediment, a number of scaling relations can be given, see e.g., Kranenburg (1994). On the basis of certain assumptions concerning the floc structure, the relations between yield stress σ_y of a volume-filling network and volume fraction, and permeability and volume fraction are given by:

$$\sigma_y \sim \phi_p^{\frac{2}{3-D}} \tag{4}$$

$$k \sim R_p^2 \phi_p^{\frac{-2}{3-D}} \tag{5}$$

The yield stress σ_y is assumed to depend on the number and strength of the connections between the particles. If we assume that a consolidating mud layer is yielding continuously, there is an argument for relating σ_y to the effective stress.

Figure 5 shows the effective stress data for T8, T13 and T24 as functions of the volume fraction of solids. Note that double-logarithmic scales are used so that (4) and (5) appear as straight lines. Ignoring effective stress levels below 20 Pa, which is the accuracy of the effective stress calculation, the data do appear as a more or less straight line. The fractal dimension can be calculated from the slopes of the lines. For these experiments the average fractal dimension was $D = 2.71 \pm 0.05$. The straight lines corresponding to $D = 2.71$ are also included in Figure 5 (dotted lines). The agreement between experimental data and theory is good. Notice that (4) and (5) are merely proportionality relationships.

Bowden (1988) did similar settling column experiments on Combwich mud. In his work he plotted the effective stress data against void ratio. In Figure 6 the effective stress data of his experiment RB18 (initial height 1.25 m, initial density 1,080 kg/m^3 and salinity 0.5 %) are plotted against volume fraction of solids on double log-scales. The figure shows that the effective stress data of that experiment also can be described by a power law. The fractal dimension, derived from the slope of the straight line approximation, equals 2.75 for experiment RB18, which is in line with the fractal dimension found in the experiments reported herein.

To indicate the validity of the fractal concept, lines corresponding to $D = 2.71$ are presented together with the permeability data in Figure 7. Although these data show significant scatter, the correspondence is satisfactory, especially for experiment T24. However, one could also suggest a dimension closer to 3 for experiment T13. The quality of the data for T8 was too poor and therefore, the data are not presented.

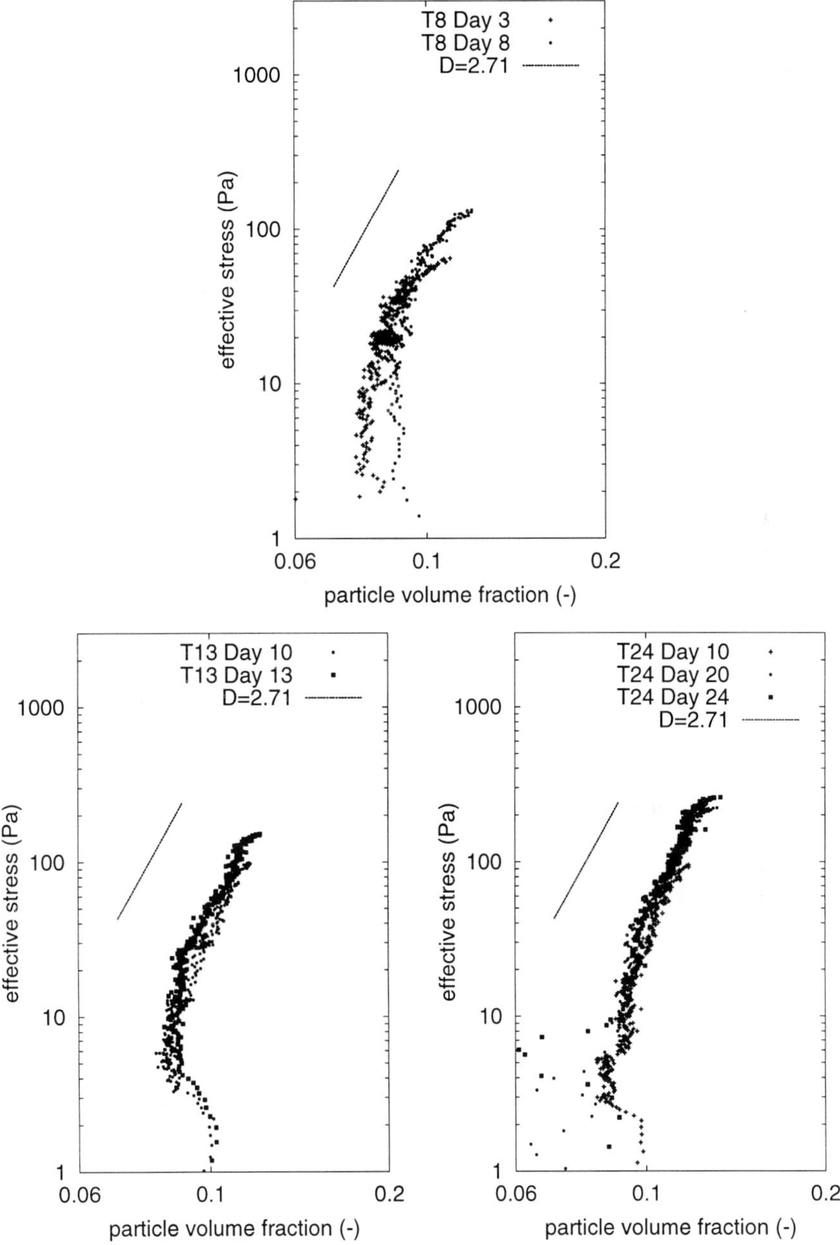

Figure 5. Effective stress data, tests T8, T13 and T24.

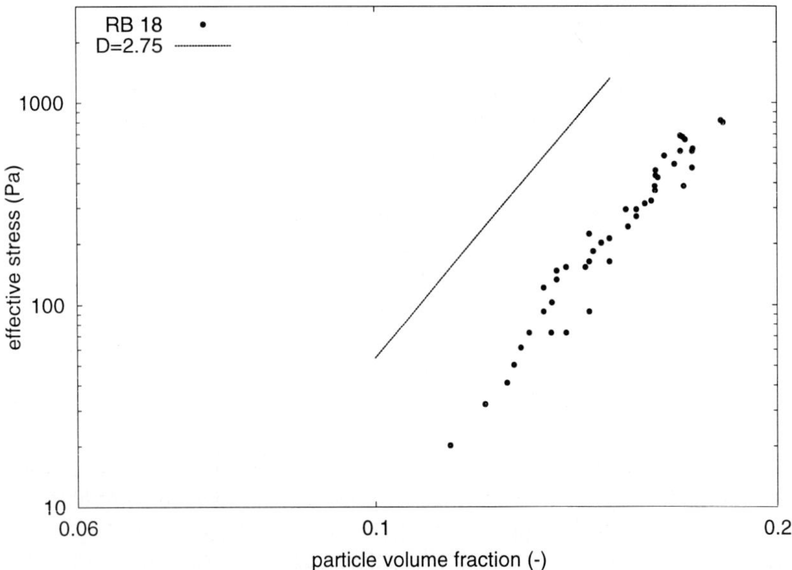

Figure 6. Effective stress data, experiment RB18 (Bowden, 1988).

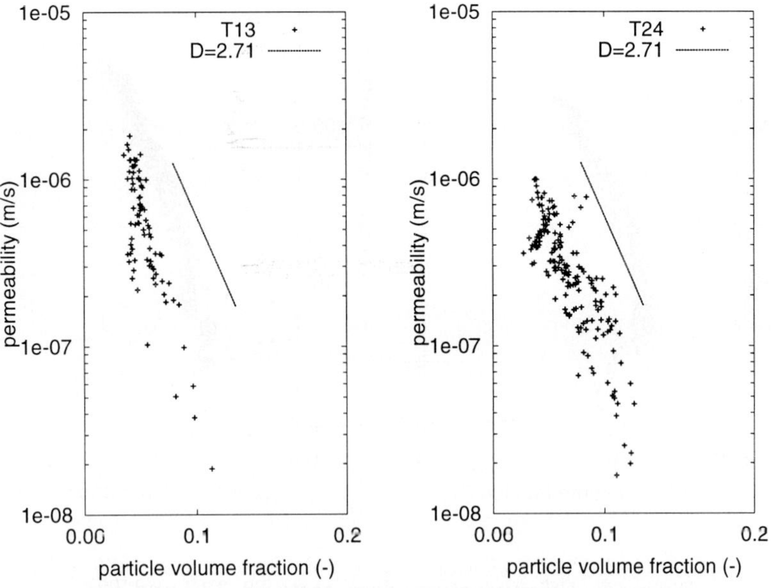

Figure 7. Permeability data, experiments T13 and T24.

5. PEAK SHEAR STRESS

Analyzing shear strength in terms of effective stress is a common approach in geotechnical engineering, and usually gives good correlations. In the vane tests, horizontal shear stresses on vertical planes were measured. The most appropriate effective stress would be the horizontal effective stress. Since the horizontal stress could not be measured, we analyzed the peak shear strength in terms of vertical effective stress; see Figure 8.

The figure shows a more or less linear relationship between the effective stress and the peak shear stress and indicates the presence of a positive peak shear stress at zero effective stress, the so-called true cohesion, τ_c. The data points of T13 and T24 can be represented by a straight line according to $\tau_{peak} = \tau_c + c\sigma'$, where $\tau_c = 24$ Pa and $c = 0.6$. The data points of T8 can also be represented by a straight line with c equal to 0.6, but with a true cohesion of approximately 15 Pa. If the increase in true cohesion between day 8 and day 13 had been significant, it would have indicated thixotropic behavior.

Similar straight line approximations were obtained by Bowden (1988). For experiments RB16-18, which were similar in all respects but for duration, the true cohesion was of order 35 Pa and the slope was about 0.45. For lower effective stress levels the true cohesion seemed to increase with time, which also indicates thixotropic effect.

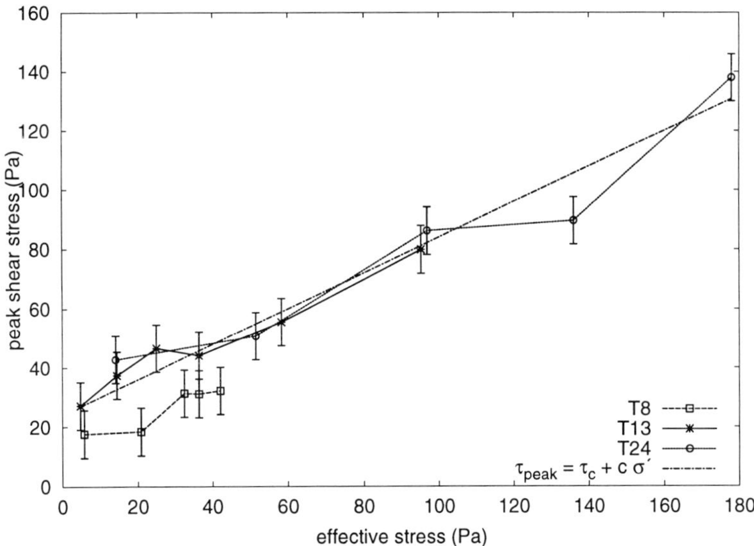

Figure 8. Peak shear stresses against effective stresses after 8, 13 and 24 days of consolidation.

6. CONCLUSIONS

We have presented three settling column experiments on the Caland-Beer Channel mud, aimed at identifying material-specific relations useful for consolidation and strength modeling. Three measurement techniques were used: non-destructive density measurements with an X-ray density profiler, pore water pressure measurements and destructive shear strength measurements with vane tests.

Effective stresses and permeabilities were obtained from density profiles and pore water pressure profiles. We selected the peak shear stress as a representative strength parameter.

Effective stress data were compared with relationships based on the concept of fractal structure as applied to cohesive sediment by Kranenburg (1994). The agreement between experiment and theory is good for effective stress levels higher than 20 Pa. The fractal dimension derived from these data is 2.71 ± 0.05. Permeability data were in line with a fractal dimension of 2.71. Thus, the concept of fractal structure provides a useful tool to relate both effective stress and permeability to volume fraction of solids through the fractal dimension. Applying the fractal concept to effective stress data of similar experiments done by Bowden (1988) also yields a good agreement and a fractal dimension of 2.75, which is close to the value found for the present experiments.

The results of the vane tests done at 8, 13 and 24 days after initiating the experiments showed that the peak shear stress increased with time and depth. A linear relationship between effective stress and peak shear stress seemed to exist. Extrapolation of the straight line approximation indicates the presence of shear strength at zero effective stress, the so-called true cohesion. True cohesion seemed to increase with the passage of time, whereas the coefficient multiplying the effective stress remained constant within the accuracies of the measurements. Bowden (1988) found similar linear relationships between effective stress and peak shear stress and a time dependent true cohesion.

The linearity between effective stress and peak shear stress might be useful for parameterizing the strength of soft mud layers, which is important in engineering practice. However, further investigation is necessary to assess the applicability of the results to different column heights, initial conditions and muds.

7. ACKNOWLEDGMENT

This research was partly financed by the Technology Foundation (STW).

REFERENCES

Been, K., 1980. Stress-strain behaviour of a cohesive soil deposited under water. *Ph.D. Thesis*, Oxford University, Oxford.

Been, K., and Sills, G. C., 1981. Self-weight consolidation of soft soils: an experimental and theoretical study. *Géotechnique*, 31(4), 519-535.

Berlamont, J., van den Bosch, L., and Toorman, E., 1992. Effective stresses and permeability in consolidating mud. *Proceedings of the 23rd International Conference on Coastal Engineering*, American Society of Civil Engineers, New York, 135-136.

Bowden, R. K., 1988. Compression behaviour and shear strength characteristics of a natural silty clay sedimented in the laboratory. *Ph.D. Thesis*, Oxford University, Oxford.

Elder, D. McG., 1985. Stress-strain and strength behaviour of very soft soil sediments. *Ph.D. Thesis*, Oxford University, Oxford.

Family, F., and Landau, D. P., 1984. *Kinetics of Aggregation and Gelation*. North-Holland, Amsterdam.

Jullien, R., and Botet, R., 1987. *Aggregation and Fractal Aggregates*. World Scientific Publications, Singapore.

Kranenburg, C., 1994. The fractal structure of cohesive sediment aggregates. *Estuarine, Coastal and Shelf Science*, 39, 451-460.

Mandelbrot, B. B., 1982. *The Fractal Geometry of Nature*. Freeman, New York.

McDermott, I. R., 1992. Seismo-acoustic investigations of consolidation phenomena. *Ph.D. Thesis*, University of Wales, Menai Bridge.

Migniot, C., and Hamm, L., 1990. Consolidation and rheological properties of mud deposits. *Proceeding of the 22nd International Conference on Coastal Engineering*, American Society of Civil Engineers, New York, 2975-2983.

Sills, G. C., 1995. Time dependent processes in soil consolidation. In: *Proceeding of the International Symposium on Compression and Consolidation of Clayey Soils*, H. Yoshikuni and O. Kusakabe eds, , A. A. Balkema, Rotterdam, 875-890.

Sills, G. C., 1997. Consolidation of cohesive sediments in settling columns. In: *Cohesive Sediments*, N. Burt, R. Parker and J. Watts eds., John Wiley, Chichester, UK, 107-120.

Terzaghi, K., 1936. The shearing resistance of saturated soils and the angles between the planes of shear. *Proceedings of the 1st International Conference on Soil Mechanics*, 1, 54-56.

van Kessel, T., 1997. Generation and transport of subacqueous fluid mud layers. *Ph.D. Thesis*, Delft University of Technology, Delft.

Winterwerp, J. C., 1998. A simple model for turbulence induced flocculation of cohesive sediment. *Journal of Hydraulic Research*, 36(3), 309-326.

Coastal and Estuarine Fine Sediment Processes
W.H. McAnally and A.J. Mehta (Editors)

A framework for cohesive sediment transport simulation for the coastal waters of Korea

D. Y. Lee[a], J. L. Lee[b], K. C. Jun[a] and K. S. Park[a]

[a]Korea Ocean Research and Development Institute, P.O. Box 29, Ansan, Kyunggi, Korea

[b]Sungkyungkwan University, Civil Engineering Department, Suwon, Kyunggi, Korea

Korea Ocean Research and Development Institute (KORDI) is carrying out an ambitious task to establish an operational cohesive sediment transport prediction system for long-term simulation of suspended fine sediment movement under strong wave and tide action. A semi-automatic system for cohesive sediment transport simulation has been established by preparing the essential data base, and also by integrating various components of coastal models including those for waves, tide, storm surge and wave-induced circulation, from which various processes in cohesive sediment transport are numerically modeled. The system allows easy testing and calibration of the model for any given area and time of interest, and can be used as a tool for further development of a more comprehensive model for engineering application. The results of a simulation test in the southwestern coastal waters of Korea are shown and compared with field measurements.

1. INTRODUCTION

There are a multitude of problems along the coast of Korea related to the transport of cohesive sediment and the absorbed pollutants, especially along the west and south coasts of Korea, which are abundant in cohesive sediments. Many large-scale coastal development projects are planned along the west coast of Korea, which borders the Yellow Sea. An operational sediment transport model is one of the urgently required tools to handle these environments.

The west coast of Korea is characterized by macro-tides resulting in wide tidal flats and strong tidal currents combined with severe waves and storm surges during the winter monsoon, and frequent passages of typhoons. This high-energy coastal environment leads to complexities and difficulties in predicting cohesive sediment transport using numerical models.

In planning and designing large-scale development projects and managing the coastal zone, predictions of the coastal response, such as morphological changes, are needed for a sustainable coastal development. Various types of cohesive sediment transport models have been developed since the early 1980s (e.g., Sheng, 1983; Hayter and Mehta, 1986; Odd, 1988; Le Hir, 1997; Hamm et al., 1996). In order to apply such predictive methods for the coastal waters of the Yellow Sea over extended periods of time, the physical environmental conditions should be properly modeled and interfaced with a sediment transport model, and the numerous empirical parameters of the model tested and calibrated using available information.

As a first part of work in developing an appropriate simulation system applicable to the evaluation of the environmental impacts of suspended sediment transport and morphological changes, an operational prediction system that can be used semi-automatically for longer intervals in any area of interest in the coastal waters of Korea has been established. This system is described next.

2. SEDIMENT TRANSPORT PREDICTION

The tide chart of the M_2 component around the Korean peninsula (Figure 1) indicates that the tidal range is large. Hence, strong tidal currents develop along the west coast (Kang et al., 1991). The tidal range along this coast is 4-9 m, producing wide tidal flats. High waves of 3-6 m occur there during the winter monsoon and the passage of typhoon in the summer. Figure 2 is an example of a wave rose for the southern part of the west coast of Korea obtained from long-term wave simulation (Lee et al., 1997).

In order to develop an operational simulation system that can be used for any site and time of interest, external forces and coastal environmental parameters must be modeled and interfaced systematically with a sediment transport model. The following three phases of system development are in progress.

i) Phase I: Development of a 2-D sediment transport prediction system by preparing the essential data bases and coastal environmental prediction models, and integration of these with a sediment transport model.
ii) Phase II: Extension of the system for quasi 3-D modeling by synthesizing the vertical profiles of flow and suspended sediment concentration at each grid point of the 2-D model.
iii) Phase III: Development of a full 3-D system by utilizing 3-D hydrodynamic and transport models.

In the first stage, emphasis is placed on the preparation of a basic data base and development of an automatic system for the generation of the depth grid and the boundary conditions for all the required models of the system, and systematic interfacing of different types of coastal environmental models at different regional scales to develop an operational sediment transport simulation system for an arbitrary location and time of interest.

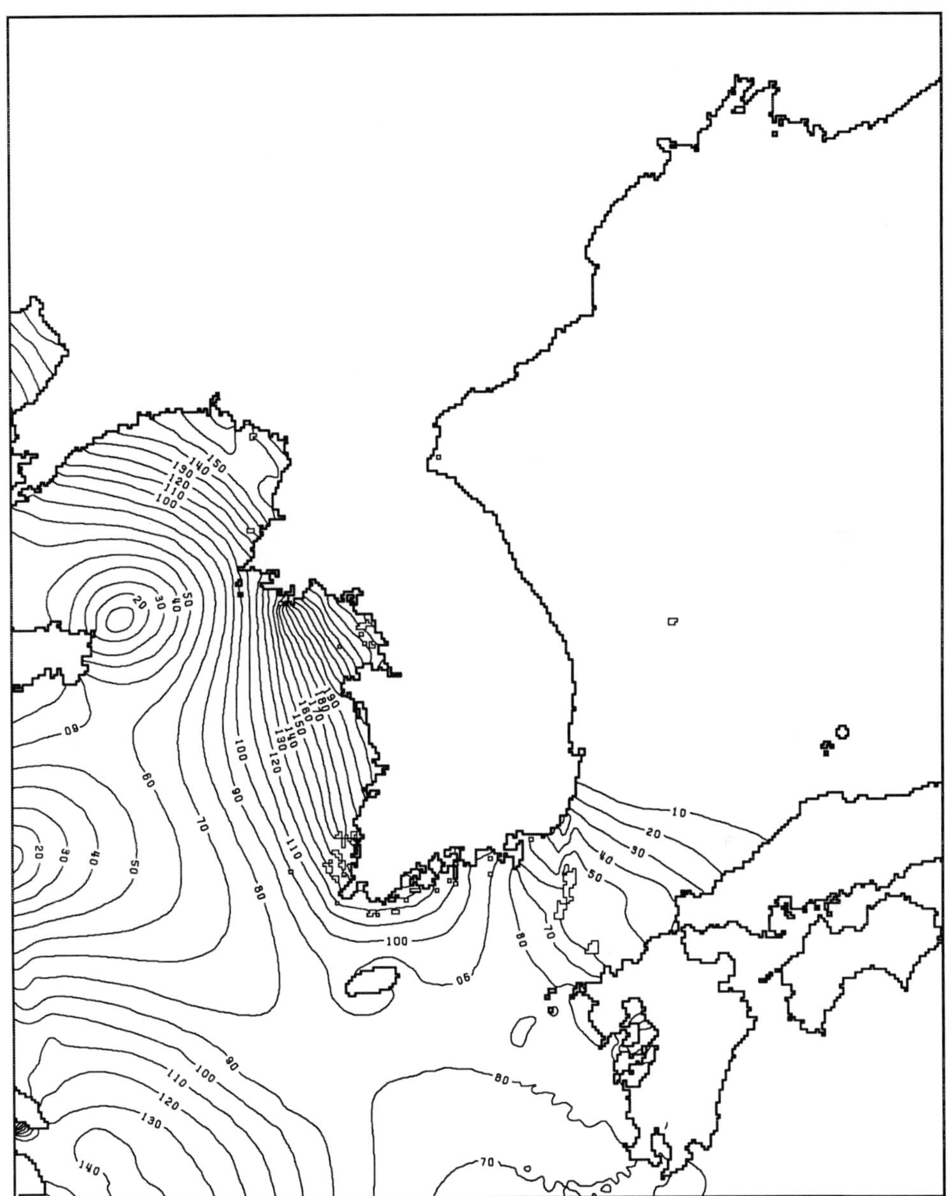

Figure 1. Tidal co-amplitude chart of the M_2 harmonic (unit: cm).

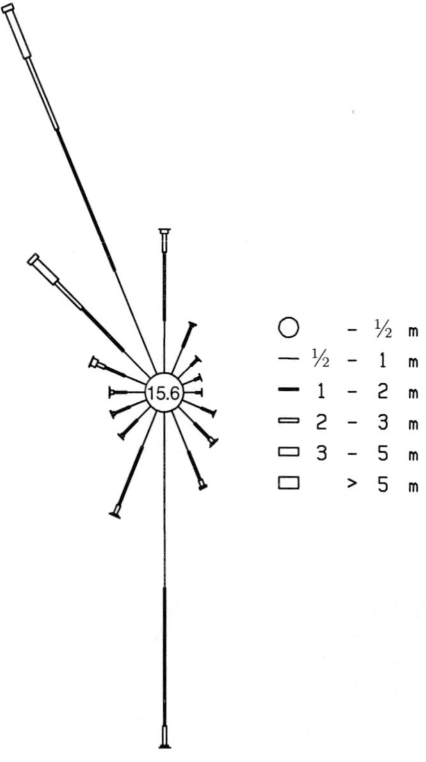

Figure 2. Example of a wave rose for the southwestern coastal waters of Korea.

3. COHESIVE SEDIMENT TRANSPORT

Suspended-load transport is described by the advection-diffusion equation for the sediment concentration. Two-dimensional horizontal models are derived by depth-integration under the assumption of vertical similarity of the advection terms:

$$\frac{\partial}{\partial t}(dC)+u\frac{\partial}{\partial x}(dC)+v\frac{\partial}{\partial y}(dC)=$$
$$\frac{\partial}{\partial x}\left(dD_{xx}\frac{\partial C}{\partial x}+dD_{xy}\frac{\partial C}{\partial y}\right)+\frac{\partial}{\partial y}\left(dD_{yx}\frac{\partial C}{\partial x}+dD_{yy}\frac{\partial C}{\partial y}\right)+S_T \tag{1}$$

where, C is the mass of sediment per unit volume of water and sediment mixture, D_{ij} is the effective sediment dispersion tensor, S_T is the sediment source/sink term, and d is the flow depth. The source/sink term can be expressed as

$$S_T = Q_e + Q_d \tag{2}$$

where Q_e is the rate of sediment addition due to erosion from the bed, and Q_d is the rate of sediment removal due to deposition. To solve (1), the source and sink in (2), i.e., Q_d and Q_e, must be estimated and the flow velocity components u, v, determined from a circulation model.

Several studies propose formulas for the deposition rate (Krone, 1962: Mehta and Partheniades, 1975). The settling velocity needs to be estimated through available parameters such as the bed shear stress, suspended concentration and salinity (Dyer, 1989; Le Hir et al., 1993). The rate of mud erosion depends on bed properties and the bed shear stress, and simple functional relationships have been derived using such variables (Partheniades, 1965).

The threshold shear stress, or shear strength, for simulating erosion is expressed empirically as a function of bed density (Migniot, 1968). Bed (dry) density and shear strength profiles are therefore specified initially for each grid point. The bed is separated into several layers according to the shear strength profile as described by Hayter and Mehta (1986). The resuspension rate is calculated through an empirical formula as a function of the bed shear stress due to combined wave and current induced motions and the shear strength of the top bed layer. For newly deposited materials, the erosion properties of bed are allowed to change with time due to consolidation.

Suspended sediment transport is determined using the above formulations by a new hybrid method (Lee, 1998). It is based on the forward-tracking particle method for advection. However, unlike the random-walk Lagrangian approach, it solves the diffusion process with respect to a fixed Euclerian grid, which requires neither an interpolating algorithm nor a large number of particles.

4. FIRST PHASE SEDIMENT TRANSPORT SIMULATION

Basic information has been developed for all the coastal waters of Korea for the cohesive sediment transport model. The depth grid system can be generated for the area of interest by interpolating the depth data retrieved from the data base comprising of 90 electronic charts for the coastal waters. Major tidal harmonic constants (M_2, S_2, K_1, O_1) have been calculated for the northeast Asian regional seas (Kang et al., 1991). Offshore wave conditions have been simulated continuously for 20 years. The boundary conditions for the local tide and wave models can be retrieved from these data bases for any arbitrary time of interest.

The coastal tidal flow and wave conditions are produced at each time step and grid point of the sediment transport model by means of systematic interfacing of various coastal models. For efficient application for long time periods, two system modes have been established: one for detailed simulation including interaction between the system processes, and the other for fast simulation using pre-set coefficients.

4.1. Coupled simulation mode

In shallow water, the interactions between tide, storm surge, waves and local wind forcing need to be considered, especially during extreme sea-states. Accordingly, the flow chart of sediment transport simulation is shown in Figure 3. The flow velocity and water level due to tide- and storm-induced combined-flow are determined by solving the depth-averaged equations of mass and momentum conservation. The hydrodynamic equations are solved by a fractional-step method in conjunction with approximate factorization techniques leading to an implicit finite difference scheme. Wave conditions are obtained using the shallow water wave spectral transformation model developed by Lee et al. (1997) based on the wave energy conservation equation.

The next step is the calculation of the bottom shear stress for the combined wave and current condition, from which deposition and erosion rates are calculated for the pre-assigned bottom bed profile as described in Hayter and Mehta (1986). Sediment advection-dispersion is calculated using a hybrid method. Different time steps are used for the hydrodynamic model, the wave model, the sediment advection-dispersion and the consolidation model. For the near-shore area, a finer grid model is nested within the outer area model, where hydrodynamics and shallow water wave models are fully interfaced to deal with wave-induced circulation.

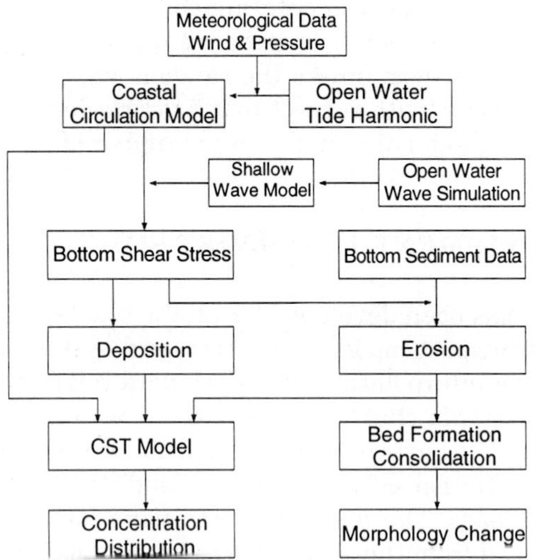

Figure 3. Flow chart of the cohesive sediment transport prediction system for simulation under extreme conditions.

4.2. Uncoupled fast simulation mode

It is time consuming to use the above described simulation system for most cases of normal sea condition. Thus, when the interactions between the processes are not significant, a fast simulation method can be used. In this mode, tide and tidal current as well as waves are calculated using pre-determined coefficients at each grid point without actually running the models. The flow chart of the fast simulation mode is shown in Figure 4.

An example of the amplitude and phase of the M_2 tidal component near a west coast area is shown in Figure 5. For a simple shallow water wave information simulation (Lee et al., 1997), the coastal wave conditions were pre-calculated for about 100 combinations of the offshore boundary conditions (wave directions, periods and heights) using the SWAN shallow water wave model (Ris, 1997), and transformation coefficients from the offshore boundary waves to those at each grid point were calculated and stored in the data base. The coefficients for a sample case are shown in Figure 6. For a given offshore wave condition, the shallow water wave condition at each grid point of the sediment transport model is produced by means of interpolation of these transformation coefficients. This mode is suitable for long-term simulations of sediment transport. However, in this fast mode, neither local effects of wind on coastal circulation and waves nor the interaction between tide and coastal waves are considered.

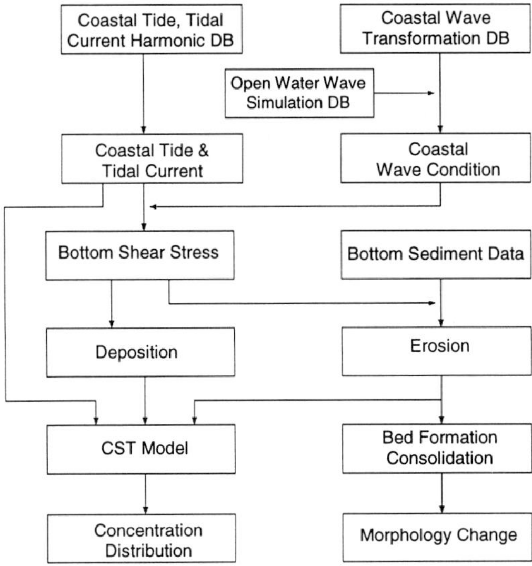

Figure 4. Flow chart of the cohesive sediment transport prediction system for long-term simulation under normal conditions.

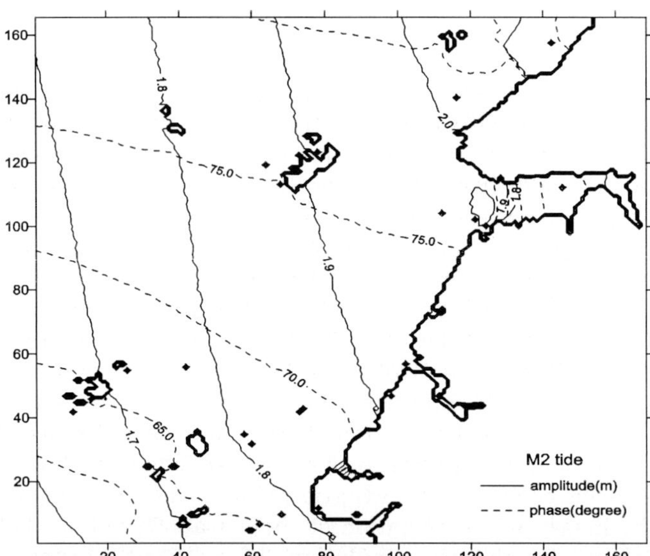

Figure 5. Example of the amplitude and phase of the M_2 tidal harmonic component from the data base for the Yeongkwang area.

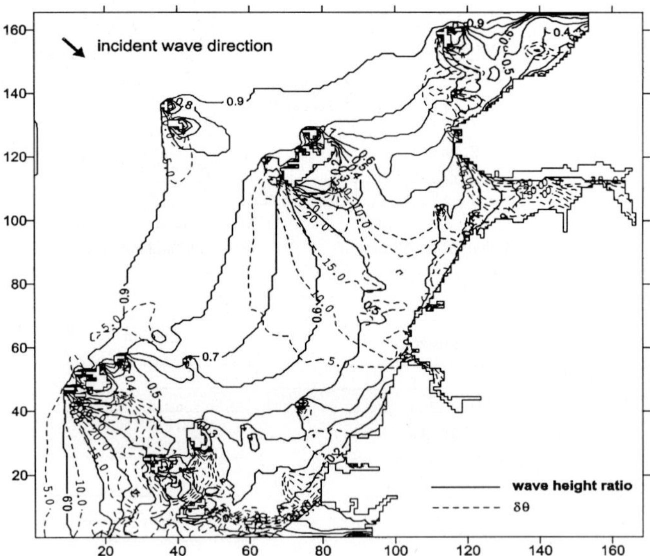

Figure 6. Example of shallow water wave transformation coefficients from data base for the Yeongkwang area.

5. SIMULATION TEST

The system was tested for the Yeongkwang area in the southwestern coastal waters. The site and the bathymetry are shown in Figure 7. The bottom surface sediment distribution is shown in Figure 8. The suspended sediment concentration was measured at 1.6 m from the sea bottom using bottom-mounted water quality sensors at location S1 in Figure 7 (Lee et al., 1992). The concentration was in the range of 200-300 mg/l when the sea state was mild. However, the concentration increased considerably with the advent of high waves, and reached a value higher than 3,000 mg/l (Figure 9).

The results of the simulation test are shown in Figure 10 together with time variations of tide and storm induced current and bottom shear stress due to the combined action of tide and waves calculated using a formula given by Tanaka and Shuto (1981). It is observed that the system provides a useful tool in the study of cohesive sediment transport modeling in the combined wave-tide zone such as the Yellow Sea. A comparison between data and simulation however indicates that the simulation system needs to be improved for storm events.

The uncertainty in prescribed bed properties and empirical parameters in estimation of erosion and deposition rates due to the wave orbital motion is

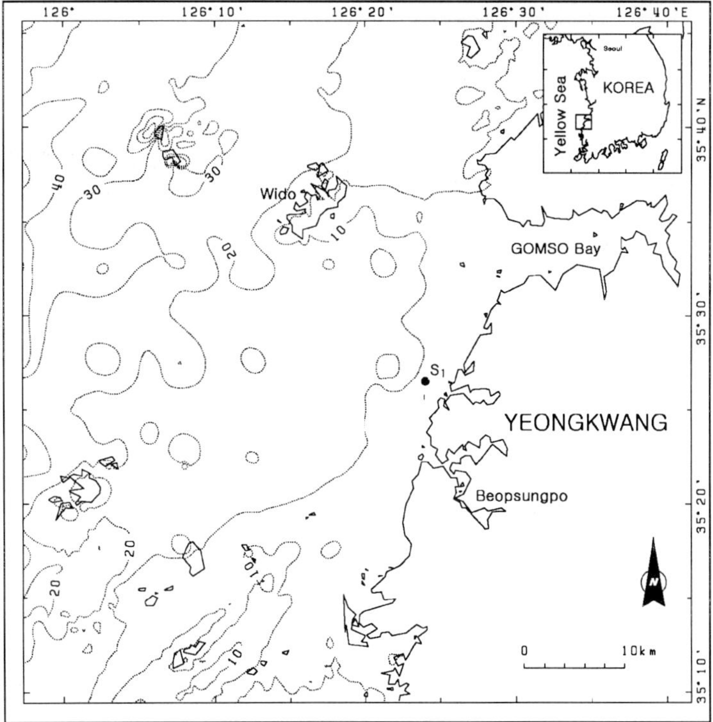

Figure 7. Site and the bathymetry of the study area.

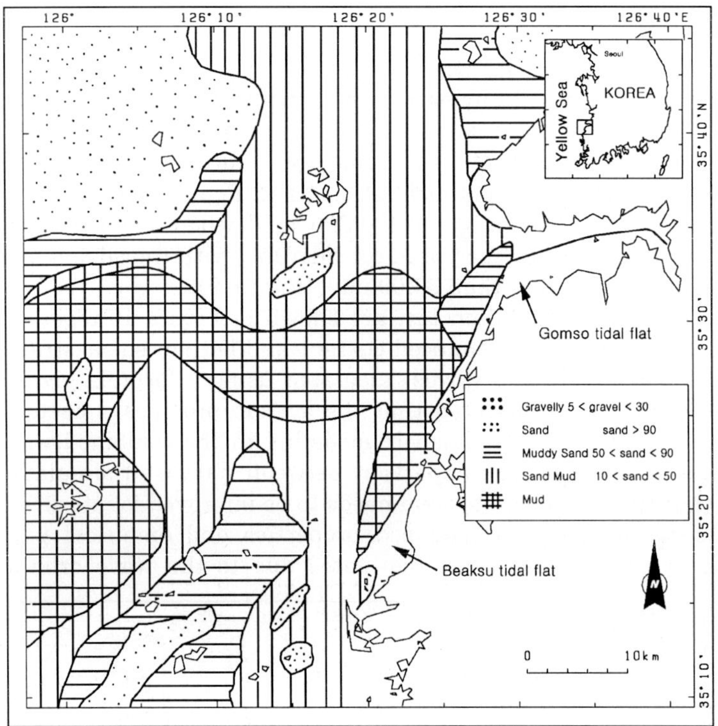

Figure 8. Bottom surface sediment distribution in the study area.

considered to be a major cause of the discrepancy observed in Figure 10. The time lag between the concentration peak and the bottom shear stress peak may be accounted for by the three dimensional character of suspended sediment dynamics and also partially by the process of bed fluidization due to wave forcing.

6. CONCLUSIONS

The developed system should serve as an efficient tool for sediment transport modeling for engineering applications. All the processes are evaluated semi-automatically except for the bottom sediment profiles and the sedimentation parameters, which must be provided from field and laboratory experiments, and adjusted by simulation testing.

The simulation test of the system produces the overall trend of suspended sediment concentration variation when compared with measurements, but does not simulate the time variation in detail. Further work is therefore necessary to improve the system. As noted, information on the deposition and erosion

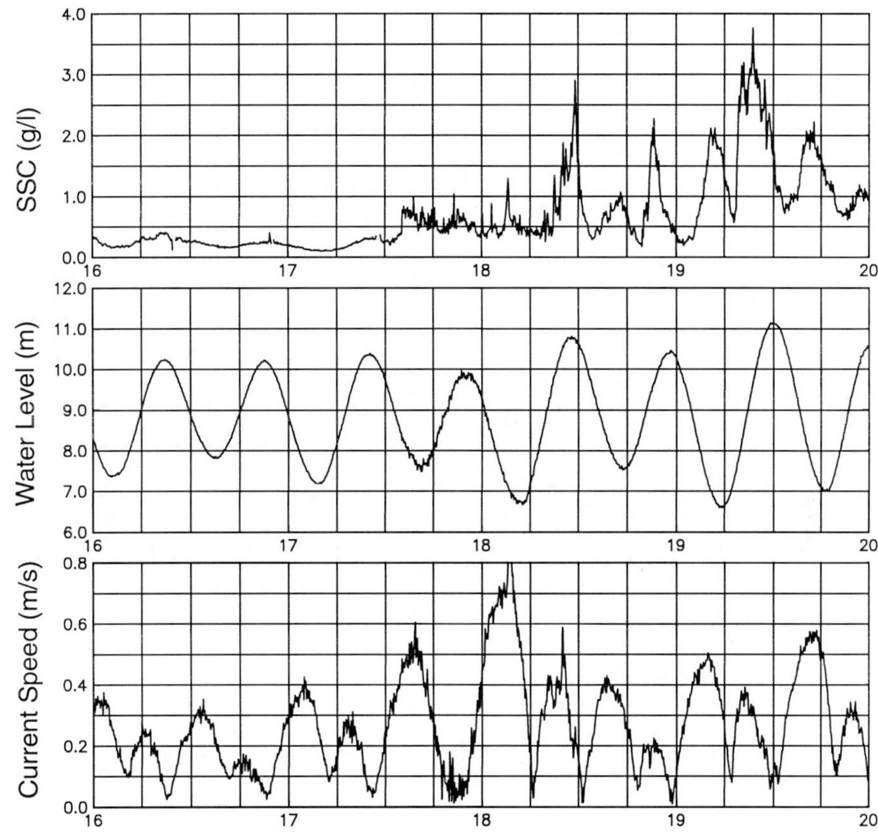

Figure 9. Time variations of the suspended sediment concentration, water level and current measured 1.6 m above sea bottom at location S1 in Figure 7.

processes due to wave-current action needs to be developed from field and laboratory experiments and simulation testing. Handling three dimensional processes with 2-D modeling is another likely cause of the discrepancy between simulation and measurement. This problem needs to be addressed by expanding the system to cover 3-D features of the sedimentation processes, especially the transport of near-bottom fluid mud.

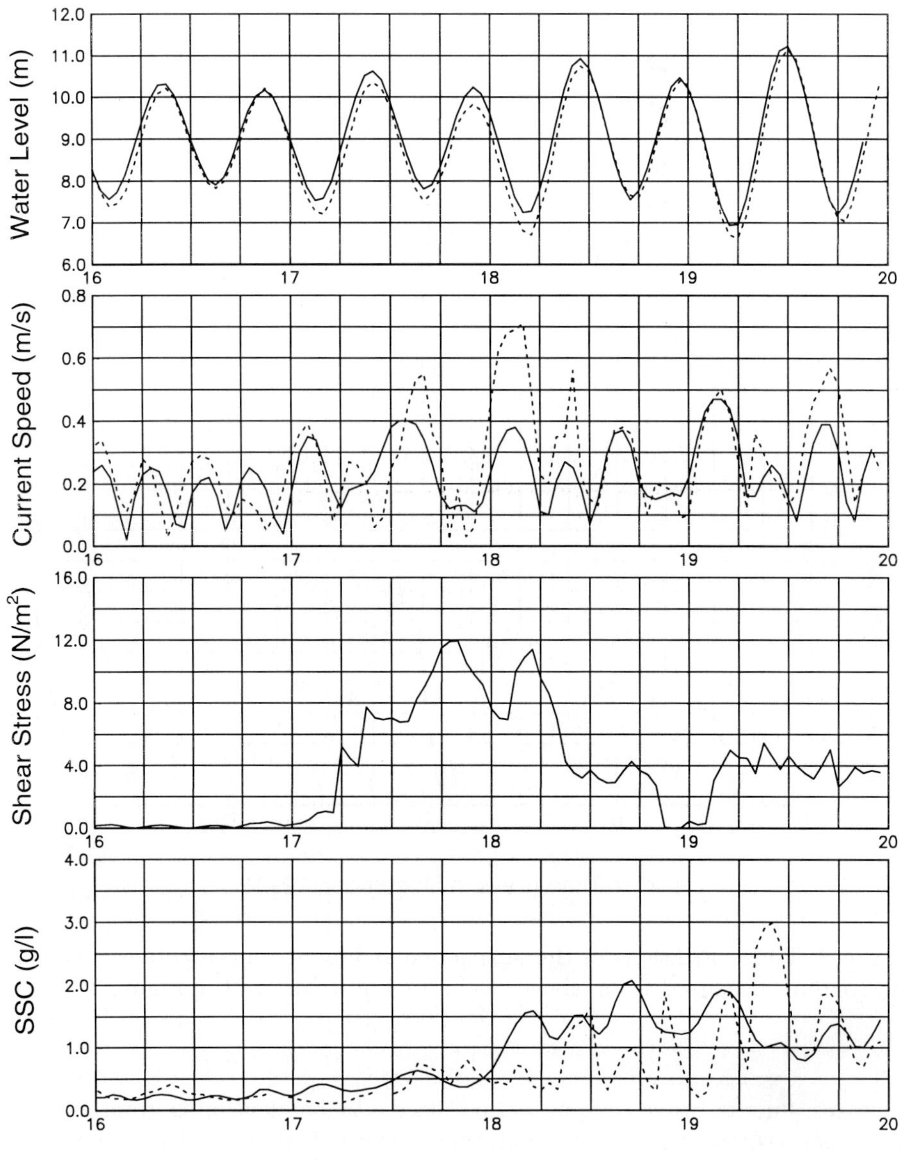

Figure 10. Comparison between simulated results and measurements of concentration and current speed together with wave-current induced bottom shear stress (dotted line: measurement; solid line: model simulation).

REFERENCES

Dyer, K. R., 1989. Sediment processes in estuaries: Future research requirements. *Journal of Geophysical Research*, 90(C10), 14,327-14,339.

Hamm, L., Chesher, T. J., Jakobsen, F., Peltier, E., and Toorman, E., 1996. Data analysis for cohesive sediment transport modelling. *Report No. 52184R7*, MAST G8M Coastal Morphodynamics, Project 4-Topic E: Mud Morphodynamic Modelling. SOGREAH, Grenoble, France.

Hayter, E. J., and Mehta, A. J., 1986. Modeling cohesive sediment transport in estuarial waters. *Applied Mathematical Modeling*, 10(4), 294-303.

Kang, S. K, Lee, S. R., and Yum, K. D., 1991. Tidal computation of the East China Sea, the Yellow Sea and East Sea. In: *Oceanography of Asia Marginal Seas*, K. Takano ed., Elsevier Oceanography Series 504, Amsterdam.

Krone, R. B., 1962. Flume studies of the transport of sediment in estuarial shoaling processes. *Final Report*, Hydraulic Engineering Laboratory and Sanitary Research Laboratory, University of California, Berkeley.

Lee, D. Y., Park, K. S., Kang, S. W., and Chu, Y. S., 1992. Studies on the numerical modelling of fine-grained sediment transport. *BSPG 00154-494-2*, Korea Ocean Research and Development Institute, Ansan.

Lee, D. Y., Jun, K. C., and Lee, J. L., 1997. Shallow water wave estimation system (IV). Project Report for Korea Ministry of Marine Affairs and Fisheries by Korea Ocean Research and Development Institute, Ansan.

Lee, J. L., 1998. A high-accuracy approach for modeling flow-dominated transport. In: *Environmental Coastal Regions*, C. A. Brebbia ed., Proceedings of Second International Conference on Environmental Coastal Regions, Cancun, Mexico, 277-286.

Le Hir, P., 1997. Fluid and sediment "integrated" modelling application to fluid mud flows in estuaries. In: *Cohesive Sediments*, N. Burt, R. Parker and J. Watts eds., John Wiley, Chichester, UK, 417-428.

Le Hir, P., Bassoullet, P., and L'Yavanc, J., 1993. Application of a multivariate transport model for understanding cohesive sediment dynamics. In: *Coastal and Estuarine Studies*, A. J. Mehta ed., 42, American Geophysical Union, Washington, DC, 504-519.

Mehta, A. J., and Partheniades, E., 1975. An investigation of the depositional properties of flocculated fine sediments. *Journal of Hydraulic Research*, 13(4), 361-381.

Migniot, C., 1968. A study of the physical properties of different very fine sediments and their behaviour under hydrodynamic action. *La Houille Blanche,* 7, 591-620 (in French, with abstract in English).

Odd, N. V. M., 1988. Mathematical modelling of mud transport in estuaries. In: *Physical Processes in Estuaries*, J. Dronkers and W. van Leussen eds., Springer Verlag, Berlin, 503-531.

Partheniades, E., 1965. Erosion and deposition of cohesive soils. *Journal of the Hydraulics Division*, ASCE, 91(1), 105-138.

Ris, R. C., 1997. Spectral modelling of wind waves in coastal areas. *Ph.D. Thesis, Report No. 97-4*, Communications on Hydraulic and Geotechnical Engineering, Delft University of Technology, Delft, The Netherlands.

Sheng, Y. P., 1983. Mathematical modeling of three-dimensional coastal currents and sediment dispersion: Model development and application. *Technical Report CERC-83-2*, Coastal Engineering Research Center, Waterways Experiment Station, Vicksburg, Mississippi.

Tanaka H., and Shuto, N., 1981. Friction coefficient for a wave-current coexistent system. *Coastal Engineering in Japan*, 24, 105-128.

Coastal and Estuarine Fine Sediment Processes
W.H. McAnally and A.J. Mehta (Editors)

Application of the continuous modeling concept to simulate high-concentration suspended sediment in a macrotidal estuary

P. Le Hir, P. Bassoullet and H. Jestin

Institut Français pour la Recherche et l'Exploitation de la Mer (IFREMER), Direction de l'Environnement Littoral, DEL-EC-TP, B.P. 70, 29280 Plouzané, France

A model for the transport of high-concentration fine sediment in estuaries is described in accordance with the *continuous approach* in which water and soft sediment are considered as a whole, without empirical exchanges (e.g., erosion, entrainment and deposition) between compartments. The model accounts for viscoplastic behavior, enabling the simulation of the Bingham threshold. Turbulence closure follows the mixing length concept, and damping parameters have been calibrated with experiments on fine sand suspensions, requiring a more rapid decrease of viscosity than classically accepted. Starting from uniform suspensions, steady-state runs lead to the formation of lutoclines and mud layers, with layer thickness dependent on the total sediment mass for a given forcing. Equilibrium concentration in the upper layers is essentially dependent on the friction velocity and the settling rate, and converges towards the so-called *saturation concentration* if damping of viscosity is neglected and the settling velocity is constant. Continuous modeling has been applied to the formation and erosion of a fluid mud layer in the area of the turbidity maximum in the Gironde estuary in France. Results are encouraging, pointing out the importance of hindered settling, turbulence damping and viscoplastic behavior for explaining the observed lutoclines and soft mud layers, with surprising velocity gradients even at spring tide. A fully continuous three-dimensional model is needed for eliminating advection-induced ambiguities.

1. INTRODUCTION

In turbid environments, suspensions of cohesive sediments can become highly concentrated, depending on the total mobile sediment available, the settling rate and the turbulent kinetic energy. For instance in estuaries, local concentrations of suspended sediment related to the so-called turbidity maxima often generate fluid mud layers near the bottom at neap tide; e.g., Inglis and Allen (1957) for the

Thames, Allen et al. (1977) for the Gironde, Gallenne (1974) for the Loire and Kirby and Parker (1983) for the Severn. Winterwerp (1998) reviewed publications that mention the occurrence of such dense suspensions he termed *High-Concentrated Mud Suspensions* (HCMS). These fluid mud layers can reach several meters in height as reported by Castaing (1981), and their concentration range is wide; Kirby and Parker (1983) distinguish mobile suspensions of up to 150 gl^{-1}, and stationary suspensions on the order of 250 gl^{-1}. Estuaries are not the only areas where fluid mud can occur; on exposed muddy coasts sediment can be resuspended by waves, sometimes as a result of liquefaction, forming fluid mud layers which interact with waves and contribute largely to mud transport (e.g., Li, 1996). Similar processes are probably involved in the formation of mud suspensions around the entrance of Rotterdam Port as mentioned by Winterwerp (1998), or at the mouth of the Seine estuary (Brenon, 1997). Finally, offshore turbidity currents that generate large sediment slumping on continental slopes (e.g., Mulder and Cochonat, 1996) can often be considered as fluid mud movements.

A better knowledge of the processes leading to fluid mud layers is needed for many applications including dredging, waterway maintenance, impact of turbidity maximum on local ecosystems, muddy coast and shoal (mudbank) evolution, offshore turbidity currents and others.

Modeling fluid mud layers is difficult, as it requires simultaneous simulation of refined hydrodynamics and complex sedimentary behavior. In addition, observations in highly turbid environments are rare and complex, as they require investigations in the intermediate range of concentrations at undeterminable vertical positions; good examples have been provided by Kirby (1988) in the Severn estuary. Recently, we investigated the formation and erosion of fluid mud layers in a macrotidal estuary (Gironde, France); simultaneous concentration and velocity profiles over a fortnightly tidal cycle were measured, providing new insights into the dynamics of soft sediment layers and the appearance of lutoclines.

Up until now, models have rarely dealt with the internal structure of fluid mud layers, assuming a sharp interface (lutocline) between the mud layer (of uniform density) and the water above, with semi-empirical exchange formulations between them for entrainment and settling. These models can account for fluid mud flows, e.g., as in the modeling work of Odd and Owen (1972) in the Thames. Le Normant (1995) coupled a 2DH fluid mud model to a fully 3D model for simulating the turbidity maximum in the Loire estuary. However, these models do not simulate the full interaction between fluid dynamics and the sediment structure. In practice, in highly turbid environments it becomes difficult to distinguish the water column from the sediment, so that classical models, which assume a strong discontinuity at the water/sediment interface and no vertical velocity gradient within the sediment, fail to reproduce the actual dynamics. Earlier, we presented an *integrated* modeling approach, which consists in simulating water- and mud-dynamics straightforwardly, solving for the mass conservation and momentum equations over the entire column (Le Hir, 1997). In the present work, the same concept is used, with the word *integrated* being replaced by *continuous*, which is probably more representative of a progressive transition between the water column and bottom sediment. Equilibrium profiles of velocity and concentration are emphasized, and model capability has been extended to viscoplastic mud behavior. Also, the model is used to simulate the behavior of the turbidity maximum in the Gironde.

2. MAIN FEATURES OF THE 1DV CONTINUOUS MODEL

In fluid mud flows, the density of particles is so high that interactions between particles and water become strong, and a full description of the mechanics needs a multiphase approach. This was achieved by Teisson et al. (1992), who reproduced the character of a fluid mud layer by using a two-phase flow model. This approach also enables a combination of hydrodynamics and soil mechanics, for which pore-water movement relative to settling particles must be considered. However, the computations are very time-consuming, and unsteady configurations cannot be simulated easily.

In most natural fine sediment transport, particles and water have similar horizontal velocity components, so that a unique set of horizontal momentum equations can be considered for the water/sediment mixture. In addition, vertical acceleration can often be neglected, and the vertical momentum equation reduces to one with hydrostatic pressure for the fluid along with an equilibrium settling velocity for each dispersed phase. As pointed out by Le Hir (1997), according to the previous assumptions, differences between single and multiphase approaches then lie in problem formulation.

In the applications of the present paper, horizontal advection is neglected by assuming horizontal uniformity of velocities and concentrations, and the dynamics is described by a one-dimension vertical model (SAM-1DV, for Simulation of Multivariable Advection). Accordingly, the set of equations can be written as:

Mass conservation:

$$\frac{\partial C}{\partial t} + \frac{\partial}{\partial z}(C\,W_s) = \frac{\partial}{\partial z}\left(K_z \frac{\partial C}{\partial z}\right) \tag{1}$$

where C is the mass concentration of sediment, W_s is the settling velocity and K_z is the vertical diffusivity.

Horizontal momentum balance:

$$\frac{\partial u_i}{\partial t} = -\frac{1}{\rho}\frac{\partial P}{\partial x_i} + \frac{\partial}{\partial z}\left(-\overline{u_i'w'} + \frac{T_{iz}}{\rho}\right) \tag{2}$$

where u_i is the horizontal velocity along the x_i direction, P is the pressure , ρ is the density of the mixture, u_i' and w' are the turbulent fluctuations of velocity and T_{iz} is the "viscous" shear stress. The main forcing is the imposed pressure gradient.

2.1. Turbulence closure

The concept of eddy viscosity is adopted and a simple mixing length based turbulence closure is used. For 1DV computations, we consider that higher orders of turbulence closure are not highly pertinent, as mean vertical velocities are not accounted for and horizontal transport is neglected. The most important process is stratification-induced turbulence damping; it is classically formulated by using an empirical relationship between the mixing length and the local Richardson number, as proposed by Munk and Anderson; e.g., in Nunes Vaz and Simpson (1994), in which a comparison between different formulations is given. The relevant formulation is

$$-\overline{u_i'w'} = \nu_t \frac{\partial u_i}{\partial z} = l_0^2 (1 + \alpha Ri)^{\beta} \sqrt{\sum \left(\frac{\partial u_i}{\partial z} \right)^2} \frac{\partial u_i}{\partial z} \tag{3}$$

where ν_t is the eddy viscosity, l_0 is the mixing length computed as $l_0 = \kappa \min(z, h-z, 0.20\, h)$, κ is the von Karman constant, z is the distance from the bottom and h is the water depth. Further, Ri is the Richardson number calculated as $-[(g/\rho)\, \partial\rho/\partial z]/[\Sigma(\partial u_i/\partial z)^2]$, and α and β are damping parameters, with selected values of 3 and -4, respectively.

2.2. Generalized viscosity

A major feature of highly concentrated suspensions is the increase in the molecular viscosity with concentration. A power law is assumed (e.g., Toorman, 1992). However, the mixture may become non-Newtonian, which can be formulated by a dependence of the apparent viscosity on the shear rate. Thus we propose a generalized viscosity expression:

$$\begin{aligned} -\overline{u_i'w'} + \frac{T_{iz}}{\rho} &= (\nu_t + \nu_m) \frac{\partial u_i}{\partial z} \\ \nu_m &= \nu_0 + k_1 C^{k_2} \left(1 + \frac{k_3}{k_4 + \sqrt{\sum \left(\frac{\partial u_i}{\partial z} \right)^2}} \right) \end{aligned} \tag{4}$$

where ν_t depends on the turbulence closure and is generally formulated as in (3), or as a function of the turbulent kinetic energy k and a turbulence length scale, or the energy dissipation rate.

In the molecular viscosity term ν_m, k_i are empirical coefficients and ν_0 is a minimum viscosity, which normally corresponds to the water viscosity. By setting k_4 very low, the resulting viscosity becomes very high for a low shear rate, and rapidly decreases when the latter increases, as shown on Figure 1. Such a parameterization accounts for the viscoplastic behavior, very close to a Bingham behavior. In particular, the product $\rho k_1 C^{k_2} k_3$ represents the shear strength of the sediment. Further, when such a behavior is selected, $k_4 = 10^{-3}$ and $k_3 = 0.1$ to 3. As for the increase in viscosity with density, the power 3 ($= k_2$) is considered to be representative of the scarce available data (Toorman, 1992), and k_1 is on the order of 2×10^{-10} according to measurements on mud samples from the Gironde, to be considered further (Migniot, 1989). The resulting viscosity is what is plotted in Figure 1.

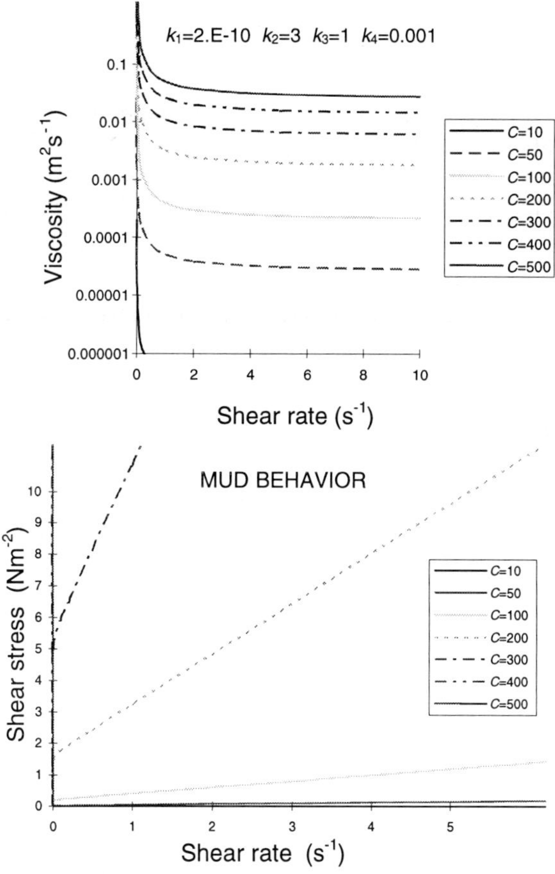

Figure 1. Simulated viscosity (top) and shear strength (bottom) of soft mud as a function of the shear rate according to (4). C values are in kg m^{-3}.

The diffusivity of suspended sediment, K_z, has a molecular component and a turbulent component. Without any guidance from the literature, considering that mixing is not enhanced in concentrated mixtures, we assume that molecular diffusivity (K_{z0}) is independent of density. On the other hand, turbulent diffusivity follows variations in the eddy viscosity, and is similarly modeled according to the mixing length concept, with a stratification-induced damping factor:

$$K_z = K_{z0} + K_{zt}$$
$$K_{zt} = l_0^2(1+\gamma Ri)^{\delta}\sqrt{\sum\left(\frac{\partial u_i}{\partial z}\right)^2} \qquad (5)$$

with selected values of $\gamma = 3$ and $\delta = -3.5$.

2.3. Settling velocity

The settling velocity is a sensitive parameter of suspended sediment transport, and remains difficult to assess as it depends on the state of flocculation of the suspension. An increase in floc size and consequently settling velocity with concentration is commonly assumed, although the floc density can decrease (e.g., van Leussen, 1994). In addition, in highly concentrated suspensions settling is hindered (e.g., Mehta, 1989). Accordingly, the following semi-empirical formulation is selected:

$$C \le C_{W\max} \Rightarrow W_s = W_{\min}(1+\alpha_1 C + \alpha_2 C^2)$$
$$C_{W\max} \le C \le C_{cr} \Rightarrow W_s = \left(1-\beta_1 C^{\beta_2}\right)^{4.65}\Re, \quad \Re = \frac{W_{\min}(1+\alpha_1 C_{W\max} + \alpha_2 C_{W\max}{}^2)}{\left(1-\beta_1 C_{W\max}^{\beta_2}\right)^{4.65}} \qquad (6)$$

where $C_{W\max}$ is the concentration at which the settling velocity is maximum ($W_{\max}$), $W_{\min}$ is the settling velocity of dispersed particles, α_1 and α_2 are empirical coefficients calculated in order to yield a given maximum settling velocity at concentration $C_{W\max}$, $\beta_1 C^{\beta_2}$ represents the volumetric concentration, and the power 4.65 is a classical value deduced from experiments of Richardson and Zaki (e.g., in Mehta, 1989). The power β_2 is taken to be equal to 0.5 in order to account for a possible densification of flocs below the critical value C_{cr}, which is conceptually close to the gel point. For concentrations larger than C_{cr}, settling results from consolidation and is assumed to vary with the sediment concentration following the power law $W_s = \gamma_1 C^{\delta_1}$.

In the following sections, settling parameters have been fitted according to the SEDIGIR experiment described later. Thus the maximum settling velocity is chosen as 2 mm s^{-1} and $W_{\min}$ 0.5 mm s^{-1}, whereas hindered settling occurs from $C_{W\max}$ = 20 kg m^{-3} up to C_{cr} = 60 kg m^{-3}, with β_1 = 0.085. Lastly, γ_1 and δ_1 are calculated so that W_s is 10^{-7} m s^{-1} for C = 500 kg m^{-3}. The resulting relationship between the settling velocity and the dry density is plotted on Figure 2, and exhibits similarity with measurements of settling velocity (e.g., Thorn and Parsons 1980, in Mehta, 1986) and/or permeability (Alexis et al., 1992).

Several particle classes can be considered, with respective advection/diffusion equations (1). However, presently, interactions between different fractions are not taken into account, except in the formulation of the settling velocity, where C represents the total mass concentration (dry density).

The formulations of the settling velocity and the generalized viscosity represent key features of the continuous modeling concept. Lastly, the set of equations is closed by the equation of state:

$$\rho = \rho_w + \sum_{cl} C_{cl} \left(1 - \frac{\rho_w}{\rho_{cl}} \right) \tag{7}$$

where ρ_w is the water density, and C_{cl} and ρ_{cl} respectively are the concentration and the grain density of the particles of class "cl".

2.4. Initial and boundary conditions

The initial and boundary conditions for (1) and (2) are expressed as zero fluxes of sediment at the surface and the bottom, i.e.,

$$\left[CW_s - K_z \frac{\partial C}{\partial z} \right]_{surf} = \left[CW_s - K_z \frac{\partial C}{\partial z} \right]_{bottom} = 0 \tag{8}$$

which means that no erosion or deposition is possible at the bottom, although deposition is permitted to occur within the near-bottom layers of the water column.

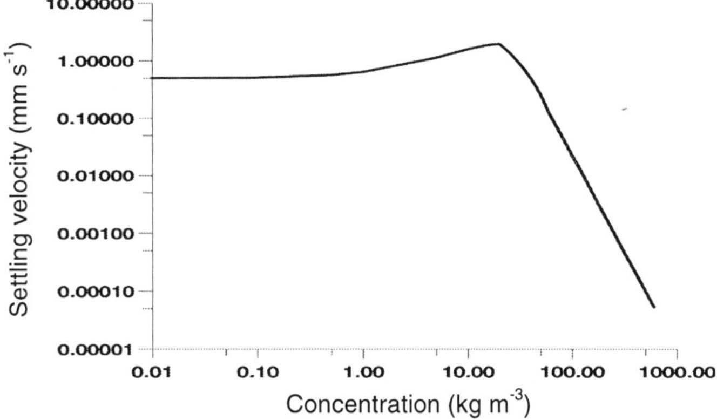

Figure 2. Simulated settling velocity as a function of sediment concentration according to (6).

For the flow, the wind-induced shear stress can be specified at the surface, whereas at the bottom a zero velocity is imposed. Such a condition avoids any modeling of the bottom shear stress which is often computed from the near-bottom velocity through a quadratic law conforming with a logarithmic velocity profile close to the boundary. This assumption, commonly used in the so-called k-ε models, is not realistic when dealing with fluid mud flows. Naturally, a no-slip condition is only possible when the near bottom discretization is very small.

As for the initiation of simulation, initial concentrations have to be specified. Most commonly the flow is assumed to be zero at the beginning.

2.5. Numerical features

The model is solved by means of a semi-implicit finite difference method, with staggered grids for u_i, W_s and C, and for Ri, ν and K_z. The mesh size can be regular or otherwise. Typically, the grid spacing gradually increases from the bottom, from 5 mm to 20 cm near the surface, for a 10 m water depth (100 meshes). In such a case the time step is on the order of 10-60 s.

3. STEADY STATE SIMULATIONS

3.1. Calibration of turbulence damping functions

Turbulence damping is introduced through the parameters α, β in (3), and γ, δ in (5). A previous inter-comparison of damping parameters had been conducted by Nunes Vaz and Simpson (1994), and tests have also been run for simulating the salinity plume of the river Seine (Cugier, 1999). Hence, we set α and γ each equal to 3. The remaining parameters have been calibrated by using an experiment of Coleman (1981) (in Galland, 1996). In a recirculating flume, Coleman simultaneously measured the vertical profiles of velocity and concentration of fine sand, and observed a reduction of flow when the volumetric fraction of particles exceeded 5×10^{-3} (Figure 3). Results of our simulation are also presented in Figure 3, and are closer to measurements than those obtained by using the Munk and Anderson damping functions ($\alpha = 10$, $\beta = -0.5$, $\gamma = 3.33$, $\delta = -1.5$). The fitting leads to $\beta = -4$ and $\delta = -3.5$. A large value of β is needed for correct velocity damping; in particular, this coefficient is required to be at least as large as δ, which leads to an unusual decrease in the turbulent Prandtl-Schmidt number, ν_z/K_z, with Ri.

3.2. Equilibrium profile

Several steady state tests were carried out for a 10 m deep water column, with single-class particles (Figure 4). The different tests were defined by the forcing (surface slope), initial uniform concentration (C_i), density-induced viscosity increase (k_1) and "Bingham behavior" (k_3). The settling velocity was not changed, nor the turbulence damping parameters α, β, γ, δ. However, the latter coefficients could be set to zero, in order to neglect turbulence damping. Examples are given in Figure 4 for a unique forcing.

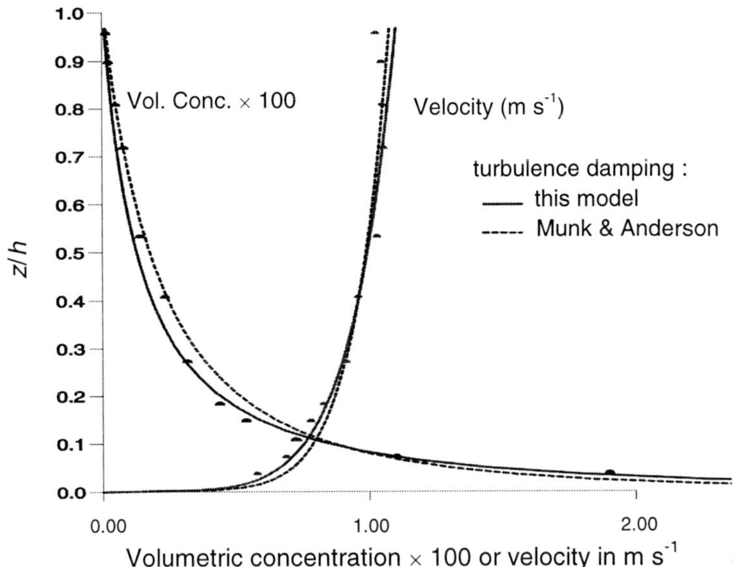

Figure 3. Simulation of experiment 18 from Coleman (1981). Conditions: mean sand diameter = 0.105 mm, mean concentration = 3.74×10^{-3} m^3 m^{-3}, mean velocity = 0.95 m s^{-1}, water depth = 0.170 m and settling velocity = 0.0087 m s^{-1} (Zanke formulation) + hindering. Measured and computed concentration and velocity profiles are shown.

In most cases, fluid mud layers occur near the bottom at equilibrium, sometimes with a strong lutocline around 50 kg m^{-3} due to strong hindered settling in this concentration range (*cf.* Figure 2). When the viscosity increase with concentration is accounted for, velocity and turbulent mixing are damped in this domain, and a soft mud layer is formed in the range of 200-300 kg m^{-3}.

Except in the case of *no turbulence damping* for which the water column is rather well-mixed, all tests show similar concentration profiles in the upper layer, regardless of the total mass and the Bingham behavior. On the other hand, viscosity increase seems to change the upper profile, even in the very low concentration range, probably because of its effect on the velocity which partly controls the concentration profile. Lastly, when the initial concentration is *very* low (0.1 gl^{-1}), the concentration does not reach an "equilibrium" value, the maximum concentration that the flow can transport for a given material characterized by its settling velocity. This can be connected to the concept of *saturation concentration* developed by Winterwerp et al. (this volume) based on the works from Teisson et al. (1992) and Galland (1996).

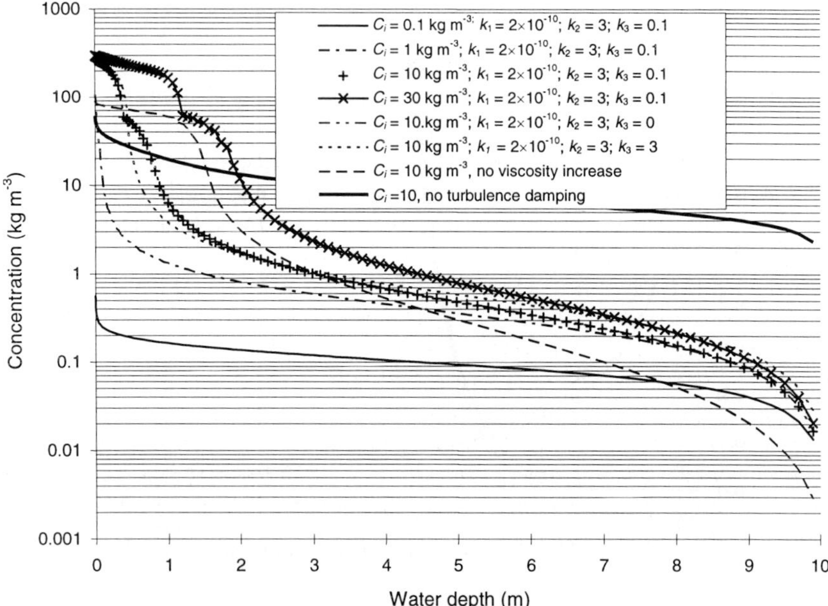

Figure 4. Equilibrium vertical profiles of concentration for different conditions and total mass. Common forcing: a surface slope of 10^{-6}, i.e., a bottom friction velocity of 0.01 m s^{-1}.

3.3. Saturation concentration

Assuming a logarithmic velocity profile and equilibrium between settling and mixing, the flux Richardson number can be stated (Teisson et al., 1992; Winterwerp et al., this volume) as:

$$Ri_{\mathrm{f}} = g\left(1-\frac{\rho_w}{\rho_s}\right)\kappa\frac{C}{\rho_w}\frac{W_s}{\left[u_*^3\left(\frac{1}{z}-\frac{1}{h}\right)\right]} \tag{9}$$

where g is the acceleration due to gravity, ρ_w and ρ_s are the density of water and grains, respectively, and u_* is the friction velocity.

Field experiments provide evidence of a critical value (≈0.15 according to Winterwerp et al., this volume) that Ri_{f} cannot exceed, leading to a maximum concentration that the flow can sustain:

$$C_{\max}(z) = Ri_{\mathrm{fmax}}\frac{\rho_w}{\left[g\left(1-\frac{\rho_w}{\rho_s}\right)\kappa\right]}u_*^3\left(\frac{\frac{1}{z}-\frac{1}{h}}{W_s}\right) \tag{10}$$

The diversity of concentration profiles shown on Figure 4 in the upper layer seems to indicate that the saturation concentration is not encountered. However, in reality the settling velocity is not constant, and, especially, the velocity profile is not logarithmic because the viscosity is damped by stratification.

In order to "validate" the saturation concept, a set of tests were performed with a constant W_s (1 mm/s) at low concentrations, a mixing length varying as $z(1-z/h)^{0.5}$, and stratification-induced damping only for the eddy diffusivity (α and β = 0, γ = 3, δ = -1.5). These tests are presented in Figure 5. Here the profiles are much more similar. In addition, the flux Richardson number, deduced from the computed gradient Richardson number and the Prandtl-Schmidt number, proves to be rather uniform in the upper layer, with a maximum value of 0.13 (it actually depends on the damping parameters γ and δ). When computing the saturation concentration profile with this critical Ri_{fmax}, predicted concentrations are very close to the modeled profiles (Figure 5).

The saturation concept is thus validated provided the velocity profile remains logarithmic. As noted, this is generally not the case in stratified waters; however, even when viscous damping is accounted for, the concentrations come together above the fluid mud layer (Figure 4).

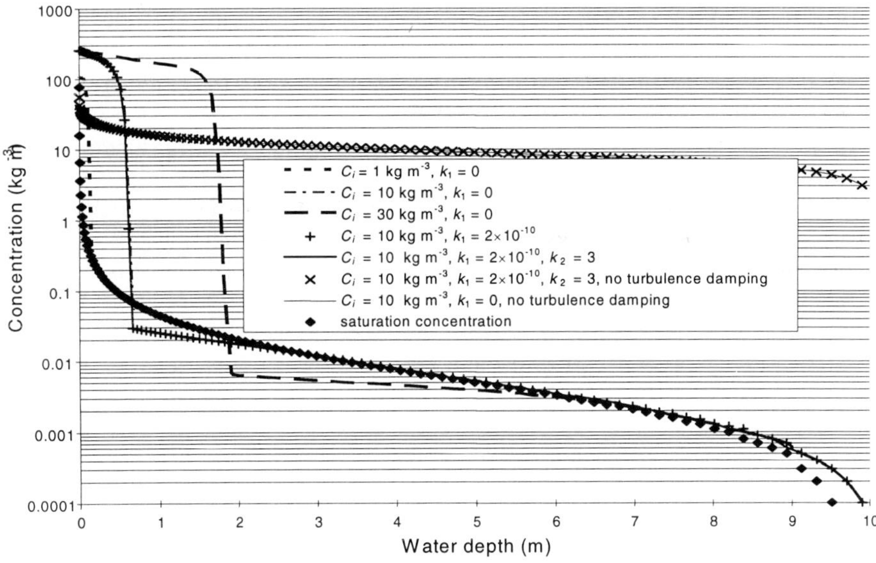

Figure 5. Equilibrium vertical profile of concentration for different conditions and total mass. See text for settling velocity parameters and turbulence damping factors (for the eddy diffusivity only). Diamonds: computed saturation concentration according to (10). Common forcing: a surface slope of 10^{-6}, i.e., a bottom friction velocity of 0.01 m s^{-1}.

4. FLUID MUD FLOW IN THE GIRONDE

4.1. The SEDIGIR experiment

The Gironde estuary is macrotidal, with a 5 m range at spring tide. The tide propagates 160 km upstream of the mouth located on the Atlantic Ocean. Mean river discharge is approximately 1000 $m^3\ s^{-1}$. The estuary is known for its well-developed turbidity maximum (Allen et al., 1980), and fluid mud layers which can attain a 2 m thickness. In order to observe fluid mud formation at neap tide, as well as its erosion when the tidal range increases, a 12 day experiment (SEDIGIR) was conducted in June 1996. Using the SAMPLE probe (Jestin et al., 1994), simultaneous vertical profiles of sediment concentration, flow velocity and direction, temperature and salinity were obtained at a fixed location, once or twice per hour during 9 tidal cycles spread over the fortnightly lunar cycle. Thanks to a double calibration curve of the OBS turbidity sensors used, and confirmed by synchronous data from an acoustic density probe, sediment concentrations were obtained nearly continuously in the range of 0.1 to 400 kg m^{-3}. As an example, a spring tidal cycle is represented in Figure 6a, and examples of concentration profiles at neap and spring tides are given in Figure 7a. Several features need to be pointed out:

- Settling exhibits a delay with respect to flow reversal, and gives rise to a relatively homogeneous fluid mud in the range of 50-70 kg m^{-3}. This range is seen throughout the fortnightly cycle, and seems to be governed mainly by settling, and also resuspension. It can probably be related to a discontinuity in the settling rate, which leads to the formation of a lutocline;
- At maximum ebb and flood, resuspension occurs quickly although flow acceleration is gradual, which implies a non-linear effect;
- When concentration exceeds several tens of kg m^{-3} turbulence is damped, as shown and partly quantified by the velocity contours;
- During flow reversals, a bottom advance is seen, especially when the suspended sediment concentration is low.

Notwithstanding some ambiguity regarding the hard bottom level, the vertically-integrated sediment mass seems to change significantly, revealing strong advective effects. For instance, around 279 hours in Figure 6a the total sediment mass increases, which necessitates an input by advection. A longitudinal survey of the estuary revealed that in actuality the core of the turbidity maximum was slightly upstream of the measurement location.

4.2. Simulation of SEDIGIR data

The 1DV continuous model was used for simulating resuspension and settling over the fortnightly tidal cycle in a similar environment as one where the Gironde experiment took place. The chosen flow intensity, water height and total mobile sediment were based on data, although height and total sediment

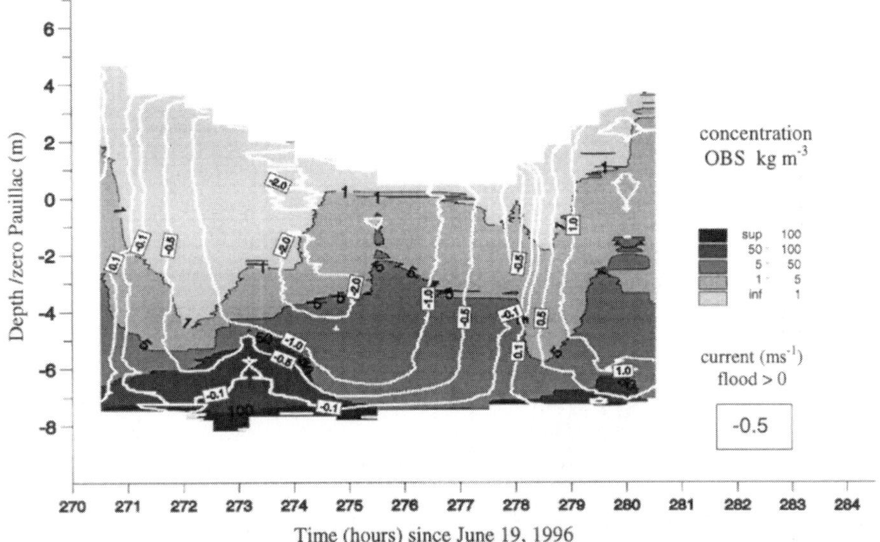

Figure 6a. Time-evolution of suspended sediment concentration and velocity at spring tide in the turbidity maximum zone of the Gironde estuary during the SEDIGIR experiment at Pauillac near Bordeaux, France.

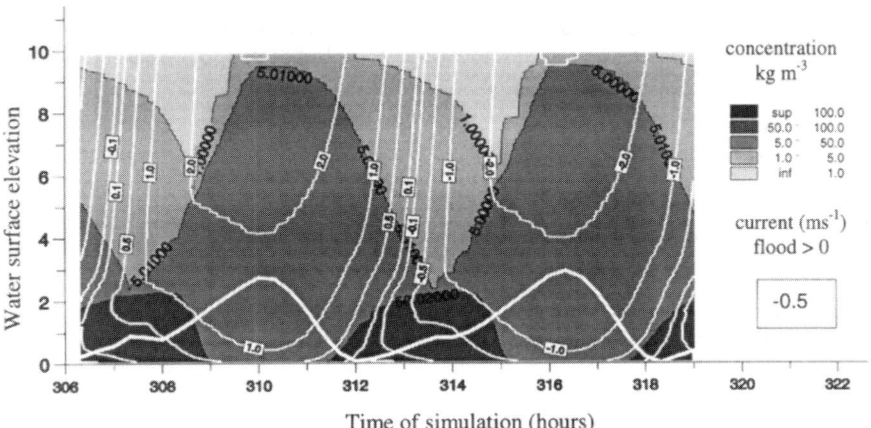

Figure 6b. Computed time-evolution of suspended sediment concentration and velocity at spring tide in a similar environment, with a continuous 1DV model (SAM-1DV). The model is forced by a sinusoidal mean flow with a superimposed spring/neap/spring tide. Thick white line: bottom shear stress (N m^{-2}).

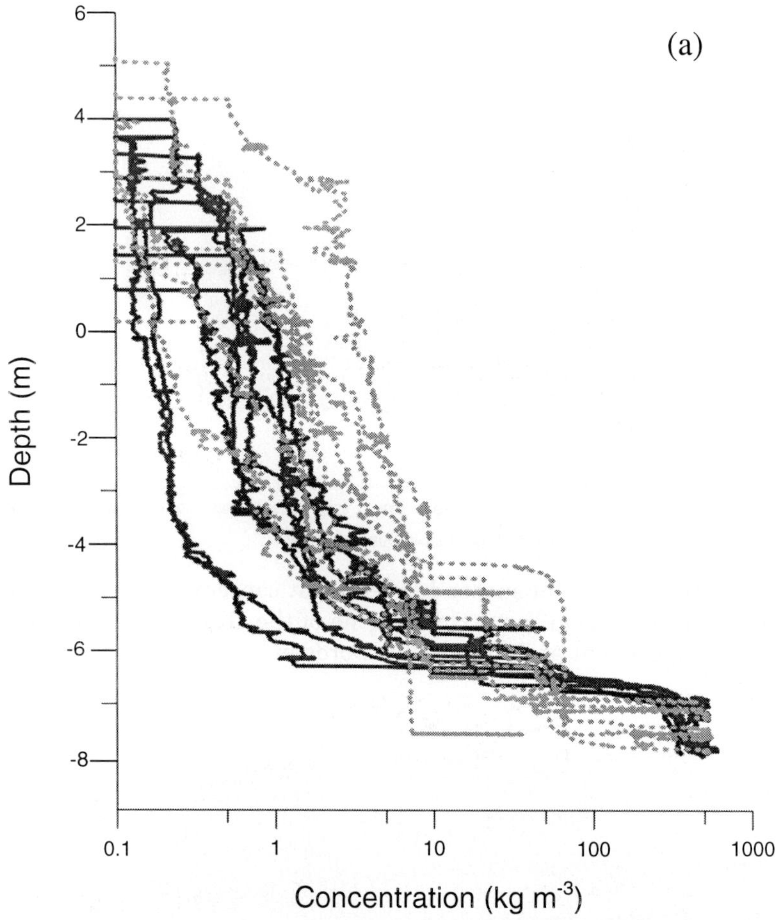

Figure 7a. Measured vertical profiles of sediment concentration in the turbidity maximum zone of the Gironde estuary in June 1996: neap tide (dark lines); spring tide (dotted lines).

were assumed to be steady throughout the cycle, for the sake of mass conservation in the 1DV formulation. The model was forced by the mean flow, with sinusoidal time-variation including a spring/neap/spring variation of tidal amplitude. The vertical resolution was about 5 cm. The main features were qualitatively reproduced (Figures 6b and 7b,c), including the large range of suspension concentration (from 0.1 kg m^{-3} to several hundreds of kg m^{-3}), decay of current intensity within fluid mud, which reflects turbulence damping, and a slight advance of the bottom layer close to flow reversal.

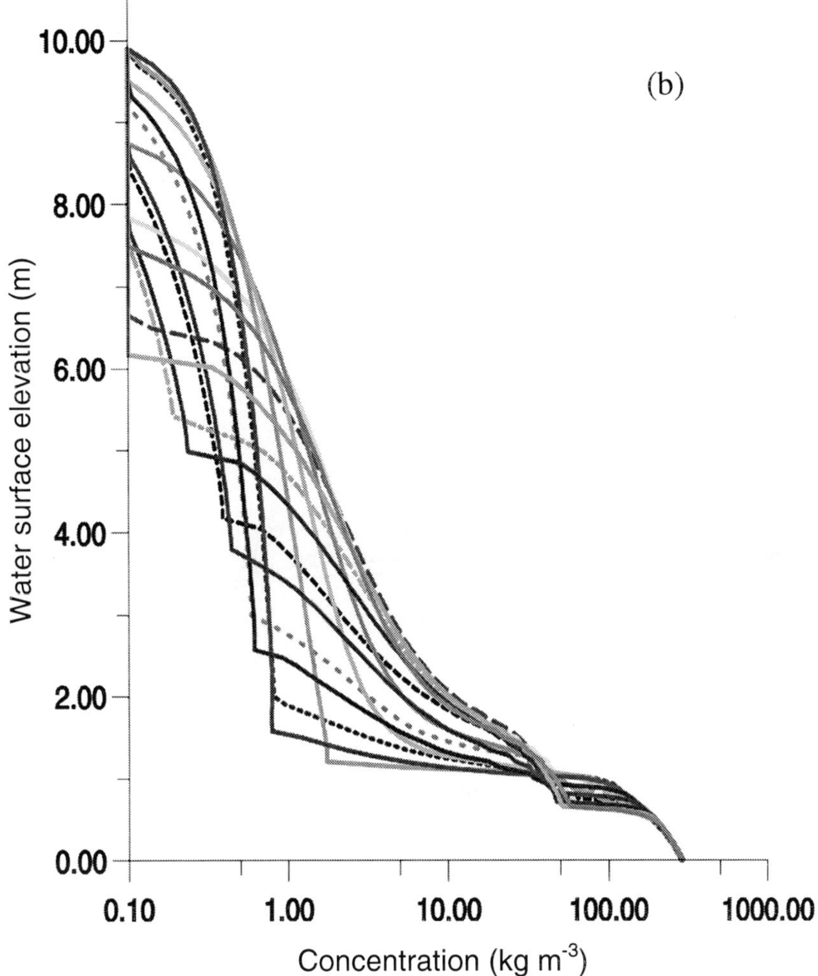

Figure 7b. Computed vertical profiles of sediment concentration in the turbidity maximum zone of the Gironde estuary during a neap tide in June 1996.

The neap/spring cycle was also simulated, with the formation of soft mud at neap tide. The contrast between the profiles at neap, characterized by a 1m thick soft mud layer below a small and variable liquid mud layer (about 50 kg m^{-3}) and turbid layers in the range of 1 kg m^{-3} on one hand, and the spring profiles, characterized by higher turbidity in the upper layer (around 5 kg m^{-3}) and a temporary high (2 m) liquid mud layer on the other hand, is well reproduced (Figure 7). However, differences are at times large; when resuspension is

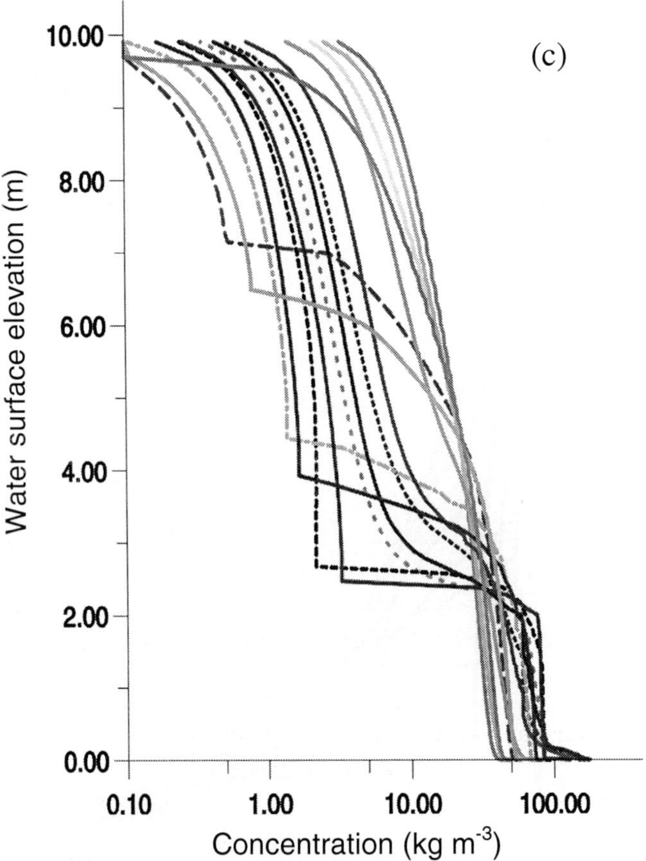

Figure 7c. Computed vertical profiles of sediment concentration in the turbidity maximum zone of the Gironde estuary during a spring tide in June 1996.

maximum at spring, the model over-predicts the concentration (Figure 6), essentially because in reality the turbid mass is largely advected. Also, the observations in the liquid mud (about 50-70 kg m^{-3}) show a large zone of null velocity in a homogeneous layer, 2 hours after reversal (hour 273 in Figure 6a). The model proved unable to reproduce such a feature. Possibly an increase in the viscosity or the Bingham threshold caused such a damping of velocity. However, settling would then have occurred within this layer, which is not observed. Alternatively, a reduction of viscosity could have enhanced the observed homogeneity of concentration. However, laminar velocity gradients would then have occurred over the entire layer. Given these sets of inferences, two explanations are possible: 1) turbulence closure may be inadequate, requiring the need to account for the transport of turbulent quantities, or 2) the observed behavior could be due to longitudinal advection of the fluid mud, linked to a possible slope of the interface with the upper layer. To explore such a case, a full 3D continuous modeling is needed. Within the 1DV framework, progress can be achieved by forcing the model with a surface slope derived from a depth-averaged model of the estuary yielding more realistic accelerations. However, the uncertainty in the vertical distribution of advective fluxes would still remain.

5. CONCLUSIONS

By using a one-dimensional vertical model based on *the continuous modeling concept*, the main processes involved in the formation of high suspended sediment concentration and fluid mud layers have been explored.

Stratification-induced turbulence damping is fundamental for explaining lutocline events in high-energy environments, such as macrotidal estuaries. For a better agreement between modeling and measurements, it appears that eddy viscosity damping should be enhanced in contrast to classical empirical damping functions. However, it is clear that such a conclusion must be confirmed by other, more elaborate turbulence closure schemes, which must be applied in the specific environment of soft mud bottom.

The settling velocity is the second important parameter which influences fluid mud layers. An effort at validation must be undertaken in the concentration range of 20-200 kg m^{-3}, within the zone of strongly hindered settling. The likely influence of horizontal and/or turbulent movements increases the difficulty in validation.

The viscous behavior of mud significantly contributes to the damping of velocity in soft mud, and a simple formulation has been proposed to account for viscoplasticity at low shear rates. The likely effect of such a laminar behavior on the mixing of material has not been considered here, but should be examined.

Naturally, calibration and comparison with measurements should be carried out. In particular, the set of data acquired during the SEDIGIR experiment in the Gironde are useful; however, a full 3D model needs to be used for improving validation. The final goal of the exercise should be not only to provide a set of calibrated parameters, but to assess the relative importance of the involved processes.

6. ACKNOWLEDGMENT

This work is the result of a collaborative effort between IFREMER and the University of Bordeaux. The SAM-1DV model was developed under the French Program National d'Océanographie Côtière, and partly funded by the Commission of the European Communities through the INTRMUD project under contract MAS3-CT95-0022. The authors would like to thank Dr. A. Sottolichio and Prof. P. Castaing of the University of Bordeaux for their contribution to the SEDIGIR experiment and for fruitful discussions.

REFERENCES

Alexis, A., Bassoullet, P., Le Hir P., and Teisson, C., 1992. Consolidation of soft marine soils: unifying theories, numerical modelling and *in situ* experiments. *Proceedings of the 23rd International Conference on Coastal Engineering,* American Society of Civil Engineers, New York, 2949-2961.

Allen, G. P., Sauzay, G., Castaing, P., and Jouanneau, J. M., 1977. Transport and deposition of suspended sediment in the Gironde estuary, France. In: *Estuarine Processes, Vol. II: Circulation, Sediments and Transfer of Material in the Estuary,* Academic Press, New York, 63-81.

Allen, G. P., Salomon, J. C., Bassoullet, P., Du Penhoat, Y., and De Grandpré, C., 1980. Effects of tides on mixing and suspended sediment transport in macrotidal estuaries. *Sedimentary Geology,* 26, 69-90.

Brenon, I., 1997. Modélisation de la dynamique des sédiments fins dans l'estuaire de la Seine. *Thèse de Doctorat,* No. 501-1997, Université de Bretagne Occidentale, France, 204p.

Castaing, P., 1981. Le transfert à l'océan des suspensions estuariennes. Cas de la Gironde. *Thèse d'Etat,* Université de Bordeaux, France, 530p.

Coleman, N. L., 1981. Velocity profiles with suspended sediment. *Journal of Hydraulic Research,* 19(3), 211-229.

Cugier Ph., 1999. Modélisation du devenir à moyen terme dans l'eau et le sédiment des éléments majeurs (N, P, Si) rejetés par la Seine en Baie de Seine. Thèse de l'Université de Caen (France).

Gallenne, B., 1974. Les accumulations turbides dans l'estuaire de la Loire. Etude de la "crème de vase". *Thèse de 3ème cycle,* Université de Nantes, France, 323p.

Galland, J.-C., 1996. Transport de sédiments en suspension et turbulence. *Rapport HE-42/96/007/A,* Laboratoire National d'Hydraulique, EDF-DER, Chatou, France, 88p.

Inglis, C., and Allen, F. H., 1957. The regimen of the Thames estuary as affected by currents, salinities and river flow. *Proceedings of the Institution of Civil Engineers,* London, 827-868.

Jestin, H., Le Hir, P., and Bassoullet, P., 1994. The "SAMPLE system", a new concept of benthic station. *Proceedings of Oceans'94,* Brest, France, 278-283.

Kirby, R., 1988. High concentration suspension (fluid mud) layers in estuaries. In: *Physical Processes in Estuaries,* J. Dronkers and W. van Leussen eds., Springer-Verlag, New York, 463-487.

Kirby, R., and Parker, W. R., 1983. Distribution and behavior of fine sediment in the Severn Estuary and Inner Bristol Channel, U.K. *Canadian Journal of Fisheries and Aquatic Sciences,* 40(1), 83-95.

Le Hir, P., 1997. Fluid and sediment "integrated" modelling: application to fluid mud flows in estuaries. In: *Cohesive Sediments,* N. Burt, R. Parker and J. Watts eds., John Wiley, Chichester, UK, 417-428.

Le Normant, C., 1995. Modélisation numérique tridimensionnelle des processus de transport des sédiments cohésifs en environnement estuarien. *Thèse de Doctorat,* No 1002-1995, Université de Toulouse, France, 236p.

Li, Y., 1996. Sediment-associated constituent release at the mud-water interface due to monochromatic waves. *Ph.D. Thesis,* University of Florida, Gainesville, 344p.

Mehta, A. J., 1986. Characterization of cohesive sediment properties and transport processes in estuaries. In: *Estuarine Cohesive Sediment Dynamics.* Lecture Notes on Coastal and Estuarine Studies, Vol. 14, A. J. Mehta ed., Springer-Verlag, New York, 290-325.

Mehta, A. J., 1989. On estuarine cohesive sediment suspension behavior. *Journal of Geophysical Research,* 94(C10), 14,303-14,314.

Migniot, C., 1989. Tassement et rhéologie des vases, 2ème partie. *La Houille Blanche,* 95-111.

Mulder, T., and Cochonat, P., 1996. Classification of offshore mass movements. *Journal of Sedimentary Research,* 66(1), 43-57.

Nunes Vaz, R. A., and Simpson, J. H., 1994. Turbulence closure modeling of estuarine stratification. *Journal of Geophyscial Research,* 99(C8), 16,143-16,160.

Odd, N. V. M., and Owen, M. W., 1972. A two-layer model of mud transport in the Thames estuary. *Proceedings of the Institution of Civil Engineers,* Supplement paper 7517S, London, 175-205.

Teisson, C., Simonin, O., Galland, J. C., and Laurence, D., 1992. Turbulence and mud sedimentation: a Reynolds stress model and a two-phase flow model. *Proceedings of the 23rd International Conference on Coastal Engineering*, American Society of Civil Engineers, New York, 2853-2866.

Thorn, M., and Parsons, J., 1980. Erosion of cohesive sediments in estuaries. An engineering guide. *Proceedings of the Third international Conference on Dredging Technology*, Bordeaux, B.H.R.A., Cranfield, Bedford, UK, 349-358.

Toorman, E., 1992. Modelling of fluid mud flow and consolidation. *Ph.D Thesis,* Katholieke Universiteit Leuven, Heverlee, Belgium, 219p.

van Leussen, W., 1994. Estuarine macroflocs and their role in fine-grained sediment transport. *Ph.D. Thesis,* University of Utrecht, Utrecht, The Netherlands, 488p.

Winterwerp, J., 1998. Sediment-fluid interactions. *Report for the MAST3 Cosinus project,* Delft Hydraulics and Technical University of Delft, Delft, The Netherlands, 24p + 4 fig.

Winterwerp J. C., Uittenbogaard, R. E., and de Kok, J. M., 2000. Rapid siltation from saturated mud suspensions. In: *Coastal and Estuarine Fine Sediment Transport Processes*, W. H. McAnally and A. J. Mehta eds., Elsevier, Oxford, UK.

Coastal and Estuarine Fine Sediment Processes
W.H. McAnally and A.J. Mehta (Editors)

Modeling of fluid mud flow on an inclined bed

R. Watanabe[a], T. Kusuda[b], H. Yamanishi[b] and K. Yamasaki[a]

[a]Department of Civil Engineering, Fukuoka University, Nanakuma 8-19-1, Jonan-ku, Fukuoka 814-0180, Japan

[b]Department of Urban and Environmental Engineering, Graduate School, Kyushu University, Hakozaki 6-10-1, Higashi-ku, Fukuoka 812-8581, Japan

Characteristics of fluid mud flow on an inclined bed were investigated in the quiescent state in an experimental flume. The slope of the inclined bed was adjustable from 0 to 1:473. Suspension concentration in the experiments ranged from 3.2 to 64 kg m^{-3}. With an increase in the initial concentration, the maximum velocity of fluid mud decreased but the highest concentration increased. Apparent viscosity in the fluid mud was three orders of magnitude larger than that of water. Based on these experimental results the constitutive equation, the rate of deposition and the dispersion coefficient in fluid mud were obtained. Using these equations discretized by the SIMPLE method, simulation of fluid mud flow on the bed was performed. The simulation results in the quiescent state agreed well with the experimental results, and accurately reproduced fluid mud flow on the inclined bed.

1. INTRODUCTION

Fluid mud formation following settling of suspended solids in muddy tidal rivers and estuaries plays an important role in narrowing river cross sections and shoaling navigation channels. Shoaling of navigation channels in estuaries has been studied intensively, and is known to be often due to horizontal transport of suspended solids as fluid mud, rather than by direct settling of suspended solids to the channels (e.g., Odd and Cooper, 1989). Thus, the formation of fluid mud flow on an inclined bed has received much attention. Although a variety of numerical models have been developed for simulating fluid mud (e.g., Kessel and Kranenburg, 1996; Ali et al., 1997; Crapper and Ali, 1997), fundamental elements of the process such as the constitutive equations and the dispersion rate in the fluid mud layer have not been fully understood yet. In this study the constitutive equation, the rate of deposition on the bed mud and the dispersion coefficient in fluid mud are introduced theoretically and experimentally. Based on these results and calculated dispersion rates in fluid mud, its movement on an inclined bed is simulated.

2. MOVEMENT OF FLUID MUD ON AN INCLINED BED

A fluid mud layer formed with a supply of suspended solids by settling is outlined in Figure 1. The layer grows with continuing supply of solids. A lutocline exists between the layer and the overlying suspension. The layer is structurally divided into two parts: a mobile fluid mud layer and a stationary fluid mud layer (Ross and Mehta, 1989), hereafter called bed mud layer. This layer grows with supply of suspended solids, that is, by deposition from the mobile fluid mud layer. The boundary between these two layers may shift vertically depending on the applied shear stress at the surface of the bed mud.

The movement of fluid mud is controlled by the rate of supply and dissipation of energy. The main driving force is gravity which causes a density current. The settling of suspended solids supplies the fluid mud with additional potential energy. The shear stress in the fluid mud is calculable from the balance of forces under supply of suspended solids.

We introduce some assumptions for simplicity of modeling the movement of fluid mud as follows:

1. The concentration of suspended solids in the upper suspension layer is kept constant (C_a);

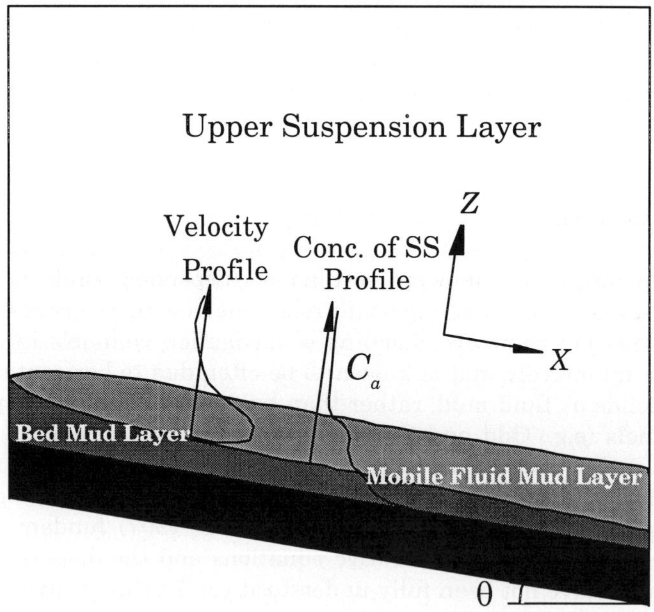

Figure 1. Fluid mud layer on an inclined bed.

2. The Boussinesq approximation is applicable to the fluid mud layer because it is thin;
3. Fluid mud flow is assumed to be quasi-uniform; and
4. The Reynolds Number of the movement of fluid mud is less than unity.

Under these assumptions, two-dimensional continuity and momentum equations for fluid mud flow are introduced as follows:

Continuity Equation:

$$\frac{\partial u}{\partial x}+\frac{\partial w}{\partial z}=0 \tag{1}$$

Momentum Equation:

$$\frac{\partial \rho_m u^2}{\partial x}+\frac{\partial \rho_m uw}{\partial z}= -\frac{\partial}{\partial x}\int_z^{H(x)} R(C-C_a)\ g\cos\theta\ dz+R(C-C_a)\ g\sin\theta+\frac{\partial \tau_{xz}}{\partial z} \tag{2}$$

$$\frac{\partial \rho_m uw}{\partial x}+\frac{\partial \rho_m ww}{\partial z}=-\frac{\partial P}{\partial z}+R(C-C_a)\ g\cos\theta \tag{3}$$

$$R=\frac{\rho_s-\rho_l}{\rho_s} \tag{4}$$

where C is the concentration of fluid mud, C_a is the concentration in the upper suspension layer, τ_{xz} is the shear stress along the inclined bed, u and w are the velocities of fluid mud in the x and z directions, respectively, P is pressure, g is the acceleration due to gravity, θ is the slope of the inclined bed, ρ_s is the density of suspended solids, ρ_l is the density of liquid, ρ_m is the ambient density of fluid mud, z is the axis perpendicular to the inclined bed, positive upward, and x is the axis along the inclined bed.

The boundary conditions to solve (1), (2) and (3) are as follows: u = 0 and w is constant at the boundary between fluid mud layer and bed mud layer; w and P are continuous at the interface between fluid mud layer and bed mud layer, and u, w and P are continuous at the interface between overlying water and fluid mud.

The mass conservation equation for the solids phase is as follows:

$$\frac{\partial C}{\partial t}+\frac{\partial uC}{\partial x}+\frac{\partial (w-w_s)C}{\partial z}=\frac{\partial}{\partial z}\left(K_z\frac{\partial C}{\partial z}\right) \tag{5}$$

The dispersion coefficient is one of the controlling factors of vertical mass flux in the fluid mud layer. Under a steady-state condition, the dispersion coefficient is calculated from (6):

$$K_z = \frac{\left[w_{sa} C_a - w_s C + \frac{\partial}{\partial x} \int_z^{\infty} uC\, dz \right]}{\frac{\partial C}{\partial z}} \tag{6}$$

When the concentration profile above and in the fluid mud changes insignificantly, that is, the settling flux of suspended solids is almost equal to the deposition flux of fluid mud, the dispersion coefficient is easily obtained by (7).

$$K_z = \frac{\left[w_s C - F_d \right]}{\frac{\partial C}{\partial z}} \tag{7}$$

where C_a and C respectively are the concentrations of suspended solids far above the fluid mud and at a point where K is calculated, W_{sa} and W_s are the settling velocities of suspended solids as a function of C_a and C, respectively, and F_d is the flux to bed mud layer.

3. EXPERIMENTAL RESULTS AND INTERPRETATION

The experimental apparatus, methods and test materials used are as follows (see Kusuda et al., 1993,1997, for further reference). The sediment was obtained from the Kumamoto Port in Ariake Bay located in the western part of Japan. The grain size distribution of the material was 48% in the clay range, 47% in the silt range, and 5% in the sand range. The mean grain diameter was 0.060 mm, grain specific gravity was 2.65 and loss on ignition was 12.2%. All experiments were conducted in salt water of specific gravity 1.025. The settling velocity of the sediment in the quiescent state was obtained from column tests using the same salt water. The column was 10 cm in diameter and 1 m high. A flume employed in the experiments was 2 m high, 2 m wide and 0.2 m thick. An inclined bed was installed in it, with slope adjustable from 0 to 1:473 (0.523 rad). Sampling tubes were installed in 4 rows, 0.30, 0.60, 0.90 and 1.20 m away from the upstream end, at levels 0.7 and 1.20 m above the bottom of the flume and from 0 to 5 cm from the inclined bed surface at a distance of 2.5 mm. Movement of fluid mud on the inclined bed was recorded by a video camera with a close-up lens through the flume wall. Fluid mud was sampled through small tubes, and the concentration of suspended solids was measured.

3.1 Settling velocity

The settling velocity of suspended solids at the initial stage was obtained as

$$w_s = 7.3 \times 10^{-4} \left(1 - 5.0 \times 10^{-3} C\right)^{6.6} \tag{8}$$

where w_s is the settling velocity (m s^{-1}) and C is the corresponding concentration (kg m^{-3}).

3.2 Constitutive equation in fluid mud

The apparent viscosity in the fluid mud layer is a function of the deformation rate D (s^{-1}) and the solids fraction in the layer, as shown in Figure 2. Accordingly, the apparent viscosity can be written as

$$\mu_a = \mu_w \left(6.1\, D^{-0.66} + 1\right) \left\{3.8 \times 10^3 \left(1 - \varepsilon\right)^{1.7} + 1\right\} \tag{9}$$

where μ_w is the viscosity of water and $1 - \varepsilon$ is the solids fraction (concentration).

3.3 Deposition rate on bed mud

A non-dimensional flux to the bed mud is shown in Figure 3. The depositional flux to the bed mud from the fluid mud layer is determined from the solids fraction profile and velocity profile in each run. These profiles were measured at intervals of 10 to 15 minutes (see Kusuda et al., 1993, for further reference). In

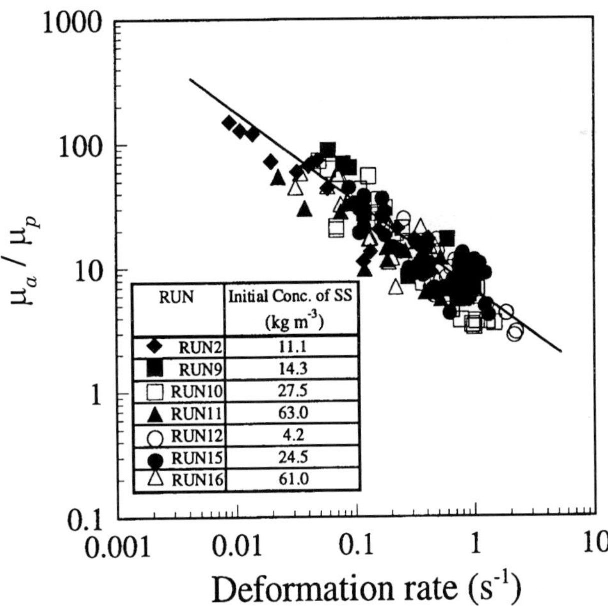

Figure 2. Apparent viscosity as a function of deformation rate.

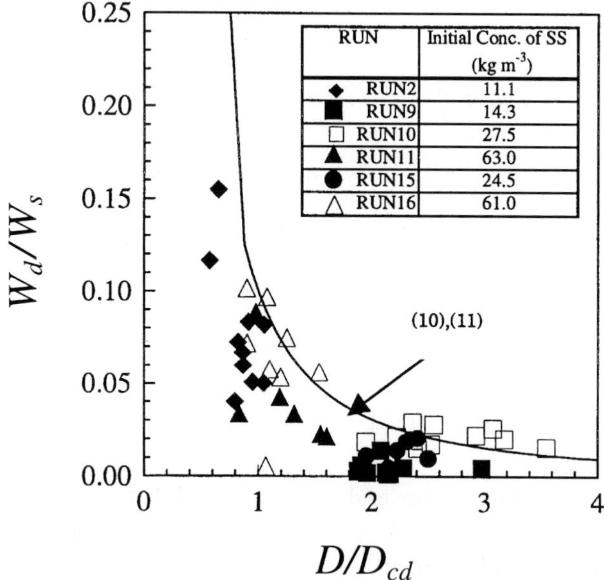

Figure 3. Relation for the non-dimensional deposition rate to bed mud.

Figure 3, two sets of data, Runs 11 and 16, in which the initial concentrations were higher than the critical value over which effective stress appears, do not fall on the line of Run 2 data because of an underestimation of the hindered settling velocity. The data in the figure, except Runs 11 and 16, are expressed as a function of deformation rate as follows:

$$\frac{w_d}{w_s} = \left(1 - \frac{D}{D_{cd}}\right) \qquad D < (1-\alpha)\ D_{cd} \tag{10}$$

$$\frac{w_d}{w_s} = \alpha\left(\frac{D}{D_{cd}}\right)^{-1.2} \qquad D_{ce} > D \geqq (1-\alpha)\ D_{cd} \tag{11}$$

where $\alpha = 0.1$, D_{cd} is nearly equal to 0.4 (s^{-1}), w_d is the deposition velocity, and w_s is the settling velocity in the quiescent state. The depositional flux is given by the product of the settling velocity and the concentration at the surface of bed mud.

3.4 Dispersion coefficient in the fluid mud

The flow regime in the fluid mud layer is schematized in Figure 4. In this figure, the lower fluid mud layer occurs between the bed mud surface and the middle layer, and the upper fluid mud layer occurs between the middle layer and

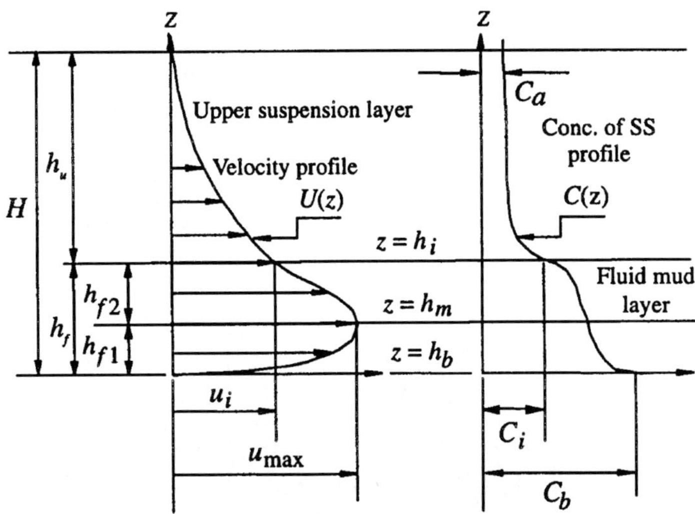

Figure 4. Flow regime in the fluid mud layer. The bed is assumed to be horizontal.

the interface. The dispersion coefficient characterizing "wall turbulence" is conventionally expressed as a function of z^2D (Hinze, 1975), where z is a distance from the bed surface. According to this expression, and considering the effects of the concentration of suspended solids in fluid mud, K_z in the lower fluid mud layer is deduced as shown in Figure 5 and is described by the equation

$$K(z) = 0.12 \ z^2 D\left(\frac{\kappa_a}{\kappa_0}\right) \tag{12}$$

where κ_0 is the von Karman constant, and κ_a is the apparent von Karman constant. Figure 6 indicates the relationship between κ_a / κ_0 and the solids fraction in the fluid mud layers. This ratio is obtained as a function of solids fraction as follows (van Rijn, 1980):

$$\frac{\kappa_a}{\kappa_0} = 1 + \left(\frac{1-\varepsilon}{1-\varepsilon_0}\right)^{0.8} - 2\left(\frac{1-\varepsilon}{1-\varepsilon_0}\right)^{0.4} \tag{13}$$

where $1-\varepsilon_0 = 0.35$.

The dispersion coefficient K_i at the interface between the fluid mud layer and the upper suspension layer, as shown in Figure 7, is given by:

$$K_i(z) = 8.8 \times 10^{-7}\left(\frac{u_m - u_i}{h_{f2}}\right)^{0.4} \tag{14}$$

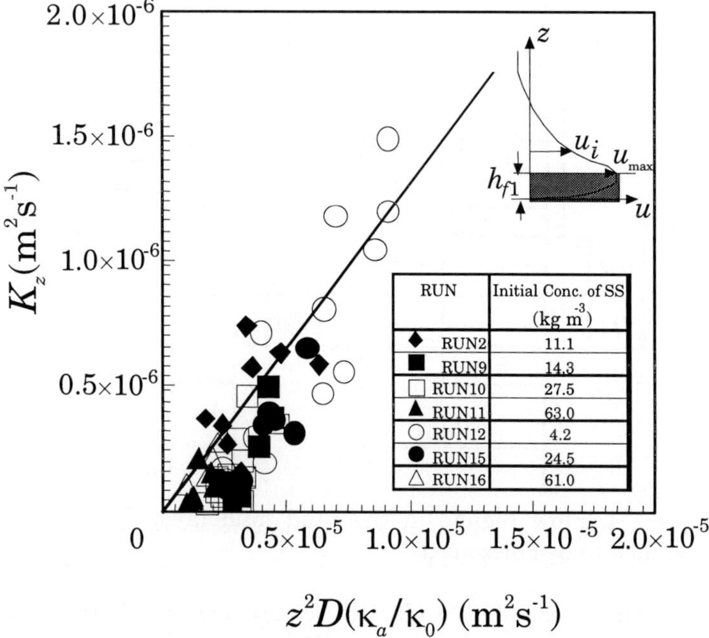

Figure 5. Dispersion coefficient in the lower fluid mud layer.

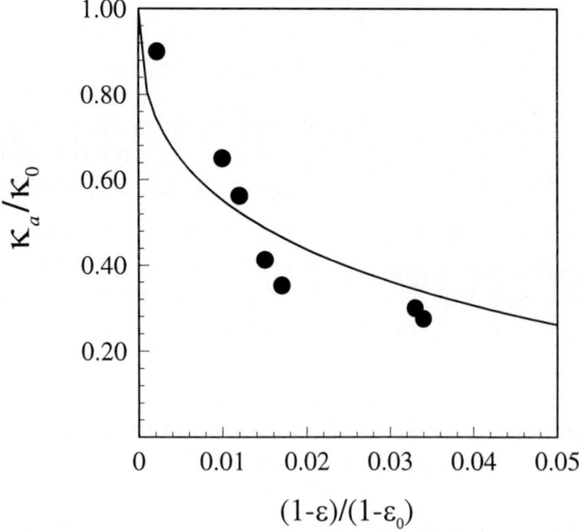

Figure 6. The relation between solids fraction and κ_a/κ_0.

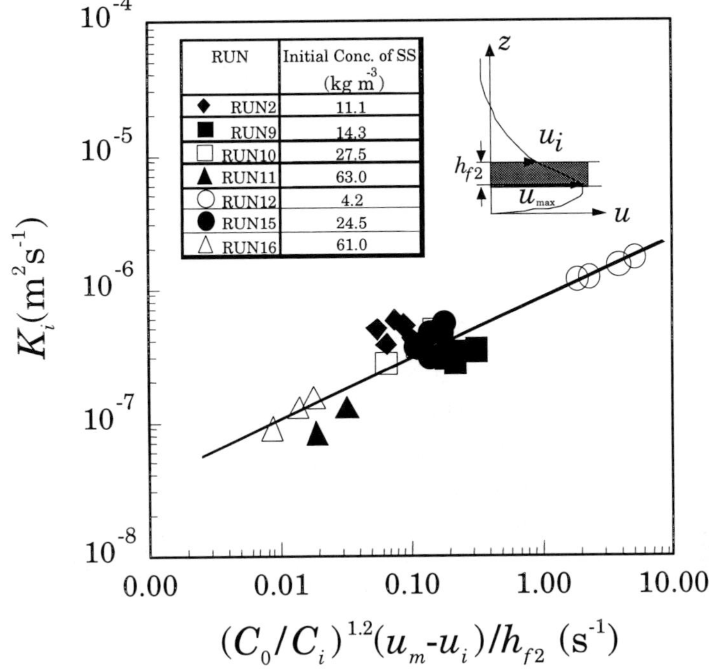

Figure 7. Dispersion coefficient in the fluid mud layer.

4. NUMERICAL SIMULATION AND DISCUSSION

The procedure for simulation is as follows. The bottom velocity (u=0 at z=0), homogeneous concentration of the upper suspension (C_a), and the slope angle are given as initial conditions. A set of the equations is discretized by the SIMPLE (Semi-Implicit Method for Pressure-Linked Equations) method. The calculational domain has a range of 5 m in the x-axis from the upper end (x = 0 m) and 0.05 m in the z-axis from the inclined bed (z = 0 m). The domain is divided into 100 divisions in both x and z-directions. The time increment is 10 (s) in each case.

Figures 8 and 9 show computed and experimental results (Run 10) for the solids fraction and velocity profiles. The slope is set at 0.196 rad and the initial solids fraction is 0.0104 (C_a = 27.5 kg m^{-3}) in the computation. In these figures, solid lines indicate calculated results, and the dots denote experimental data. The calculated velocity and solids fraction profiles along the x-direction increase with distance from the upper end, and the fluid mud layer gradually becomes thicker along the slope. These simulation results reproduce the experimental data reasonably well.

Figures 10 and 11 show calculated and experimental results for Run 15. The slope is set at 0.245 rad and the initial solids fraction is 0.0092 (C_a = 24.5 kg m^{-3}). These results exhibit similar trends as those for Run 10.

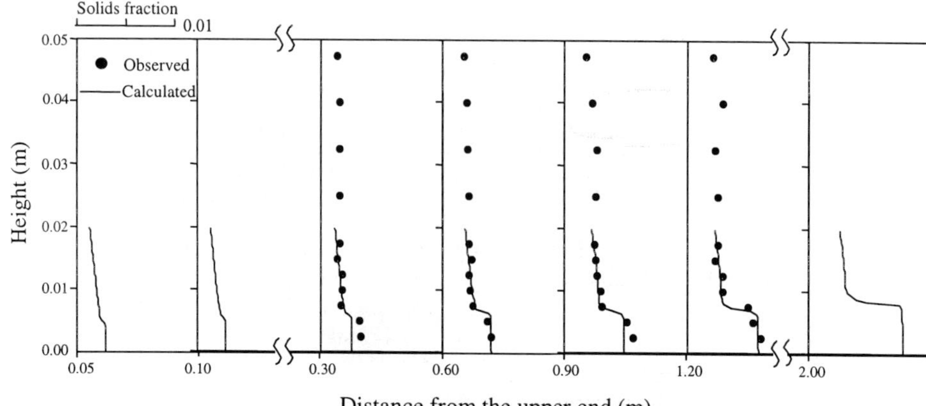

Figure 8. Figure 8. Solids fraction profiles along the x-axis. Run 10 (initial solids fraction 0.0104 and slope 0.196 rad).

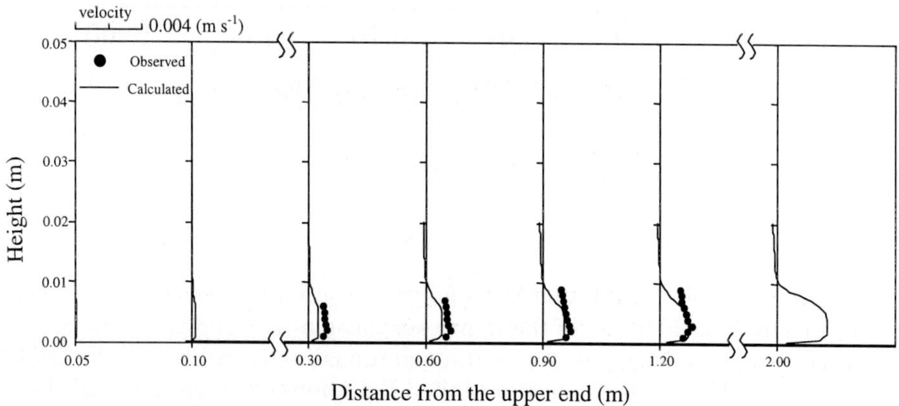

Figure 9. Velocity profiles along the x-axis. Run 10 (initial solids fraction 0.0104 and slope 0.196 rad).

Figure 12 shows the calculated deposition rate of suspended solids on to the bed. W_d is the rate of deposition of suspended solids from the fluid mud layer to the bed mud layer. In the cases of 0.05, 0.10, and 0.15 rad, W_d attained constant values. These results show that the settling rate of suspended solids from the overlying layer to the fluid mud layer is almost equal to the depositional rate to the bed mud layer. In other words, the thickness of fluid mud layer is unvarying. In the other cases shown, the deposition rate gradually approached zero, that is, the depositional flux also approached zero. These results mean that the thickness of fluid mud layer increased toward the downstream.

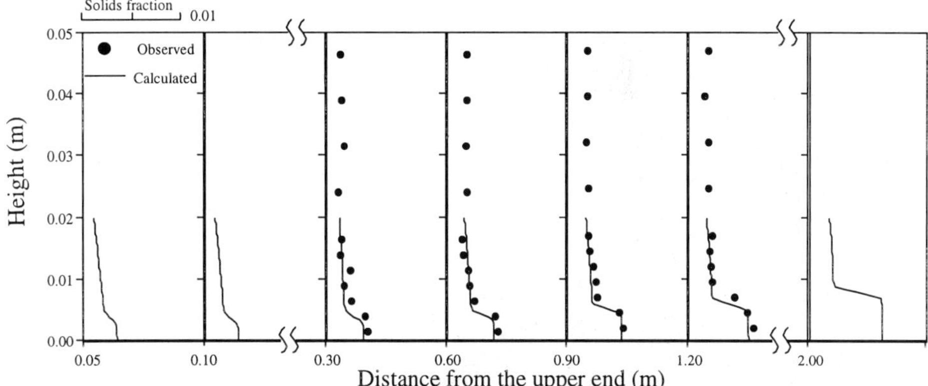

Figure 10. Solids fraction profiles along the x-axis. Run 15 (initial solids fraction 0.0092 and slope 0.245 rad).

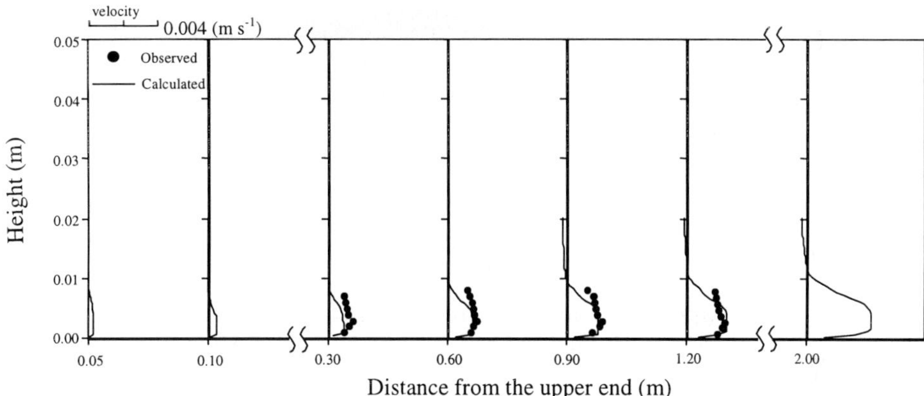

Figure 11. Velocity profiles along the x-axis. Run 15 (initial solids fraction 0.0092 and slope 0.245 rad).

5. CONCLUSIONS

Model simulated results consistently explain the velocity and solids fraction profiles in the fluid mud layer. Also, using this model, the depositional flux to the bed on an inclined bed can be predicted. Further work on the micro-structure of fluid mud flow, especially related to the floc behavior, is necessary for improving predictive accuracy.

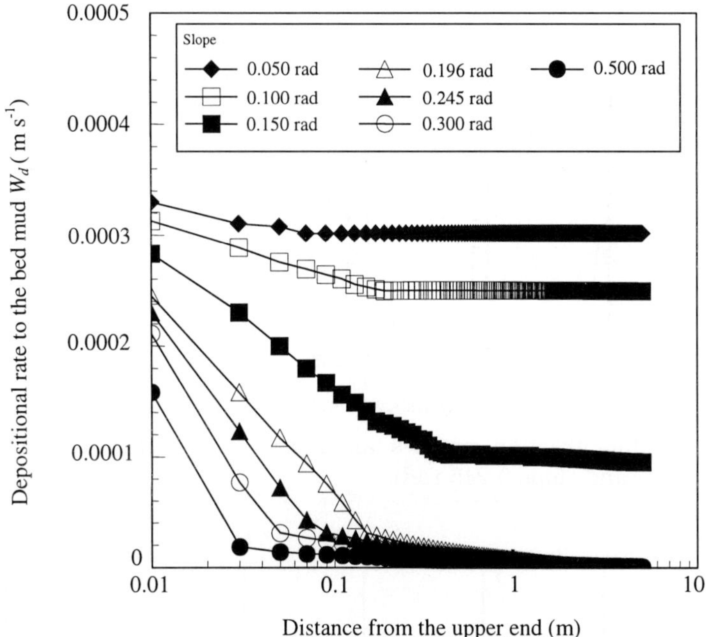

Figure 12. Variation of the depositional rate W_d with distance.

REFERENCES

Ali, K. H. M., Crapper, M., and O'Connor, B. A., 1997. Fluid mud transport. *Proceedings of the Institution of Civil Engineers, Water, Maritime, and Energy,* 124, London, 64-78.

Crapper, M., and Ali, K., 1997. A laboratory study of cohesive sediment transport. In: *Cohesive Sediments*, N. Burt, R. Parker and J. Watts eds., John Wiley, Chichester, UK, 197-211.

Hinze, J. O., 1975. *Turbulence*. McGraw-Hill, New York.

Kessel, T. V., 1997. Generation and transport of subaqueous fluid mud layers. *Report No.97-5*, Delft University of Technology, Delft, The Netherlands.

Kessel, T. V., and Kranenburg, C., 1996. Gravity current of fluid mud on sloping bed. *Journal of Hydraulic Engineering*, 122(12), 710-717.

Kusuda, T., Watanabe, R., Futawatari, T., and Yamanishi, H., 1993. Fluid-mud movement on an inclined bed. In: *Nearshore and Estuarine Cohesive Sediments Transport*, A. J. Mehta ed., American Geophysical Union, Washington, DC, 42, 218-294.

Kusuda, T., Watanabe, R., and Yamanishi, H., 1997. Mass fluxes in fluid-mud layers on an inclined bed. In: *Cohesive Sediments*, N. Burt, R. Parker and J. Watts eds., John Wiley, Chichester, UK, 407-415.

Odd, N. V. M., and Cooper, A. J., 1989. A two-dimensional model of the movement of fluid mud in a high energy turbid estuary. *Journal of Coastal Research*, SI5, 185-193.

Patankar, S. V., 1980. *Numerical Heat Transfer and Fluid Flow.* Hemisphere Publishing Corporation, Washington, DC.

van Rijn, L. C., 1984. Sediment transport, part 2: suspended load transport. *Journal of Hydraulic* Engineering, 110(11), 1613-1641.

Ross, M. A., and Metha, A. J., 1989. On the mechanics of lutoclines and fluid mud. *Journal of Coastal Research*, SI5, 51-62.

Coastal and Estuarine Fine Sediment Processes
W.H. McAnally and A.J. Mehta (Editors)

Predicting the profile of intertidal mudflats formed by cross-shore tidal currents

W. Roberts[a] and R. J. S. Whitehouse[b]

[a]16 McMullan Close, Wallingford, Oxfordshire OX10 0LQ, United Kingdom

[b]HR Wallingford, Howbery Park, Wallingford, Oxfordshire OX10 8BA, United Kingdom

A schematic modeling approach is presented, which allows calculation of the equilibrium profile of intertidal mudflats that experience cross-shore currents, but no wave action or long-shore directed currents. This situation is representative of a class of mudflats. The equilibrium is defined as zero net sediment transport at each point along the profile over a chosen period, either a tidal cycle or a spring-neap cycle. Equilibrium profiles are derived for forcing by a variety of tidal ranges, sediment input conditions and for a spring-neap cycle.

1. INTRODUCTION

There is at present a lack of reliable methods for predicting the effect of natural or engineering changes on mudflat morphology, both of which are of widespread concern. Manifestations of climate change, such as sea level rise or increased storminess are likely to alter existing mudflats; and the effects of engineering works in estuaries, such as reclamations, dredging, barrage construction and port developments are not presently fully understood.

The aim of this work is to investigate the hydrodynamic and sediment factors that control the morphology of mudflats and to establish how a change in the forcing that acts on the mudflat may lead to changes in the gross morphology. This improved understanding will lead to improved methods of predicting the effects of climate change and engineering works.

The topography of mudflats can be examined at a number of length scales. This work addresses primarily the large scales and seeks to predict the overall mudflat profile. Typically mudflats also have smaller scale features, such as drainage gullies and/or ridge and runnel systems (Dyer, 1998). At an even smaller scale, the surface roughness and porosity of the mudflat can be affected by burrows and trails of invertebrates and by patches of algae. These smaller scale effects may be influential on the local hydrodynamics and hence have a link to the larger scale topography. This is discussed in more detail below, but is essentially disregarded in the approach developed in this paper on the grounds that it is a secondary effect.

2. FORCING ON MUDFLATS

The principal input of energy to mudflats comes from tidal currents and from wind generated surface waves. The effect of tidal currents can be sub-divided into cross-shore currents (perpendicular to the shore) and long-shore currents (parallel to the shore). Cross-shore currents arise mainly from the rise and fall of the tide, which requires a volume of water to move onto the mudflat during the flood phase of the tide and to drain from the mudflat during the ebb. Long-shore currents on mudflats also arise due to tidal water level variations, but in this case the primary movement of the water is along the estuary or coastline where the mudflat is situated.

In this paper, we restrict our attention to forcing by cross-shore tidal currents. This means that the analysis described below is applicable to the class of mudflats where cross-shore tidal currents are the most important source of energy. For this to occur, the mudflat must normally be protected somehow from strong shore-parallel currents, for example by the existence of hard promontories, so that the principal current direction is cross-shore. Also, if the mudflat is exposed to strong winds and/or long fetches then large waves can be formed, which will normally dominate over tidal currents; therefore we restrict ourselves also to relatively sheltered mudflats. Cross-shore currents are rapid when the tidal range is large and the width of the mudflat is large, so that the vertical change in water surface level is associated with a rapid horizontal movement of the water's edge. Many of the issues discussed below also apply to mudflats dominated by long-shore tidal currents and/or waves, but different equations must be used to describe the hydrodynamics.

An example of a mudflat dominated by cross-shore tidal currents is Spurn Bight, near the mouth of the Humber Estuary on the east coast of England. This is a macrotidal mudflat, with a mean spring tidal range of approximately 7 m. The mouth of the Humber is constricted by Spurn Head, a sandy spit extending from the northern shore. This shelters the Spurn Bight area both from strong long-shore currents and from the sometimes severe waves in the North Sea. Figure 1 shows mathematical model calculations of the depth-averaged spring tide current pattern at Spurn Bight, both on the flood and the ebb. The modeling software used was TELEMAC-2D (Hervouet, 1994), which uses a finite element method to solve depth-averaged shallow water equations. It can be seen that the currents are directed mainly perpendicular to the shore except at the lowest part of the mudflat where they merge with the shore-parallel currents in the main estuarine channel. At High Water when the currents are relatively slack, a weak clockwise eddy forms over the mudflat, leading to a greater directional variability in currents. The dominant forcing currents can be regarded as being in the cross-shore direction. We will return to the behavior of the Spurn Bight mudflat later.

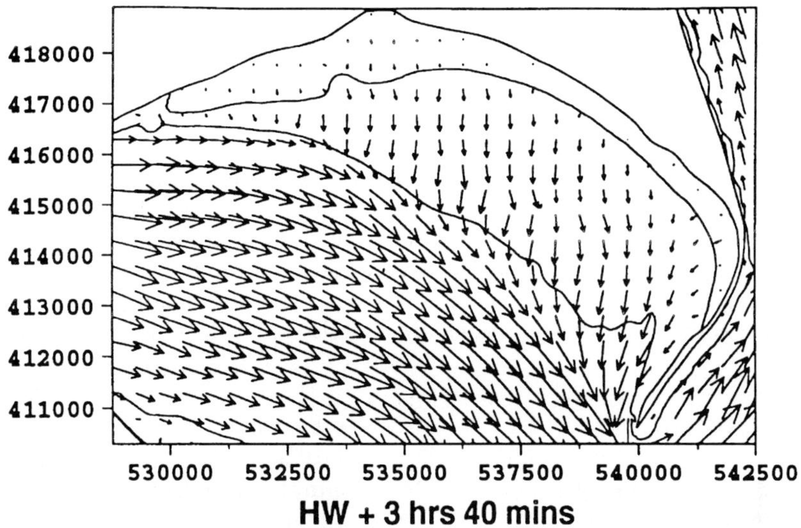

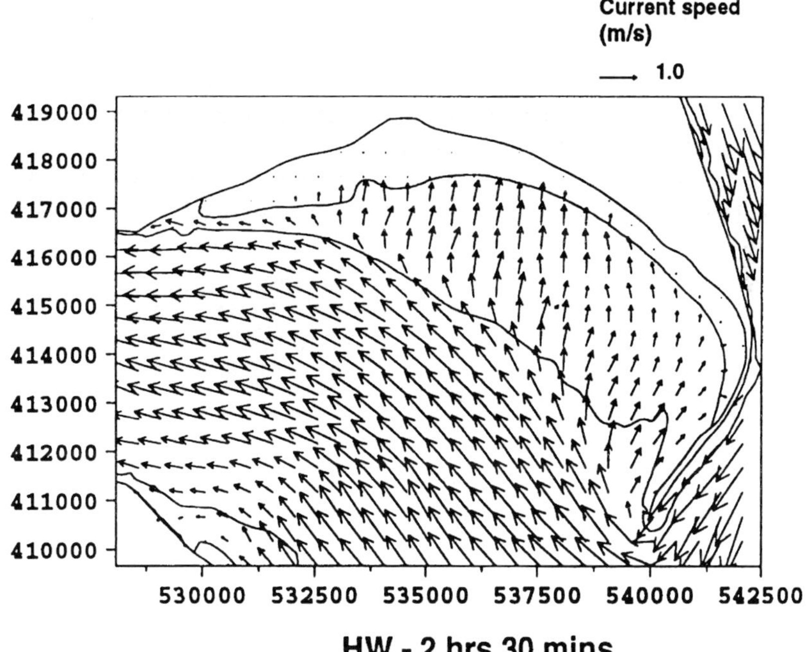

Figure 1. Current vectors on the Spurn Bight mudflat, Humber Estuary UK. Bathymetry contours shown are LWS and MWL.

3. TIME AND SPACE SCALES

The difficulty of modeling the morphodynamics of mudflats, arises principally because of the large range of time scales and spatial scales at which important processes occur. In spatial terms, these range from interactions between the grains of the sediments, occurring at micron scales, through to tidal processes occurring on the scale of the whole estuary. Time scales of interest extend from turbulent fluctuations with periods of a fraction of a second, through individual wave processes on a time scale of seconds, the tidal cycle occurring twice daily, the spring-neap cycle on the scale of several days, through to infrequent storm events occurring perhaps once every few years. The significant time and space scales to consider for mudflat systems are as follows.

Time scales	
• turbulence	< 1 second
• waves	1-10 seconds
• depth/settling velocity	minutes – hours
• tidal cycle	12 hours
• consolidation of sediment	days
• time between wave events	days – months
• spring-neap cycle	2 weeks
• biological activity	annual seasonal cycle
• seasonal wave climate	annual seasonal cycle
• time between major storms	> 1 year
• time between new engineering works	1 – 10 years
• relative mean sea level change	> 100 years
• major climate change	> 10,000 years
Space scales	
• primary grain size	microns
• particle aggregate	0.1 – 1 mm
• microtopography of mudflat	1-10 mm
• drainage channels	0.1 – 10 m
• ridge/runnel dimensions	0.1 – 10 m
• tidal range	1 – 10 m
• mudflat width	50 m – 5 km
• estuary dimensions	10 – 100 km

De Vriend (1991, 1998) presents a clear discussion of the issues raised by this large range of scales and summarizes a number of possible techniques for addressing them. Firstly, he notes that time and space scales tend to be linked, so that a phenomenon that occurs on a large spatial scale also has a long time scale: similarly for small space and time scale features. Also, a phenomenon on a certain scale interacts most strongly with other similar scale phenomena. Processes on a much smaller scale can be treated as "noise" or at least their effects can be combined into a single average influence. Processes acting on a much larger scale can be treated as "extrinsic conditions", which will be unaffected by changes in the system under study.

In this work the main interest is the mudflat profile shape and how this might be influenced by engineering works or natural changes, on a time scale of roughly 1 – 50 years. Therefore small scale bedforms are not explicitly represented and the influence of turbulence is only included in a parameterized way through the usual formulation of bed shear stress, τ, as a function of depth averaged tidal currents. Sediment and hydrodynamic processes occurring within a tidal cycle are taken into account.

4. EQUILIBRIUM PROFILE

The idea of equilibrium of a mudflat, or other sediment-water system, is that given sufficient time and constant forcing the morphology will adjust to a stable form, which does not change when viewed over a suitable time scale. This can be defined as

$$\int_{t^-}^{t^+} deposition(x,t)\ dt = \int_{t^-}^{t^+} erosion(x,t)\ dt \tag{1}$$

for all points x on the mudflat, where (t^-, t^+) is an appropriate time interval.

Intuitively this is a straightforward idea, but to obtain a precise definition, in a situation where the forcing on the mudflat is varying on various time scales, becomes much more complicated. As a simple first case, we address the situation of a uniform tidal range, i.e., a system where every tide is the same. In this case the time interval in (1) is the tidal period and the definition of equilibrium allows variation of mudflat level during the tide, but no net change over a tidal cycle. Thus periods of erosion must be matched by periods of deposition. The magnitude of these intratidal fluctuations (e.g., Whitehouse and Mitchener, 1998) would be expected to vary according to sediment properties and sediment supply. More complex situations will be addressed later.

Friedrichs and Aubrey (1996) and Kirby (1992) make the distinction between the cross-shore profile and hypsometry. The hypsometry of a mudflat is the relationship between the intertidal area and elevation, whereas the profile is simply a cross-section through the topography. Friedrichs explains how a curving shoreline can mean that two mudflats with the same profile may have a different hypsometry. In this work, we restrict ourselves to mudflats that are uniform in the longshore direction and in this case the hypsometry and cross-sectional profile are essentially equivalent.

To evaluate the expressions in (1) requires knowledge of hydrodynamics and sediment transport rates. The approach taken in this work is to use simple formulations of the water and sediment behavior, to see which aspects of mudflat form can be shown to arise from the basic elements of the water and sediment dynamics. More complex representations of behavior can be introduced later if deficiencies of the simplest approach are identified.

The hydrodynamics of the flow of tidal currents across the mudflat have been represented by solving an equation for the conservation of water volume, with a sinusoidally varying water level imposed at the seaward boundary of the modeled area. As the water level rises and falls, water must flow onto the mudflat during the rising tide and off the mudflat during the falling tide. The conservation of momentum equation is ignored, which means that the water surface is always horizontal. Therefore, any shallow water, inertia and frictional effects and thus any effects of tidal asymmetry on residual sediment transport are not represented by this approach. This is a reasonable approximation as long as the width of the mudflat is small compared with the tidal wavelength, or equivalently, that the Froude number, $Fr = u/\sqrt{(gh)}$, of flow across the mudflat is small, where u is the depth-averaged velocity and h is the water depth. This approximation may break down when the slope of parts of the mudflat becomes very small (i.e., very flat) and this is discussed further below.

The current speeds are calculated from the volume flux per unit width divided by the local water depth. A sinusoidally varying water level has been applied. The main influences on the current speed are the rate of change of water level, fastest at mid-tide, and the slope of the bed. Shallow bed slopes mean that the water's edge must move quickly as the water level changes and this leads to rapid cross-shore currents.

$$\frac{\partial h}{\partial t} + \frac{\partial (uh)}{\partial x} = 0 \tag{2}$$

where h is the water depth, u is the depth-averaged velocity and x is the cross-shore distance. This equation of volumetric continuity is solved with a boundary condition for water depth at the offshore boundary of the model domain, i.e.,

$$h_{bnd} = 0.5R\cos\left(\frac{2\pi t}{T}\right) \tag{3}$$

where R is the tidal range and T is the tidal period, and

$$\frac{\partial (h + z_b)}{\partial x} = 0 \tag{4}$$

where z_b is the local elevation of the sea bed, i.e., the water surface is forced to be horizontal.

To calculate the sediment transport, again a simple approach was taken. The conservation of sediment is expressed as a depth-averaged equation for the advection of suspended sediment, with source and sink terms representing the exchange of sediment between suspension and the bed. The equation is solved with a boundary condition for suspended sediment concentration (SSC) on inflow, representing the external supply of sediment to the mudflat.

$$\frac{\partial(ch)}{\partial t}+\frac{\partial(uch)}{\partial x}=Q_e-Q_d \tag{5}$$

where c is the depth-averaged concentration, Q_e is the flux of material from the bed into suspension by erosion and Q_d is the flux of suspended material depositing on the bed. Q_e and Q_d are calculated using the Partheniades and Krone formulations respectively (e.g., Dyer, 1986):

$$Q_e=m_e\left(\frac{\tau}{\tau_e}-1\right) \tag{6}$$

$$Q_d=cw_s\left(1-\frac{\tau}{\tau_d}\right) \tag{7}$$

where m_e is the erosion rate, τ_e is the critical bed shear stress for erosion, w_s is the settling velocity (assumed constant) and τ_d is the critical bed shear stress for deposition. Note that the usual longitudinal diffusion term has not been explicitly included in (5), but the numerical method used to solve the advection equation introduces a certain amount of numerical diffusion. Large concentration gradients can occur in the shallow water, but the numerical method for advection makes use of the unidirectional currents to maintain conservation and non-negativity, with an acceptably small amount of numerical diffusion. The boundary condition can be expressed as

$$c_{bnd}=c_0,\ u<0 \tag{8}$$

where u is positive in the offshore direction. For a pure advection equation, no boundary condition on c is required on outflow. In real estuarine situations, the suspended sediment concentration will normally vary throughout the tidal cycle. However, rather than represent the full complexity of the natural variations, the above schematic approach was taken, to investigate the gross effect of high or low sediment supply.

The bed shear stress was calculated from the depth-averaged velocity, as follows:

$$\tau=\rho c_D u^2 \tag{9}$$

where c_D is a drag coefficient, assigned a constant value in the simulations of 0.002.

Having established a method for calculating the expressions on either side of (1), an algorithm is now required for adjusting the mudflat profile so that (1) is satisfied. Two approaches have been used: one is the simulated annealing method, described in the next section; the other is a morphodynamic approach, described in a later section.

5. SIMULATED ANNEALING METHOD

The problem of finding the equilibrium profile can be expressed as the minimization of a function that depends on the profile shape and the forcing (waves, currents etc). For example, in the case where the equilibrium is defined as a situation of no net change in bed level, the function would be the net change (erosion or accretion) over the chosen time period and minimization of the function (hopefully to a zero value) corresponds to finding the equilibrium solution.

There are a number of well-established methods for minimization of functions. Many of them rely on the use of derivatives of the function. In general, for the kind of problem we are considering, it is not possible to write down the derivative analytically, and is probably prohibitively time-consuming to establish the derivative numerically. Therefore, consideration has been limited to methods that require only evaluation of the function itself. Also, the aim is to find a global rather than a local minimum of the function. The function will not be a linear in the independent variables. To try all of the possible shapes of the mudflat increases as the factorial of the number of points representing the profile and so a "brute force" approach would be impractical.

The most promising method is known as simulated annealing (Press et al., 1989, Metropolis et al., 1953). This method is based on an analogy with the cooling of liquids and the formation of low-energy crystalline states as liquid metals solidify, a process known as ‘annealing’. If metal is cooled sufficiently slowly, it will reach the minimum energy crystalline state, which corresponds to the minimization of a function in the method of simulated annealing. If a liquid metal is cooled quickly, known as ‘quenching’, it may not form the lowest-energy state, because the atoms have not had the opportunity to align themselves correctly. This is analogous to finding a local minimum of a function, which may not necessarily be the global minimum.

The following function is used to represent the "distance" of the profile from equilibrium:

$$f = \int_{profile} \int_{tide} (erosion(x,t) - deposition(x,t))\, dt\ dx \qquad (10)$$

The function f depends on the profile shape $\{z_i,\ i = 1,N\}$, the tidal range, the boundary condition for SSC and the parameters used to define the sediment properties. It is minimized with respect to $\{z_i,\ i = 1,N\}$ by application of the simulated annealing process as described below.

The method works as follows:

1. Choose an initial condition of the system that corresponds to the state of a number of variables, in this case the elevation of the mudflat at different distances from the shore. The result of the minimization should be independent of the initial condition, but convergence will be more rapid if the initial "guess" for a solution is a good one. Calculate the value of the function f for this configuration, by solving the hydrodynamic and sediment transport equations for a tidal cycle.
2. Calculate the local gradient of the mudflat at each point,
   $m_i = (z_{i+1} - z_i) / dx$
3. Choose a random value of i between 1 and N.
4. At this point, make a random change to the local gradient.
   $m_i \rightarrow m_i + \Delta m$
   where Δm is chosen from a Gaussian distribution. Reject any change that would lead to the local gradient being zero or positive, to maintain a monotonic decrease in mudflat elevation away from the shore.
5. Recalculate $\{z_i, i = 1,N\}$ from the gradients, starting from the fixed high water elevation at the shore (i=0). Note that a change in gradient at one location affects the mudflat elevation at all points further from the shore.
6. Calculate the function f for the new mudflat profile. If the change leads to a reduction in the value of f (i.e., closer to equilibrium) then accept this change as the new state of the system. If the change causes an increase in the function, then accept the change with a probability $\exp\{-(\Delta f)/kT\}$ where Δf is the change in the function, T is the analogue of temperature and k is a constant. Thus there is some probability of accepting changes in the system that temporarily worsen the process of minimizing the function. This is essential to avoid homing in on a local minimum, which may not be the global minimum.
7. After a certain number of iterations of steps 2 - 6, decrease the value of T, to make adverse changes in the function less likely to be accepted. Carry out a large number of further iterations. Steadily decrease T until a constant value is reached. The schedule for decreasing T ('cooling') tends to require some trial and error. Too rapid cooling can cause convergence to a local rather than global minimum. Too slow cooling is inefficient, and this method tends to require a large number of iterations, so it is desirable to avoid carrying out more calculations than necessary. The lower the value of T, the less likely it is that an adverse change to the system will be accepted.

The assumption of monotonicity seems to be a true reflection of the majority of observed mudflats (Dyer, 1998), disregarding the local effect of drainage channels or small-scale bedforms. It is useful in reducing the number of possible solutions for the equilibrium mudflat shape and also allows the use of a simple and efficient numerical method for the solution of the hydrodynamic and sediment transport equations.

When considering random changes to the mudflat level as part of the optimization process, any change that would break the monotonicity constraint was rejected.

6. MORPHODYNAMIC APPROACH

The morphodynamic approach used for this work was to start from an initial profile, solve (2) and (5) over a tidal cycle and use the calculated quantities of erosion and deposition to adjust the bed elevation at the end of the tidal cycle. The new bathymetry is then used for the next tidal cycle. Changes in bed level within the tidal cycle are generally small enough that there is no significant effect of "saving" the changes until the end of the cycle.

In principle, there is no guarantee that the morphodynamic approach will give a stable equilibrium profile, although for the system modeled here, convergence was smooth. One advantage of this technique is that it can offer some information on the rate of approach to equilibrium, because changes in the profile can be related to numbers of tidal cycles. This is not the case with the simulated annealing method.

With simple equations and a small number of grid points as in this case, there is no need to extrapolate the bed changes forward in time, removing some of the stability problems associated with certain morphodynamic approaches. It is quite practical to calculate 100 years of 700 tides per year, for example, which was the approach taken for the tests reported here. However, the number of iterations required to reach equilibrium will depend on the initial condition. As in the simulated annealing approach, the constraint was imposed that the mudflat slope should remain monotonic. The solution of the simplified equations of flow requires this. Without this assumption, more complex numerical methods are required. This assumption is particularly useful in the very shallow water near the high water line, where only a small change in bed level would be required to pierce the water surface and therefore move the high water line seaward.

In the next section, results of the calculations by both the simulated annealing method and the morphodynamic method are presented. The comparison between the results of the two methods is discussed.

7. CALCULATED PROFILES

The following influences on equilibrium mudflat profile shape were considered:

- tidal range
- sediment supply (as represented by boundary suspended sediment concentration)
- sediment properties, in particular the critical shear stress for erosion

The results for the effect of tidal range on the mudflat profile are presented in Figures 2 and 3. Figure 2 shows calculated results using the simulated annealing method and Figure 3 shows results of the morphodynamic method. For each of these series of tests the following parameters were used:

Boundary suspended sediment concentration:	0.1 g/l
Critical shear stress for erosion:	0.2 Pa
Critical shear stress for deposition:	0.1 Pa
Settling velocity:	1 mm/s
Erosion rate:	5.0×10^{-5} kg/m^2/s

These were chosen to be typical of observed values in UK mudflat areas. A grid of 40 points was used for the simulations, with a grid spacing of 250-300 m.

To quantify the comparisons, the shapes of the calculated profiles have been characterized by simple parameters: the average slope of the intertidal part of the mudflat, and the position of the mean water level as a proportion of the overall mudflat width (see Table 1).

The mean slope has been calculated by dividing the tidal range by the distance between the high water and low water lines. The mean water (MW) position is expressed as the ratio of the distances from the high water line to the mean water and low water points. A value greater than 0.5 indicates a generally convex upward profile and a value less than 0.5 indicates a generally concave upward profile.

It can be seen that the calculated profile shapes are convex upwards in the upper part of the profile and more linear in the lower part of the profile. This is consistent with the theoretical analysis of Friedrichs and Aubrey (1996), which involved deriving the profile shape that gives a spatially uniform value of peak

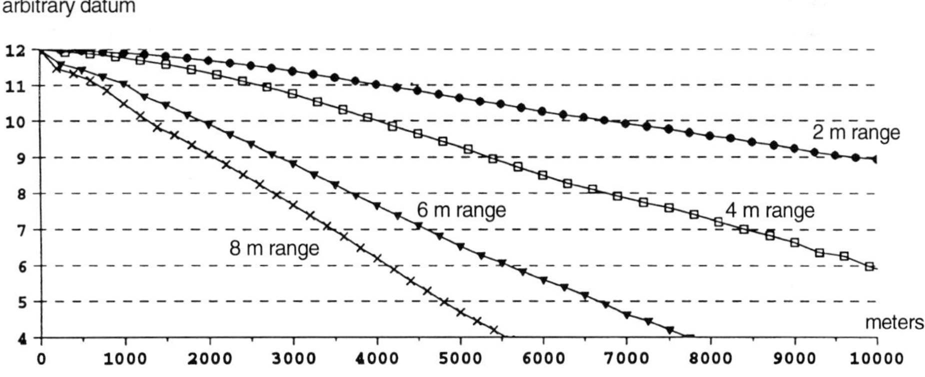

Figure 2. Equilibrium mudflat profile for different tidal ranges – simulated annealing method.

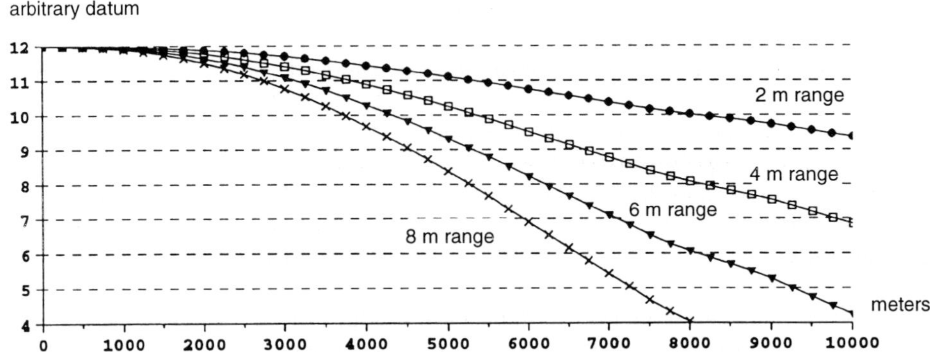

Figure 3. Equilibrium mudflat profile for different tidal ranges – morphodynamic method.

Table 1
Effect of tidal range on profile shape, boundary concentration = 0.1 g/l

	Simulated annealing approach		Morphodynamic approach	
Tidal range (m)	Mean slope	MW position	Mean slope	MW position
2	2.95×10^{-4}	0.60	2.48×10^{-4}	0.66
4	5.92×10^{-4}	0.59	4.92×10^{-4}	0.66
6	1.08×10^{-3}	0.51	7.40×10^{-4}	0.65
8	1.41×10^{-3}	0.52	9.95×10^{-4}	0.66

bed shear stress over the mudflat profile. The mean slope of the profile is inversely related to the tidal range, with the total mudflat width, defined by the horizontal distance between the low water and high water positions, being approximately independent of tidal range. Note that the curves in Figures 2 and 3 extend beyond the low water position into the sub-tidal area and that they have been plotted so that the high water levels coincide at 12 m above datum. Therefore the low water elevations are 10 m, 8 m, 6 m and 4 m for tidal ranges 2 m, 4 m, 6 m and 8 m, respectively. These all lie approximately 8,000 m horizontally from the high water point. The average slopes of the profiles calculated by the simulated annealing method are systematically steeper than those calculated using the morphodynamic approach. The difference occurs mainly at the uppermost part of the profile, where the morphodynamic approach generates a wide upper mudflat with a very shallow slope. Analysis of the "distance from equilibrium" at different parts of the profile, as measured by the net erosion or accretion during the tidal cycle, shows that the uppermost part of the profile is most sensitive to small variations in slope. It appears to be the case that the simulated annealing approach is tending towards a similar solution to the morphodynamic approach but is less well converged in this crucial upper section.

Figures 4 and 5 show the calculated peak bed shear stress of the equilibrium profiles, calculated by the simulated annealing and morphodynamic approaches respectively, for the different tidal range scenarios. The peak shear stress values turn out to be largely independent of tidal range, as predicted analytically by Friedrichs and Aubrey (1996), depending only on the boundary suspended sediment concentration and the parameter values for sediment properties. The smaller tidal ranges are associated with flatter bed slopes, so that the maximum cross-shore velocity (and hence bed shear stress) remains approximately independent of tidal range. Both methods give similar results for the peak shear stress in the lower part of the profile, approximately 0.3 N/m^2 for the choice of input parameters listed above. In Figure 4, the peak shear stress in the upper part of the profile shows a lot of variability between the different curves and within each curve, suggesting that the profile has not fully converged to an equilibrium state. The curves in Figure 5 are much more consistent and show a steady increase in peak shear stress until very close to the high water line. This is associated with the very low gradient of the bed in the upper part of the profiles.

The dependence of the equilibrium profile shape on the imposed boundary suspended sediment concentration was investigated. The results are summarized in Table 2 and shown graphically in Figures 6 and 7. Larger boundary concentration corresponded with less steep average profile slopes. In Figure 6, the profiles for C=0.025 g/l and C=0.1 g/l are very similar and this may be because the upper part of the profiles have not fully converged to the equilibrium solution. The reason for this is that the larger supply of sediment increases the accretion rate during deposition periods, thus requiring higher current speeds to

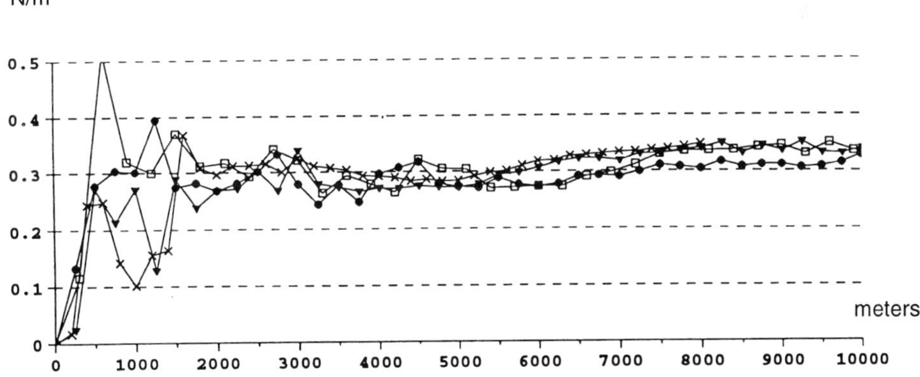

Figure 4. Peak shear stress for different tidal ranges – simulated annealing method.

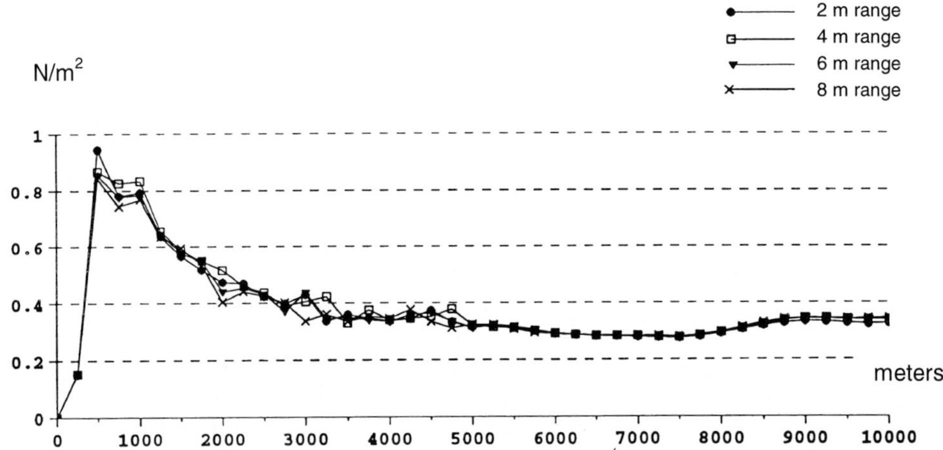

Figure 5. Peak shear stress for different tidal ranges – morphodynamic method.

Table 2
Effect of boundary concentration on profile shape, tidal range = 8 m

Boundary SSC (g/l)	Simulated annealing approach		Morphodynamic approach	
	Mean slope	MWL position	Mean slope	MWL position
0.025	1.50×10^{-3}	0.54	1.17×10^{-3}	0.63
0.1	1.41×10^{-3}	0.52	9.95×10^{-4}	0.66
0.2	1.22×10^{-3}	0.55	8.89×10^{-4}	0.65
0.4	1.08×10^{-3}	0.54	7.69×10^{-4}	0.65

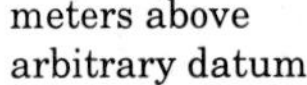

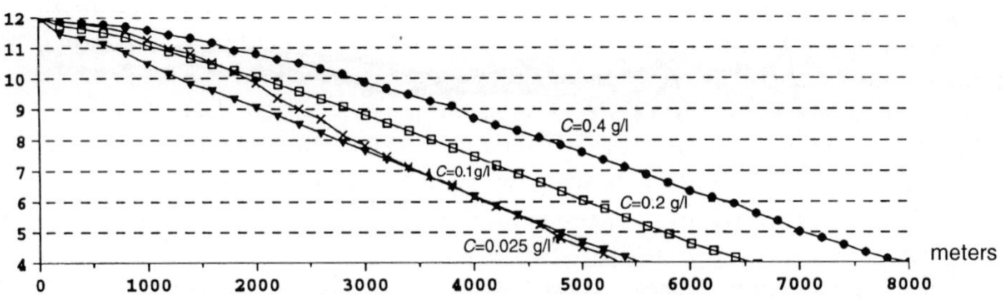

Figure 6. Equilibrium mudflat profile for different boundary SSC – simulated annealing method.

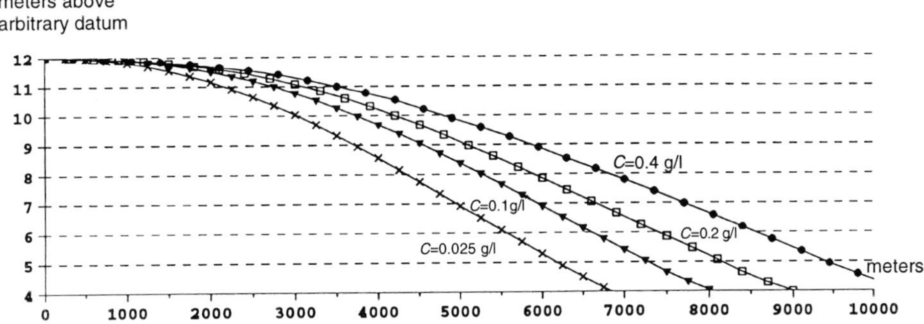

Figure 7. Equilibrium mudflat profile for different boundary SSC – morphydynamic approach.

increase the capacity for erosion to maintain a balance. The peak shear stress occurring on the mudflat thus increases as a function of the boundary concentration, so that mudflats in a very turbid environment could be expected to be wider, with higher current speeds, than those in a low concentration environment.

If the boundary concentration is set to zero, then no net accretion of the profile is possible, and so the final profile obtained becomes dependent on the initial condition. If the initial profile is sufficiently steep that peak shear stresses are less than the critical shear stress for erosion, then no change to the initial profile occurs. If the initial state has high peak shear stresses, then the profile erodes until the peak bed shear stress becomes equal to the threshold value at each point of the mudflat.

Figure 8 illustrates the typical pattern of erosion and deposition through the tidal cycle, at three locations across the mudflat. The mudflat profile was at equilibrium for an 8 m tidal range and a boundary suspended sediment concentration of 0.1 g/l. Note that there are two periods of erosion, appearing as periods of decreasing bed level in Figure 8, one during the flood and one during the ebb, which must be matched by single period of deposition (increasing bed level) around high water slack. As the High Water line is approached, the time of submergence tends to zero, concentrations are generally non-zero, so the deposition at slack water must still be balanced by periods of erosion. This tends to lead to a solution where the periods of erosion and deposition become very short and a very flat mudflat is required to achieve flow velocities high enough to maintain the sediment in suspension. This can lead to short lived high Froude number flows, which violate the assumption of low Froude number made in the simplification of the hydraulic equations. A more physical way of achieving zero net sediment transport on the upper part of the mudflat appears to be for deposition to occur in calm conditions around high water, balanced either by erosion when waves are present, or by sediment transport on the ebb tide in very

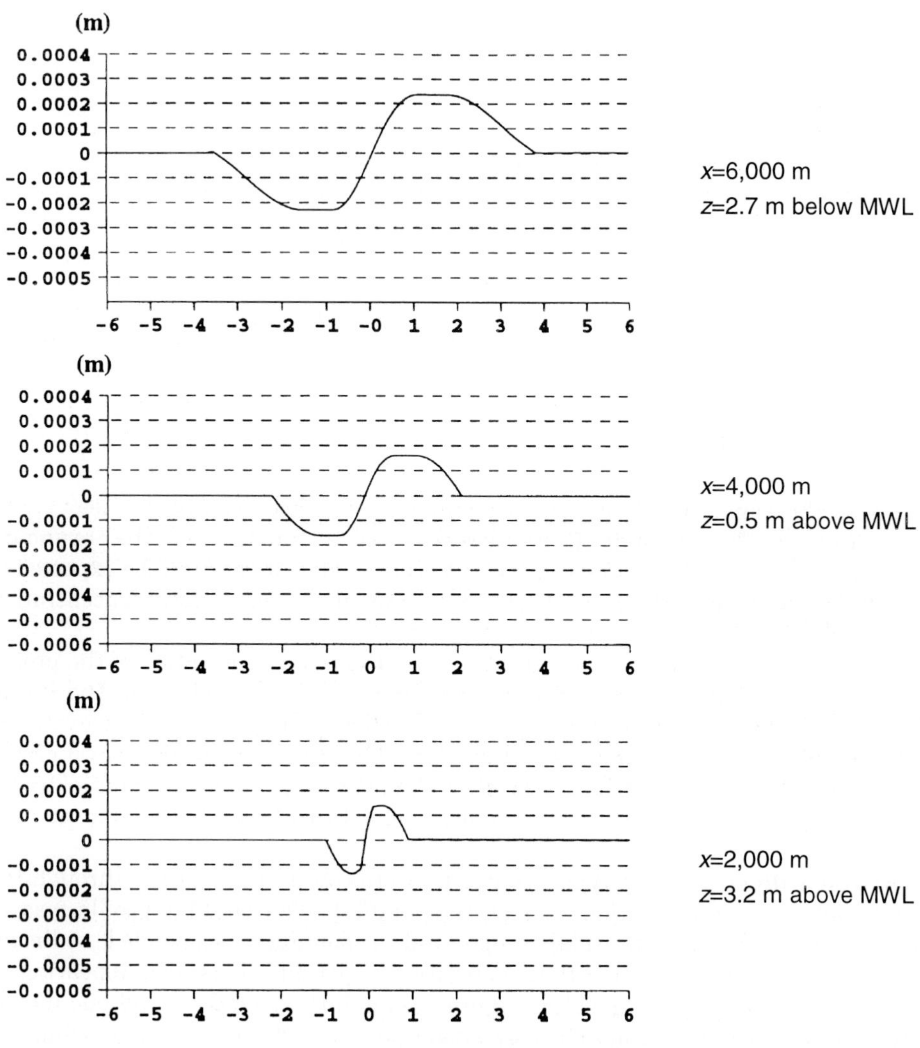

Figure 8. Net change in bed level over a tidal cycle. Tidal range 8 m, boundary SSC 0.1 g/l.

shallow water. Observations on mudflats show that turbid water continues to drain through the network of channels in the mudflat surface, long after the water line has receded. These channels are not resolved in the modeling approach presented here and so this type of flow is not represented. Further work is in progress to investigate these aspects.

8. COMBINATION OF CONDITIONS

The previous section has considered the equilibrium profiles that arise from the continuous application of a single tidal range. However, in nature, the tidal range undergoes a continuous variation at a number of frequencies. The most important variation for most applications is the spring-neap cycle. A mudflat profile that is in equilibrium for a spring tide for example will be significantly out of equilibrium for a neap tide. The changes in bed level arising due to erosion and accretion within a spring-neap cycle will generally have a small or negligible effect on the overall profile shape. Observations of bed level on a mudflat in the Severn Estuary (Whitehouse and Mitchener, 1998) found the variation within the spring-neap cycle to be approximately 20-25 mm. Therefore the mudflat profile must assume a shape where erosion during the rapid flows on spring tides is balanced by deposition during the more tranquil conditions of neap tides. The resulting profile will be a response to the overall combination of conditions.

The morphodynamic model was run for a large number of spring-neap cycles, each consisting of 28 tides, until a stable profile was reached, defined as zero net sediment transport over the spring-neap cycle at each point along the profile. The tidal range varied between a maximum of 8 m and a minimum of 4 m. The boundary suspended sediment concentration on inflow was set to be proportional to the tidal range, equal to 0.1 g/l on the largest tide and 0.05 g/l on the smallest tide.

The resulting profile is illustrated in Figure 9, together with the profiles obtained for 8 m tides only and for 4 m tides only. It can be seen that the average slope of the spring-neap cycle profile lies between that arising from spring tides only and from neap tides only. However, the profile calculated for the spring-

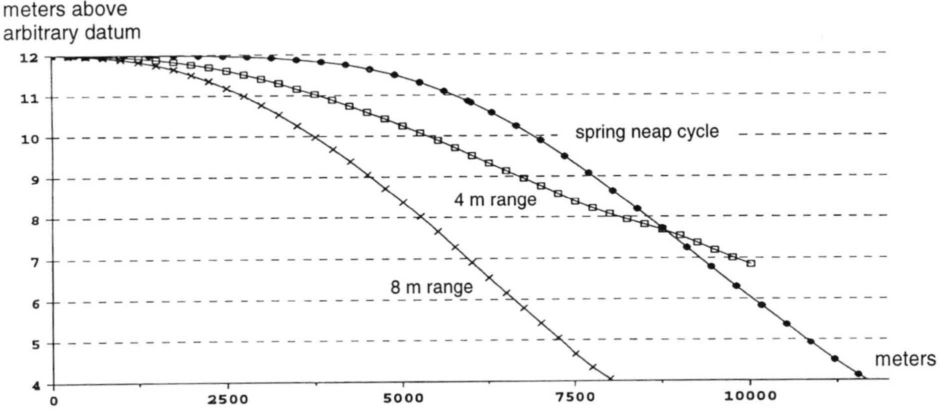

Figure 9. Profile arising from spring-neap cycle, compared with profiles from 8 m and 4 m tides only.

neap cycle has a very wide and flat upper section. This arises from the requirement for sufficient erosion in the short immersion period to balance the deposition. This section of the profile is only covered on spring tides when the generally fast current speed leads to a high concentration of material in suspension and therefore a potentially high deposition flux during the high water slack period. Because of the shallow water depth, the Froude number becomes very high and the method for calculating velocity, using the volume equation only becomes unrealistic. This is part of the reason that such an apparently unphysical solution is obtained.

9. DISCUSSION

The methods presented above consider the behavior of mudflats in an idealized way. Real mudflats behave in a much more complicated manner and respond to a much more complicated set of forcing conditions. Nonetheless, this type of approach can give some useful insight into the way mudflats behave, and by considering which aspects of observed mudflats are well represented and which are not, it may be possible to make deductions about what are the important processes on mudflats.

The type of forcing considered in this paper, that of cross-shore currents, is relevant to one class of mudflat, where wave action and long-shore currents are insignificant. The Spurn Bight mudflat in the Humber Estuary, UK is as an example of a mudflat where cross-shore currents dominate over long-shore currents. Figure 10 shows a cross-section through the bathymetry of Spurn Bight, based on surveys carried out as part of the LISP project (Black and Paterson, 1998). The tidal heights at this location are as follows:

(all heights in meters above Chart Datum (CD))	
Mean High Water Spring (MHWS)	7.0
Mean High Water Neap (MHWN)	5.5
Mean Low Water Neap (MLWN)	2.7
Mean Low Water Spring (MLWS)	1.0

Comparing the profile with these heights, it can be seen that it is convex upward up to approximately 5.3 m CD, just below the level of MHWN, with an average slope close to 1:1000 (i.e., 1.0×10^{-3}). Above 5.3 m CD, the profile is much steeper, approximately 1:100, and is concave upwards. The mudflat extends 200-300 m further landward from this apparent crossover point in the convex/concave parts of the profile. At a level of around 5.7 m above CD, the mudflat becomes a salt marsh, with no noticeable cliff at the interface. The salt marsh zone is of the order of 100 m wide where it meets a flood defense embankment. The general shape and average slope of the convex section of the profile are in broad

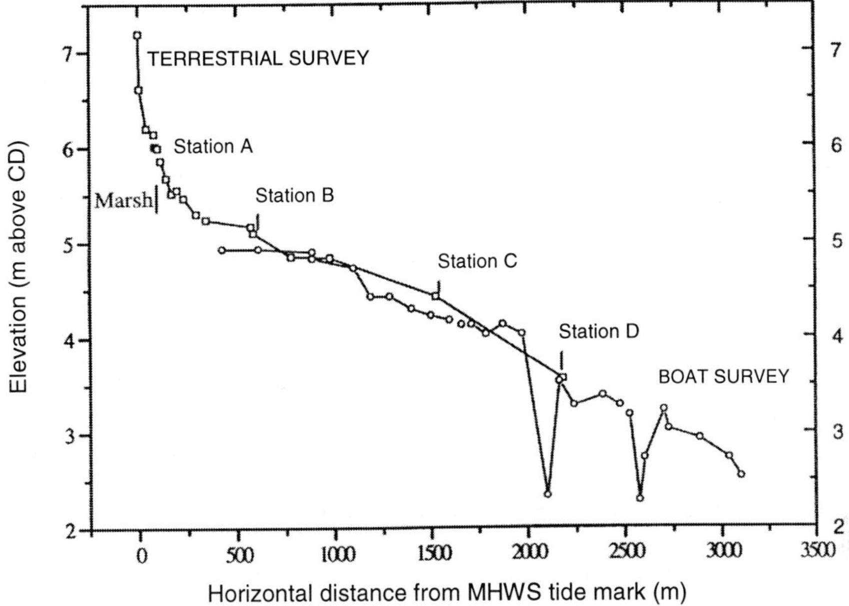

Figure 10. Bathymetric cross-section of the Spurn Bight mudflat. Reproduced from Black and Paterson (1998) by permission of the Geological Society, London.

agreement with the results of the modeling work, with the significant difference that in the model results, the convex profile extends all the way up to the level of high water on spring tides.

Observed peak current speeds are smaller in the upper part of the mudflat than the lower part (Christie and Dyer, 1998), in contrast to the model results, where peak current speeds tend to be uniform or higher, though short-lived, in the upper part of the mudflat. This is intimately linked with the mudflat profile and it is therefore inevitable that if the modeled and observed water depths are different that the current speeds will also differ. However, the model does successfully represent the observed fact that the highest suspended sediment concentrations at a given location occur when the water is very shallow.

Using the formulation of hydrodynamics and sediment transport presented above, the net sediment transport for a profile such as that observed on Spurn Bight mudflat would lead to deposition in the upper mudflat, i.e., net onshore transport in the upper region. Rapid accretion of the upper mudflat is not observed and this demonstrates the presence of an additional process not represented in our schematic approach. This additional process (or processes) must lead to an offshore net transport of sediment.

The most likely option appears to be the action of waves, which tend to be largest around High Water when fetch lengths are at their longest. The overall profile will be a response to a mixture of wavy and calm conditions, associated with net erosion and net accretion of the upper mudflat. Friedrichs and Aubrey (1996) present an analysis of wave-dominated mudflats, which concludes that the effect of waves alone leads to a concave upward profile. If waves dominate over currents in the upper part of the profile, but currents dominate over waves in the mid part of the profile, that may lead to an explanation of the observed shape, but this has not yet been investigated in detail. Another possibility is that there is some transport of sediment in very shallow water during the ebb phase of the tide, when friction (not represented in our simplified hydrodynamics) extends the time required for the mudflat to drain. Work is in progress to investigate the latter two items: the influence of the salt marsh on the intertidal profile is an interesting question for future work.

Another feature of mudflats not included in this schematic modeling approach is the spatial (and possibly temporal) variability in sediment properties. Measurements made at Spurn Bight during the INTRMUD project (Mitchener et al., 1998, INTRMUD, 1998) show that the mudflat exhibits a fining of grain size with a decreasing proportion of sand towards the upper mudflat, increasing water content of the bed sediments and increasing organic content with proximity to the upper edge of the mudflat. The detailed surface topography of the mudflat varies across it, with a relatively featureless surface at the uppermost part, with a few drainage channels, giving way to distinct ridge and runnel features in the mid mudflat and deeply incised drainage channels in the lower mudflat. The higher the bed elevation, the longer is the exposure to the atmosphere and sunlight, affecting drying consolidation and types and abundance of flora and fauna.

Kirby (1992) discusses the Spurn Bight mudflat and his hypsographic curve for the area is similar to the elevation profile presented in Figure 10, although it does not extend to such high elevations. Kirby notes that Spurn Bight is a site of long term accretion, which raises the question of whether it can be regarded as being in equilibrium with the hydrodynamic forcing. Relative sea level in the area has been rising at a rate of approximately 1 mm/year in the last few hundred years. Possibly more significant is the extensive reclamation of upper intertidal areas in the estuary over a period of centuries. A careful examination of time scales of response to these changes in forcing is required to understand the behavior of the Humber mudflats and that is outside the scope of this work.

The results of the modeling have been compared with the mudflat profile shapes derived analytically by Friedrichs and Aubrey (1996) for mudflats dominated by cross-shore currents. The approach presented here has in common with Friedrichs and Aubrey the use of only the mass conservation equation to calculate the currents and water depths on the mudflat. However, Friedrichs and Aubrey assumed that the equilibrium profile shape was such that the peak shear stress was uniform at each point of the mudflat. Our approach is to introduce in addition a sediment transport equation and to calculate the profile shape that corresponds to zero net sediment transport. Our calculations are in agreement with many aspects of the Friedrichs and Aubrey analysis: a generally convex upward shape for this type of mudflat, becoming approximately linear below mean water level and with peak shear stress independent of tidal range. However, the calculated peak shear stress on our equilibrium profiles is not uniform, as discussed in Section 7. Because of the inclusion of a sediment transport equation, the approach presented here allows the peak shear stress on the mudflat to be calculated as a function of the sediment supply and sediment properties. Larger sediment supply and/or greater resistance to erosion lead to a higher peak shear stress and hence wider mudflat for a given tidal range.

10. CONCLUSIONS

A method has been developed for calculating the equilibrium cross-shore profile of intertidal mudflats, where the equilibrium is defined as zero net cross-shore sediment transport at each point of the mudflat. It has been applied to mudflats where the hydrodynamic forcing is dominated by cross-shore tidal currents. This leads to the following conclusions about the behavior of this type of mudflat:

- the equilibrium shape is convex upwards in the upper mudflat and approximately linear in the lower mudflat
- the peak shear stress arising from the tidal currents at equilibrium is independent of tidal range, because a smaller tidal range leads to a less steep mudflat slope
- the width of the mudflat is approximately independent of tidal range
- a large supply of suspended sediment will lead to a wider (and hence less steep) mudflat
- stronger, less erodible sediment will lead to a wider and less steep mudflat
- a mudflat that is in equilibrium under tidal conditions modulated by the spring-neap cycle is not in equilibrium for any single tidal range; spring tides lead to net erosion and neap tides lead to net accretion; the uppermost part of the mudflat is only affected by spring tides

Comparison of the model predictions with observations of the Spurn Bight mudflat in the Humber Estuary, UK, show many similarities but also some significant differences. The influence of waves on this mudflat, not considered in the modeling presented here, is the most likely reason for the difference. Further work is ongoing to examine the effect of waves within our modeling framework.

11. ACKNOWLEDGMENT

Thanks are due to Pierre Le Hir and Ashish Mehta for useful suggestions. This work was funded by the UK Ministry of Agriculture Fisheries and Food as part of its program of flood and coastal defense research; the UK Environment Agency as part of its R & D program; and by the Commission of the European Communities Directorate General for Science Research and Development as part of the INTRMUD collaborative research program, under Contract Number MAS3-CT95-0022.

REFERENCES

Black, K. S., and Paterson, D. M., 1998. LISP-UK Littoral investigation of sediment properties: an introduction. In: *Sedimentary Processes in the Intertidal Zone*, K. S. Black, D. M. Paterson and A. Cramp eds., Geological Society, London, Special Publications, 139, 1-10.

Christie, M., C., and Dyer, K. R., 1998. Measurements of the turbid tidal edge over the Skeffling mudflats. In: *Sedimentary Processes in the Intertidal Zone*, K. S. Black, D. M. Paterson and A. Cramp eds., Geological Society, London, Special Publications, 139, 45-56.

de Vriend, H. J., 1991. Mathematical modeling and large-scale coastal behaviour. *Journal of Hydraulic Research*, 29(6), 727-753.

de Vriend, H. J., 1998. Prediction of aggregated-scale coastal evolution. Proceedings Coastal Dynamics '97, Plymouth, U.K., June 1997 E. B. Thornton ed., American Society of Civil Engineers, 644-653.

Dyer, K. R., 1986. *Coastal and Estuarine Sediment Dynamics*. John Wiley, New York.

Dyer, K. R., 1998. The typology of intertidal mudflats. In: *Sedimentary Processes in the Intertidal Zone*, K. S. Black, D. M. Paterson and A. Cramp eds., Geological Society, London, Special Publications, 139, 11-24.

Friedrichs, C. T., and Aubrey, D. G., 1996. Uniform bottom shear stress and equilibrium hypsometry of intertidal flats. In: *Mixing in Estuaries and Coastal Seas. Coastal and Estuarine Studies*, C. Pattiaratchi ed., American Geophysical Union, Washington, DC, 50, 405-429.

Hervouet, J.-M., 1994. Computation of 2D free surface flows: the state of the art in the TELEMAC system. *Proceedings of the First International Conference on Hydroinformatics*, Delft, The Netherlands,

INTRMUD, 1998. The morphological development of intertidal mudflats: *Second Annual Report to the European Commission*. MAST III Contract MAS3-CT95-0022.

Kirby, R., 1992. Effects of sea-level rise on muddy coastal margins. In: *Dynamics and Exchanges in Estuaries and the Coastal Zone*, D. Prandle ed., American Geophysical Union, Washington DC, 313-334.

Metropolis, N., Rosenbluth, A., Rosenbluth, M., Teller, A., and Teller, E., 1953. Equation of state calculations by fast computing machines. *The Journal of Chemical Physics*, 21, p1087.

Mitchener, H. J., Lee, S. V., and Whitehouse, R. J. S., 1998. Erodibility measurements at Skeffling, Humber Estuary, *3-9 July 1997: Data Report TR52*, HR Wallingford, Wallingford, UK.

Press, W. H., Flannery, B. P., Teukolsky, S. A., and Vetterling, W. T., 1989. *Numerical Recipes: The Art of Scientific Computing*. Cambridge University Press.

Whitehouse, R. J. S., and Mitchener, H. J., 1998. Observations of the morphodynamic behaviour of an intertidal mudflat at different timescales. In: *Sedimentary Processes in the Intertidal Zone*, K. S. Black, D. M. Paterson and A. Cramp eds., Geological Society, London, Special Publications, 139, 255-272.

Coastal and Estuarine Fine Sediment Processes
W.H. McAnally and A.J. Mehta (Editors)

Monitoring of suspended sediment concentration using vessels and remote sensing

J.-Y. Jin, D.-Y. Lee, J. S. Park, K. S. Park and K. D. Yum

Coastal and Harbor Engineering Research Center, Korea Ocean Research and Development Institute, Ansan P.O. Box 29, Seoul 425-600, Korea

Two methods for continuous monitoring of the surface suspended sediment concentration (SSSC) along routine ship routes are considered; one by deploying the monitoring instrument on the underwater part of a ferryboat, and the other by pumping surface water through the instrument located on ship. These methods have been field tested for different vessels including a chartered fishing boat, a ferry operating near the coast and an open-water car ferry. Experience derived from the tests and implementation of the methodologies for a semi-automated observation system for SSSC are described, and the problems of each method are discussed. The monitoring of SSSC using ferryboats is cost-effective when properly integrated with satellite remote sensing in collecting valuable information for initial and boundary conditions of sediment transport models, and also for understanding sedimentation processes in coastal waters with high waves and strong tidal currents.

1. INTRODUCTION

The sedimentation process in the western coastal waters of Korea is forced by high waves and strong tides, and the variation of the suspended sediment concentration is large both in space and time. In that regard, a long-term, routine, surface suspended sediment concentration (SSSC) monitoring program to cover a wide area in a relatively short period would be very useful in understanding the sedimentation process.

The interpretation of the data from a bi-monthly or seasonal monitoring program using designated vessels along the west coast of Korea is difficult because of the rather rapid temporal variation of SSSC with wave and tidal forcing. Point measurements from a buoy or an observation tower provide important information on the sedimentation processes. However, in that case it is not feasible to cover a wide area to monitor the long-term variability of SSSC.

Continuously monitored SSSC data can yield the open boundary condition for sediment transport models, and also for model validation. Remote monitoring of the SSSC using satellites such as SeaWiFs, NOAA, and Landsat can cover large areas. However, the algorithms necessary to determine the suspended sediment concentration from satellite remote sensing are presently not fully established. Hence, *in situ* data from different locations must be used to provide ground truth information for the calibration of satellite measurements. To obtain SSSC data over a large area accurately, instruments need to be deployed at various points, which can be extremely costly. Perhaps the most cost-effective method for long-term monitoring of SSSC is the use of a ship routinely operating (SRO) such as a ferryboat or a merchant ship. When periodic monitoring by using an SRO is integrated with satellite remote sensing, it should be possible to obtain the distribution of SSSC economically.

Most of the existing underway measurements using an SRO have been carried out by oceanographers. For example, Harashima (1993,1994) measured seawater temperature, salinity, pH and nutrients using an automated system installed on a car ferry crossing the Korea Strait. Cooper et al. (1998) measured the partial pressure of carbon dioxide using a fully automated instrument installed on a merchant ship traveling between the U.K. and the Caribbean. MOMAF (1999) have developed a more elaborate system to measure various chemical and biological parameters and have successfully tested it along the 15-hour long ferry route between Inchon and Cheju Island.

The Korea Ocean Research and Development Institute (KORDI) is establishing an operational cohesive sediment transport prediction system for the coastal waters of Korea. A continuous monitoring system for SSSC is essential in obtaining the initial and boundary conditions of the sediment transport model, and also in calibrating and testing the model. A cost-effective system for the long-term monitoring of SSSC along the west coast of Korea is needed to test and implement the prediction system for practical applications as well as to improve our understanding of the complex sedimentation processes in the Yellow Sea. Accordingly, a systematic monitoring of SSSC using a routinely operating ferryboat and remote sensing has been tested and field implemented in the western coastal waters of Korea.

2. INSTRUMENTATION

2.1. Instruments

Some automated methods for monitoring SSSC are available. A relatively simple method is underway-sampling of seawater by using a pump or automated *in situ* samplers such as the AUTTLE (Jin et al., 1999) or the RAS 3-48N of MacLane Inc. of USA. The WTS6-24FH, also of MacLane Inc. collects suspended particulate matter by *in situ* filtering; however its discrete sampling capacity is a limitation. The time consuming process for filtering of the sampled seawater may not be appropriate for automated continuous monitoring. Optical backscatter sensors (OBS) may help to accomplish the purpose of the present study, although they also have an inherent limitation in that the OBSs need to be reliably calibrated.

Figure 1. 6000UPG OBS retrieved after 39-day deployment in the Lavaca Bay, USA.

A multi-parameter water quality monitor, the 6000UPG of YSI Inc. of USA, was selected for this study. This device can optionally measure temperature, salinity, water depth, dissolved oxygen, pH, turbidity, and ammonia in direct reading, self-recording or telemetering mode. Its main advantage is a self-cleaning wiper to prevent bio-fouling of the OBS, which is essential for long-term SSSC monitoring (Figure 1).

2.2. Instrument set-up in a ferry

There are two methods of instrument set-up for continuous measurement of SSSC using an SRO: one is by pumping the surface water into a shipboard measuring (SM) system, and the other is *in situ* measurement (IM) by attaching the instrument to the underwater part of the ferryboat. Both methods have advantages and disadvantages.

Before the field implementation test to determine the optimal monitoring method, the SM was selected because it has the advantage of the instrument's security. Laboratory experiments were conducted with the flow cell that YSI Inc. supplied for the SM. The pumped water flows in and out of the cell that includes the measuring sensors. The tests indicated that there was a problem of air bubble formation in the cell when the flow was strong. To avoid air bubble formation, which can interfere with the measurement of SSSC, the flow rate must be adjusted by trial and error to assure accurate measurement.

For the field experiments using the two methods, a system for SM (Figure 2a) and a guard frame (Figure 2b) for IM were constructed. The dimensions of the frame for SM were 70cm×50cm×31cm. The hose nozzle for the SM and the OBS for the IM were located in 1 m water depth.

3. IMPLEMENTATION TESTS

In order to evaluate these methods to decide on the optimal monitoring procedure, different field tests and test implementations were carried out. The first test was done with two 6000UPGs on November 14, 1997 near Inchon harbor. Another test was done using a chartered fishing boat in the coastal waters near Youngkwang (TCML in Figure 3). Surface seawater was sampled with a bucket for the calibration of the OBSs during the experiments.

Implementation experiments with the SSSC monitoring system were carried out in two cases: one with a near-coast ferry route, and the other with an open water operating car ferry. The ferryboat (KA-line in Figures 3,4) operates daily in

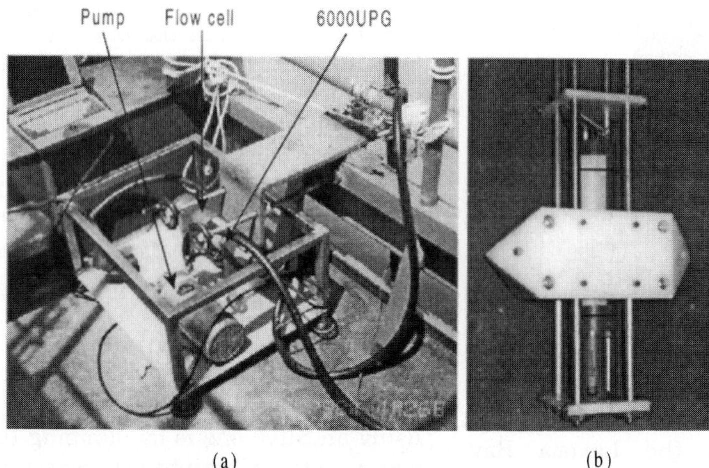

Figure 2. Configuration of shipboard measurement (a), and guard frame for *in situ* measurement (b).

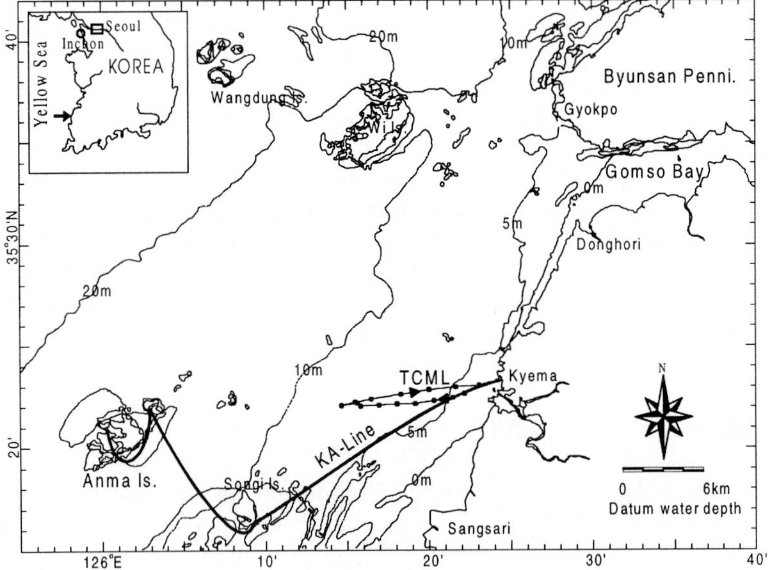

Figure 3. Test cruise monitoring line (TCML) and passenger ship route between Keyma harbor and Anma Island (KA-line) in the Youngkwang area.

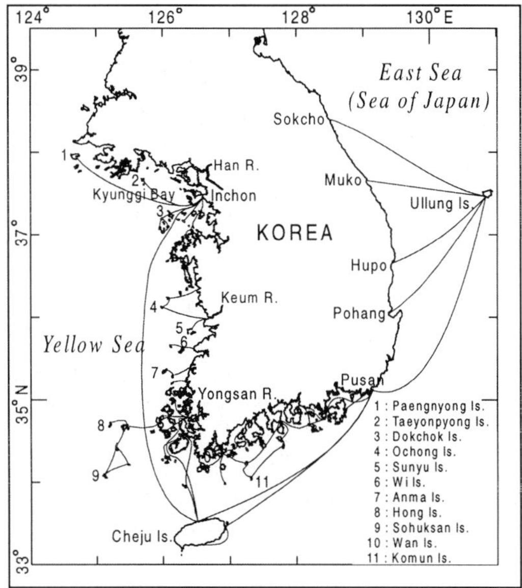

Figure 4. Representative passenger ship routes in Korea.

the coastal waters between Keyma harbor and Anma Island. According to the classification of Hayes (1979), the area around the route is a low macrotidal regime with mean spring and neap tidal ranges of about 5.4 m and 2.2 m, respectively (Korea Electric Power Company, 1994). The sedimentary processes in the area largely depend on strong wave action due to the winter monsoon from Siberia. The route of the second implementation test in open water between Inchon and Cheju is shown in Figure 4, covering the offshore region along the west coast of Korea.

During the test using a chartered fishing boat near Inchon harbor, both methods (SM and IM) were simultaneously tested. Figure 5 shows the correlation between manually measured SSSC by means of filtering of the sampled water and the output of the 6000UPG obtained by the two methods. In this test, both methods showed similar calibration characteristics. However, it was found that in the SM, fine sands were deposited in the flow cell at the end of the 2-hour cruise. This implies that it is possible that the pumping method could cause some error due to the accumulation of sediment in the flow cell. The SM is thus advantageous with regard to instrument security, while the IM can remove any uncertainty of data contamination that can be caused by pumping.

The calibration curves are shown to be quite different at different test sites as indicated by Figure 6, which means that reliable calibration must be conducted because the gain (volts per mg/l) of an OBS varies by a factor of 200 with particle

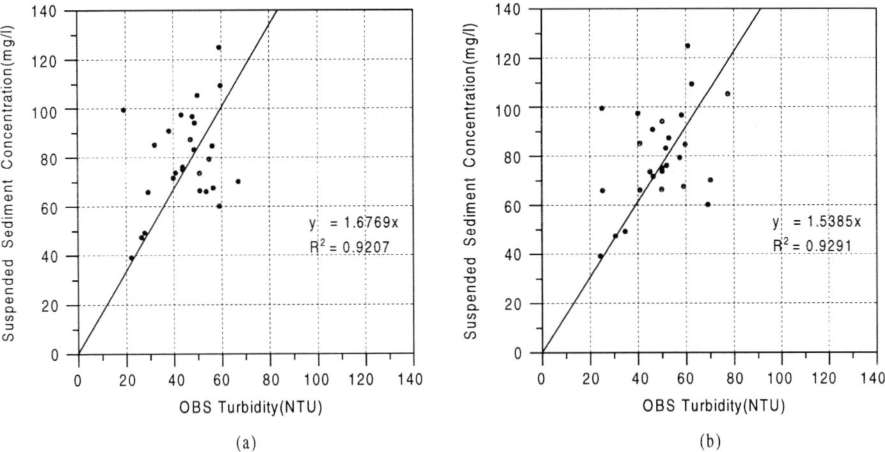

Figure 5. Correlation between measured concentration and OBS turbidity near Inchon harbor: *In situ* measurement (a), and shipboard measurement (b).

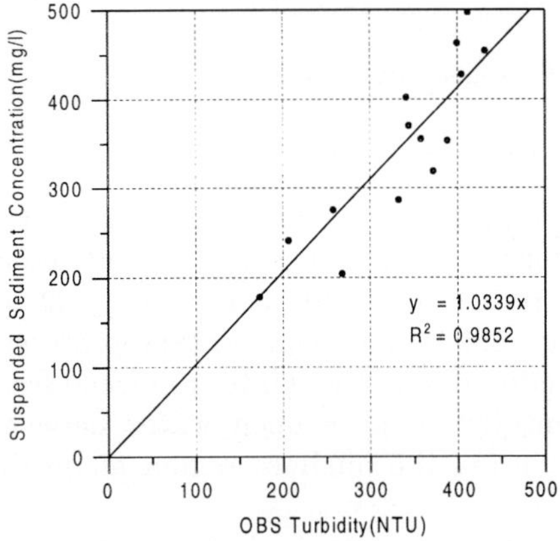

Figure 6. Correlation between measured concentration and OBS turbidity along the test cruise monitoring line near Keyma harbor.

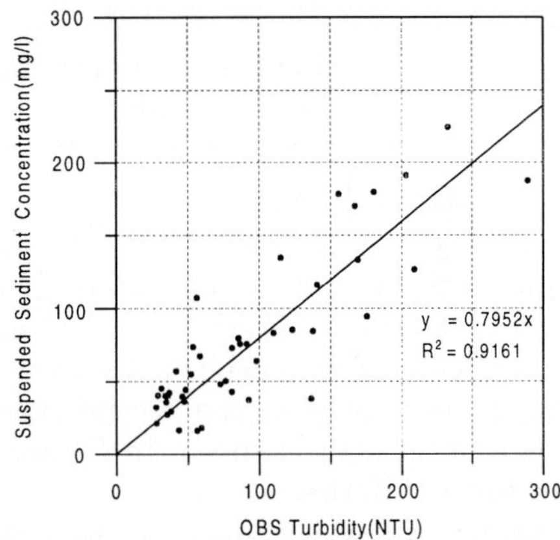

Figure 7. Correlation between measured concentration and OBS turbidity along KA-line near Keyma harbor.

size (D&A Instrument Company, 1991). The 6000UPG requires two kinds of calibrations because its output unit is NTU instead of voltage. After NTU calibration using a standard solution in the laboratory, *in situ* calibration was carried out on a fishing boat along a test cruise monitoring line (TCML in Figure 3) on December 14, 1997. As shown in Figure 6, the slope of the regression line is gentler and the determination coefficient is higher than those around the Inchon area (Figure 4). From the characteristic of the OBS with a gain that has an inverse relationship with particle size, it can be seen that the fraction of fine-grained sediment in the Youngkwang area is higher than that near Inchon.

In situ self-recording measurements were conducted for a month beginning March 6, 1998, from the car ferry, the *Sinhae IX*, which plies once a day between Keyma harbor and Anma Island (KA-line in Figures 3 and 4). The weight, length and width of the *Sinhae IX* are 154 tons, 49.5 m and 8.4 m, respectively. Its average cruising speed is 13 knots. During each cruise, the crew sampled surface seawater at six sites. However, they encountered the problem of instrument security after two weeks of implementation. Because of high cruising speed, the pipes of the guard frame were bent, the connecting part was broken and the OBS was lost.

The OBS calibration obtained from two-week monitoring is shown in Figure 7. The slope of the regression line was smaller than that along TCML. This might be due to the milder sea state and the greater water depth along the KA-line. With these conditions, the average concentration became low and the fine sediment fraction in suspension became large. Sample variations in the sea surface temperature (SST), salinity and SSSC for each cruise (Figure 8) show that there is a distinct boundary between high and low SSSC areas.

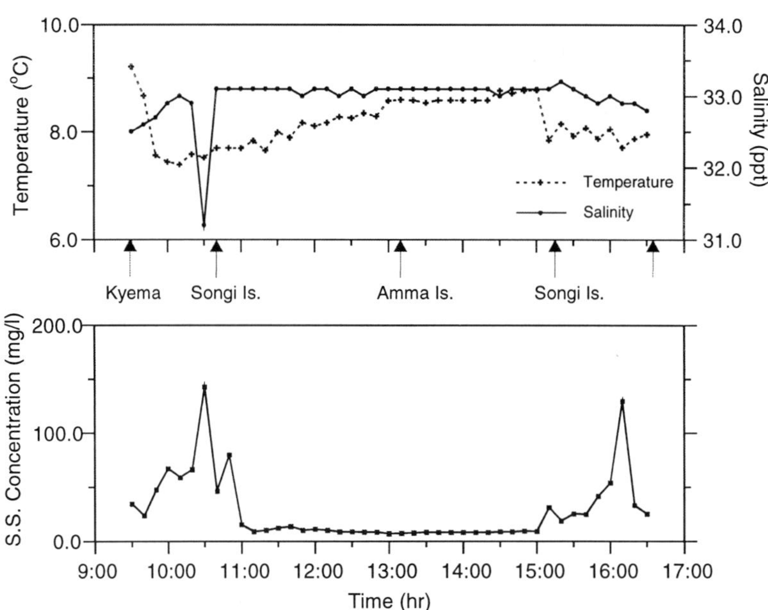

Figure 8. Sample variation in sea surface temperature, salinity, and SSSC obtained along the KA-line by *in situ* monitoring.

The SM was tested again after modifying the flow cell to prevent sediment accumulation and bubble formation. However, another mode of data contamination occurred as shown in Figure 9. The variation in salinity was unreasonable. The most probable cause of this variation was the formation of air bubbles in the flow cell, which was not observed visually. A flashlight held very close to the cell indicated that there were a great number of micro-bubbles circulating in it. It was also found that micro-bubbles gathered on the conductivity probe, and then vanished after reaching a critical size. The periodic fluctuation of salinity essentially represents gathering and vanishing of bubbles. Due to the possible effect of bubbles on the OBS, the concentrations in Figure 9 may not represent *in situ* values, although the discrepancy may be somewhat lower than that for salinity, due to the self-cleaning wiper.

The utility of SSSC monitoring on a relatively long route and the performance of the system developed by MOMAF (1999) were jointly tested on the 3,872-ton *Semo Ferry I* plying three times a week between Inchon and Cheju Island. *Semo Ferry I* left Inchon in the evening of September 11, 1998 and arrived at Cheju the following morning. The sensing interval was 5 minutes, and pumped seawater samples were taken every 30 minutes for OBS calibration. The calibration result and the variations in SST, salinity and SSSC are shown in Figures 10 and 11, respectively.

An interesting result was obtained when the ferry passed through the high-turbidity tidal mixing zone in the southwest region. The SST from near Dokchok Island (ND) to the west of Anma Island (WA) remained almost constant at 25°C. The temperature decreased markedly from 26.6°C at WA to 17.4°C halfway

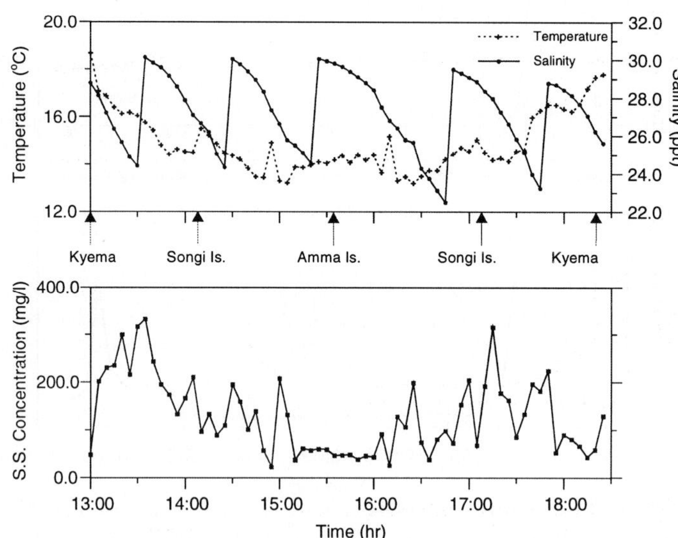

Figure 9. Test result of shipboard monitoring along the KA-line.

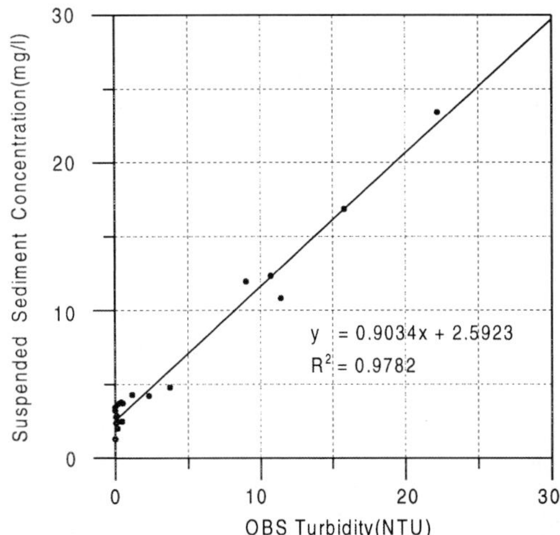

Figure 10. Correlation between measured concentration and OBS turbidity along the Inchon-Cheju route.

between Chin Island and Taehuksan Island (CCT), and then increased rapidly to 26.1°C southwest of Chin Island (SWC). The temperature gradient in the south frontal zone, the second half between CCT and SWC, was 0.3°C/km.

Because of heavy clouds on the test day, the clearest NOAA-14 SST image 4 days before the test was compared with the measured SST in order to estimate the extent of the mixing zone (Figure 12). Tidal mixing zones have developed around Seohan Bay, near Jangsan cape and the southwest tip (SWT) region. By considering vertical mixing (Simpson and Hunter, 1974), Wells (1988) has discussed turbidity fronts 25-50 km offshore of the depth of 20-80 m in the SWT region.

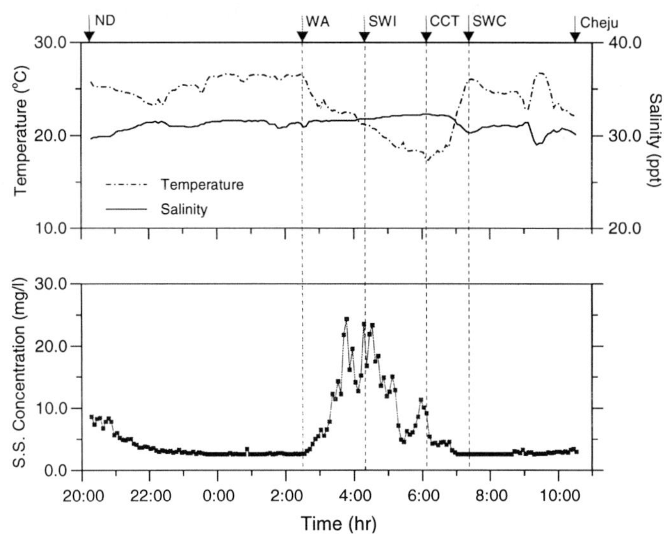

Figure 11. Results showing tidal mixing zone in the SWT region of Korea. ND; near Dokchok Is., SWI; southwest of Imja Is., CCT; center of Chin Is. and Taehuksan Is., SWC; southwest of Chin Is.

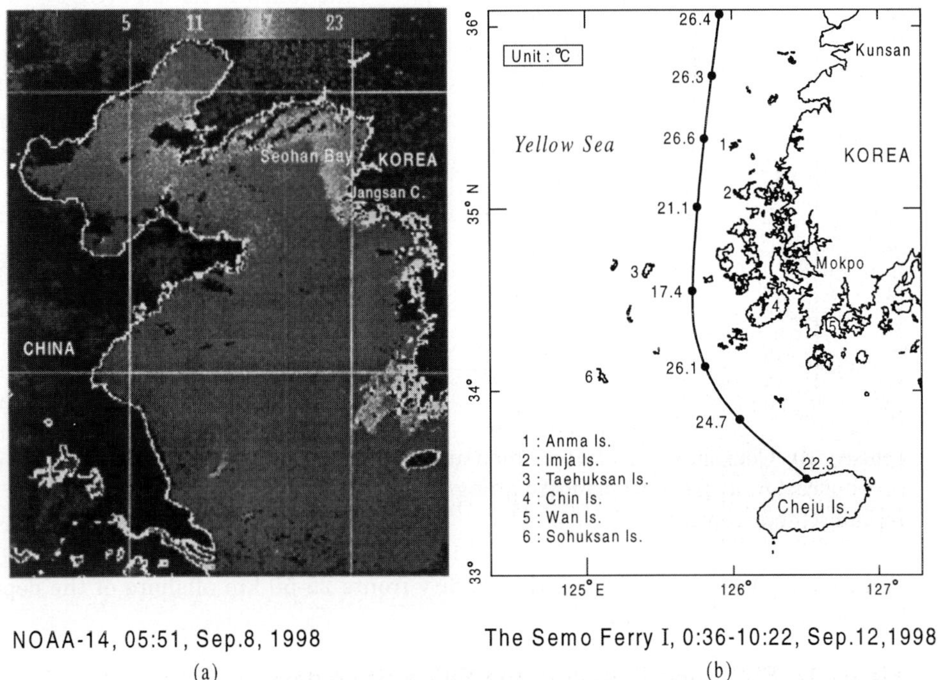

Figure 12. SST distribution in the Yellow Sea 4 days before the Inchon-Cheju test (a) and measured values in the SWT region (b).

The NOAA-14 image taken on a clear day 4 days after the test was calibrated to generate SSSC distribution in the Yellow Sea. The measured values along the ship route and the horizontal distribution of SSSC mapped from the satellite data are shown in Figure 13. It can be seen that the general pattern of SSSC in the Yellow Sea can be identified using the satellite remote sensing data. A high turbidity zone around the Changjiang River and the mouth of old Huanghe is distinctly observed. An isolated turbid zone south of the Shandong Peninsula tip indicating high sediment accumulation rate in this area (Alexander et al., 1991) is shown in the map. A tidal mixing zone along the west coast of Korea is also shown.

4. CONCLUSIONS

Methods for continuous monitoring of the SSSC along a routinely operating ferry route were tested. The IM, by deployment of the instrument into the sea surface water, provides more reliable information than the SM by pumping surface water in which contamination occurred due to bubble formation and

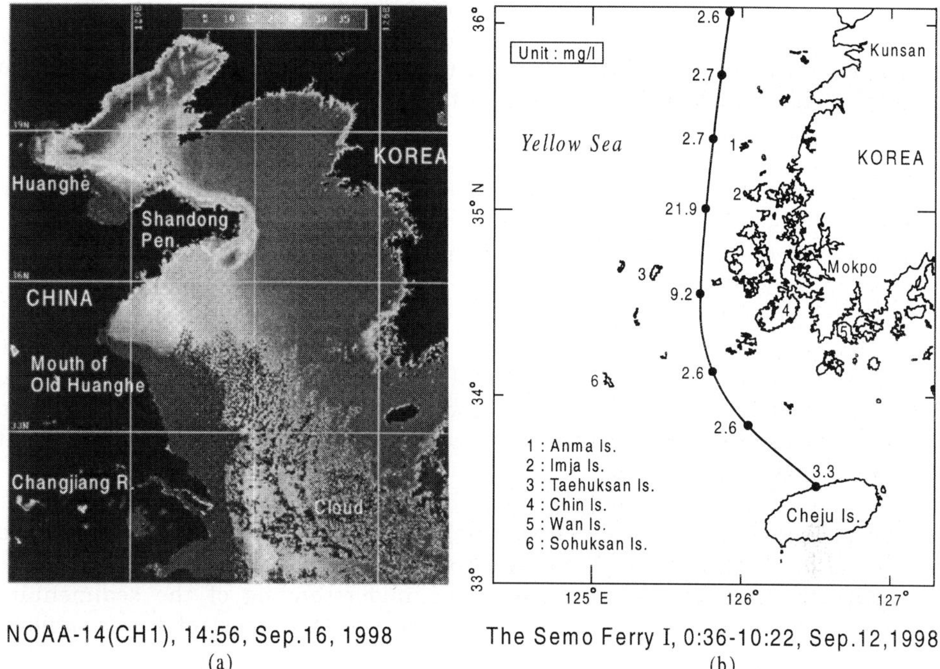

NOAA-14(CH1), 14:56, Sep.16, 1998
(a)

The Semo Ferry I, 0:36-10:22, Sep.12,1998
(b)

Figure 13. SSSC distribution in the Yellow Sea 4 days after the Inchon-Cheju test (a), and measured values in the SWT region (b).

siltation in the chamber. The SM is convenient as an automated system for relatively long routes; however, care must be taken to minimize the errors caused by such problems. A careful adjustment of the flow rate into the measuring chamber to minimize such errors is needed. The correlation between SSSC measured by filtering of the water sample and the YSI instrument is shown to change with sediment characteristics which vary with location and sea condition. Manual measurements of SSSC by filtering of the water samples are needed together with continuous instrument monitoring once in a while to obtain reliable SSSC values with the OBS.

The usefulness of the monitoring of SSSC was tested along two ferry routes. The SSSC in the Yellow Sea was quantitatively mapped with measured concentrations along the Inchon-Cheju Island route. Since remote sensing calibration algorithms are not available for the suspended sediment concentration, the IM of SSSC using routinely operating ferries in the Yellow Sea is needed. The two routes chosen in this study (Keyma-Anma and Inchon-Cheju lines) are important in understanding the sedimentary processes in the Yellow Sea, especially along the west coast of Korea. The Keyma-Anma route lies along the northern boundary of this high-turbidity zone. The net sediment flux across the route calculated from the long-term SSSC data and sediment transport modeling will yield important information on the sediment transport patterns along the west coast of Korea. By means of satellite remote sensing, when properly calibrated with ferry data from these two lines, it should be possible to monitor the time and space variations of the SSSC. The transport of suspended sediment in the Yellow Sea, sediment transport near the tidal mixing zone, formation of eddies and entrapment of suspended sediment can also be understood from such data.

In predictive modeling of sediment transport for the west coast of Korea, the long-term transport of the sediments of the Yellow Sea system must be also understood. For that purpose, a simulation system for cohesive sediment transport is being established by KORDI. The offshore boundary condition of the cohesive sediment transport model will be obtained from an integration of satellite remote sensing and ferryboat monitoring. The synthesized data should serve as a major source of data in the calibration and testing of the sediment transport model.

There are several coastal passenger ship routes along the coast of Korea as shown in Figure 4. Once the systems for routine SSSC monitoring are installed on ships operating along the important routes and the space and time variations of SSSC are monitored continuously, the understanding of the sedimentary processes in the Yellow Sea, especially along the west coast of Korea, will be significantly improved. To understand the sedimentation processes in the Yellow Sea, such a program needs to be extended to the international car ferry routes in the Yellow Sea. Cooperation between China and Korea will be required in the monitoring and prediction of suspended sediment transport in the Yellow Sea.

5. ACKNOWLEDGMENT

Guidance provided by Dr. Dana Kester in the early stages of this study is greatly appreciated. This study was supported by Ministry of Marine Affairs and Fisheries under Grant No. PM98015. We are grateful to Captain Il-Rae Kim and the crew of the *Sinhae IX*. We are also indebted to Dr. Sung-Hyun Kahng and Mr. Youn Gyoun Lee for their help in the test along the Inchon-Cheju Island route.

REFERENCES

Alexander, C. R., DeMaster, D. J., and Nittrouer, C. A., 1991. Sediment accumulation in a modern epicontinental-shelf setting: The Yellow Sea. *Marine Geology*, 98, 51-72.

Cooper, D. J., Watson, A. J., and Ling, R. D., 1998. Variation of PCO2 along a north Atlantic shipping route (U.K. to the Caribbean): A year of automated observations. *Marine Chemistry*, 60, 147-164.

D&A Instrument Company, 1991. Instruction manual of OBS-1 & 3. Port Townsend, WA, 41 p.

Harashima, A., 1993. High temporal-spatial resolution marine biogeochemical monitoring from Japan-Korea ferry – 1991 results. Monitoring report on global environment, Center for Global Environmental Research, Japan.

Harashima, A., 1994. High temporal-spatial resolution marine biogeochemical monitoring from Japan-Korea ferry – 1992-1993 results. Monitoring report on global environment, Center for Global Environmental Research, Japan.

Hayes, M. O., 1979. Barrier island morphology as a function of tidal and wave regime. In: *Barrier Islands*, S. P. Leatherman ed., Academic Press, New York, 1-27.

Jin, J.-Y., Hwang, K. C., Park, J. S., Yum, K. D., and Oh, J. K., 1999. Development of new time-selective and self-triggering water sampler. *Journal of Korean Society of Oceanography*, 34(4), 200-206.

Korea Electric Power Company, 1994. A study on the reduction of thermal discharge effects around nuclear power plants. 1st interim report (Youngkwang). *Report No. 92-802*, 336p (in Korean).

MOMAF, 1999. Marine environmental monitoring by using ship-of-opportunity. *Report No. BSPM98001-00-1118-2*, Ministry of Maritime Affairs and Fisheries, Seoul, Korea (in Korean).

Simpson, J. H., and Hunter, J. R., 1974. Fronts in the Irish Sea. *Nature*, 59, 699-710.

Wells, J. T., 1988. Distribution of suspended sediment in the Korea strait and southeastern Yellow sea: Onset of winter monsoons. *Marine Geology*, 83, 273-283.

Coastal and Estuarine Fine Sediment Processes
W.H. McAnally and A.J. Mehta (Editors)

Seasonal variability of sediment erodibility and properties on a macrotidal mudflat, Peterstone Wentlooge, Severn estuary, UK

H. J. Mitchener[a] and D. J. O'Brien[b]

[a]Defence Evaluation Research Agency, Underwater Sensors and Oceanography, Winfrith Technology Centre, Dorchester, Dorset DT2 8XJ, United Kingdom

[b]Hyder Consulting Ltd, PO Box 4, Pentwyn Road, Nelson, Mid Glam CF 46 6YA, United Kingdom

A comprehensive data set of sediment erodibility and properties has been obtained from the Peterstone Wentlooge mudflat on the north shore of the Severn estuary, UK. Three monitoring stations were visited on a monthly basis during full moon spring tide low water conditions from April 1997 to September 1997, with an additional survey in January 1998 to represent winter conditions. Measurements of *in situ* erosion thresholds were taken using SedErode, and a range of physical and biological parameters were determined from surface scrape sediment samples. Most of the SedErode deployments were completed on recently deposited sediment which had high water contents between 100% to 330%. Erosion thresholds over the entire period ranged between 0.10 and 0.46 Pa, and there was a general decrease in surface strength during the spring/summer period which was mainly attributable to decreasing sediment density. However, although the surface strength decreased during the period (1 - 2 mm depth), vane shear strength, which is a measure of the strength of the sediment to a depth of 5 cm, showed a clear seasonal trend of increasing strength with mean monthly temperature. Although there was considerable biological activity at the site and the production of extracellular polymeric substances (identified through measured levels of colloidal carbohydrate), there was no evidence of biostabilization. Therefore although the mudflat demonstrated a seasonal trend of development in the top 5 cm, the surface strength was dependent on large diurnal variations in density.

1. INTRODUCTION

Muddy coastlines are important habitats (Ferns, 1983) and can provide natural shoreline protection due to their ability to attenuate wave energy. At present the factors controlling cohesive sediment stability are not fully understood and this makes predicting sediment transport difficult. Previous studies (Kirby and Parker, 1983; Williamson and Ockenden, 1996; Whitehouse and Mitchener, 1998) have indicated that the stability of the mudflats on the southern shore of the Severn estuary exerts an important control on the turbidity of the upper estuary.

Recent observations have shown that the mudflats on the northern shore of the Severn estuary also have important implications for the turbidity of the estuary as they experience periodic and episodic development at seasonal as well as shorter timescales (O'Brien et al., 2000). Most previous studies have been designed to determine the spatial variability in surface sediment strength and have consequently been based over short time periods (Williamson and Ockenden, 1996, Amos et al., 1998; Widdows et al., 1998). Some studies have addressed the temporal variation in sediment stability but have used standard geotechnical techniques, e.g., vane shear strength (VSS), to determine sediment stability (Kraeuter and Wetzel, 1986; Amos et al., 1988). Derived values of VSS are at best only weakly correlated with *in situ* erosion thresholds (Amos et al., 1996).

Kraeuter and Wetzel (1986) suggested that sediment stability for a mudflat in Virginia, USA, was due to biostablization and was controlled by the seasonal variation in temperature. Amos et al. (1988) showed dramatic increases in sediment stability for a mudflat in the Bay of Fundy, Nova Scotia, during the summer when the tidal flat exposure coincided with midday and these increases were attributed to desiccation due to solar radiation (Anderson and Howell, 1984). Measurements made by Frostick and McCave (1979), in the Deben estuary, UK indicated that the summer period between April and September was associated with accretion and the winter months with erosion. They noted that the accretion in summer coincided with algal growth but was also coincident with naturally lower windspeeds (i.e., lower wave activity). The mudflats on the north shore of the Severn estuary also experience this same seasonal development (O'Brien et al., 2000) but it is uncertain whether biostabilization is a major control on sediment stability.

Many studies have shown the stabilizing effects of micro-organisms including benthic diatoms and bacteria (Yallop et al., 1994). The two main reasons for the stabilizing effect of these organisms is the binding of their secretions, i.e., extracellular polymeric substances (EPS), with surface sediment and the reduction of surface roughness also due to EPS (Paterson, 1997). The effect of the density of macrofauna is less clear on stability of surface sediment (Jumars and Nowell, 1984). Whereas some macrofauna stabilize the sediment through compaction due to tube building (Meadows and Tait, 1989), others weaken sediment by burrowing and also increase surface roughness (Widdows et al., 1998).

The aim of many contemporary studies is to derive predictive relationships between *in situ* sediment stability and the density of benthic organisms. However, these relationships are complex as it is the mechanism of stabilization/turbation that is important and not necessarily the biomass (Paterson, 1997). Bioturbation/stabilization due to macrofauna is usually linked to the descriptors of erodibility through organism density (Meadows and Tait, 1989; Widdows et al., 1998), whereas chemical indicators are used to quantify the stabilizing effects of phytobenthos and bacteria. Colloidal carbohydrate has proved to be a useful indicator of EPS (and therefore sediment stability) and has even been correlated with biomass (Underwood et al., 1995). A predictive relationship between photo-pigment content (BChl a, indicating purple sulfur bacteria) and the critical erosion friction velocity (U_{*cr}) was found by Grant and Gust (1987). Quantitative relationships between colloidal carbohydrate (indicating EPS) and the descriptors of erodibility are given by Sutherland (1996) and Black (1997). A recent study found a relationship between the critical erosion stress and a combination of wet weight bulk density and colloidal carbohydrate (Amos et al., 1998).

Tidal modulation within the spring-neap cycle has been found to produce a periodic variation in mudflat bed level on the southern shore of the Severn estuary (Whitehouse and Mitchener, 1998). These authors also showed that the phasing of calm conditions within the spring-neap cycle was an important control on the sediment exchange between the water column and the mudflat surface. There are currently no data sets which have attempted to isolate the seasonal variations in stability from those that might occur over short time periods due to tidal forcing.

This paper describes measurements of sediment erosion thresholds and properties made over a 10 month period at Peterstone Wentlooge, Severn estuary, UK. The measurements were undertaken systematically so that deployments were made at the same locations on the mudflat, at the same tidal state (spring tides) and with the same relative exposure time. The resulting data set (Mitchener et al., 1998) allows trends in sediment properties and the factors that control the strength of surface sediment to be ascertained on a seasonal basis.

2. SITE DESCRIPTION

The study area was situated on the northern shore of the Severn estuary, seaward of the village of Peterstone Wentlooge on the intertidal mudflats that extend between the River Rhymney at Cardiff and the River Usk at Newport (Figure 1). The Severn estuary is one of the largest in Europe and is characterized by a large tidal range (between 10 m and 14 m during springs) and associated strong tidal currents. The intertidal area faces south-east and has a width of 1 km and a slope of 1:94 and is backed by a salt marsh and a sea defense bund (Figure 2). The mudflat is protected from swell waves entering the estuary from the west but can experience erosional events due to short period wind waves generated within the estuary (O'Brien et al., 2000).

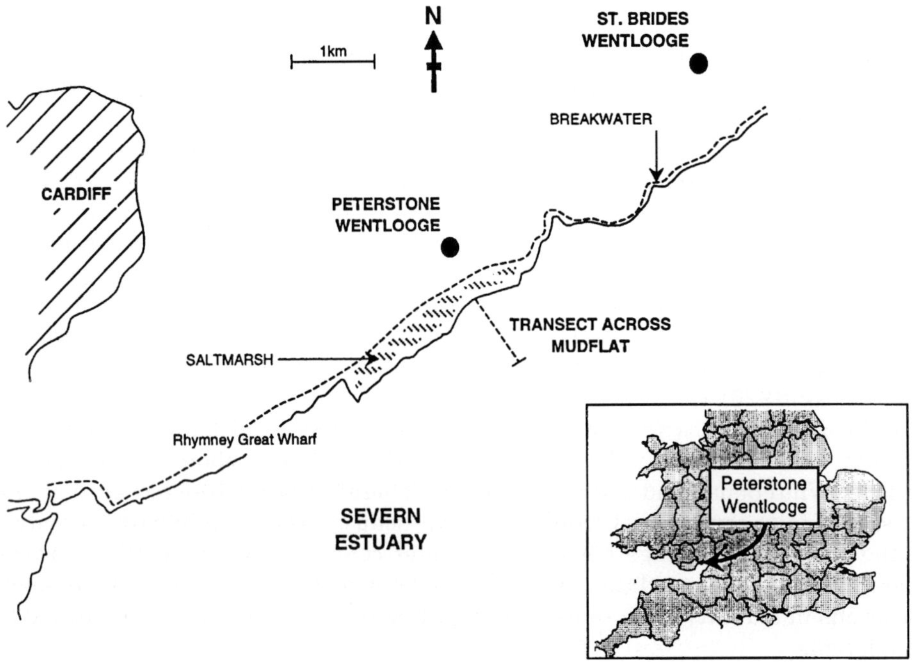

Figure 1. Peterstone Wentlooge mudflat general location plan.

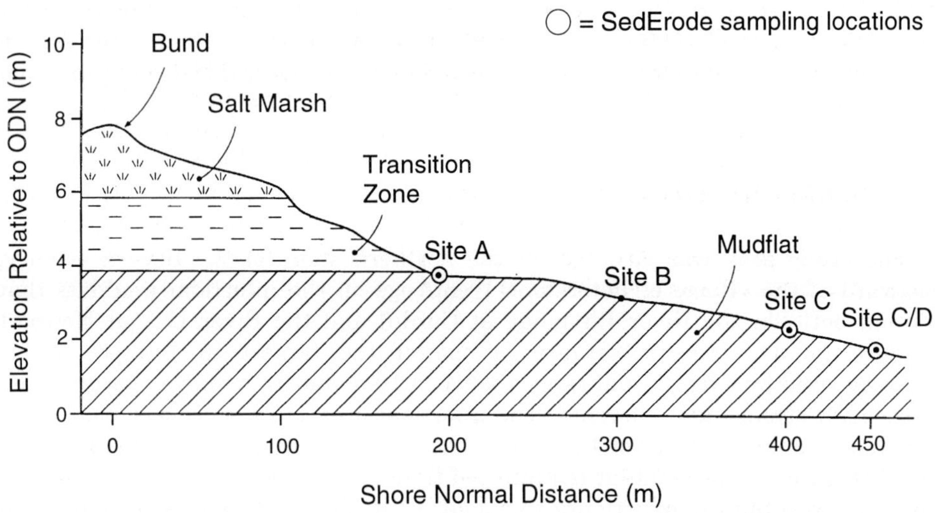

Figure 2. Erodibility seasonal survey sampling locations.

The mudflat consists of two Holocene layers: a surface layer of modern sediment which has a depth of between 0 and 30 cm and an underlying relict deposit. The relict deposit is composed of a stiff grey clay and is part of the Wentlooge Formation. The surface layer has a median grain size of between 1 - 2 μm and the water content typically exceeds 100% (O'Brien, 1998).

A ridge and runnel system (aligned shore-normal) is commonly found between 300 m and 500 m offshore, with typical ridge widths of 40 - 70 cm and heights of 5 -10 cm.

3. MONITORING STRATEGY

Three locations were selected for *in situ* measurements of erodibility and sediment properties and were situated on a shore-normal transect of monitoring stations extending seawards from about Mean High Water Neaps. Figure 2 shows a diagram of the shore-normal transect, indicating the different zones of the mudflat and the sampling locations. The stations had been previously set up in 1996 as part of a wider study (O'Brien, 1998). The selected locations were monitoring stations A, C and C/D (which were 200 m, 400 m and 450 m from the onshore bund, respectively) and modern sediment was present at all stations except during January 1998. The site was visited once a month during full moon spring tide conditions. The aim of this deployment strategy was to minimize the variations in surface sediment properties that could occur due to different tidal conditions (Whitehouse and Mitchener, 1998). Six monthly surveys were conducted between April and September 1997 with an additional day's monitoring in January 1998 to represent winter conditions. During the January visit the underlying Wentlooge Formation was exposed between stations B and C. During this site visit, measurements were taken at sites A and B/C which represented the modern sediment and the exposed Wentlooge Formation, respectively.

The sites were monitored in the same order on each measurement day, sites A, C and C/D, respectively, progressing offshore throughout the period of the falling tide. The systematic progression down the mudflat meant that exposure times at each station were similar to those from the other measurement days and any time-related increases in strength due to exposure were minimized (Paterson et al., 1990). Measurements were only taken on ridge topography in order to minimize the small-scale spatial variability in sediment strength and properties due to morphology.

4. MEASUREMENTS

At each station, three repeat measurements of τ_{cr} were made within a 5 m^2 area using SedErode (Mitchener et al., 1996a, 1998). In addition, 5 'Pilcon' vane shear strength measurements (over the surface 5 cm of sediment) were taken at each SedErode deployment along with measurements of surface sediment temperature (top 5 mm), detailed surface observations and photographs. Three surface scrape samples of about 5 ml were taken from the top 1-2 mm of sediment in the vicinity of the SedErode sites. One sample was used for colloidal carbohydrate (CC) analysis, one was used for the analysis of physical properties and one was archived. The sediment sample used for CC analysis was stored on ice until it could be frozen in the laboratory prior to analysis. The sediment samples were analyzed for water content, bulk density, loss on ignition, grain size and CC content. Whereas the physical analyses were completed for each SedErode deployment (and therefore in triplicate for each station) the CC analyses were only undertaken on the sample from the second deployment at each station.

5. METHODS

5.1. The measurement of *in situ* erosion thresholds

SedErode instrument. SedErode is the successor to ISIS (Instrument for shear Stress *In Situ*) and is a portable, fully contained instrument designed for use on intertidal mudflats and other cohesive sediments (Mitchener et al., 1996a). Williamson and Ockenden (1996) give a full technical description of ISIS. The instrument consists of 2 units: a head unit and a control unit. Figure 3 shows the major components of the SedErode head unit. It is a mini radial flume which is used to subject a sediment surface of 0.09 m diameter to increasing fluid shear stresses. The principle of SedErode measurements is that known shear stresses are applied in increments to the sediment surface and the bed response (erosion) is monitored through measurements of increasing turbidity within a 1 liter recirculating system.

Deployment method. A representative 9 cm diameter flat mud surface was selected. The site was photographed and visually analyzed for surface characteristics, and the surface sediment temperature readings were taken. SedErode was then positioned on the sediment surface and the recirculating pipework was filled with local clear (settled) seawater. At the start of each test the nephelometer turbidity sensor was zeroed to allow relative increases in turbidity relating to sediment erosion from the bed to be monitored. Throughout the tests shear stress and turbidity signals were logged at two second intervals. The lowest shear stress setting (about 0.2 Pa) was applied for the first three minutes to allow mixing of the recirculating water and to establish a baseline turbidity. The applied shear stress was then increased in controlled steps (about 0.05 Pa) at one minute intervals until the sediment surface eroded and the turbidity reading significantly increased. SedErode was then removed, drained and cleaned before the next deployment.

Each erosion test resulted in a raw time-series of voltage output from the nephelometer (turbidity sensor) and pump motor. This data was then processed via calibration functions (Mitchener et al., 1996a) to produce time-series of

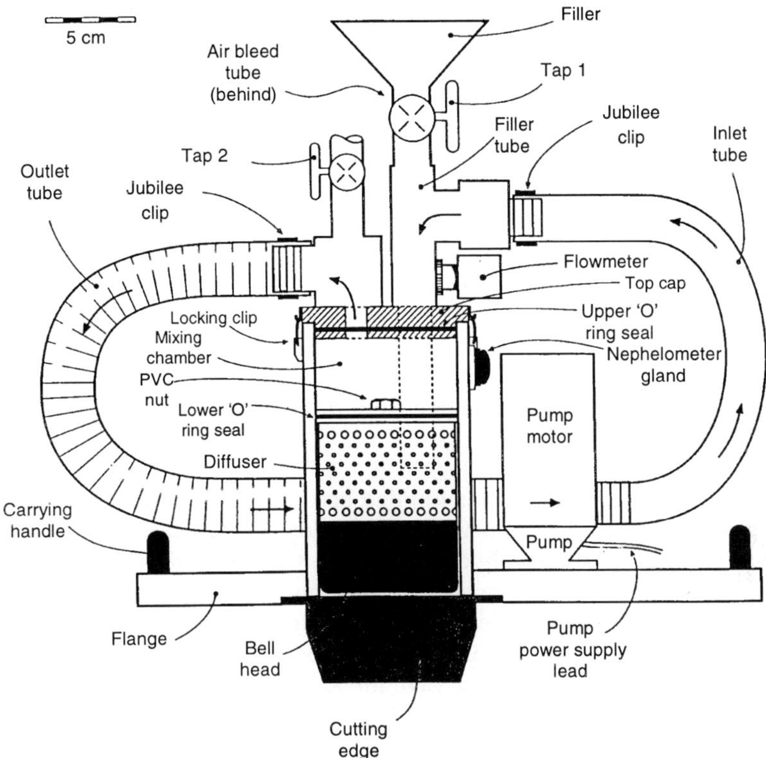

Figure 3. SedErode head unit.

applied shear stress and suspended sediment concentration (SSC) within the recirculating system. Figure 4 shows an example of the output data obtained from each deployment. The time-series plots were analyzed to establish the point at which surface erosion started. The incipient point of erosion, τ_{cr}, corresponded to a divergence in the system concentration from the baseline which indicated that material had been removed from the bed and mixed into the recirculating system. The further increases in shear stress produced higher concentrations and confirmed that surface erosion had been initiated. On the time-series plots shear stress steps τ_A and τ_B were established between which surface erosion had occurred and τ_{cr} was then calculated as the mean of τ_A and τ_B. For runs where erosion started at the lowest applied shear stress, τ_B was set at this value and τ_A was set at 0.

5.2. Surface sample analysis

Water content (the percentage ratio of the mass of water and dry sediment) was determined as the sample mass loss after drying at 110°C for 24 hours

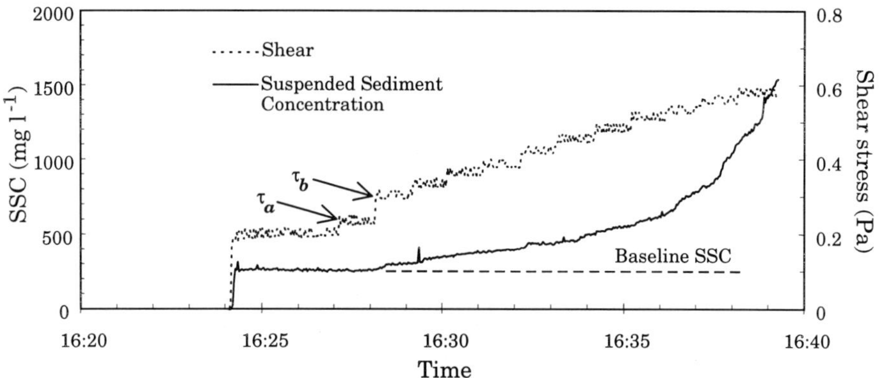

Figure 4. Example of output data from SedErode.

(Head, 1980). A salt correction was applied according to Noornay (1984) using an interstitial water salinity of 25 psu. Wet bulk density was calculated according to Lee et al. (1987) use a previously derived mean value of sediment specific gravity of 2.50 (O'Brien, 1998). The percentage loss on ignition was determined using INTRMUD protocols (1997) which involved sample ignition at 450°C for 4 hours. LOI is an overestimation of organic content (Mook and Hoskin, 1982) and organic carbon was approximately 30% of LOI for samples taken previously from this location (O'Brien, 1998).

The grain size distribution of the samples was determined by wet sieving through a 90 μm sieve and analysis of the < 90 μm fraction at 1/4 phi intervals with a Micrometrics Sedigraph™ 5100 instrument. The CC values were determined using a spectrophotometric assay (Dubois et al., 1956). Defrosted, hydrated samples were centrifuged and the supernatant treated with aqueous phenol to produce a carbohydrate-phenol mixture. Concentrated sulfuric acid (H_2SO_4) was added which produced an exothermic reaction, and resulted in the production of colored compounds. The colored samples were then scanned at 486.5 nm using a scanning spectrophotometer to measure the absorbency. The absorbencies were calibrated against a standard curve of absorption versus glucose concentration to produce the CC results as microgram glucose equivalents.

6. RESULTS

6.1. Temporal variability, April to September 1997

Table 1a,b gives a summary of the data obtained during the 7 deployment days. Meteorological conditions varied considerably on measurement days. The daily maximum temperature at Rhoose Airport shows a range of between 8.4°C for deployment 7 (winter measurements) to 25.8°C for the August deployments.

Table 1a
Averaged data for the Cardiff seasonal survey

Date	Deployment	Daily max. temp. / rainfall (°C/mm)	Sediment temp. (°C)	Vane shear strength (kPa)	Water content (%)	Bulk density (kgm^{-3})	LOI (%)
Site A							
18/04/97	1A	12.6/-	16.7	0.5	231.2	1242	9.81
22/05/97	2A	10.0/-	10.9	0.8	150.0	1337	8.05
19/06/97	3A	15.6/2.9	14.8	0.9	216.3	1254	8.12
22/07/97	4A	24.2/-	28.3	1.0	200.5	1275	10.06
19/08/97	5A	25.8/-	29.7	1.3	279.2	1209	12.16
18/09/97	6A	21.9/-	23.0	0.9	333.4	1181	11.05
14/01/98	7A	8.4/-	8.2	5.6	219.5	1251	9.72
Site B/C							
14/01/98	7BC	8.4/-	8.3	16.8	82.7	1516	6.64
Site C							
18/04/97	1C	12.6/-	16.7	0.1	231.3	1241	10.98
22/05/97	2C	10.0/-	11.7	0.3	180.5	1292	9.45
19/06/97	3C	15.6/2.9	18.6	0.6	207.1	1263	8.73
22/07/97	4C	24.2/-	27.0	0.7	169.9	1311	9.42
19/08/97	5C	25.8/-	28.8	0.9	220.9	1250	11.17
18/09/97	6C	21.9/-	23.2	1.0	246.0	1230	10.65
Site CD							
18/04/97	1CD	12.6/-	21.9	0.4	206.6	1264	10.39
22/05/97	2CD	10.0/-	10.9	1.1	159.2	1321	9.51
19/06/97	3CD	15.6/2.9	19.7	1.2	139.0	1355	8.98
22/07/97	4CD	24.2/-	/	1.8	120.4	1394	9.11
19/08/97	5CD	25.8/-	26.3	1.6	146.1	1342	10.40
18/09/97	6CD	21.9/-	22.5	1.4	142.6	1349	8.97

Triplicate data for surface sediment properties except for carbohydrate where there is one determination for each station and vane shear strength which is the mean of 15 observations. Daily maximum air temperature and daily rainfall were obtained from the Meteorological Office's synoptic weather station at Rhoose (Cardiff Airport).

Although there were some showers on deployment days 1, 2 and 6, the only significant rainfall occurred during the June deployments. During deployment 1 it was observed that the SedErode filling method was causing sediment erosion prior to the start of experiments so the method was changed for subsequent deployments. It is uncertain to what extent the filling method affected the results of deployment 1, so these results have been excluded from any interpretation.

The critical erosion shear stress values, τ_{cr}, were in the range 0.10 Pa to 0.46 Pa. The temporal variability of mean τ_{cr} for each station is shown in Figure 5 for each station when 3 deployments were completed and demonstrates that there was a general decrease in the erosion threshold from around 0.30 Pa to 0.15 Pa during the spring/summer period. There was the opposite trend for surface water content which was in the range 150% to 330% and generally increased during the period (Figure 6).

Table 1b
Averaged data for the Cardiff seasonal survey

Date	Deployment	Colloidal carbohydrate (mgg^{-1})	Median size (μm)	Erosion threshold 1 (Pa)	Erosion threshold 2 (Pa)	Erosion threshold 3 (Pa)	Mean erosion threshold (Pa)
Site A							
18/04/97	1A	2.56	2.95	/	/	/	/
22/05/97	2A	0.49	5.09	0.13	0.38	0.30	0.27
19/06/97	3A	1.04	3.00	0.13	0.40	0.30	0.28
22/07/97	4A	5.47	1.81	0.11	0.29	0.33	0.24
19/08/97	5A	3.75	2.98	0.10	0.25	0.21	0.19
18/09/97	6A	1.83	1.66	0.13	0.13	0.13	0.13
14/01/98	7A	0.49	4.65	0.15	0.24	0.20	0.20
Site B/C							
14/01/98	7BC	0.54	4.81	0.23	0.24	0.41	0.30
Site C							
18/04/97	1C	2.62	2.11	/	/	/	/
22/05/97	2C	1.17	3.02	0.43	0.39	0.31	0.38
19/06/97	3C	1.09	2.22	0.31	0.28	0.27	0.29
22/07/97	4C	6.48	1.88	0.32	0.27	0.26	0.28
19/08/97	5C	4.07	2.69	0.30	0.10	0.26	0.22
18/09/97	6C	2.21	1.92	0.14	0.13	0.12	0.13
Site CD							
18/04/97	1CD	2.76	2.20	/	/	/	/
22/05/97	2CD	0.71	2.88	0.46	0.40	0.30	0.39
19/06/97	3CD	1.67	2.32	0.36	0.45	/	/
22/07/97	4CD	1.72	2.37	0.24	/	/	/
19/08/97	5CD	2.21	2.12	0.37	0.36	/	/
18/09/97	6CD	2.81	2.67	0.39	0.31	0.27	0.32

Triplicate data for surface sediment properties except for carbohydrate where there is one determination for each station and vane shear strength which is the mean of 15 observations. Daily maximum air temperature and daily rainfall were obtained from the Meteorological Office's synoptic weather station at Rhoose (Cardiff Airport).

Bulk density had an inverse relationship with water content and therefore followed the decreasing trend of the erosion thresholds with values between 1,180 and 1,395 kg m^{-3} (Table 1a,b). Although the erosion thresholds generally decreased during the period (Figure 5), mean vane shear strength (which is a measure of the strength of the surface 5 cm of sediment) increased from around 0.1 kPa in April 1997 to over 1.0 kPa by August 1997 (Figure 7). These values reflect a gradual consolidation of the surface 5 cm of the mudflat during the spring/summer period.

The surface sediment temperature was in the range 10.9 to 29.7°C and exhibited a sinusoidal temporal trend at all 3 stations which was similar to the maximum daily air temperature (Figure 8). In general the surface sediment temperature observed at each station was 2 - 4°C greater than the daily maximum air temperature (Table 1a,b). For stations A and C the temporal trend

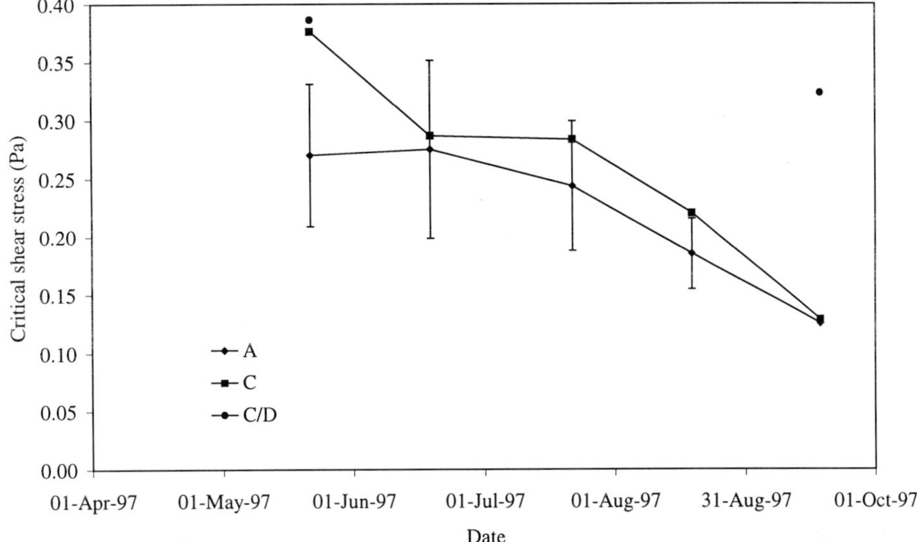

Figure 5. Time-series of τ_{cr} between May 1997 and September 1997, mean values ± 1 standard deviation for station A.

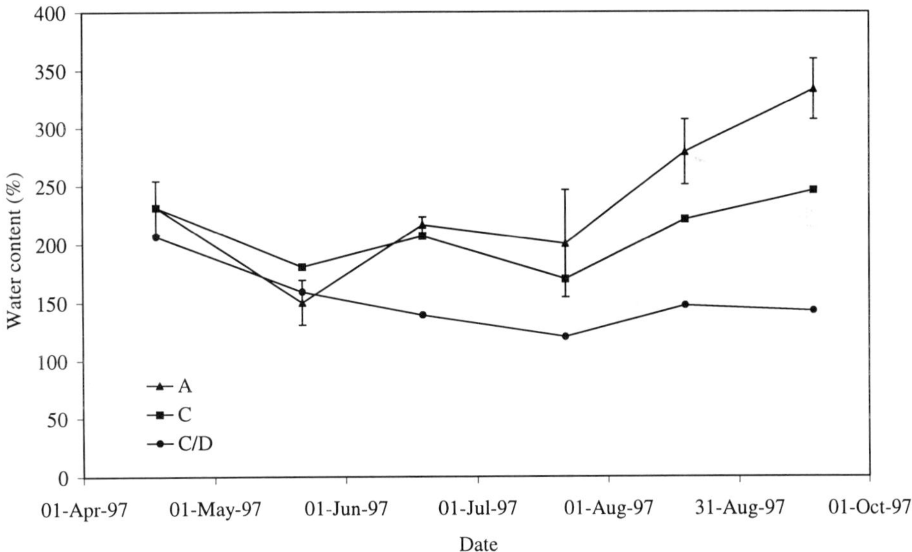

Figure 6. Time-series of surface water content between April 1997 and September 1997, mean values ± 1 standard deviation for station A.

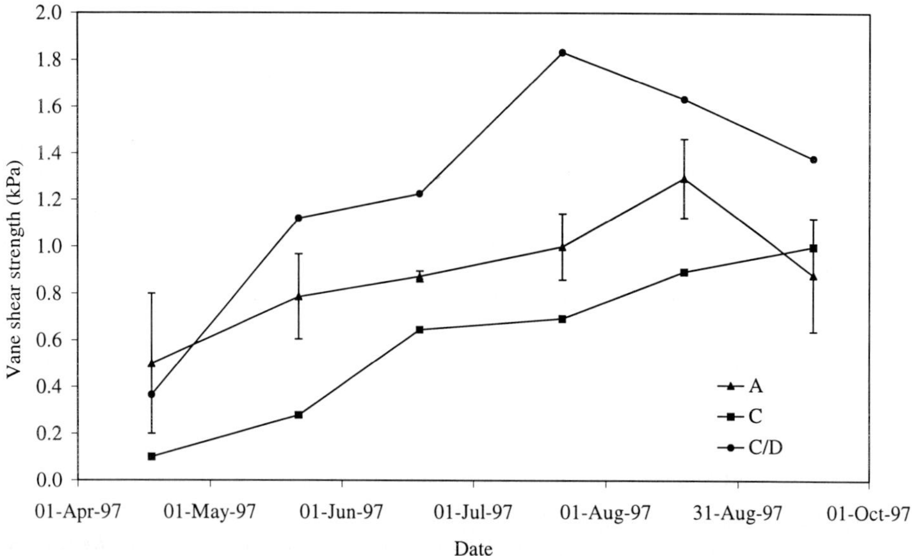

Figure 7. Time-series of vane shear strength between April 1997 and September 1997, mean values ± 1 standard deviation for station A.

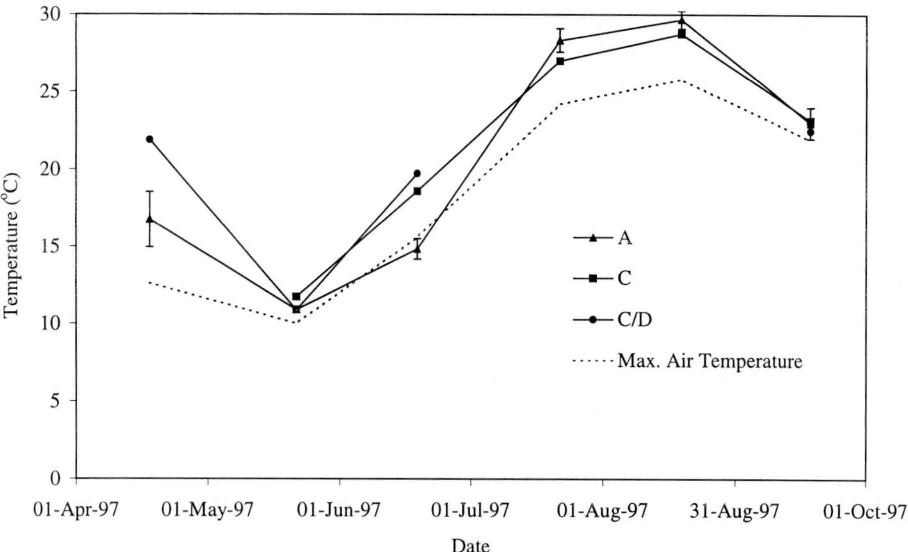

Figure 8. Time-series of surface sediment temperature and daily maximum air temperature between April 1997 and September 1997, mean values ± 1 standard deviation for station A.

of colloidal carbohydrate followed closely that of air and sediment temperature (Figure 9) ranging from values below 1mg^{-1} glucose equivalent in May to over 5 mg g^{-1} in July 1997. The reason why station C/D exhibited a different trend than the other stations in the months of July and August is uncertain. The results for LOI showed a decrease in values between April and June of approximately 2% to 8 - 9% before increasing to maximum values of between 10 - 12% in August (Figure 10). It is thought that the increase in LOI during July to September was due to the substantial increases in the number of macrofauna including the polychaete *Nereis diversicolor* and gastropod *Hydrobia ulvae* during the same period.

The grain size distribution of the surface sediment samples was very uniform during the survey period and there was no temporal variation observed (Table 1a,b). The median grain size had a narrow range between 1.6 microns and 6.1 microns, with slightly higher values at site A. The dominant sediment fraction on the survey transect was silt (45% to 78%), followed by clay (22% to 54%) and then sand which was consistently very low (< 4%).

6.2. Spatial variability, April to September 1997

The mean erosion threshold for deployments 2 - 6 increased with distance seaward with values of 0.22 (sample size n = 15), 0.26 (n = 15) and 0.36 Pa (n = 11) for stations A, C and C/D, respectively. Mean surface water content for stations A, C and C/D for deployments 1 - 6 was 235%, 209% and 152%,

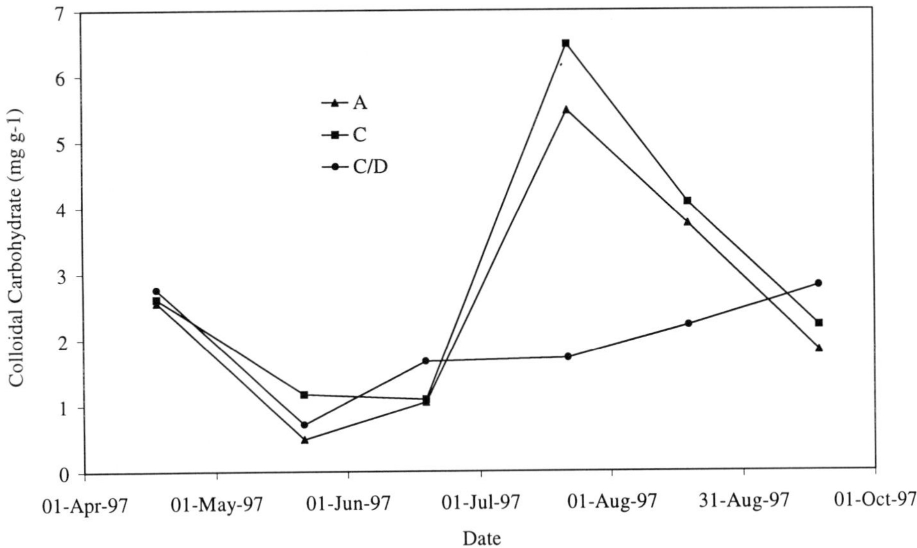

Figure 9. Time-series of surface colloidal carbohydrate between April 1997 and September 1997.

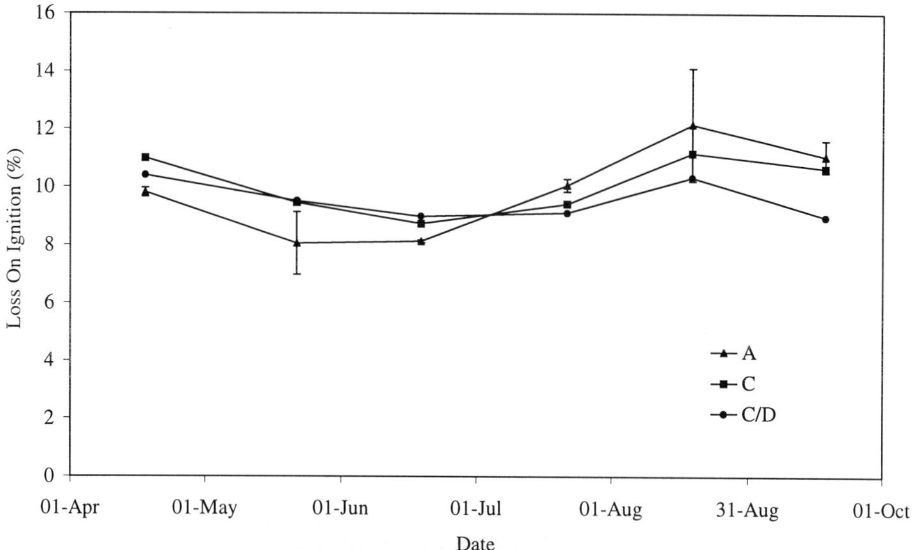

Figure 10. Time-series of loss on ignition between April 1997 and September 1997, mean values ± 1 standard deviation for Station A.

respectively. Although there was a steady decrease in mean surface water content with distance seaward, the same trend was not found for vane shear strength where station C had the lowest mean value. The mean vane shear strength for stations A, C and C/D was 0.9, 0.6 and 1.3 kPa, respectively. There was little spatial variation in surface sediment temperature, colloidal carbohydrate (except for station C/D during the July and August deployments), LOI and grain size.

6.3. Winter deployment erosion thresholds and sediment properties

Between September 1997 and January 1998 there was a period of sustained erosion and the modern sediment had been removed from the area between stations B and C (Figure 2) leaving the Wentlooge Formation exposed. The three deployments at station A had erosion thresholds of 0.15, 0.24, and 0.20 Pa giving a mean value of 0.20 Pa. These low erosion threshold values were associated with high sediment water contents of between 215% and 223%. The second set of triplicate SedErode measurements on the 14th January 1998 were made on the Wentlooge Formation between stations B and C. The mean of the three erosion threshold measurements was 0.30 Pa.

6.4 Intercomparisons

Quantitative relationships between τ_{cr} and sediment parameters were derived through least squares regression analysis. The data used in the regressions was the mean of each triplicate results at all stations. All the data was used from deployments 2 - 7 (when available in triplicate) with the exception of the data from station B/C during deployment 7 which represented the Wentlooge Formation. Water content, wet weight bulk density and loss on ignition were the only sediment properties that were found to correlate with τ_{cr} at the 95% level and above. These relationships are as follows:

$$\tau_{cr} = -0.0012wc + 0.51 \qquad (r^2 = 0.66, P < 0.001) \qquad (1)$$

$$\tau_{cr} = 0.0013\gamma - 1.4 \qquad (r^2 = 0.63, P < 0.005) \qquad (2)$$

$$\tau_{cr} = -0.041\text{LOI} + 0.65 \qquad (r^2 = 0.36, P < 0.050) \qquad (3)$$

where τ_{cr} is the mean surface erosion threshold from each triplicate deployment (Pa), wc is the mean water content (%), γ is the mean bulk density (kgm^{-3}) and LOI is the mean loss on ignition (%). The derived relationship between τ_{cr} and wc is shown in Figure 11 and demonstrates that there is a negative correlation between the two parameters. As γ is calculated from wc the r^2 (coefficient of determination) is similar for both (1) and (2). The relationship between τ_{cr} and LOI is just significant at the 95% level and (3) accounts for only 36% of the measured variation in τ_{cr}. Note that in (1), (2) and (3) P represents the significance level of the correlation.

Least squares regression analysis of parameters measured on a daily basis revealed that there was also a strong relationship between sediment temperature and colloidal carbohydrate (CC). Although a linear regression was significant at

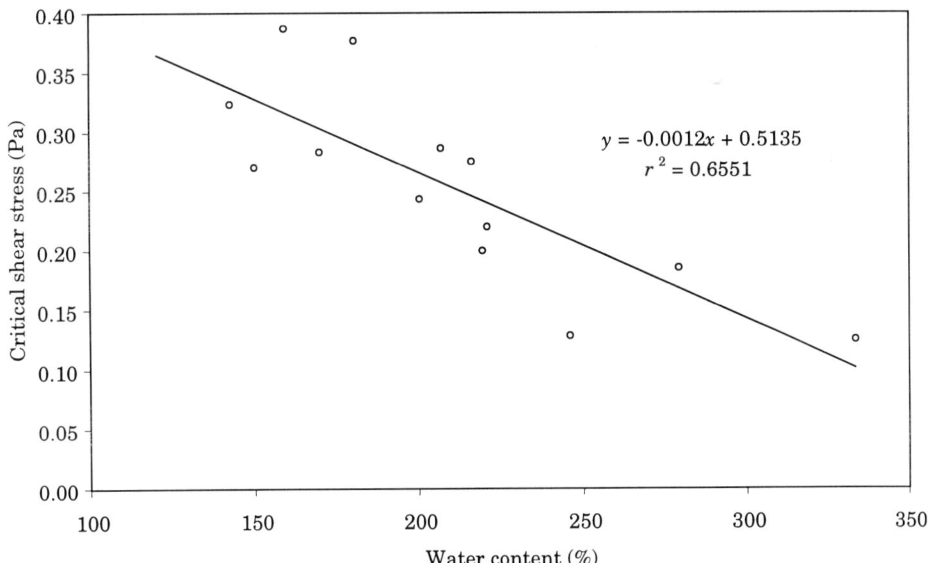

Figure 11. τ_{cr} against surface water content.

the 99% level, the most significant relationship was between sediment temperature and the natural log of colloidal carbohydrate (Figure 12). The following relationship is positively correlated and accounts for 84% of the measured variation in colloidal carbohydrate:

$$\ln(CC) = 0.11T_s - 1.48 \qquad (r^2 = 0.84, P < 0.001) \qquad (4)$$

where CC is colloidal carbohydrate (mg g^{-1}) and T_s is the surface sediment temperature (°C).

As the SedErode measurements and sampling were made near the end of each month (Table 1a,b) each measured parameter was correlated with mean monthly temperature at Rhoose Airport to determine any significant relationships. The only significant relationship determined was between mean monthly temperature and vane shear strength at each station. These relationships are shown in Figure 13 are given below:

$$\text{Station A} \quad VSS = 0.076\, T_{mm} - 0.19 \qquad (r^2 = 0.94, P < 0.005) \qquad (5)$$

$$\text{Station C} \quad VSS = 0.088\, T_{mm} - 0.63 \qquad (r^2 = 0.75, P < 0.025) \qquad (6)$$

$$\text{Station C/D} \quad VSS = 0.136\, T_{mm} - 0.66 \qquad (r^2 = 0.86, P < 0.010) \qquad (7)$$

where VSS is the mean vane shear strength at each station (kPa, n = 15) and T_{mm} is the mean monthly temperature (°C).

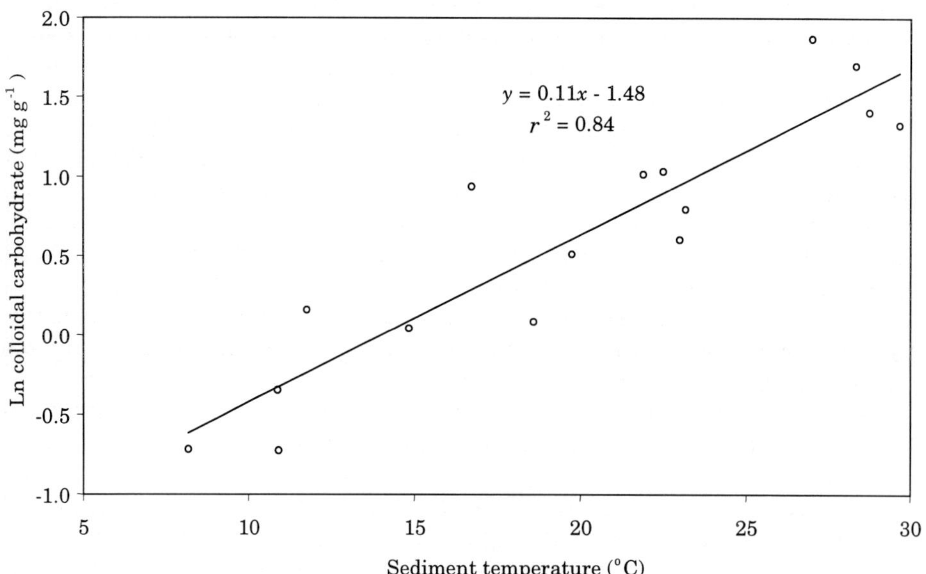

Figure 12. Surface colloidal carbohydrate against sediment temperature.

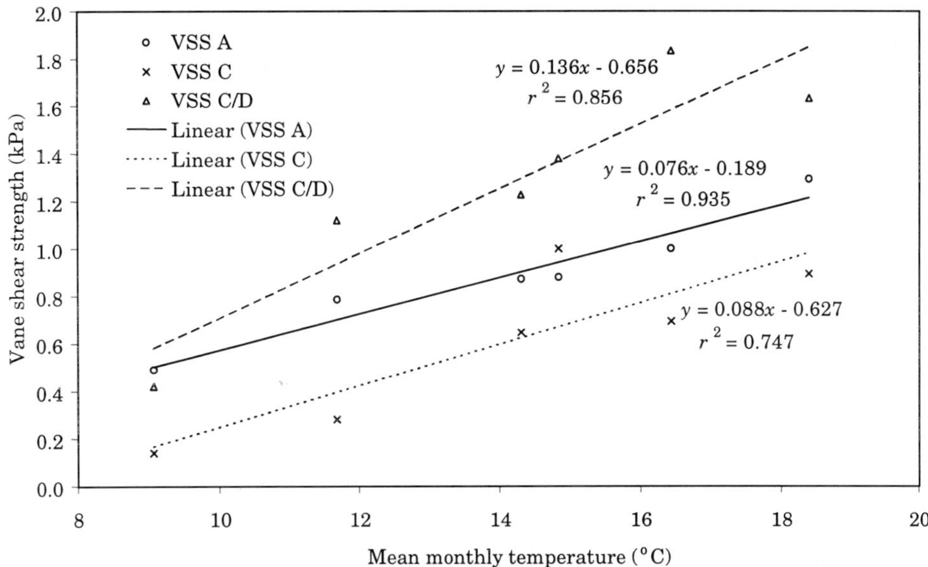

Figure 13. Vane shear strength against monthly mean air temperature.

7. DISCUSSION

The erosion threshold results were between 0.10 - 0.46 Pa. These compare well with values from a previous *in situ* erosion study at Portishead on the southern shore of the Severn estuary (Williamson and Ockenden, 1996). There was no significant seasonal trend in erodibility but there was a general decrease in the erosion thresholds during the study period (April to September 1997) which related to an increase in water content. Kraeuter and Wetzel (1986) also found an increase in the sediment water content at an intertidal mudflat over summer months although their results could have been influenced by sampling subtidally. There was a good relationship between the erosion thresholds and surface sediment compaction as represented by water content and bulk density [(1) and (2), respectively]. This result is similar to other studies which have demonstrated that at some sites, τ_{cr} was strongly dependent on the bulk density and water content (Yallop and Paterson, 1994; Mitchener et al.,1996b).

The only true measure of seasonality (and to some degree erodibility) is vane shear strength. Vane shear strength was the only parameter to have a significant relationship with monthly temperature [(5), (6) and (7)]. This indicates the fundamental difference between surface measurements of erodibility and those measured over depth. The modern sediment layer as a whole (to a depth of 5 cm) was consolidating and increasing in shear strength, whereas the erosion threshold actually decreased during the period due to increasing surface water content. This dichotomy poses a problem for modeling mudflat behavior at this location, accurate predictions of erosion rates will only be produced if the surface sediment is fully characterized over a depth of several centimeters.

The negative correlation between erosion threshold and loss on ignition suggests that the sediment may have been weakened due to bioturbation during the period (3), however the relationship is weak and is barely significant at the 95% level. If the loss on ignition results do represent bioturbation then it is likely that this is due to the surface dwelling gastropod *Hydrobia ulva*. *Hydrobia* is regarded as a bioturbator (Widdows et al., 1998) and this is mainly due to the increased roughness associated with the organism's tracks.

The Wentlooge Formation is more consolidated than the modern layer with a vane shear strength of 16.8 kPa compared with the modern layer which was in the order of 1 kPa. It was thought that this compaction would lead to a threshold of erosion that was considerably higher than that of the modern layer. There were two reasons for the low erosion threshold (0.30 Pa). Firstly, although the surface was scoured, some modern sediment did remain in small desiccation cracks and it was thought that this sediment was eroded prior to any erosion of the Wentlooge Formation. The presence of the modern layer was demonstrated quantitatively by the surface scrape sample having a bi-modal grain size distribution (Mitchener et al., 1998). Secondly, the first millimeter of the surface of the Wentlooge Formation was observed to have a higher water content than had been previously observed through the collection of cores (O'Brien, 1998). It was thought that there was some weakening of the surface sediment due to exposure.

In some other areas there is a large change in grain size distribution across the width of the intertidal with grain size increasing with distance seaward (Amos, 1995). The erosion threshold is only significantly affected by small changes in composition when the sediment consists of a mud/sand mixture e.g., if the 63 μm fraction is increased in a predominantly sandy sediment (Mitchener et al., 1996b). No such composition changes occur on the Peterstone intertidal as the sand fraction is negligible so at this location grain size is not a control on erodibility.

The erosion threshold was not significantly correlated with colloidal carbohydrate, sediment temperature, grain size distribution and vane shear strength. The lack of a relationship between the erosion threshold and colloidal carbohydrate is surprising as many authors have shown that colloidal carbohydrate is a good indicator of EPS and therefore sediment stability (Yallop et al., 1994; Sutherland, 1996; Black, 1997; Underwood et al., 1995; Amos et al., 1998). The values of colloidal carbohydrate from this study (2.38 ± 1.63 mg g^{-1}) are similar to those from a previous study made at Portishead on the southern shore of the estuary (Yallop et al., 1994; 1.46 ± 0.79 mg g^{-1}). However, tidal currents at Peterstone are greater than at Portishead leading to greater resuspension and more rapid changes in bed level (O'Brien et al., 2000). Therefore any stabilizing effects by microphytobenthos and bacteria are secondary to the rapid surface density changes that can occur over short timescales during springs (< 24 hours). Due to the measured concentrations of colloidal carbohydrates it is conceivable that the surface sediment may experience biogenic stabilization at other stages of the spring-neap tidal cycle and during periods when there is little variation in bed density.

It is concluded that at this site during spring tide conditions that the inherent sediment matrix (the water content) is the major controlling factor in governing the surface strength and that biological processes exert only a secondary influence. The major controlling influence on the surface sediment characteristics was the degree of consolidation of sediment layers deposited on the intertidal profile. This process was strongly affected by short term weather patterns. O'Brien et al. (2000) found that the surface of the study transect was unstable and were susceptible to even low wave activity on a variety of timescales ranging from daily to seasonal. The reduction in τ_{cr} from April to September 1997 may reflect the calm conditions generally experienced over the summer period allowing the soft fluid mud sheet deposited at slack tide to persist on the surface. The stability measurements were made at springs where deposition was likely to be at its greatest in the absence of wind waves (Whitehouse and Mitchener, 1998, O'Brien et al., 2000).

The mean station values of erosion threshold increased with distance offshore and this could be in part due to exposure time. Each set of three SedErode measurements took approximately 1 hour and 45 minutes and stations A, C and C/D were visited sequentially. If the time for the ebb tide to pass from station A to C/D is taken into account, station C and C/D measurements had 1.5 hours and 3 hours greater exposure time than station A, respectively. Increases in stability during exposure are primarily due to drainage and desiccation (Anderson and Howell, 1984) and have been demonstrated *in situ* by Paterson et al. (1990).

The results have demonstrated the value of taking long term measurements. However, to enable seasonal trends in surface stability to be resolved at such an active site, it is recommended that measurements are made at more frequent intervals to eliminate fluctuations due to weather and the exact position within the spring-neap cycle, and also to increase the statistical significance of the results obtained.

8. CONCLUSIONS

1. A comprehensive data set of sediment surface erodibility and sediment properties has been obtained at Peterstone Wentlooge site for a 10 month period (April 1997 - January 1998) under full moon spring tide conditions.
2. The mudflat had two sediment layers: underconsolidated modern fine-grained sediment overlying a stiff layer of Wentlooge Formation clay. The Wentlooge Formation was exposed in January 1998 after substantial erosion during the autumn of 1997.
3. The erosion threshold of the modern layer, τ_{cr}, ranged from 0.10 to 0.46 Pa and generally decreased during the spring/summer period primarily due to an increase in water content (and decrease in wet bulk density). Predictive relationships were derived between τ_{cr} and water content/wet bulk density.

4. Although the surface sediment (1 - 2 mm) weakened during the spring/summer period, vane shear strength (which is a measure of sediment strength to a depth of 5 cm) showed a clear seasonal trend of increasing strength with mean monthly air temperature.
5. Although the Wentlooge Formation was considerably more consolidated than the modern layer (as determined by vane shear strength) it had an erosion threshold of 0.30 Pa. This suggested that the surface of the formation had become weakened due to exposure.
6. The spatial and temporal variability of grain size was found to be very consistent and was not a control on erodibility of the modern layer.
7. There was a strong relationship between the natural log of colloidal carbohydrate and sediment temperature although there was no evidence that EPS production was a control on erodibility of surface sediments during spring tide conditions.

9. ACKNOWLEDGMENT

D. J. O'Brien acknowledges funding from Natural Environment Research Council, CASE Award Ref. GT4/95/305/MAS, in association with National Power Plc. and Hyder Plc. The work at HR Wallingford Ltd. was part funded by the Commission of the European Communities Directorate General for Science, Research and Development as part of the INTRMUD collaborative research program, under contract number MAS3-CT95-0022, and also by the UK Ministry of Agriculture, Fisheries and Food as part of its program of flood and coastal defense research and the UK Environment Agency as part of its Research and Development Programme. Provision of Meteorological data by the Met. Office is gratefully acknowledged. The authors would like to thank Mr. Stephen Shayler of the Department of Earth Sciences, University of Cardiff for carrying out the CC analyses. The authors acknowledge the help given by Dr. Richard Whitehouse during the course of the research and comments on the data interpretation.

REFERENCES

Amos, C. L., 1995. Siliciclastic tidal flats. Geomorphology and sedimentology of estuaries. In: *Developments in Sedimentology 53*, G. M. E. Perillo ed., Elsevier Science B.V., New York, 273-306.

Amos, C. L., Van Wagoner, N. A., and Daborn, G. R., 1988. The influence of subaerial exposure on the bulk properties of fine-grained intertidal sediment from Minas basin, Bay of Fundy. *Estuarine, Coastal and Shelf Science*, 27(1), 1-13.

Amos, C. L., Sutherland, T. F., and Zevenhuizen, J., 1996. The stability of sublittoral, fine-grained sediments in a subartic estuary. *Sedimentology*, 43(1), 1-19.

Amos, C. L., Brylinsky, M., Sutherland, T. F., O'Brien, D. J., Lee, S., and Cramp, A., 1998. The stability of a mudflat in the Humber estuary, South Yorkshire, U.K. In: *Sedimentary Processes in the Intertidal Zone*, K. S. Black, D. M. Paterson, and A. Cramp eds., Geological Society, London, 25-43.

Anderson, F. E., and Howell, B. A., 1984. Dewatering of an unvegitated muddy tidal flat during exposure - desiccation or drainage?, *Estuaries*, 7(3), 225-232.

Black, K. S., 1997. Microbiological factors contributing the erosion resistance in natural cohesive sediments. In: *Cohesive Sediments*, N. Burt, R. Parker, and J. Watts eds., John Wiley, Chichester, UK, 231-244.

Dubois, M., Gilles, K. A., Hamilton, J. K., Rebers, P.A., and Smith, F., 1956. Colorimetric method for determination of sugars and related substances. *Analytical Chemistry*, 28(3), 350-356.

Ferns, P. N., 1983. Sediment mobility in the Severn estuary and its influence upon the distribution of shorebirds. *Canadian Journal of Fisheries and Aquatic Science*, 40(1), 331-340.

Frostick, L. E., and McCave, I. N., 1979. Seasonal shifts of sediment within an estuary mediated by algal growth. *Estuarine and Coastal Marine Science*, 9(5), 569-576.

Grant, J., and Gust, G., 1987. Prediction of coastal sediment stability from photopigment content of mats of purple sulfur bacteria. *Nature*, 330(6145), 244-246.

Head, K. H., 1980. *Manual of Soil Laboratory Testing Volume 1: Soil Classification and Compaction Tests*. Pentech Press, Plymouth, UK, 339p.

INTRMUD Protocols, 1997. Guide document compiled for INTRMUD partners, SERG (Severn Estuary Research Group), Gatty Marine Laboratory, St Andrews, Scotland, UK.

Jumars, P. A., and Nowell, A. R. M., 1984. Effects of benthos on sediment transport: difficulties with functional grouping. *Continental Shelf Research*, 3(2), 115-130.

Kirby, R., and Parker, W. R., 1983. Distribution and behaviour of fine sediment in the Severn estuary and Inner Bristol Channel, UK. *Canadian Journal of Fisheries and Aquatic Science*, 40(1), 83-95.

Kraeuter, J. N., and Wetzel, R. L., 1986. Surface sediment stabilisation-destabilisation and suspended sediment cycles on an intertidal mudflat. In: *Estuarine Variability*, D. A. Wolfe ed., Academic Press Inc., San Diego, CA, 203-222.

Lee, H. J., Chough, S. K., Jeong, K. S., and Han S. J., 1987. Geotechnical properties of sediment cores from the southeastern Yellow Sea: Effects of depositional processes. *Marine Geotechnology*, 7(1), 37-52.

Meadows, P. S., and Tait, J., 1989. Modification of sediment permeability and shear strength by 2 burrowing invertebrates. *Marine Biology*, 101(1), 75-82.

Mitchener, H. J., Whitehouse, R. J. S., Soulsby, R. L., and Lawford, V. A., 1996a. Estuarine morphodynamics - instrument development for mud erosion measurements. Development and testing of SedErode - sediment erosion device. *HR Report No. TR 16*, HR Wallingford, Oxon, UK.

Mitchener, H. J., Torfs, H., and Whitehouse, R. J. S., 1996b. Erosion of mud/sand mixtures. *Coastal Engineering*, 29, 1-25.

Mitchener, H. J., O'Brien, D. J., and Whitehouse, R. J. S., 1998. Seasonal erodibility measurements at Peterstone Wentlooge, Cardiff, UK, April 1997 to January 1998. *HR Report No. TR 57*, HR Wallingford, Oxon, UK.

Mook, D. H., and Hoskin, C. H., 1982. Organic determinations by ignition: caution advised. *Estuarine, Coastal and Science*, 15(6), 697-699.

Noornay, I., 1984. Phase relations in marine soils. *Journal of Geotechnical Engineering*, 110(4), 539-543.

O'Brien, D. J., 1998. The sediment dynamics of a macrotidal mudflat on varying timescales. *Unpublished Ph.D. Thesis*, Department of Earth Sciences, University of Wales, Cardiff.

O'Brien, D. J., Whitehouse, R. J. S., and Cramp, A., 2000. The cyclic development of a macrotidal mudflat on varying timescales. *Continental Shelf Research*, 20(12-13), 1593-1619.

Paterson, D. M., 1997. Biological mediation of sediment erodibility: ecology and physical dynamics. In: *Cohesive Sediments*, N. Burt, R. Parker, and J. Watts eds., John Wiley, Chichester, UK, 215-229.

Paterson, D. M., Crawford, R. M., and Little, C., 1990. Subaerial exposure and changes in the stability of intertidal estuarine sediments. *Estuarine, Coastal and Shelf Science*, 30(6), 541-556.

Sutherland, T. F., 1996. Biostabilization of estuarine subtidal sediments. *Unpublished Ph.D. Thesis*, Dalhousie University, Nova Scotia, Canada, 184p.

Underwood, G. J. C., Paterson, D. M., and Parkes, R. J., 1995. The measurement of microbial carbohydrate exopolymers from intertidal sediments. *Limnology and Oceanography*, 40(7), 1243-1253.

Whitehouse, R. J. S., and Mitchener, H. J., 1998. Observations of the morphodynamic behaviour of an intertidal mudflat at different timescales. In: *Sedimentary Processes in the Intertidal Zone*, K. S. Black, D. M. Paterson, and A. Cramp eds., Geological Society, London, 225-271.

Widdows, J., Brinsley, M., and Elliot, M., 1998. The use of in situ flume to quantify particle flux (biodeposition rates and sediment erosion) for an intertidal mudflat in relation to changes in current velocity and benthic macrofauna. In: *Sedimentary Processes in the Intertidal Zone*, K. S. Black, D. M. Paterson, and A. Cramp eds., Geological Society, London, 225-271.

Williamson, H. J., and Ockenden, M. C., 1996. ISIS: An instrument for measuring erosion shear stress *in-situ*. *Estuarine, Coastal and Shelf Science*, 42(1), 1-18.

Yallop, M. L., and Paterson, D. M., 1994. Survey of the Severn estuary. In: *Biostabilisation of Sediments*, W. E. Krumbein, D. M. Paterson and L. J. Stal eds., BIS Publishers, Oldenburg, Germany, 279-326.

Yallop, M. L., De Winder, B., Paterson, D. M., and Stal, L. J., 1994. Comparative structures, primary production and biogenic stabilisation of cohesive and non-cohesive marine sediments inhabited by microphytobenthos. *Estuarine, Coastal and Shelf Science*, 39(6), 565-582.

Coastal and Estuarine Fine Sediment Processes
W.H. McAnally and A.J. Mehta (Editors)

Observations of long and short term variations in the bed elevation of a macro-tidal mudflat

M. C. Christie, K. R. Dyer and P. Turner

Institute of Marine Studies, The University of Plymouth, Plymouth, Devon, PL4 8AA, U.K

A study was carried out to examine the variations in suspended sediment flux and bed level across a macro-tidal mudflat in the Humber estuary (U.K.). Measurements were obtained for one year of flow velocity, water depth, salinity, suspended sediment concentration (SSC), bed level and water and air temperatures. Mean current velocities exceeded 0.4 ms^{-1} during early flood inundation and final ebb retreat (i.e., water depths below 1.0 m). At these times, under calm conditions, SSC was closely correlated to mean velocities. However, wave activity significantly increased the erosion of surface sediment. Under calm conditions, the bed level usually increased a few millimeters around slack water, due to deposition of suspended sediment eroded during the earlier flood tide. The newly settled material was often resuspended during the ebb by the accelerating tidal flow. Storms and waves modified this behavior, by enhancing erosion of the mudflat during immersion and preventing deposition during slack water. Wave activity governed the residual sediment flux, producing an onshore transport during calm periods, and an opposing offshore transport under stormy conditions. A decline in surface biology and increase in wave activity produced several centimeters of erosion during the autumn and winter. The local hydrodynamics drive the sediment transport, with tidal forcing and wave activity controlling the immediate erosion response of the sediment. The cumulative effect of the sediment flux variations controlled the seasonal behavior of the mudflat.

1. INTRODUCTION

In-situ measurements of the response of intertidal mudflats to changing environmental conditions are vital to our understanding of the behavior of these shallow water locations. Local hydrodynamics drive the system, with tidal forcing and wave activity often determining the physical and biological characteristics of the mudflat. Even small waves in the shallow water can suspend material and carry it shorewards as the tide rises (Anderson et al., 1981). The duration of wave attack at a level on the mudflat is the result of wave attenuation and the relationship between the mudflat slope and the tidal curve (Kirby, 1986) and there is a complex feedback situation. Storms can erode significant volumes of sediment while calm conditions often favor the gradual accretion and recovery of sediment over the mudflat (Whitehouse and Roberts, 1999). The shallow water sediment fluxes are often markedly different over the spring-neap cycle. The floral and faunal distributions can affect sediment transport by either stabilizing or destabilizing the surface sediment (Paterson, 1997; Widdows et al., 1998). There are likely to be considerable seasonal variations in the erosion response of a mudflat as the plants and animals grow and die.

Many mudflats appear to be accreting at the same rate as sea level rises, suggesting an equilibrium exists between the dynamic forcing and the sedimentary response. However, the timescales and lags involved in this are unknown. The response of mudflats to changing sea level, climate and anthropogenic pressure is an increasing problem for environmental protection and management. As part of the MAST III project INTRMUD, a 12 month study was carried out to examine the changes in sediment fluxes and elevation across the macro-tidal Skeffling mudflats. This paper uses *in-situ* measurements of velocity, concentration and bed elevation to identify the dominant hydrodynamic processes controlling the erosion and deposition of sediment across the intertidal zone.

1.1. Tidal flat characteristics

The Skeffling intertidal area is roughly 5 km wide, with a mean slope of 1:1000, resulting in peak cross-shore flow speeds of about 0.4 ms^{-1}. The mudflat was exposed to a 15 km fetch and prevailing south-westerly winds were able to generate significant wave activity. Four sampling stations, A, B, C and D were defined along a shore normal transect (Figures 1 and 2). The upper mudflat

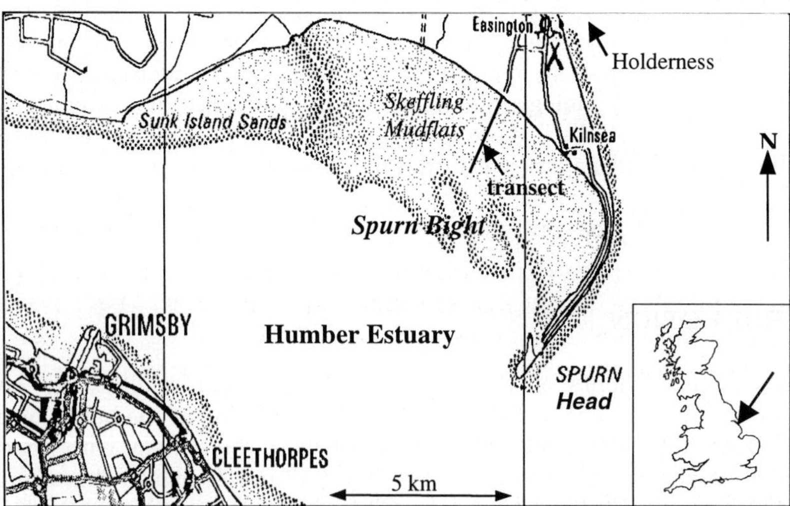

Figure 1. Location of the Humber Estuary (U.K.) and study transect on the Skeffling Mudflats.

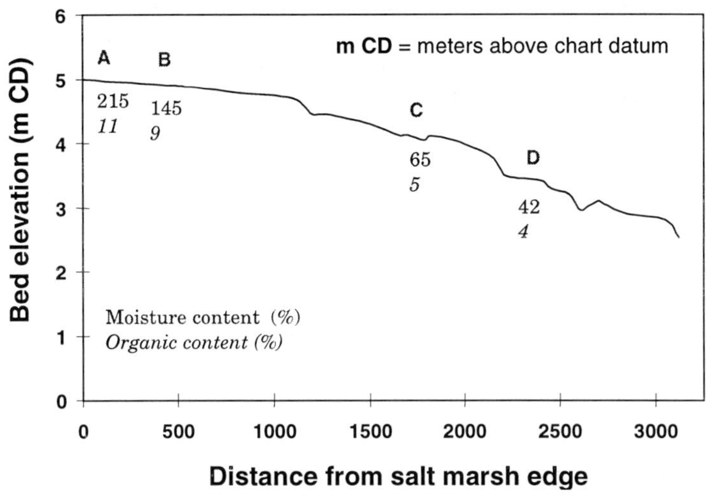

Figure 2. Cross-shore tidal flat profile (data from Black, 1995).

around station A was flat and smooth, stations B and C had a convex profile, while the lower mudflat appeared to be more erosive with a characteristic concave profile around station D. Shore normal aligned bedforms developed with distance offshore, increasing in scale from ripple features around station B, to larger gullies and channels around stations C and D. The largest channels were about 10 m wide and 2 m deep.

Percentage sand content varied with distance offshore, increasing from about 4% at station A, to 96 % around the mean low water mark. Moisture and organic contents decreased with distance offshore (Figure 2). Paterson (1997) shows there is often a correlation between chlorophyll *a* and diatom biomass, and that the microbiology stabilizes the sediment surface. Data from July 1997, showed that chlorophyll *a* levels were about 40.7 $\mu g\ g^{-1}$ near station A and decreased with distance offshore, reaching minimum values at D of 4.2 $\mu g\ g^{-1}$. The upper mudflat thus appeared to have more diatoms and hence potentially higher erosion threshold (τ_c) values than the lower sediments. Amos et al. (1996) measured the surface τ_c in April 1995 and found τ_c values of 0.48 Nm^{-2} on the upper mudflat, and 0.3 Nm^{-2} the lower mudflat. More recently, Mitchener et al. (1998) measured τ_c values of about 0.3 Nm^{-2} at stations A, B and C during July 1997. The variation in biological and physical properties appeared to produce relatively uniform τ_c values across the mudflat. In this case, erosion potential was considered to depend upon the properties of the tidal flow and the associated bed shear stresses. Erosion was likely to occur during shallow water periods around the mid-tide level, when tidal currents were greatest. Tidal flows were much weaker at stations A and B, with an immersion period around the time of slack water.

Christie and Dyer (1998) show the movement of the shallow water's edge across the mudflat dominates the local sediment fluxes. It appeared that the upper mudflat was likely to favor the deposition and consolidation of suspended sediment, and erosion processes would be limited to the mid-tide level.

2. METHODOLOGY

2.1. Instrumentation

The measurements were obtained over a period of 12 months, about 10 km from the mouth of the estuary. Instruments were deployed from March 9, 1997 to March 28, 1998 at station D (Figure 2) corresponding to a mid-tide level on the mudflats. The measuring equipment was serviced every six weeks to maintain sensor calibration and minimize the effects of biological fouling. Single point values of velocity, depth, temperature and salinity were measured every 15 minutes at 20 cm above the bed, using an Environmental Multi-Sensor Monitoring Probe (EMP2000) from Applied Microsystems Limited. The horizontal components of velocity were measured using an electro-magnetic flow sensor to an accuracy of ± 5 cms^{-1}. Single turbidity readings were obtained every five minutes by an optical backscatter sensor (OBS) located at 22 cm above the bed. The OBS was connected to an Endeco YSI6000 for 6 months, and then to a second EMP2000 for the last part of the study. Turbidity was measured to ± 1.5 NTU and converted to concentration data by repeated calibration with local surface sediment during the service periods. Changes in mudflat elevation were measured every five minutes during immersion by an ARX II acoustic scour monitor from Scan Group Ltd (accurate to ± 1 mm). A horizontal datum was set up at the beginning of the study to calibrate the acoustic scour data. Careful measurements were made of the bed level changes with respect to this datum. Air temperature was recorded during mudflat exposure, and tidal heights and meteorological data were measured every 10 minutes by the Humber Observatory (Hardisty and Rouse, 1996). Wave heights were not measured directly, however wave activity was inferred from the wind data and the variations in water depth around high water.

2.2. Calculated bed shear stresses

Bed shear stresses (τ_b) were calculated using two methods. The shear stress due to the mean flow was calculated from a logarithmic velocity profile approach, using a value of 0.2 mm for the typical roughness length of mud as suggested by Soulsby (1998). During stormy periods, shear stresses due to waves and currents were calculated using a parameterization of the Huynh-Tanh and Temperville (1991) kinetic energy and mixing length model.

2.3. Suspended load

It was possible to calculate a suspended load value (c_{tot}) every 15 minutes, for the locally eroded fraction of the suspended sediment by making three assumptions. Firstly, erosion could only occur when $\tau_b > \tau_c$. For this study, the mean 0.3 Nm^{-2} value from Mitchener et al. (1998) was assumed to represent τ_c. This threshold is plotted in Figures 7 and 11. Secondly, not all of the suspended sediment was supplied by local erosion of the bed. A proportion of the suspended material formed a slowly varying background concentration that was continually generated within the estuary. This background level (c_{bck}) was taken to be the minimum concentration measured during each immersion period, usually around the time of high water. Lastly, an assumption had to be made regarding the nature of the vertical SSC profile. During calm periods, the measured concentration was considered to decrease linearly towards the water surface, so that at high water, the surface value was assumed equal to the background concentration. For all other depths, the surface value decreased proportionally from the measured value as a function of h/h_{max} (defined below). Thus, for calm conditions the suspended load was calculated as:

$$c_{tot} = ch - c_{bck}h - \left[\frac{h}{h_{max}} \frac{1}{2} \left(ch - c_{bck}h \right) \right] \tag{1}$$

where c = concentration (mg cm^{-3}), h = depth (cm), and h_{max} = depth at high water (cm). However, for stormy conditions, the vertical concentration profile was simply considered to be homogeneous, being well mixed by wave action. In this case c_{tot} was calculated as:

$$c_{tot} = ch - c_{bck}h \tag{2}$$

Taking the time derivative of the resulting c_{tot} time series produced a time series of erosion and deposition rates. Erosion rates were calculated as the difference between successive suspended load values, during periods of increasing suspended load.

No data has been found to quantify the role of the large channels to the cross-shore sediment transport though these features must have some role in the overall sediment transport. A crude estimate shows that during the uncovered, low tide period the channels move less than 1% of the total water volume distributed across the mudflat at high tide. The volume of water across the transect at high tide was conservatively estimated to be about 8×10^7 m^3, taking the mudflat slope to be 1:1000, and the intertidal area to be 4 km×10 km. There were about five large channel systems within this area, each channel having a cross sectional area of about 20 m^2 (i.e., 2 m deep×10 m wide) at the seaward end. During the six hour immersion period each channel could drain approximately 9×10^4 m^3 of water off the mudflat (i.e., 20 $m^2\times0.2$ $ms^{-1}\times$ 21600 s). Thus the total drainage flow would be about 4.5×10^5 m^3, which is less than 1% of the total volume of water at high tide. However, sediment concentrations are highest in shallow water and the water movements at this time dominate the sediment fluxes (Christie and Dyer, 1998). In this respect, the 1% volume value underestimates the importance of the channels. Nevertheless, the volume of suspended sediment moved by the tide during immersion was probably much greater than that moved by the channels alone. For this study, it was considered reasonable to ignore the role of the channels in the subsequent sediment transport estimates.

3. RESULTS

3.1. Calm conditions

A typical calm period is represented by time series from July 4, 1997. Wind and pressure data are summarized in Figures 3 and 4. Data were obtained during a period of rising (high) pressure, and low offshore wind speeds (less than 5 ms^{-1}). The weather produced a calm sea surface with no wave activity, and the data were collected following about two days of low wind speeds. The antecedent conditions favored deposition and consolidation of sediment. The tidal range was 5.6 m.

Figure 5 shows depth, velocity and SSC measured every 15 minutes during this calm period. The instruments were covered shortly after 15:30 hrs, and high water was at about 18:30 hrs. Current velocities were greatest at the beginning and end of immersion as the rate of depth change was greatest at these times. High concentrations (typically > 0.1 gL^{-1}) were measured in the first few minutes of immersion and appeared to be due to the resuspension of the low density, diatom rich sediment, that had accumulated on the mudflat surface during exposure (Paterson, 1997). The mean sediment fluxes (i.e., $\overline{u}\,\overline{c}$) were calculated for each immersion period and typical results for calm conditions are shown in Figure 6.

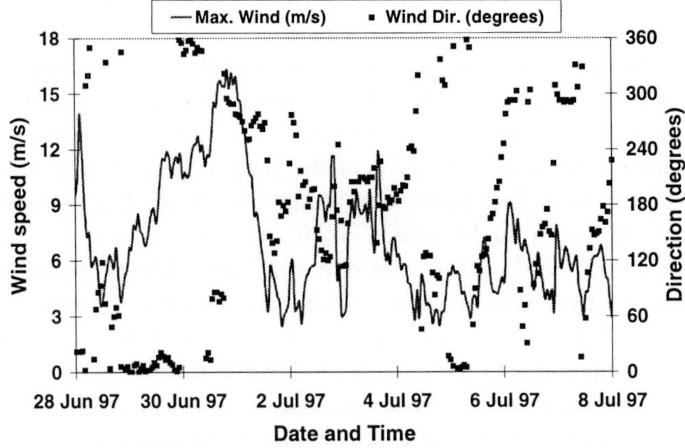

Figure 3. Wind data June 28 - July 8, 1997.

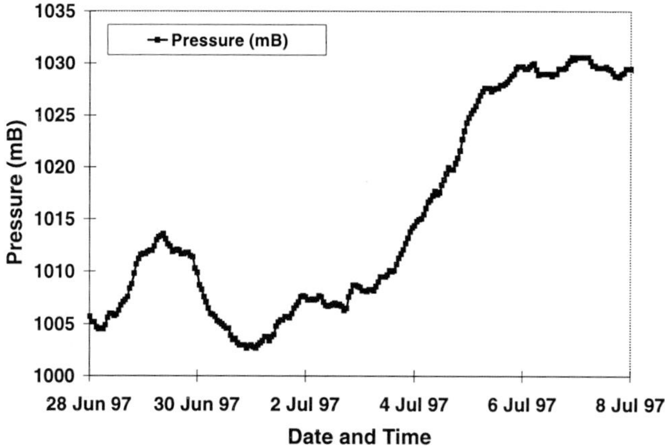

Figure 4. Atmospheric pressure June 28 - July 8, 1997.

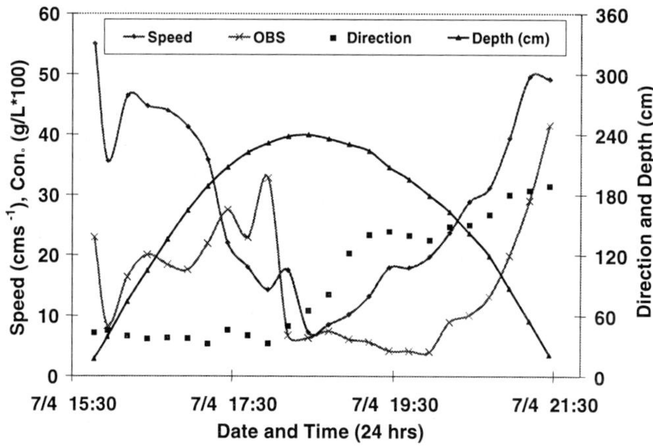

Figure 5. Depth, velocity and SSC time series July 4, 1997.

Christie and Dyer (1998) show the oscillatory fluxes were about two orders of magnitude less than the mean fluxes and thus, oscillatory fluxes have not been considered in this study. There was a marked onshore flux of sediment during the early part of the flood tide. The bulk of the suspended load was thus moved onto the upper mudflat. During calm periods the high concentrations appeared to be able to settle during slack water onto the upper part of the mudflat. Fennessy (personal communication) measured mean settling velocities of about 0.3 mms^{-1} at Skeffling during calm conditions. At this rate the majority of the suspended

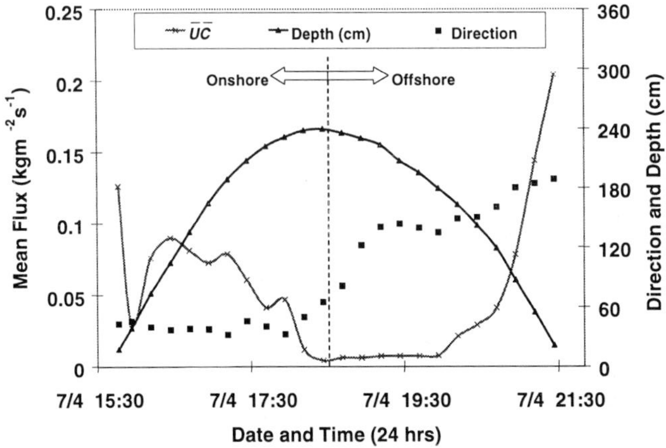

Figure 6. Mean fluxes and depth July 4, 1997.

sediment at stations A and B could settle to the bed within 1 hour during the slack water period, when $h \approx 1$ m. During the first two hours of the ebb phase, velocities were less than 0.3 ms^{-1} and water depths exceeded 1.5 m. Concentrations were low, presumably because bed shear stresses remained below the critical erosion threshold. This implies, under calm conditions, the majority of sediment that settled onto the upper mudflat over slack water would remain deposited during the early ebb tide. But, sediment deposited lower down the transect, where current velocities reached maximum values around mid-tide, could be resuspended when $\tau_b > \tau_c$. This type of mechanism would account for the concentration peak in Figure 5 at the end of the immersion period. The tidal flow was aligned predominantly cross-shore.

Flood and ebb flux values were simply defined according to the sign of the cross shore direction component, positive for flood, negative for ebb, and thus a residual flux was calculated. For the data illustrated in Figure 6 the net flux was +0.152 kgm^{-2}s^{-1} onshore. This was the result of the shorter, intense flood phase coupled with scour lag processes around high water. It was possible to quantify the fluxes of sediment across the bed/water interface by measuring the actual changes in bed elevation during immersion. Figure 7 shows the variations in bed elevation at station D during the "calm" tide, and the corresponding bed shear stress due to the tidal current. The scour data were simply a measure of the distance between the mudflat surface and an acoustic sensor. An increase in scour indicates that sediment has been eroded from the surface of the mudflat.

The bed level increased from about 309 to 311 mm during the first few minutes of covering. This change in level corresponds to the rapid resuspension of the thin surface biofilm. Half an hour after covering, scour values decreased, indicating sediment deposition in the trailing edge of the high concentration zone. Deposition

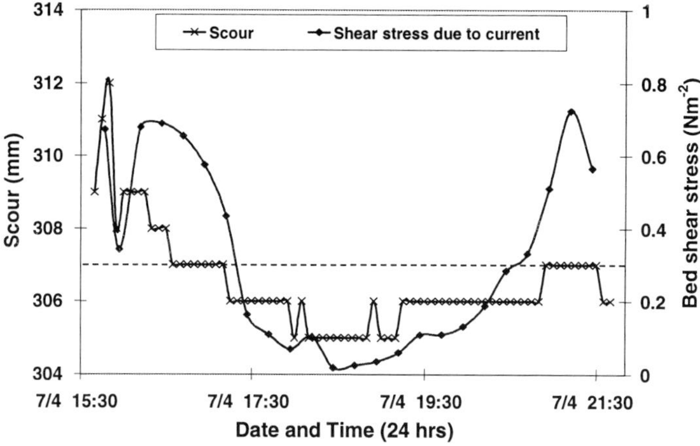

Figure 7. Change in bed level and τ_b July 4, 1997.

was likely as the bed shear stresses would be reduced by the high concentrations, and accordingly the suspended flocs would tend to grow and settle faster.

The suspended sediment in the leading water's edge would settle while being advected onto the upper mudflat, and would deposit around slack water, particularly once τ_b decreased below the assumed 0.3 Nm^{-2} τ_c threshold (measured by Mitchener et al. , 1998). Minimum scour readings were 305 mm , indicating that a 5 mm layer of sediment was deposited at D during the last two hours of the flood. Bed shear stresses increased following slack water as the flow accelerated and depth decreased. Scour readings increased slightly when bed shear stresses exceeded about 0.1 Nm^{-2} as some of the weaker newly settled sediment was resuspended. Because of scour lag effects only about 2 mm of the freshly deposited sediment was eroded on the ebb, leaving behind a new 3 mm thick layer of sediment.

It was assumed that erosion only occurred when $\tau_b > \tau_c$, i.e. from about 15:30 to 17:30 hrs at the start of the flood, and from about 20:45 hrs onwards on the ebb. No erosion was considered to occur around high water. The mean erosion rate from 15:30 to 17:30 hrs was 3.16 mgcm^{-2}h^{-1}, which was similar to the erosion rates obtained by Sanford et al. (1991). Erosion rates of this scale are associated by Sanford et al. (1991) with regular tidal resuspension processes that affect a thin, unconsolidated layer of flocculent material on the sediment surface. Visual observations confirmed that a thin (order mm) biofilm, that had accumulated during exposure, was eroded during the early part of the flood tide.

3.2. Storm conditions

Typical stormy conditions are represented by time series from January 3, 1998. Wind and pressure data are summarized in Figures 8 and 9 respectively. The data were obtained during a period of falling pressure and strong south-

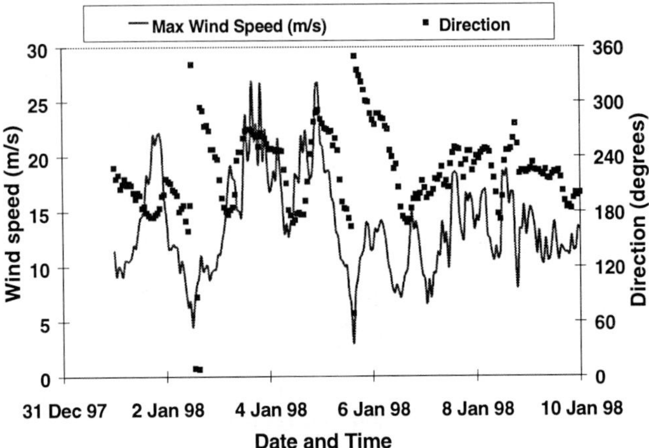

Figure 8. Wind data December 31, 1997 - January 10, 1998.

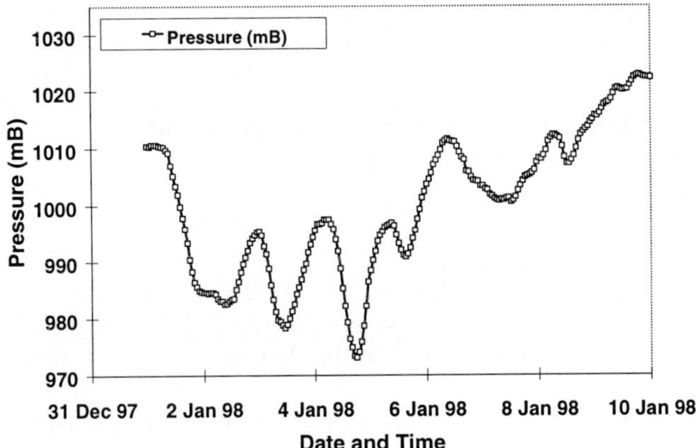

Figure 9. Atmospheric pressure December 31, 1997 - January 10, 1998.

westerly winds exceeding 15 ms^{-1}. These onshore winds produced significant wave activity as the tidal range increased towards springs.

Depth, velocity and SSC data are illustrated in Figure 10. Data recording started at 06:00 hrs when the leading water's edge reached the instruments. High water was at about 08:30 hrs. Significant wave height was conservatively estimated to be about 0.3 m, from the measured fluctuations in depth around high water. Current velocities were greatest in shallow water, with a maximum speed of 0.8 ms^{-1} recorded about 30 minutes from the start of the record.

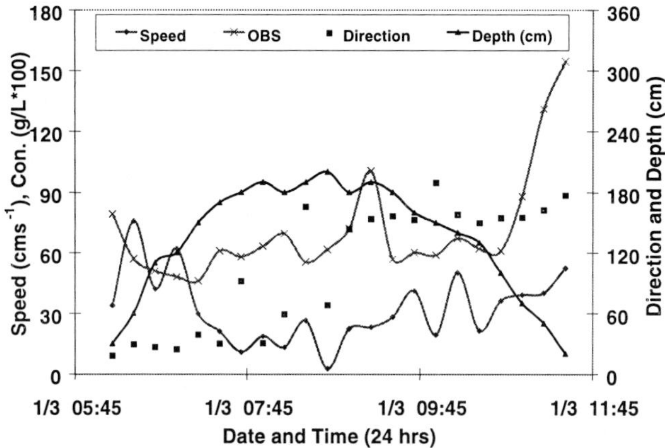

Figure 10. Depth, velocity and SSC time series January 3, 1998.

Around high water, c_{bck} was about 0.6 gL^{-1}, being roughly an order of magnitude greater than the corresponding "calm" values in Figure 5. This difference was attributed to the stormy conditions and high rainfall of the previous few days. Peak concentrations were again measured in shallow water, with values about 0.85 gL^{-1} at the start of the flood and a maximum value of 1.5 gL^{-1} in the final moments of uncovering.

However, there was also a peak in concentration of about 0.9 gL^{-1} in deeper water, at about 09:00 hrs. The concentration time series was related to local changes in bed elevation by examining the corresponding variations in scour and bed shear stress due to waves and currents (see Figure 11). About 6 cm of surface sediment was removed during the first 2 hours of immersion (i.e., $\tau_b >> \tau_c$)

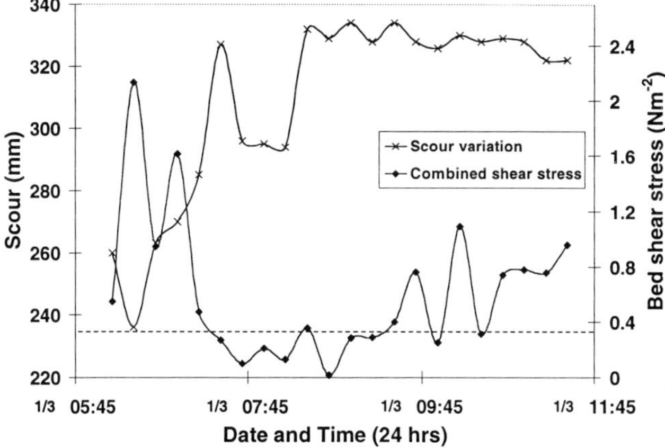

Figure 11. Change in bed level and τ_b January 3, 1998.

because of the combined effects of the relatively large waves and high flow speeds in the shallow water (i.e., depth < 1.0 m). The surface sediment shear strength would have increased as erosion exposed the underlying over-consolidated sediment, and this would limit the erosion process. This type of mechanism would restrict erosion to the first part of the flood tide or until a weaker layer was deposited on top of the newly exposed bed. Measurements by Amos et al. (1996) and Mitchener et al. (1998) suggest that it is reasonable to use the 0.3 Nm^{-2} value to define τ_c during exposure. However, there is no data available for τ_c variations during immersion, making it difficult to provide a thorough explanation of the shear and scour data. From about 07:30 hrs, τ_b decreased below the assumed 0.3 Nm^{-2} τ_c threshold and erosion was expected to cease. Turbulence and mixing processes would be less intense as the water depth increased and the surf zone moved onshore. The high concentrations of sediment throughout the water column would tend to settle towards the bed and produce a near-bed concentration gradient.

In 1995, Fennessy (personal communication) measured settling velocities of around 1 mms^{-1} during south westerly breezes, and Amos et al.(1996) measured peak *in-situ* settling rates of 2.46 mms^{-1} during the same period. In still conditions, with these settling rates, it would only take about 15 minutes for a surface floc to settle to the bed in water 1.4 m deep. There was obviously some vertical mixing but there was probably a strong settling flux during the latter part of the flood phase. This settling potential would tend to produce a concentration gradient, similar to that measured by Christie and Dyer (1998). The decrease in scour in Figure 11, from about 06:45 to 08:30 hrs, was considered to be due to high concentrations close to the bed. Laboratory tests showed the acoustics were sensitive to such density gradients and it was considered that the lower scour values represent a 3-4 cm thickness mud fluid layer. Current speeds peaked to 0.3 ms^{-1} immediately before slack water (τ_b) and it appears that this velocity rapidly resuspended the fluid mud layer. Bed shear stresses were almost zero at high water, but flow speeds increased above 0.3 ms^{-1} soon afterwards and vertical mixing appeared to maintain the sediment suspension. The peak in concentration immediately after high water could be a lag effect of the resuspended "fluid mud" or the result of advection. Since there was no correlation with the scour data, this concentration peak was attributed to advection.

Following high water, current speeds and wave turbulence intensified and concentrations increased further, especially during the last hour. Following our assumption that erosion would occur when $\tau_b > \tau_c$, erosion should have recommenced during the latter stages of the ebb but the scour data suggest otherwise. An explanation is that the shear strength of the over-consolidated sediment (exposed by the flood erosion) was greater than the 0.3 Nm^{-2} value measured by Mitchener et al. (1998). Indeed, the scour reading decreased slightly in the final minutes of immersion, which represents the high concentrations (> 1.2 gL^{-1}) at this time.

A mean erosion rate of 17.64 $mgcm^{-2}h^{-1}$ was obtained for the during the first hour of the flood tide. This value was an order of magnitude greater than measured by Sanford et al. (1991) and it is indicative of massive erosion resulting from the superposition of tidal and wind driven currents and waves (Ward, 1985). This erosion rate corresponds well with the measured scour data. Mean flux data are presented in Figure 12. There was a net offshore directed flux of -0.579 $kgm^{-2}s^{-1}$, being about four times greater than the onshore flux value under calm conditions. In this case, the velocity dominated the sediment flux. The higher background concentrations for this tide formed the bulk of the total suspended load.

3.3. Seasonal considerations

The longer term effects of residual tidal flux variations maybe quantified by examining the whole dataset. Figures 13 and 14 summarize the wind conditions recorded during the whole study. The modal wind speeds were about 6 ms^{-1}, predominantly from a southwest direction.

The seasonal effects of storm and calm periods on the cross-shore sediment transport were investigated by examining the variation of the tidal residual flux with maximum wind speed (Figure 15). The maximum wind speed was selected from the 12 hour period before final uncovering. This period included waves generated further offshore during low tide exposure, because light winds during mudflat immersion would not necessarily correspond with calm sea conditions. Local seas required several hours to become calm, because variations in the wave field lagged behind the variations in wind velocity. The maximum wind speed was selected because this value would have the greatest affect on sea conditions. The cross-shore component of wind speed was calculated, with a positive speed relating to an onshore direction. Calm sea conditions resulted from low wind speeds and/or offshore directed winds.

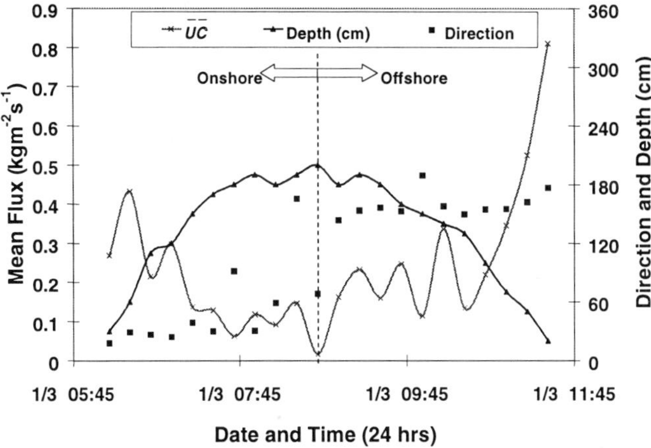

Figure 12. Mean fluxes and depth January 3, 1998.

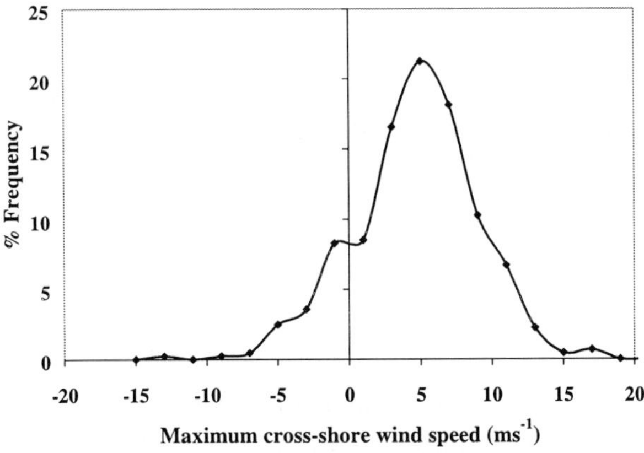

Figure 13. Distribution of wind speed.

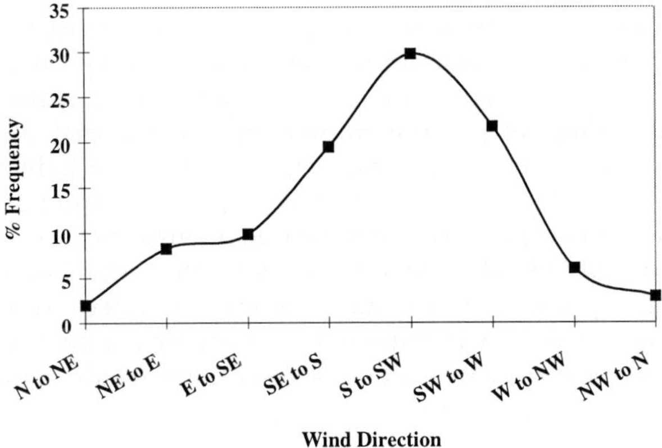

Figure 14. Distribution of wind direction.

During calm periods, variations in the residual sediment flux were typically between –0.5 to +0.5 $kgm^{-2}s^{-1}$, with a negative flux being directed offshore. The residual fluxes were not correlated to wind velocity under calm conditions, but there appears to be an apparent threshold for wind and wave influence on the sediment flux at about 6 ms^{-1}. Above this limit, there was a negative correlation between wind velocity and residual flux. As onshore wind speeds increased, the residual offshore flux also increased, reaching maximum values around -3.5 $kgm^{-2}s^{-1}$ (measured during very stormy conditions in January 1998). A possible regression line is plotted in Figure 15; however the correlation coefficient was < 0.3 and more data are required to quantify the relationship between wind speed and flux. A multiple regression approach including tidal range and river runoff should produce a statistically more meaningful result. For onshore winds in excess of 6 ms^{-1}, the linear relationship was simply:

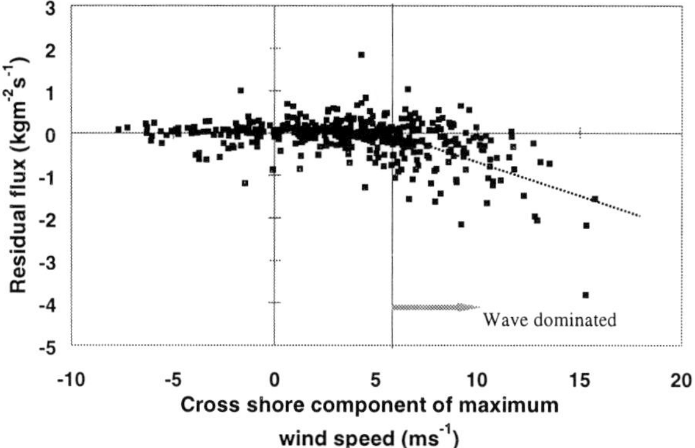

Figure 15. Residual flux vs wind speed.

$$Rf = 0.972 - 0.157\ W \approx 1 - 0.16\ W \tag{3}$$

where: Rf = residual flux (kgm^{-2}s^{-1}), W = cross-shore component maximum wind speed (ms^{-1}). it must be noted that (3) is a tentative relationship and is not statistically significant.

Figure 16 shows the cumulative flux plotted alongside mean variations in scour for the study period. The early scour data was very noisy because the acoustic sensor was mounted on a vertical scaffold pole that was driven about 2 m into the sediment. The tidal currents caused excessive scour around the base of the pole, and there were unusual variations in the scour data because the sensor

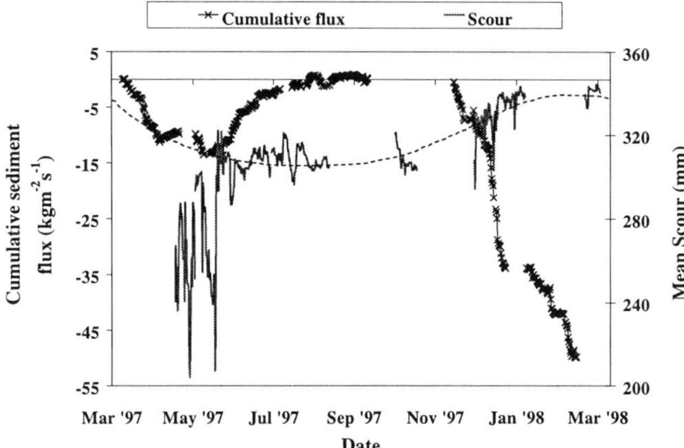

Figure 16. Variations in cumulative flux and scour with season.

was measuring the uneven erosion close to the sides of the pole. This problem was solved at the end of May 1997, by repositioning the scour sensor on a cross bar over undisturbed sediment. Variations in scour were now attributed to erosion or deposition. The early scour data, until 1 June 1997, needs to be treated with caution. There were also some gaps in both the scour and sediment flux time series. The scour data is best examined from June 1997 to the end of March 1998. The mean scour at the start of this period was about 315 mm. Positive fluxes were measured during June and July 1997 and correlate with a decrease in scour. Tidal variations in scour were about ±1 cm during this summer period, but the mean scour was about 310 mm at the start of August. There appears to have been about 5 mm of sediment deposited during the early part of the summer. Unfortunately, scour data were not recorded during August, but there was about three weeks of data in September, with final scour values of about 305 mm. This indicates about 1 cm of sediment was deposited at station D during the summer of 1997. Visual observations showed several centimeters of sediment were deposited onto the upper mudflat around stations A and B during the same time interval.

The greatest change in scour occurred during the autumn 1997, and this change correlates with the greatest change in cumulative flux. About 3 cm of sediment were removed during continuous westerly storms at the end of December, and there were correspondingly large negative fluxes. If the trends from this data set are typical, there will be long term erosion of sediment at station D at a rate of about 2 $cmyr^{-1}$. However, this erosion rate is contradicted by results from Pejrup et al. (1998) who dated sediment cores with Pb^{210}, and show that sediment has been accumulating on the middle mudflats at about 1 $mmyr^{-1}$ for roughly the last 40 years.

It is unfortunate that the scour data from March to May 1997 were unreliable, because data from this period may explain the discrepancy. The dotted line in Figure 16 shows an assumed trend in mean scour for the whole sample period, with the result that the net change in bed level is very small. This type of annual cycle in bed level would produce very small net changes in bed level (i.e., a few mm per year) which corresponds better with the results from the sediment core analysis.

4. DISCUSSION

Calm conditions favored the deposition and accretion of sediment, particularly on the upper mudflat. The upper mudflat was exposed to the air throughout high water neaps and sediment deposited during earlier spring tides could consolidate over this neap tide immersion period and thus resist further erosion. Sediment deposited at the salt marsh edge was liable to be trapped within the salt marsh vegetation. Several millimeters of soft, low density sediment could accumulate on the mudflat surface after a few days of calm weather. The advancing flood water was able to resuspend the thin diatom rich biofilm that accumulated on the lower mudflat surface during exposure. Typically, when the water depth exceeded about 1 m during current dominated conditions, $\tau_b \leq \tau_c$, and erosion would cease. The resuspended sediment was advected onshore and over slack water, this sediment could settle onto the upper mudflat, forming a layer a few millimeters thick. Under calm conditions, this layer could strengthen and thus resist erosion during the ebb cycle, and typically there was a net onshore flux of sediment. The deposited material could experience biological stabilization, especially during the summer months.

Large waves modified this transport behavior, by maintaining the sediment suspensions over high water, and producing a net offshore directed sediment flux. The waves enhanced the bed shear stresses in shallow water such that $\tau_b >> \tau_c$, and thus several centimeters of sediment could be eroded during the flood phase. Wind generated waves affected the balance between erosion and deposition and altered the direction of the tidal residual sediment transport.

The apparent 6 ms^{-1} wind speed threshold in Figure 16 coincides with the modal wind velocity. The mudflat profile responded continuously to the local forcing environmental conditions. The residual sediment flux fluctuated in response to deviations from modal wind conditions, and the mudflat profile is in dynamic equilibrium. Commonly occurring wind conditions produced a relatively stable transport situation, with low fluxes across the mudflat. Unusual or "extreme" wind conditions consisted of either, strong onshore winds (and large waves) or offshore winds and/or low wind speeds (i.e., calm conditions). Stormy periods served to transport large amounts of sediment offshore. However, the storms generally only last a few days and longer periods of calm weather allowed the profile to slowly recover. The mudflat profile was continuously adjusting because the forcing environmental conditions were never constant, and thus it is unrealistic to consider a steady state profile.

Other environmental factors must be responsible for the variations in residual flux when wind speeds were less than 6 ms^{-1}. The most important factors seem to be tidal range and river runoff. A mathematical model of the Skeffling transect by Whitehouse and Roberts (1999) shows there is a spring/neap signal in the cross-shore sediment movement under calm conditions. The model results show that spring tides produce a net onshore transport of sediment, resulting from the flood dominated tidal curve that is set up across the mudflat. Neap tides have less distortion, and favor an offshore sediment transport. It appears for this study, that spring tides and calm conditions produced accretion on the upper mudflat, while neap tides and storms resulted in erosion of the lower mudflat. Figure 17 shows a sketch of the response of the profile to wind velocity and tidal range.

Wind velocity and tidal range are essential to the cross-shore flux of sediment, with rainfall, river runoff and season modulating the basic mudflat dynamics. Generally, periods of negative flux correlated well with increasing scour (i.e., erosion and offshore transport of sediment) and positive fluxes corresponded to accretion of sediment, particularly during the calmer summer months (Figure 16). Microbiology can enhance accretion during summer months, but the biology

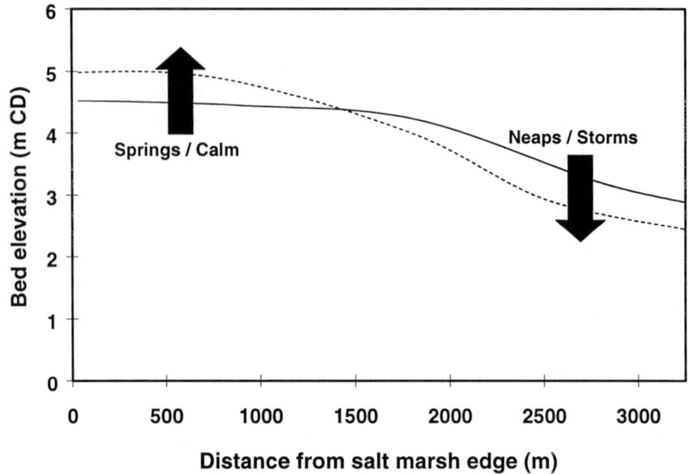

Figure 17. Conceptual response of profile to wind/tides.

will tend to die back during the winter, accompanied by a decrease in the bed shear strength and increase in erosion potential. Figure 16 shows that at the end of the study there was a net erosion of about 2 cm of sediment from station D, however, it is not clear if this behavior was typical. This net erosion is considered part of a longer period cycle and there will be some sort of recovery of sediment on the middle mudflat. Visual observations indicated that there was accretion of sediment around stations A and B during the study, and historical records show the upper mudflat to be accreting (De Boer, 1979). It is postulated that when there was a positive sediment flux at D, the suspended sediment was transported onshore by the flood tide and deposited onto the upper mudflat during slack water. There was not a significant accumulation of sediment around station D under calm conditions. Without waves, the mudflat level at D remained approximately constant until the next storm event removed more material from D and transported this sediment offshore. Indeed, almost 2 cm of sediment were removed in the first week in January 1998 during a particularly long period of extremely stormy conditions. This type of mass erosion may not fit the typical environmental conditions, and it is possible we have measured an episodic event that occurs at much longer intervals than the annual cycle. It seems unlikely that the steady erosion of sediment from D, as measured during the study, will continue.

Accretion and erosion of sediment should be self-limiting under steady forcing conditions. Sediment deposition produces shallower water depths and thus enhances the bed shear stress (particularly due to wave action) and accretion rates will then be reduced. Conversely, eroded parts of the transect result in deeper water and lower bed shear stresses that favor sediment accumulation.

5. CONCLUSIONS

- Maximum velocities generally occurred at station D during the first hour of the flood phase, exceeding 0.5 ms^{-1} during spring tides. The tidal curve at this location was flood dominated.
- Waves significantly influenced the residual fluxes when onshore winds exceeded about 6 ms^{-1}. Calm conditions were produced by offshore winds and wind speeds less than $6ms^{-1}$.
- Peak bed shear stress values due to currents in shallow water were about 0.4 Nm^{-2}. Wave activity enhanced this value, and shear stresses exceeded 2.0 Nm^{-2} for combined flows.
- There was offshore directed sediment transport under wave dominated conditions, with a residual tidal flux between 0.5 to 3.5 $kgm^{-2}s^{-1}$. Waves were able to erode a few centimeters of sediment per tide. A large proportion of this suspended material was carried offshore from the mudflat by the ebbing tide.
- Net onshore transport occurred under calm conditions, with residual fluxes around 0.1 to 0.5 $kgm^{-2}s^{-1}$.
- During calm periods, surface sediments were generally poorly consolidated and characterized by low densities. Erosion was limited to shallow water periods (i.e., $h < 1.0$ m). The strong onshore flows (> 0.4 ms^{-1}) were able to erode a millimeter thick, low density biofilm which formed on the sediment surface during exposure.

6. ACKNOWLEDGMENT

This work was funded by MAST III contract MAS3-CT95-0022 INTRMUD. We are grateful to Plymouth Marine Laboratory and Dr. Robin Howland for the loan of the EMP2000's. The Humber Observatory (Hardisty and Rouse, 1996) kindly provided the meteorological data.

REFERENCES

Anderson, F. E., Black, L., Watling, L. E., Mook, W., and Mayer, L. M., 1981. A temporal and spatial study of mudflat erosion and deposition. *Journal of Sedimentary Petrology*, 51, 729-736.

Amos, C. L., Brylinski, M., Lee, S., and O'Brien, D., 1996. Littoral mudflat stability monitoring, the Humber estuary, S. Yorkshire, England. LISPUK, April 1995. *Open File Report # 3214*, Geological Survey of Canada.

Black, K. S., 1995. Data from LISPUK experiment 1995 at http://www.st-andrews.ac.uk/~ksb2/lisp/LISP_Reports/hydrography/.

Christie, M. C., and Dyer, K. R., 1998. Measurements of the turbid tidal edge over the Skeffling mudflats. In: *Sedimentary Processes in the Intertidal Zone,* K. S. Black, D. M. Paterson, and A. Cramp eds., The Geological Society, London, 45-55.

De Boer, G., 1979. History and general features of the estuary. In: *The Humber Estuary: a selection of papers on present knowledge of the estuary and it's future potential given at two symposia arranged by the Humber Advisory Group and the University of Hull*. NERC Publications Series C, Number 20, 1-4.

Hardisty, J., and Rouse, H. L., 1996. The Humber Observatory: Monitoring & Management for the Coastal Environment. *Journal of Coastal Research*, 12(3), 683-690.

Huynh Tannh, S., and Temperville, A., 1991. A numerical model of the rough turbulent boundary layer in combined wave and current interaction. In: *Euromech 262 - Sand Transport in Rivers, Estuaries and the Sea*, R. L. Soulsby and R. Bettees eds., Balkema, Rotterdam. ISBN 90 6191 186 9.

Kirby, R., 1986. Suspended fine cohesive sediment in the Severn estuary and inner Bristol Channel, U.K. Report to United Kingdom Atomic Energy Authority under contract no: E/5A/CON/4042/1394. By Ravensrodd Consultants Ltd., Taunton, U.K.

Mitchener, H. J., Lee, S. V., and Whitehouse, R. J. S., 1998. Erodibility measurements at Skeffling, Humber Estuary, 3-9 July 1997. *Report TR52* January 1998, HR Wallingford.

Paterson, D. M., 1997. Biological mediation of sediment erodibility: ecology and physical dynamics. In: *Cohesive Sediments*, N. Burt, R. Parker, and J. Watts eds., John Wiley, Chichester, UK, 215-229.

Pejrup, M., Andersen, T., Mikkelsen, O., and Møller, A., 1998. Dating of cores from Skeffling, Humber Estuary, 1997. In: *The Second Annual Report*, for INTRMUD, to the European Commission Directorate General XII. MAST III contract MAS3-CT95-0022.

Sanford, L. P., Panageotou, W., and Halka, J. P., 1991. Tidal resuspension of sediments in northern Chesapeake Bay. *Marine Geology*, 97, 87-103.

Soulsby, R. L. (1998). *Dynamics of Marine Sands: A Manual for Practical Applications*. Published by Thomas Telford, 250 p.

Ward, L. G. 1985. The influence of wind waves and tidal currents on sediment resuspension in middle Chesapeake Bay. *Geo-Marine Letters*, 5, 71-75.

Whitehouse, R. J. S., and Roberts, W., 1999. Intertidal mudflats Final Report: Predicting the morphological evolution of intertidal mudflats. *Report SR538 Environment Agency R&D Technical Report W191*. February, 1999.

Widdows, J, Brinsley, M., and Elliot, M., 1998. Use of *in situ* flume to quantify particle flux (biodeposition rates and sediment erosion) for an intertidal mudflat in relation to changes in current velocity and benthic macrofauna. In: *Sedimentary Processes in the Intertidal Zone,* K. S. Black, D. M. Paterson, and A. Cramp eds., The Geological Society, London, 85-99.

Coastal and Estuarine Fine Sediment Processes
W.H. McAnally and A.J. Mehta (Editors)
2001 Elsevier Science B.V.

Influence of salinity, bottom topography, and tides on locations of estuarine turbidity maxima in northern San Francisco Bay

D. H. Schoellhamer

U.S. Geological Survey, Placer Hall, 6000 J Street, Sacramento, California 95819-6129, USA

Time series of salinity and suspended-solids concentration measured at four locations and vertical profiles of salinity and suspended-solids concentration measured during 48 water-quality cruises from January 1993 to September 1997 are analyzed to describe the influence of salinity, bottom topography, and tides on locations of estuarine turbidity maxima in northern San Francisco Bay, California. Estuarine turbidity maxima form when salinity is present but they are not associated with a singular salinity. Bottom topography enhances salinity stratification, gravitational circulation and estuarine turbidity maxima formation seaward of sills. The spring/neap tidal cycle affects locations of estuarine turbidity maxima. Salinity stratification in Carquinez Strait, which is seaward of a sill, is greatest during neap tides, which is the only time when tidally averaged suspended-solids concentration in Carquinez Strait was less than that observed landward at Mallard Island. Spring tides cause the greatest vertical mixing and suspended-solids concentration in Carquinez Strait. Therefore, surface estuarine turbidity maxima always were located in or near the Strait (seaward of Middle Ground) during spring tide cruises, regardless of salinity. During neap tides, surface estuarine turbidity maxima always were observed in the landward half of the study area (landward of Middle Ground) and between 0–2 practical salinity units.

1. INTRODUCTION

A feature of many estuaries is a longitudinal maximum of suspended-solids concentration (SSC), called an estuarine turbidity maximum (ETM). Several processes can contribute to the formation of ETMs (Jay and Musiak, 1994). Gravitational circulation or tidal asymmetry of velocity and SSC can cause convergent fluxes of suspended solids and form ETMs (Hamblin, 1989; Jay and Musiak, 1994; Wolanski et al., 1995). Schubel and Carter (1984) state that the origin and maintenance of ETMs once was attributed to flocculation, but that nontidal (gravitational) circulation primarily is responsible. A cycle of local deposition, bed storage, and resuspension also can contribute to the formation of ETMs (Hamblin, 1989; Uncles et al., 1994; Wolanski et al., 1995; Grabemann et al., 1997). Suppression of turbulence by salinity stratification increases settling and trapping of fine sediment and may be a more effective trapping mechanism than gravitational circulation (Hamblin, 1989; Geyer, 1993).

Because gravitational circulation and stratification are dependent on the salinity field, ETMs often are located near a particular salinity. In the Tamar Estuary, an ETM is observed at the freshwater-saltwater interface or landward, and in the Weser Estuary an ETM is observed at 6 practical salinity units (psu) (Grabemann et al., 1997). While ETMs typically are located at low salinities, they also can be located at much greater salinities. ETMs in Lorient and Vilaine Bays are at salinities of about 18 psu and 25-30 psu, respectively (Le Bris and Glemarec, 1996). Freshwater flow into an estuary affects the salinity field and, therefore, can affect the position of ETMs (Uncles and Stephens, 1993; Grabemann et al., 1997).

ETMs also may be located at longitudinally fixed locations, independent of salinity. Nonlinear interactions of first-order tides in channels with constrictions or decreasing depth in the landward direction may induce landward residual currents and convergence in part of the water column (Ianniello, 1979), which results in a fixed, topographically controlled ETM (Jay and Musiak, 1994). Wind-wave resuspension in shallow water and ebbtide transport to deeper channels also can create fixed locations for ETMs, such as in the Tay Estuary (Weir and McManus, 1987).

Data from synoptic field measurements confirm the existence of an ETM in Suisun Bay (Figure 1), the most landward subembayment of northern San Francisco Bay (Meade, 1972; Conomos and Peterson, 1977; Arthur and Ball, 1978). Ten synoptic measurements collected at high slack tide during spring and summer from 1974 to 1977 indicate that an ETM typically exists in the surface salinity range of 1-6 psu (Arthur and Ball, 1978). ETM formation was attributed to gravitational circulation and flocculation. The 2-psu bottom isohaline is used as a habitat indicator to regulate freshwater flow to the Bay because it is believed to be an easily measured indicator of the location of the ETM and a salinity preferred by many estuarine species (Jassby et al., 1995).

The purpose of this paper is to describe the influence of salinity, topography, and tides on the locations of ETMs in northern San Francisco Bay. The reasons for revisiting the issue of locations of ETMs in northern San Francisco Bay include availability of large data bases of vertical profiles and time series of salinity and SSC, evaluating the applicability of previously described ETM formation mechanisms, and improving the scientific basis for regulation of freshwater flow to the Bay.

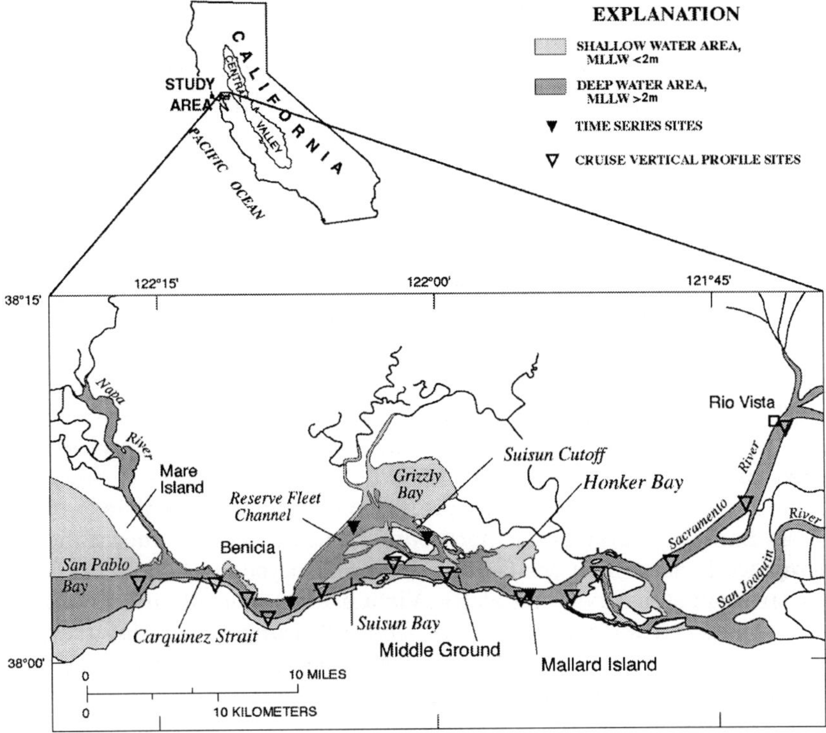

Figure 1. Northern San Francisco Bay, study area. MLLW, mean lower low water.

2. STUDY AREA

The region containing ETMs in northern San Francisco Bay is shown in Figure 1. Suisun Bay is the most landward subembayment of northern San Francisco Bay. The Sacramento and San Joaquin Rivers deliver freshwater to Suisun Bay, primarily during the winter rainy season and during the spring snowmelt and reservoir releases. The annual discharge of the Sacramento River is about six times greater than that of the San Joaquin River. Precipitation is negligible during late spring and summer. Suisun Bay is a partially mixed estuary that has extensive areas of shallow water that are less than 2 m deep at mean lower-low water (MLLW), including the subembayments of Grizzly and Honker Bays. Channels that are about 9-11 m deep are parallel to the southern and northwestern shores and are between Grizzly and Honker Bays. Carquinez Strait is a narrow channel about 18 m deep that connects Suisun Bay to San Pablo Bay, to the rest of San Francisco Bay, and to the Pacific Ocean. Tides are mixed diurnal and semidiurnal and the tidal range varies from about 0.6 m during the weakest neap tides to 1.8 m during the strongest spring tides. Freshwater inflow typically first encounters saltwater in northern San Francisco Bay, defined here as the lower rivers, Suisun Bay, and Carquinez Strait. The salinity range in northern San Francisco Bay is about 0-25 psu and depends on the season and freshwater inflow. Suspended and bed sediment in Suisun Bay is predominately fine and cohesive, except for sandy bed sediment in some of the deeper channels (Conomos and Peterson, 1977). The typical SSC range in northern San Francisco Bay is about 10-300 mg/L and sometimes up to about 1,000 mg/L at an ETM.

3. VERTICAL PROFILE AND TIME-SERIES DATA

Vertical profile and time-series data are used to analyze the influence of salinity, bottom topography, tides, and water column position on observation of ETMs in northern San Francisco Bay. Vertical profiles of SSC and salinity are measured during approximately monthly water-quality surveys of San Francisco Bay (Edmunds et al., 1997). Data are grouped into bins with a 1-m vertical resolution. Sampling sites in northern San Francisco Bay are shown in Figure 1. Cruises typically start near the time of slack before flood in Carquinez Strait and proceed landward (east), ending at Rio Vista a little more than 4 hours later, so the data are not truly synoptic. Data from 48 cruises from January 1993 to September 1997 for which the 2-psu surface isohaline was located in the study area are analyzed in this paper. Bottom salinity at Rio Vista, the most landward site, was less than or equal to 0.13 psu for all but two cruises. Thus, almost all cruises extend landward into virtually freshwater.

In addition to the monthly vertical profile data, time series of SSC at fixed stations have been measured in Suisun Bay. SSC and salinity were measured at 10-minute intervals at a height of 0.6 m above the bed in the Reserve Fleet Channel and Suisun Cutoff in September and October 1995 (J. Cuetara, U.S. Geological Survey, written communication, 1998). SSC is measured at 15-minute intervals at the Benicia Bridge at 1 m and 16 m below MLLW and at Mallard Island 1 m below the water surface and 6 m below MLLW (Buchanan and Schoellhamer, 1998, Figure 1). Data collected at the surface from May to August 1996 are presented in this paper because a large range of salinities was measured as salt returned to Suisun Bay during the period.

Time series of salinity are measured by the National Oceanic and Atmospheric Administration (National Oceanic and Atmospheric Administration, 1998) at 6-minute intervals at the same elevations as SSC measurements at Benicia and by the California Department of Water Resources (DWR) (California Department of Water Resources, 1998) 1 m below the water surface at Mallard Island at 1-hour intervals. The salinity sensor at the bottom of the water column at Benicia failed in mid July 1996. NOAA also measures velocity profiles with an acoustic Doppler current profiler at 6-minute intervals at Benicia and DWR measures water surface elevation at hourly intervals at Mallard Island.

The time-series data were low-pass filtered to remove tidal frequencies and to provide a tidally averaged analysis of salinity and SSC. All low-pass filtering was performed with a Butterworth filter with a cutoff frequency of 0.0271 per hour. The strength of the spring/neap cycle at Benicia was quantified by calculating the low-pass root-mean-squared (RMS) water speed by squaring the measured water velocity 16 m below MLLW, low-pass filtering, and taking the square root. A RMS water-surface elevation at Mallard Island was calculated to quantify strength of the spring/neap cycle. For cruise data, NOAA tidal-current predictions for Carquinez Strait were used to delineate flood and ebb tides and to quantify the spring/neap cycle by taking the mean value of the four predicted maximum flood and ebb current speeds that were temporally centered on each cruise ($\bar{U}_{max}$).

4. RESULTS: CRUISE DATA

Salinity and maximum cruise SSC are related near the water surface but not near the bed in northern San Francisco Bay. Maximum surface SSC, 1 m below the water surface, was located between 0-6 psu at the bottom for 67 percent of the cruises (Figure 2). Maximum SSC 1 m above the bed, however, was located over a wide range of bottom salinity. Only 23 percent of maximum bottom SSC was located between 0-6 psu at the bottom. Similar results were found at both elevations when surface salinity was considered. Seventy-one percent of the cruise data were collected during a predicted flood tide in Carquinez Strait. The effect of this flood tide sample bias is discussed later in this article.

Maximum SSC 1 m above the bed of northern San Francisco Bay, while not related to bottom salinity, is related to longitudinal position in the estuary. Maximum bottom SSC was located in Carquinez Strait during 67 percent of the cruises (Figure 3). Maximum surface SSC, however, was distributed throughout northern San Francisco Bay. Only 27 percent of maximum surface SSC were located in Carquinez Strait. Empirical-orthogonal-function analysis of the cruise data (not shown) produced similar results.

5. RESULTS: TIDALLY AVERAGED TIME-SERIES DATA

During the dry season of spring and summer, when freshwater flows to the estuary decrease, salinity returns to northern San Francisco Bay. Tidally averaged surface salinity at Benicia, the seaward boundary of Suisun Bay, increased from 0 psu in late May to 10 psu in June 1996 (Figure 4). At Mallard Island, the landward boundary of Suisun Bay, tidally averaged surface salinity increased from 0-2 psu in late June 1996.

Tidally averaged surface SSC did not show any maxima associated with a particular tidally averaged salinity as salinity increased in Suisun Bay in 1996

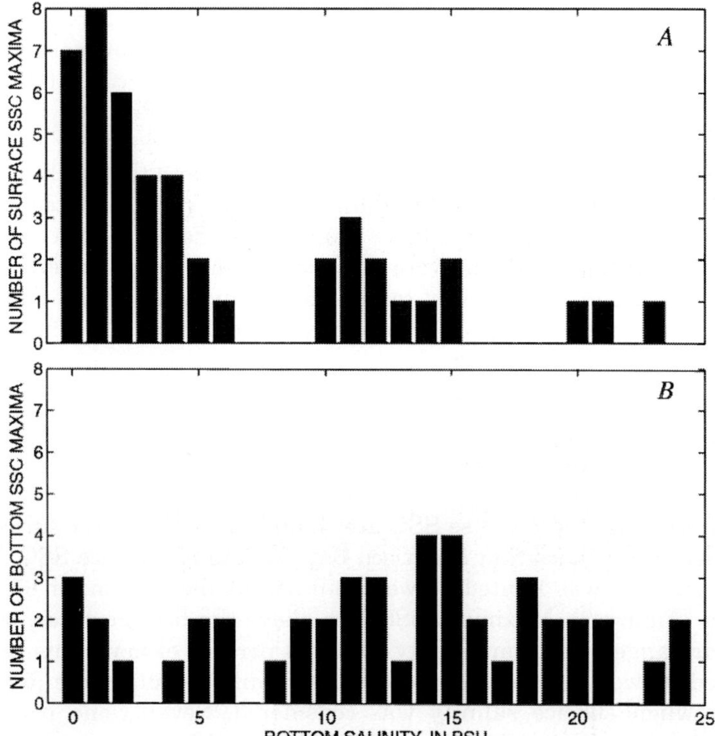

Figure 2. Positions of maximum cruise suspended-solids concentrations (SSC) at the surface *A* and bottom *B*, relative to salinity, northern San Francisco Bay.

(Figure 4). This result differs from the cruise data for which surface ETMs usually were observed between 0-6 psu bottom salinity. This discrepancy is discussed later in this article. Tidally averaged surface SSC almost always was greater at Benicia than at Mallard Island. SSC at Benicia varied with the spring/neap cycle, with minima during neap tides and maxima during spring tides. Tidally averaged SSC was slightly greater at Mallard Island only during weaker neap tides in late May and late June.

Similar results were found near the bed at two sites that are closer together as salinity returned to Suisun Bay after a wet rainy season in 1995. Tidally averaged salinity increased in the Reserve Fleet Channel (seaward site) and Suisun Cutoff (landward site) in October 1995, ranging from about 1-10 psu (Figure 5). Tidally averaged SSC always was greater in the Reserve Fleet Channel than in Suisun Cutoff, varied with the spring/neap cycle at both sites, and did not show any maxima associated with a particular salinity.

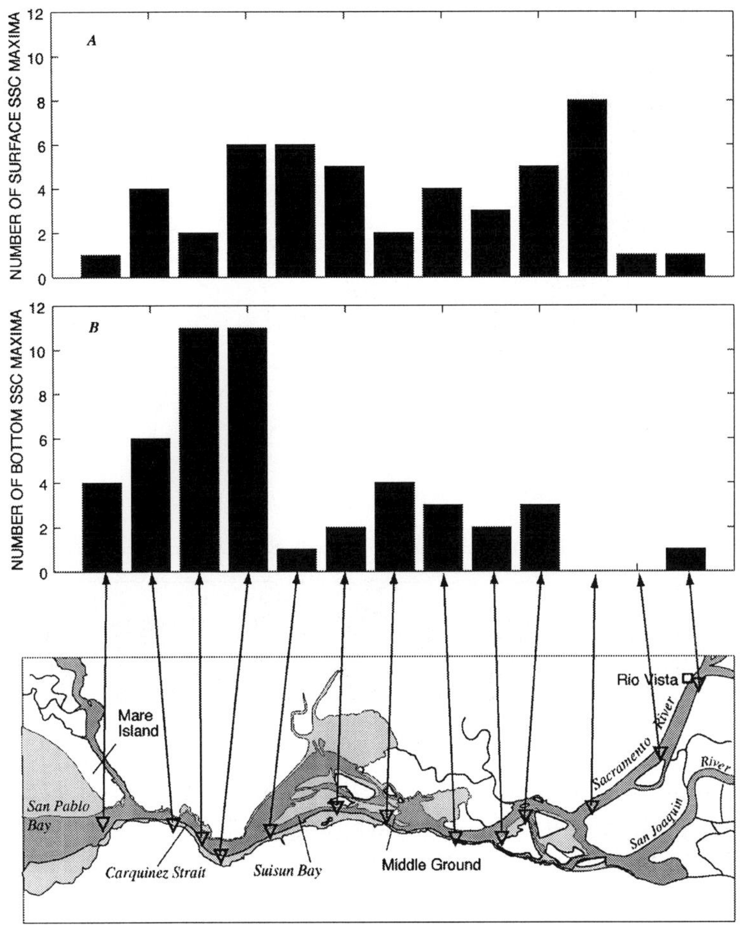

Figure 3. Position of maximum cruise suspended-solids concentration (SSC) at the suface *A* and bottom *B*, relative to sampling site, northern San Francisco Bay.

6. DISCUSSION

SSC maxima in northern San Francisco Bay are not associated with a singular salinity. Cruise data from northern San Francisco Bay indicate that there is an ETM at low salinity (0-6 psu) 1 m below the water surface. This is a larger salinity range for ETM location, however, than is observed in estuaries with a salinity-dependent ETM (Le Bris and Glemarec, 1996; Grabemann et al., 1997). The processes that account for a salinity-dependent ETM, gravitational circulation, salinity stratification, and bed storage, occur in northern San Francisco Bay and are modified by bottom topography and tides.

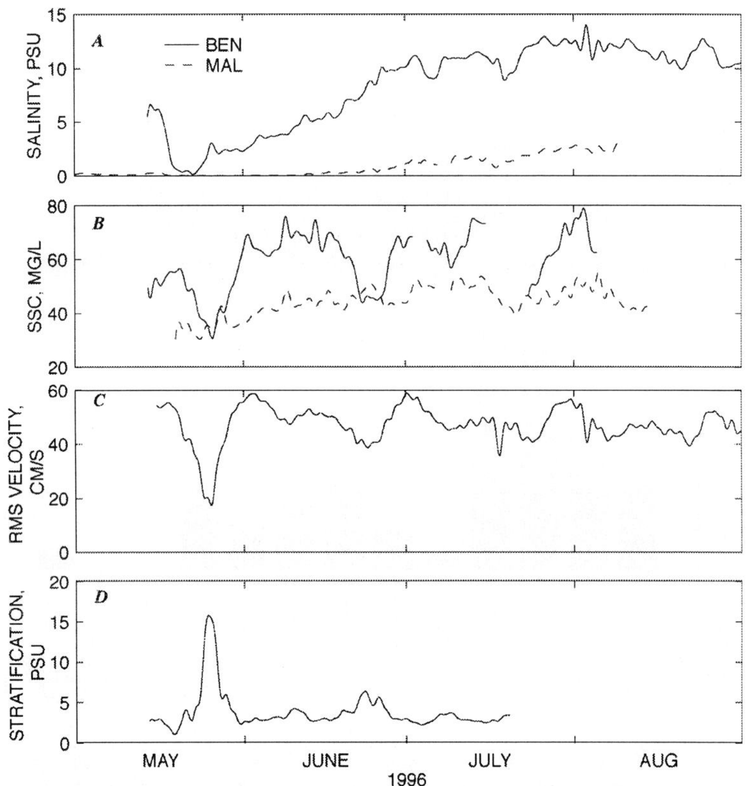

Figure 4. Tidally averaged surface salinity *A* and suspended-solids concentration (SSC) *B* at Benicia and Mallard Island, root-mean-squared (RMS) bottom velocity at Benicia *C*, and tidally averaged salinity stratification at Benicia *D*, northern San Francisco Bay.

6.1. Gravitational circulation

A natural sill is located at the boundary of Carquinez Strait and Suisun Bay slightly landward (east) of the Benicia sampling site. There is a decrease in MLLW depth from 18-11 m in the landward direction at the sill. This topographic control places an upstream limit on gravitational circulation (Jay and Musiak, 1994; Burau et al., 1998) that traps particles and creates an ETM in eastern Carquinez Strait. At the sill, the channel also bifurcates and the width of the southern channel is constricted, which also may limit gravitational circulation (Armi and Farmer, 1986). Sites in Carquinez Strait had the greatest occurrence of SSC maxima at the bottom of the water column for the cruise data (Figure 3). Tidally averaged SSC almost always was greater at Benicia than at Mallard Island as salinity increased in Suisun Bay (Figure 4).

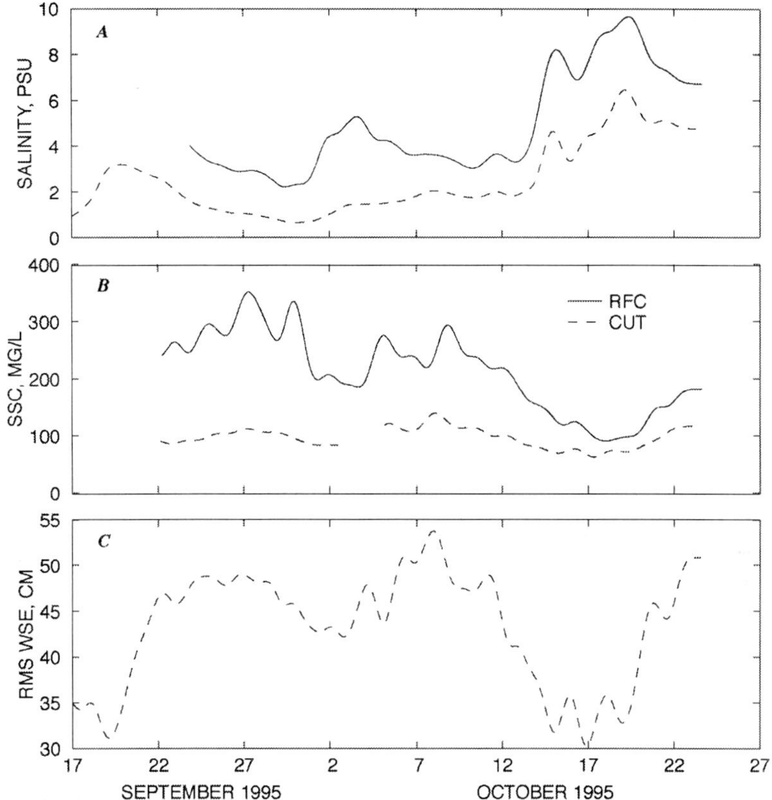

Figure 5. Tidally averaged bottom salinity *A* and suspended-solids concentration (SSC) *B* at the Reserve Fleet Channel and Suisun Cutoff, and root-mean-squared water-surface elevation (RMS WSE) at Suisun Cutoff *C*, northern San Francisco Bay.

Another sill that supports the formation of an ETM in Suisun Bay is between the Reserve Fleet Channel and Suisun Cutoff sites. Two topographic features that place an upstream limit on gravitational circulation at the sill are a decrease in MLLW depth from 9-5 m in the landward direction at the sill and constriction of the channel in Suisun Cutoff (Burau et al., 1998). This topographic control traps particles in the Reserve Fleet Channel. Tidally averaged SSC always was greater in the Reserve Fleet Channel than in Suisun Cutoff as salinity returned to Suisun Bay in 1995 (Figure 5). Gravitational circulation is driven by the longitudinal salinity gradient (Hansen and Rattray, 1965), so these ETMs are dependent on the presence of a nonzero salinity gradient, not a particular salinity.

6.2. Salinity stratification

Turbulence suppression by salinity stratification is most likely to occur in Carquinez Strait, where water depths in the study area are greatest (Burau et al., 1998). Stratification is greatest during neap tides, which reduces vertical mixing, increases deposition, and decreases SSC (Figure 4). Neap tides are the only times when tidally averaged surface SSC is less at Benicia than at Mallard Island. For example, during a neap tide in late May 1996, tidally averaged stratification of almost 16 psu between surface and bottom measurements (15 m apart) account for the smallest tidally averaged SSC measured at Benicia during the study period. In addition, bottom velocities were negligible during ebb tides during this period (not shown), increasing the duration of slack tide to hours and enhancing deposition. Deposition during neap tides creates a supply of easily erodible sediment on the bed in Carquinez Strait. When tidal energy increases as the subsequent spring tide is approached, this sediment is resuspended and gravitational circulation keeps a portion of the sediment in Carquinez Strait, creating an ETM. Thus, deposition associated with salinity stratification and subsequent resuspension is a contributor to the ETM observed in Carquinez Strait.

6.3. Bed storage

Increased sediment deposition associated with stratification is a source of suspended solids for bed storage in Carquinez Strait, especially at the subtidal time scale during neap tides. On the tidal time scale, slack tide in northern San Francisco Bay typically is only a few minutes in duration, which permits few of the suspended solids to deposit during a slack tide. In addition, the duration of high and low water slack tides in northern San Francisco Bay is symetric. In contrast, high water slack in the Tamar Estuary lasts 2-3 hours, during which time suspended solids deposit on the bed (Uncles and Stephens, 1993). Assymetry of slack tide duration in the Tamar Estuary also helps create an ETM at or landward of the freshwater/saltwater interface, which is not observed in northern San Francisco Bay.

Deposition and erosion cycles are more aligned with the spring-neap cycle than the tidal cycle. About one-half of the variance of SSC is caused by the spring-neap cycle, and SSC lags the spring-neap cycle by about 2 days (Schoellhamer, 1996). The relatively short duration of slack water limits the duration of deposition of suspended solids and consolidation of newly deposited bed sediment during the tidal cycle, so suspended solids accumulate in the water column as a spring tide is approached and slowly deposit as a neap tide is approached. Tidally averaged SSC in northern San Francisco Bay is similar to the spring/neap cycle (Figures 4 and 5). This observation is especially true at Benicia and the Reserve Fleet Channel where stratification and deposition are greatest at neap tide.

Wind-wave resuspension of bed sediment in shallow water subembayments is another topographically controlled source of suspended solids in northern San Francisco Bay. Major subembayments that are less than 2-m deep at MLLW are Grizzly and Honker Bays within Suisun Bay and San Pablo Bay, to the west of

Carquinez Strait (Figure 1). An annual cycle of deposition and resuspension begins with a large influx of sediment during winter, primarily from the Central Valley, and much of this material deposits in San Pablo and Suisun Bays (Krone, 1979). Stronger winds during spring and summer cause wind-wave resuspension of bed sediment in these shallow waters and increase SSC (Krone, 1979; Schoellhamer, 1996, 1997). Gravitational circulation in Carquinez Strait transports suspended solids from San Pablo Bay to Suisun Bay (Conomos and Peterson, 1977).

6.4. Tides and surface ETM observation

There commonly is an ETM at low salinities (0–6 psu) near the surface, according to the cruise data (Figure 2), but not according to the tidally averaged data (Figures 4 and 5). Cruise data can be described as the sum of a tidally averaged component and a component representing the instantaneous deviation from the tidally averaged value. Thus, the instantaneous deviation component represents a tidal time-scale process that accounts for this discrepancy.

The cruise data are susceptible to biasing by tidal time-scale processes because 71 percent of the measurements during the cruises were made during a predicted flood tide in Carquinez Strait. Because of this sampling bias, the difference between the tidally averaged and cruise observations of the surface ETM location could be caused by a slack tide process that reduces SSC in Carquinez Strait or a flood tide process that increases SSC at low salinities.

Salinity-dependence and longitudinal position of the surface ETM observed during cruises is modulated by the spring/neap cycle. During neap tides ($\bar{U}_{max} < 0.98\, m/s$), the surface ETM was located at salinities from 0-2 psu landward of Middle Ground in Suisun Bay (Figure 6). During spring tides ($\bar{U}_{max} > 1.15\, m/s$), the surface ETM was seaward of Middle Ground (toward Mare Island) at salinities from 0.3–20.4 psu. Thus, the process that accounts for the difference in the surface ETM observations between the tidally averaged and cruise data is more pronounced during neap tides than during spring tides. Whether or not the spring/neap modulation is present during ebb tides cannot be determined from Figure 6 because no cruises were conducted during neap ebb tides.

One process that could account for the difference between the tidally averaged and cruise observations of the surface ETM location is baroclinically driven pulses at the beginning of flood tide where the longitudinal baroclinic forcing is greatest (Burau et al., 1998; Schoellhamer and Burau, 1998). This location is where the longitudinal salinity gradient is greatest, which usually is where salinity is 0-2 psu. Pulses increase bed shear stress and vertical mixing and, thus, surface SSC. These pulses are strongest during neap tides and have been observed to cause tidally averaged landward transport of suspended solids, despite seaward water transport at mid-depth at Mallard Island during neap tides (Tobin et al., 1995). The water-quality cruise data usually are collected during flood tide and, therefore, soon after a pulse.

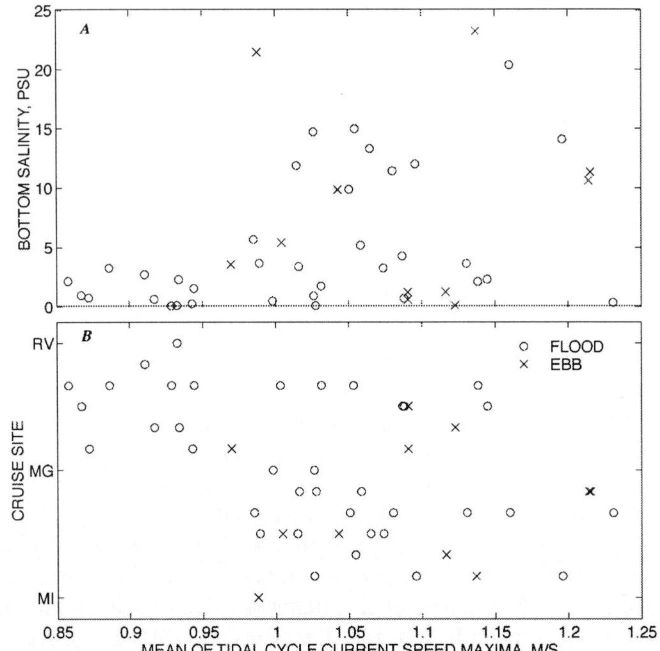

Figure 6. Bottom salinity *A* and sampling site *B* of the maximum cruise surface suspended-solids concentrations relative to the mean of the predicted tidal cycle current speed maxima ($\bar{U}_{max}$) in Carquinez Strait, northern San Francisco Bay. Predicted tidal currents in Carquinez Strait, at the time the maximum SSC for a cruise was sampled, were used to determine tidal phase. RV, Rio Vista; MG, Middle Ground; MI, Mare Island.

Another process that could account for the difference between the tidally averaged and cruise observations of the surface ETM location is particle settling during slack tide in Carquinez Strait, which reduces surface SSC. Stratification in Carquinez Strait is greatest during neap tides, which reduces vertical mixing, enhances settling, and decreases tidally averaged surface SSC (Figure 4). During spring tides, stratification is reduced, so vertical mixing, bed-shear stress, and surface SSC are greater. During cruises, the surface ETM always was located in or near Carquinez Strait during spring tides, independent of the salinity (Figure 6).

Observed surface ETM location, relative to longitudinal position and salinity, is affected by semidiurnal and diurnal tides and the spring/neap tidal cycle. Flood- tide sampling bias during cruises, baroclinically driven pulses during flood tide, and spring/neap modulation of particle settling at slack tide in Carquinez

Strait probably account for the discrepancy between the surface ETM location derived from tidally averaged data and cruise data.

7. CONCLUSIONS

Salinity, bottom topography, and tides affect the locations of estuarine turbidity maxima (ETM) in northern San Francisco Bay. ETMs are not associated with a singular salinity. Bottom suspended-solids concentration (SSC) during cruises and tidally averaged SSC did not show any maxima associated with a particular salinity. Observation of a surface ETM at 0-6 psu bottom salinity during 67 percent of the water-quality cruises probably is a result of (1) 71 percent of the cruise data being collected on flood tide, (2) baroclinically driven pulses during flood tide, and (3) spring/neap modulation of salinity stratification and particle settling at slack tide in Carquinez Strait. The longitudinal salinity gradient, not salinity, creates gravitational circulation and ETMs.

Bottom topography, especially sills in the channels, is another factor controlling the location of ETMs in northern San Francisco Bay. Locations of ETMs are related to bottom topography because salinity stratification and gravitational circulation are enhanced seaward of sills. Maximum bottom SSC was located in Carquinez Strait during 67 percent of the cruises, and tidally averaged SSC was greater in Carquinez Strait and the Reserve Fleet Channel, which are both seaward of sills, compared with more landward sites. Vertical mixing and SSC are greatest in Carquinez Strait during spring tides. Therefore, surface ETMs were always located in or near the Strait (seaward of Middle Ground) during spring tide cruises, regardless of salinity. Wind-wave resuspension of bed sediment in shallow water subembayments is another topographically controlled source of suspended solids.

The spring/neap tidal cycle affects the locations of ETMs. Bottom shear stress and SSC are greatest during spring tides and smallest during neap tides. Salinity stratification in Carquinez Strait is greatest during neap tides, which is the only time when tidally averaged SSC in Carquinez Strait was less than at a landward site at Mallard Island. Baroclinically driven pulses also are strongest during neap tides, when a surface ETM always was observed in the eastern half of the study area (landward of Middle Ground) and between 0-2 psu during cruises.

Observations of an ETM can be affected by sample timing, relative to the tidal cycle. The design of estuarine water-quality sampling programs should consider variability caused by diurnal and semidiurnal tides and the spring/neap cycle, whenever practical.

8. ACKNOWLEDGMENT

I thank Richard Bourgerie, James Cloern, and DWR for providing some of the data presented in this paper; Paul Buchanan, Jon Burau, Jay Cuetara, Robert

Sheipline, and Brad Sullivan for collecting most of the time-series data; and Jon Burau, James Cloern, Brian Cole, Wim Kimmerer, Ray Krone, Kathryn Kuivila, Ashish Mehta, and Fred Nichols for useful discussions and comments about this work. Collection of most of the time-series data and data analysis were supported by the San Francisco Regional Water Quality Control Board and by the U.S. Geological Survey Federal/State Cooperative and San Francisco Bay Ecosystem Programs.

REFERENCES

Armi, L., and Farmer, D. M., 1986. Maximal two-layer exchange through a constriction with barotropic net flow. *Journal of Fluid Mechanics*, 164, 27-51.

Arthur, J. F., and Ball, M. D., 1978. Entrapment of suspended materials in the San Francisco Bay-Delta estuary. U.S. Bureau of Reclamation, Sacramento, CA, 106p.

Buchanan, P. A., and Schoellhamer, D. H., 1998. Summary of suspended-solids concentration data, San Francisco Bay, California, water year 1996. U.S. Geological Survey Open-File Report 98-175, 59p.

Burau, J. R., Gartner, J. W., and Stacey, M. T., 1998. Results from the hydrodynamic element of the 1994 entrapment zone study in Suisun Bay, in Kimmerer, W. ed., Report of the 1994 entrapment zone study. *Technical Report 56*, Interagency Ecological Program for the San Francisco Bay/Delta Estuary. 55p.

California Department of Water Resources, 1998. California data exchange center. http:/ /cdec.water.ca.gov.

Conomos, T. J., and Peterson, D. H., 1977. Suspended-particle transport and circulation in San Francisco Bay an overview. In: *Estuarine Processes*, M. Wiley ed., Academic Press, New York, 2, 82-97.

Edmunds, J. L., Cole, B. E., Cloern, J. E., and Dufford, R. G, 1997. Studies of the San Francisco Bay, California, Estuarine Ecosystem. Pilot Regional Monitoring Program Results, 1995. U.S. Geological Survey Open-File Report 97-15, 380p.

Geyer, W. R., 1993. The importance of suppression of turbulence by stratification on the estuarine turbidity maximum. *Estuaries*, 16(1), 113-125.

Grabemann, I., Uncles, R. J., Krause, G., and Stephens, J. A., 1997. Behaviour of turbidity maxima in the Tamar (U.K.) and Weser (F.R.G.) estuaries. *Estuarine, Coastal and Shelf Science*, 45, 235-246.

Hamblin, P. F., 1989. Observations and model of sediment transport near the turbidity maximum of the Upper Saint Lawrence Estuary. *Journal of Geophysical Research*, 94(C10), 14419-14428.

Hansen, D. V., and Rattray, M., 1965. Gravitational circulation in straits and estuaries. *Journal of Marine Research*, 23, 104-122.

Ianniello, J. P., 1979. Tidally induced residual currents in estuaries of variable breadth and depth. *Journal of Physical Oceanography*, 9, 962-974.

Jassby, A. D., Kimmerer, W. J., Monismith, S. G., Armor, C., Cloern, J. E., Powell, T. M., Schubel, J. R., and Vendlinski, T. J., 1995. Isohaline position as a habitat indicator for estuarine applications. *Ecological Applications*, 5(1), 272-289.

Jay, D. A., and Musiak, J. D., 1994. Particle trapping in estuarine tidal flows. *Journal of Geophysical Research*, 99(C10), 20445-20461.

Krone, R. B., 1979. Sedimentation in the San Francisco Bay system - San Francisco Bay. In: *The Urbanized Estuary*, T. J. Conomos ed., Pacific Division of the American Association for the Advancement of Science, San Francisco, 85-96.

Le Bris, H., and Glemarec, M., 1996. Marine and brackish ecosystems of south Brittany (Lorient and Vilaine Bays) with particular reference to the effect of the turbidity maxima. *Estuarine, Coastal and Shelf Science*, 42, 737-753.

Meade, R. H., 1972. Transport and deposition of sediments in estuaries. *Geologic Society of America*, Memoir 133, 91-120.

National Oceanic and Atmospheric Administration, 1998. San Francisco Bay PORTS. ftp://ceob-g30.nos.noaa.gov/pub/ports/sanfran/screen.

Schoellhamer, D. H., 1996. Factors affecting suspended-solids concentrations in South San Francisco Bay, California. *Journal of Geophysical Research*, 101(C5), 12087-12095.

Schoellhamer, D. H., 1997. Time series of SSC, salinity, temperature, and total mercury concentration in San Francisco Bay during water year 1996: *1996 Annual Report*, Regional Monitoring Program for Trace Substances, 65-77.

Schoellhamer, D. H., and Burau, J. R., 1998. Summary of findings about circulation and the estuarine turbidity maximum in Suisun Bay, California. *U.S. Geological Survey Fact Sheet FS-047-98*, 6p.

Schubel, J. R., and Carter, H. H., 1984. The estuary as a filter for fine-grained suspended sediment. In: *The Estuary as a Filter*, V. S. Kennedy ed., Academic Press, New York, 81-105.

Tobin, A., Schoellhamer, D. H., and Burau, J. R., 1995. Suspended-solids flux in Suisun Bay, California. *Proceedings of the First International Conference on Water Resources Engineering*, San Antonio, Texas, August 14-18, 1995, 2, 1511-1515.

Uncles, R. J., and Stephens, J. A., 1993. The freshwater-saltwater interface and its relationship to the turbidity maximum in the Tamar Estuary, United Kingdom. *Estuaries*, 16(1), 126-141.

Uncles, R. J., Barton, M. L., and Stephens, J. A., 1994., Seasonal variability of fine-sediment concentrations in the turbidity maximum region of the Tamar Estuary. *Estuarine, Coastal and Shelf Science*, 38, 19-39.

Weir, D. J., and McManus, J., 1987. The role of wind in generating turbidity maxima in the Tay Estuary. *Continental Shelf Research*, 7(11/12), 1315-1318.

Wolanski, E., King, B., and Galloway, D., 1995. Dynamics of the turbidity maximum in the Fly River Estuary, Papua New Guinea. *Estuarine, Coastal and Shelf Science*, 40, 321-337.

Coastal and Estuarine Fine Sediment Processes
W.H. McAnally and A.J. Mehta (Editors)

Boundary layer effects due to suspended sediment in the Amazon River estuary

S. B. Vinzon[a] and A. J. Mehta[b]

[a]Programa de Engenharia Oceânica, COPPE - Universidade Federal do Rio de Janeiro, CEP 21945-970, Rio de Janeiro, RJ, Brazil

[b]Department of Civil and Coastal Engineering, University of Florida, Gainesville, Florida 32611, USA

The effect of suspended sediment in modulating the vertical velocity structure in the Amazon River estuary is examined. Based on simple analyses and previously made measurements on the Brazilian shelf at the mouth of the estuary, viscous damping of turbulence is shown to be a significant feature of the bottom hyperpycnal suspension layer of 2 m notional height. This layer is nested within the lower flow boundary layer on the order of 4 m thickness. For modeling sediment discharge in this estuary, it will be essential to account for the strong interaction between flow and sediment dynamics that occurs within the layered water column.

1. INTRODUCTION

Turbulence collapse within the zone of high suspended sediment concentration (SSC) is a well-known phenomenon, and in muddy rivers the sediment yield can be significantly influenced by the manner in which SSC modulates the boundary layer velocity structure. The estuarine segment of the Amazon River is a natural candidate for examination in this context, given that this river debouches, on the average, $11\text{-}13\times10^8$ tons of sediment annually onto the northern Brazilian shelf (Meade et al., 1985). Accordingly, using an approximate hydromechanical approach we have examined the significance of the viscous effect due to high SSC in influencing the vertical velocity profiles during the tidal cycle in the seaward zone of this large estuary. This examination, which is presented in detail elsewhere (Vinzon, 1998), is briefly described here.

2. VELOCITY AND SSC STRUCTURES

A comprehensive data set on flow and sediment dynamics in the region of the mouth of the Amazon was collected under AMASSEDS (A Multidisciplinary Amazon

Shelf Sediment Study) during 1989-91. Figure 1 shows the area studied and the measurement sites. Considering the oceanic boundary of the estuary to be as shown by a dotted line, we observe that anchor stations for data collection included the river mouth landward of the line, as well as the shelf seaward of this line. A detailed description of the overall study objectives and measurement strategy is found in Kineke (1993). It suffices to note that due to the large river outflow, shelf dynamics up to the ~100 m depth contour (Figure 1) is dominated by flow-sediment interactions that are essentially estuarine. The suspended sediment is largely fine-grained; two samples analyzed as part of the present study indicated the median (dispersed) grain size to be about 3 μm. The predominant clays were kaolinite and smectite, and non-clay minerals were mainly quartz and mica. About 3% (by weight) material was organic. The cation exchange capacity of the clayey fraction was ~26 meq/100 g (Vinzon, 1998).

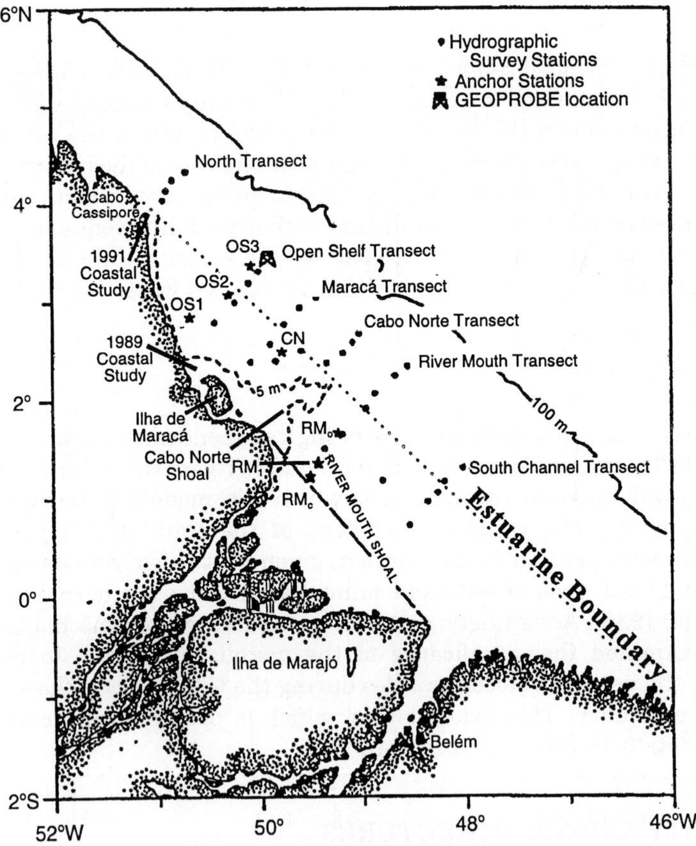

Figure 1. Amazon estuary and proximal northern Brazilian shelf along with locations of AMASSEDS measurement sites (adapted from Kineke, 1993).

Figure 2 shows illustrative plots of instantaneous (and synchronous) profiles of velocity, SSC and the (horizontal) sediment flux at anchor station OS2 (Figure 1). The lowest measurement elevation was 0.25 m above the level where the data profiler rested. Here, this level will be consistently chosen as the bottom datum. Observe that while the SSC and the flux are both high below about 2 m in comparison with the upper column, the velocity profile in this near-bottom hyperpycnal suspension layer is distinctly non-logarithmic, and in fact mimics a linear trend. Such a trend combined with high SSC are characteristic indicators of highly damped turbulence in the lower layer (Ross and Mehta, 1989). The upper bound of this layer is the lutocline, which typically oscillated between 1 and 4 m above bottom at the measurement sites depending on the location and the tidal stage (Kineke, 1993). Selecting the hyperpycnal layer as one characterized by SSC between ~10 g/L and ~75 g/L (Krone, 1962; Odd and Cooper, 1989), in Figure 2 the lutocline defines, albeit approximately, the upper level of this layer at an elevation between 1.5 and 2 m. Observe also that 75 g/L occurred at about 1 m above the datum. Since this is the nominal value of the concentration beneath which the soil tends to have a measurable shear strength against erosion, it follows that the bottom datum was ~1 m below what would be called the bed surface; bed being defined by a particle-supported sedimentary matrix (Ross and Mehta, 1989).

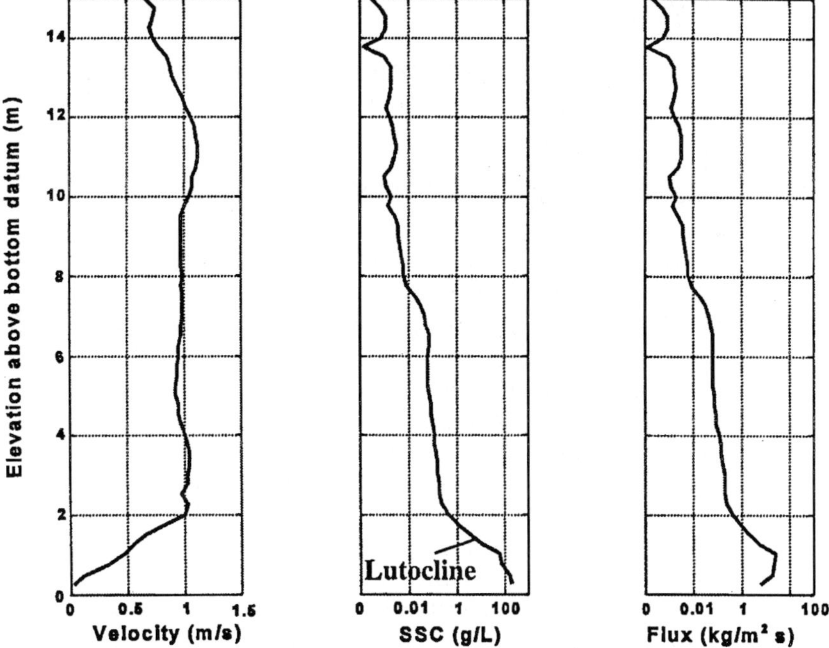

Figure 2. Examples of instantaneous (and synchronous) profiles of velocity, SSC and sediment flux during high river discharge at anchor station OS2 (after Kineke, 1993).

Similar to the Severn estuary in UK (e.g., Parker, 1987), in the high energy environment of the Amazon with a tidal range as high as ~5 m, lutocline dynamics seems to correlate strongly with tidal flow velocity, with secondary influences from the vertical gradients in salinity and temperature. This point can be illustrated by the typical salinity and temperature profiles in Figure 3, which shows that the halocline and the thermocline were considerably elevated in comparison with typical elevations of the lutocline (Figure 2). To examine further the relative effects of SSC, salinity and temperature, consider the gradient Richardson number

$$Ri = \frac{-\dfrac{g}{\rho}\dfrac{\partial \rho}{\partial z}}{\left(\dfrac{\partial u}{\partial z}\right)^2} \tag{1}$$

where

$$\rho(S,\theta,\mathrm{SSC}) = \mathrm{SSC}\frac{[\rho_s - \rho(S,\theta)]}{\rho_s} + \rho(S,\theta)$$

$$\rho(S,\theta) = 1000\frac{1 + A_1}{A_2 + 0.698A_2}$$

$$A_1 = 5890 + 38\theta - 0.375\theta^2 + 3S$$

$$A_2 = 1779.5 + 11.25\theta - 0.0745\theta^2 - (3.8 + 0.01\theta)S$$

In the above relations $u(z,t)$ is the horizontal velocity, z is the vertical coordinate originating at the bottom datum, t denotes time, g is the acceleration due to gravity, $\rho(S,\theta,\mathrm{SSC})$ is the fluid density, ρ_s is the granular density of the sediment, S is the salinity (psu) and θ is the temperature (°C). The dependence of fluid density on S, θ and SSC follows the formula of Eckart (1958).

Using measured profiles of u, SSC, S and θ at anchor station OS1, representative tide-averaged profiles of Ri are shown in Figure 4 for two cases--without (solid line) and with (dashed line) the effect of SSC. Observe that in the bottom ~2 m the two profiles diverge significantly. For convenience of interpretation of further results we will consider 2 m to be the notional height of the hyperpycnal layer. It is evident that the effect of SSC on the dominance of buoyancy relative to flow inertia, as embodied in Ri, cannot be ignored within this layer. There, the large values of Ri inclusive of the SSC effect imply the existence of comparatively stable lutoclines over the tidal cycle (Parker, 1987).

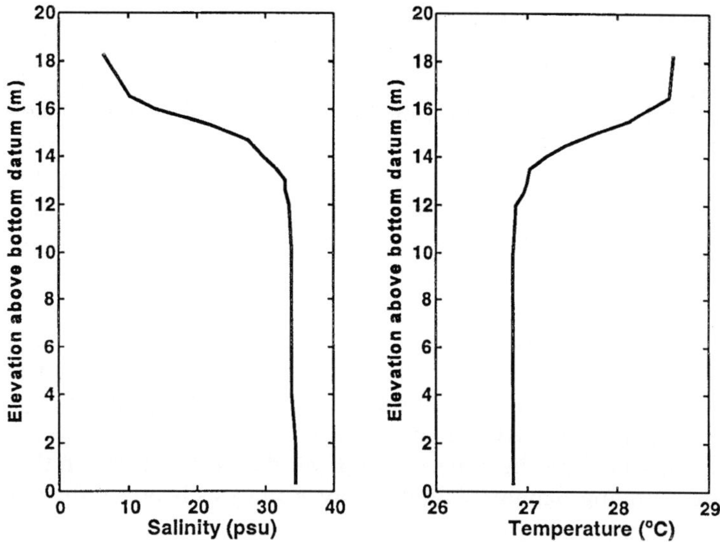

Figure 3. Examples of instantaneous (and synchronous) salinity and temperature profiles at anchor station OS2 (after Kineke, 1993).

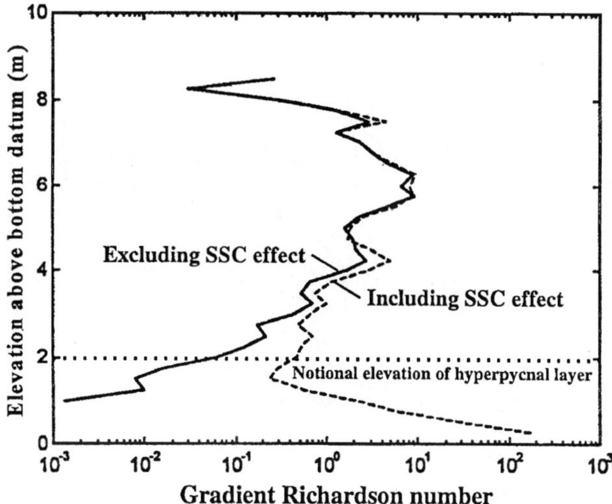

Figure 4. Representative (tide-averaged) profiles of the gradient Richardson number based on data at anchor station OS1 considering density variation due to salinity and temperature (solid line), and due to salinity, temperature and SSC (dashed line).

3. BOUNDARY LAYER CHARACTER

3.1. Velocity

Examples of tide-varying (instantaneous) velocity profiles in the lower boundary layer at anchor station OS1 shown in Figure 5 exhibit a noteworthy non-logarithmic character, as in Figure 2. Several of the profiles also show concavity($\partial^2 u/\partial z^2 > 0$) in the hyperpycnal layer. Following Geyer (1995), who considered the lower boundary layer in the study region to be confined between 3 to 5 m for neap and spring tides respectively, the top of this layer will be chosen to be notionally at a height δ =4 m.

Ross and Mehta (1990) modeled the horizontal velocity profile within the hyperpycnal layer in the Avon River (UK), which also showed a concavity, as an unsteady Couette-type flow driven predominantly by shear stress imposed at the level of the lutocline. They attributed this trend to Rayleigh flow effect arising from momentum diffusion into the hyperpycnal layer due to shear-driven flow above the lutocline. Here an approximate boundary layer analysis is presented to explain the velocity structure, as follows.

We will consider flow in the lower, oscillatory boundary layer to be essentially horizontal. For this layer, ignoring advective acceleration which can be shown to be small in comparison with temporal acceleration for the problem at hand, the equation of motion can be stated as (Nielsen, 1985)

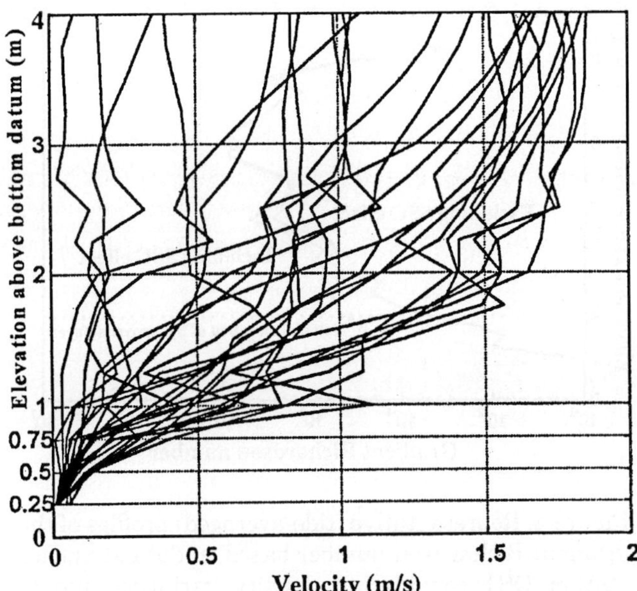

Figure 5. Tide-varying (instantaneous) velocity profiles in the lower boundary layer at anchor station OS2.

$$\rho \frac{\partial}{\partial t}(u - u_\delta) = \frac{\partial \tau}{\partial z} \tag{2}$$

where u_δ is the flow velocity at the top of the lower boundary layer, and τ is the shear stress at any elevation z. In obtaining (2) the pressure distribution is assumed to be hydrostatic, and the shear stress is considered to vanish at the top of the lower boundary layer at $z = \delta$. Equation (2) can also be written as

$$\frac{\partial(u - u_\delta)}{\partial t} = \frac{\partial}{\partial z}\left(\nu(z)\frac{\partial u}{\partial z}\right) \tag{3}$$

where $\nu(z)$ is a characteristic viscosity representing momentum diffusion within the lower boundary layer. In the viscous flow regime, ν is the kinematic viscosity, an exclusive property of the fluid. For turbulent flow ν represents the eddy viscosity, which is dependent on the flow itself.

Following Nielsen (1985) and expressing the tidal velocity profile as $u(z,t) = u(z)e^{i\sigma t}$, where σ is the tidal frequency and $i = \sqrt{-1}$, (3) becomes

$$\frac{d}{dz}\left(\nu(z)\frac{du}{dz}\right) = i\sigma(u - u_\delta) \tag{4}$$

The usual functional form of viscosity for turbulent boundary layers in a Newtonian fluid is

$$\nu(z) = \kappa u_* z \tag{5}$$

where κ is the Karman constant (which is nominally equal to 0.4) and u_* is the friction velocity. Accordingly, with the boundary conditions u=0 at z=0 (i.e., no-slip at the bottom) and $u = u_\delta$ at $z \to \infty$ (effectively at the top of the boundary layer), the general solution of (4) is

$$u = u_\delta\left[1 + A \ker\left(2\sqrt{\frac{\sigma z}{\kappa u_*}}\right)\right] \tag{6}$$

where

$$A = \frac{-1}{\ker\left(2\sqrt{\frac{\sigma}{\kappa u_*}}\right)}$$

and ker denotes a Kelvin function. For small arguments of ker this velocity profile approaches the logarithmic form, as would be expected for a turbulent boundary layer.

3.2. Viscosity

To proceed further with an interpretation of (4) in conjunction with the non-logarithmic lower boundary layer, it is necessary to examine the nature of the viscosity, ν, inasmuch as (5) is not applicable, especially in the hyperpycnal layer, which nests within the lower boundary layer. Accordingly, we will define a flow viscosity, ν_{flow}, and a fluid viscosity, ν_{fluid}, as follows.

Based on the rheological work of Faas (1986), which highlighted the pseudoplastic flow behavior of the high SSC fluid in the study area, ν_{flow} can be expressed by a power-law relation akin to that of Sisko (1958) according to

$$\nu_{flow} = \mathrm{SSC}\ \exp\left(-0.78\frac{\partial u}{\partial z} - 10.24\right); \quad \text{for } \frac{\partial u}{\partial z} < 3.9\,\mathrm{Hz} \tag{7a}$$

$$\nu_{flow} = \mathrm{SSC}\ \exp\left(-0.017\frac{\partial u}{\partial z} - 12.95\right); \quad \text{for } \frac{\partial u}{\partial z} \geq 3.9\,\mathrm{Hz} \tag{7b}$$

As for ν_{fluid}, one can determine it directly from (4) using measured velocity profiles according to

$$\nu_{fluid} = \frac{\frac{1}{\rho}\int_z^{\delta}\frac{\partial}{\partial t}(u - u_{\delta})\,dz}{\frac{\partial u}{\partial z}} \tag{8}$$

Representative profiles of ν_{flow} and ν_{fluid} thus obtained by using synchronous velocity and SSC data from several anchor stations are plotted in Figure 6. Observe that with increasing depth ν_{fluid} increases monotonically in the mean, while the corresponding ν_{flow} decreases somewhat at first, then increases and converges to ν_{fluid} within the hyperpycnal layer. Overall, ν_{flow} is comparatively uniform within the lower boundary layer. Since it is derived from viscous rheometry, it follows that the hyperpycnal layer is dominated by viscous effects that are attributable to the high SSC and associated turbulence damping.

The above finding with respect to the effect of SSC on the flow structure can be corroborated by profiles of the characteristic Reynolds number corresponding to Figure 6. It is shown elsewhere (Vinzon, 1998) that with increasing depth in the lower boundary layer, values of this number drop rapidly below 10^4, the limit below which turbulence is damped in flows of non-Newtonian clay suspensions (Wang and Plate, 1996).

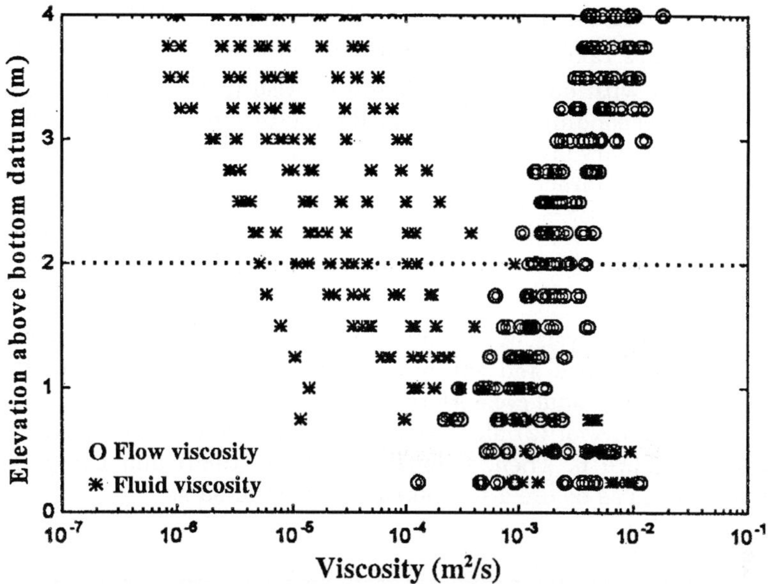

Figure 6. Representative profiles of calculated flow viscosity and fluid viscosity over the tidal cycle based on data from several anchor stations.

3.3. Velocity profile

From Figure 6, considering the fluid viscosity to decrease very approximately in a linear manner with elevation above the bottom according to $\nu_{\text{fluid}}(z) = c_1 - c_2 z$, where c_1 and c_2 are empirical constants, (4) can be solved (with $\nu = \nu_{\text{fluid}}$), along with the same boundary conditions as those used to obtain (6), as (Vinzon, 1998)

$$u = u_\infty \left[1 + B \,\text{bei}\left(2\sqrt{\frac{\sigma(c_1 - c_2 z)}{c_2^2}} \right) \right] \tag{9}$$

where

$$B = \frac{-1}{\text{bei}\left(2\sqrt{\frac{\sigma c_1}{c_2^2}} \right)}$$

and bei is a Kelvin function. Using conveniently chosen values of $\sigma = 1.45\times10^{-4}$ rad/s (nominal frequency of the prevailing semi-diurnal tide), $u_\delta = 1$ m/s, and $u_* = 0.02$

m/s, the velocity profile resulting from (9) is shown in Figure 7. Values of $c_1 = 0.001$ and $c_2 = 0.0001$ are taken to approximately represent the mean trend of fluid viscosity variation with elevation in Figure 6. Also included for a comparative purpose is the corresponding profile obtained from (6). Observe that starting at the top of the lower boundary layer, with increasing depth (9) deviates increasingly from (6). Also, (9) shows a characteristic concavity within the hyperpycnal layer.

4. TIDAL VELOCITY

Given the above analysis, it is instructive to examine the way in which the vertical profile of velocity and the bottom shear stress vary within the lower boundary layer over the tidal cycle. Here, a simple solution for the velocity is sought assuming local validity of (2) along with a uniform flow viscosity based on the observation in Figure 6. Then, following Nielsen (1985) and using the same boundary conditions as before a solution of (4) is obtained as

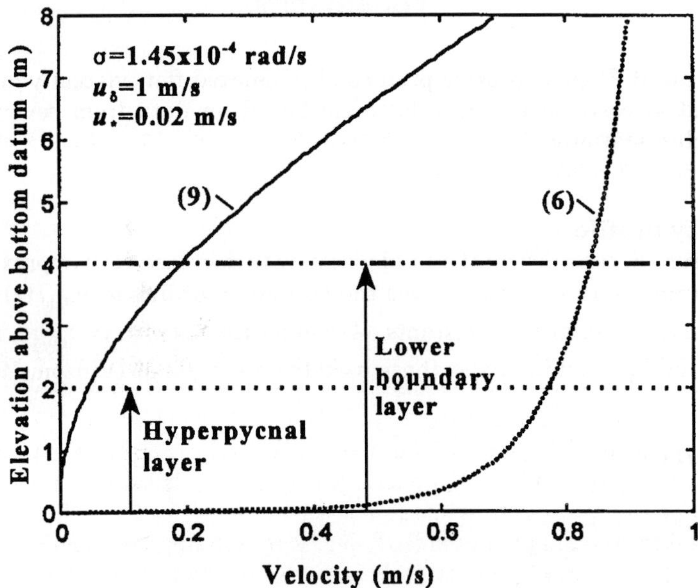

Figure 7. Examples of calculated boundary layer velocity profiles with viscosity decreasing with elevation (solid line) and increasing with elevation (dotted line). The profiles have been conveniently extended into the upper boundary layer in order to illustrate their evolution with elevation above the lower boundary layer.

$$u(z,t) = u_\delta \left[1 - \exp\left(-[1+i]\frac{z}{\sqrt{2\nu/\sigma}} \right) \right] e^{i\sigma t} \tag{10}$$

which yields the bottom shear stress as

$$\tau_b(t) = u_\delta\,(1+i)\frac{\nu\rho}{\sqrt{2\nu/\sigma}}\, e^{i\sigma t} \tag{11}$$

In (10) and (11) we will conveniently select $\nu = \nu_{flow}$ as a function of the local mean SSC and shear rate, $\partial u/\partial z$, using (7a,b).

Figure 8 compares (10) with observed hourly velocity profiles over two tidal cycles at anchor station OS1. A measure of the agreement between the calculated and the observed profiles is given in Figure 9, in which the tide-averaged absolute error is calculated from $\mathrm{er}(z) = |\hat{u}(z) - u(z)|$, where $\hat{u}$ is obtained from (10), and u is the measured velocity. For comparison, the corresponding er(z) is also calculated for the logarithmic velocity fit, i.e.,

$$u = \frac{u_*}{\kappa}\ln\frac{z}{z_o} \tag{12}$$

where z_o is the height of the bottom roughness. For solving (12), u_* was obtained from the observed velocity profiles using

$$u_* = \sqrt{\frac{1}{\rho}\int_0^\delta \rho\frac{\partial}{\partial t}(u - u_\delta)dz} \tag{13}$$

For calculation of u_* from (13) for each measured velocity profile, a crude estimate of δ was obtained as the height at which the first local maximum in the flow speed occurred. Then, knowing u_*, z_o was obtained from (12) by equating u in (12) with the corresponding measured velocity at z=δ. In Figure 9 we observe that in the hyperpycnal layer er(z) begins to decrease (below 1.8 m) when (10) is used for the velocity profile, and increases when (12) is used.

It follows from the above that since (10) is a reasonable predictor of the velocity profile in the lower boundary layer, (11) may be used to calculate the bottom shear stress, as examined next.

5. BOTTOM SHEAR STRESS

Integrating (2) between z and δ, and recognizing that $\tau|_{z=\delta} = 0$, the shear stress can be calculated from

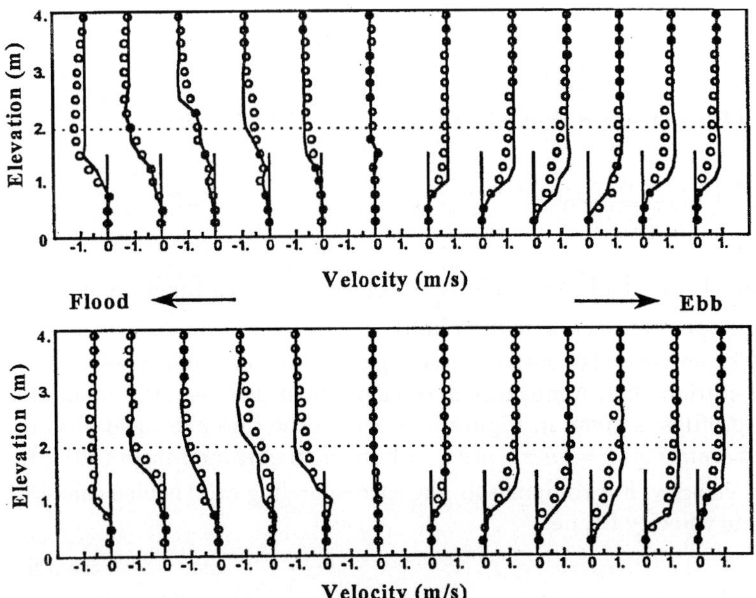

Figure 8. Comparison between calculated (lines) and observed (circles) velocity profiles over two tidal cycles at anchor station OS1, low river discharge.

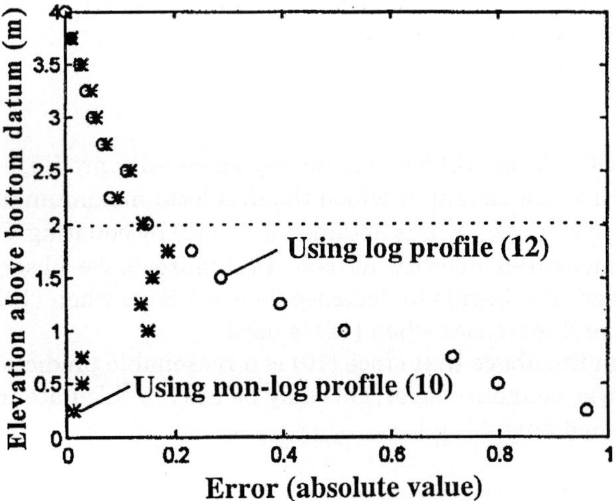

Figure 9. Absolute value of the difference (error) between measured and calculated velocity as a function of elevation, using non-logarithmic profile (10) (asterisks) and logarithmic profile (12) (circles).

$$\tau\big|_z = \int_z^{\delta} \rho \frac{\partial}{\partial t}(u - u_\delta)dz \qquad (14)$$

Thus, (14) can be used to determine the bottom shear stress, $\tau_b = \tau(z = 0.25 \text{ m})$, 0.25 m being the lowest elevation at which velocity data were obtained.

It is instructive to compare τ_b obtained from (14) with that calculated from (11). This is done in Figure 10 over a tidal cycle at anchor station CN. Observe that (11), which makes use of the measured velocity at elevation δ only, shows a more damped response to boundary layer events than (14), which accounts for the velocity structure between 0.25 m and δ. Likewise, (11) is phase-lagged from (14) by ~1 hr, presumably reflecting the time-scale of momentum exchange over the height of the boundary layer.

6. CONCLUDING COMMENTS

The high SSC near the bottom in the Amazon estuary and the associated enhanced fluid viscosity tend to inhibit turbulence development in the near-bed hyperpycnal layer. This effect markedly influences the lower boundary layer structure in a region of extreme importance to sediment yield evaluation in this estuary. Most of the sediment transport occurs over the first two meters from the bottom, where the concentrations tend to be ~10 g/L and higher. Thus it appears that a notional height of 2 m of the hyperpycnal layer, which nests within the lower boundary layer on the order of 4 m height, is characteristic of the segment of the estuary examined by Kineke (1993). For modeling sediment discharge in this estuary, it will be essential to account for the strong interaction between flow and sediment dynamics within the lower boundary layer.

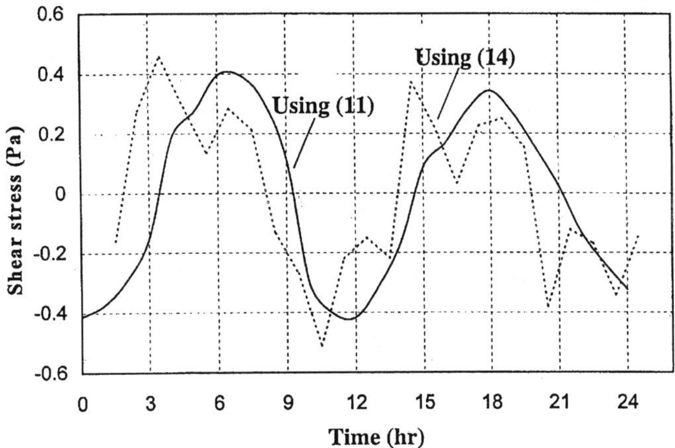

Figure 10. Shear stress calculated using (11) (solid line) and (14) (dashed line) over a tidal cycle at anchor station CN.

7. ACKNOWLEDGMENT

The authors deeply appreciate the help provided from Dr. Gail Kineke of Boston College, Chestnut Hill, MA for supplying relevant data files from AMASSEDS. Support provided by CAPES, Brazil, enabled the first author to start this study at the University of Florida.

REFERENCES

Eckart, C., 1958. Properties of water, part III. The equation of state of water and sea water at low temperatures and pressures. *American Journal of Science*, 256, 225-240.

Faas, R. W., 1986. Mass-physical and geotechnical properties of superficial sediments and dense nearbed sediment suspensions on the Amazon continental shelf. *Continental Shelf Research*, 6(1/2), 189-208.

Geyer, W. R., 1995. Tide-induced mixing in the Amazon frontal zone. *Journal of Geophysical Research*, 100(C2), 2341-2353.

Kineke, G. C., 1993. Fluid muds on the Amazon continental shelf. *Ph.D. Thesis*, University of Washington, Seattle, 259p.

Krone, R. B., 1962. Flume studies of the transport of sediment in estuarial shoaling processes. *Final Report,* Hydraulic Engineering Laboratory and Sanitary Engineering Research Laboratory, University of California, Berkeley, 118p.

Meade, R. H., Dunne, T., Richey, J. E., Santos, U. D., and Salati, E., 1985. Storage and remobilization of suspended sediment in the lower Amazon river of Brazil. *Science*, 228, 488-490.

Nielsen, P., 1985. Oscillatory boundary layers: a different problem. *Proceedings of the 21st Congress of IAHR*, Melbourne, Australia, 156-159.

Odd, N. V. M., and Cooper, A. J., 1989. A two-dimensional model of the movement of fluid mud in a high energy turbid estuary. *Journal of Coastal Research*, SI 5, 185-193.

Parker, W. R., 1987. Observations on fine sediment transport phenomena in turbid coastal environments. *Continental Shelf Research*, 7(11/12), 1285-1293.

Ross, M. A., and Mehta, A. J., 1989. On the mechanics of lutoclines and fluid mud. *Journal of Coastal Research*, SI 5, 51-61.

Ross, M. A., and Mehta, A. J., 1990. Transport of high concentration flocculated suspensions in estuaries. In: *Flocculation and Dewatering*, B. M. Moudgil and B. J. Scheiner eds., Engineering Foundation, New York, 529-538.

Sisko, A. W., 1958. The flow of lubricating greases. *Industrial Engineering Chemistry*, 50, 1789-1792.

Vinzon, S. B., 1998. A preliminary examination of Amazon shelf sediment dynamics. *Engineer Degree Thesis*, University of Florida, Gainesville, 168p.

Wang, Z., and Plate, E., 1996. A preliminary study on the turbulence structure of flows with non-Newtonian fluids. *Journal of Hydraulic Research*, 34(3), 345-361.

Coastal and Estuarine Fine Sediment Processes
W.H. McAnally and A.J. Mehta (Editors)

Modeling mechanisms for the stability of the turbidity maximum in the Gironde estuary, France

A. Sottolichio[a], P. Le Hir[b] and P. Castaing[a]

[a]Université Bordeaux I, Département de Géologie et Océanographie, UMR 5805 EPOC, Avenue des Facultés, 33405 Talence cedex, France

[b]Institut Français pour la Recherche et l'Exploitation de la Mer (IFREMER), Direction de l'Environnement Littoral, DEL-EC-TP, B.P. 70, 29280 Plouzané, France

Two numerical models are applied to the Gironde estuary in order to study basic mechanisms of turbidity maximum formation. A 2D depth-averaged model reproduces the turbidity maximum location for different river flow conditions, wherein tidal wave propagation is the only hydrodynamic forcing. Yet, excessive sediment escape to the ocean and associated loss of mass of the tidal turbidity maximum are observed. A 3D model is applied in order to incorporate vertical gradients and density stratification. The turbidity maximum is better reproduced, especially with respect to lateral gradients. The introduction of salinity gradients leads to less seaward dispersion, without any modification of the structure and location of the turbidity maximum. Results suggest that the turbidity maximum in the Gironde is exclusively tidally-induced, while density induced residual circulation imparts stability to the suspended sediment mass in the estuary. This conclusion is preliminary, as sedimentary processes are simplified in the model, and validation of sedimentary patterns is qualitative. In addition, the observation regarding the salinity effect needs to be verified by longer simulations with the 3D model, in order to provide an assessment of seaward fluxes over long periods.

1. INTRODUCTION

In estuarine environments an understanding of fine sediment transport, in particular turbidity maxima, is necessary in order to predict the fate of pollutants and to design dredging strategies for navigational channels. Many processes have been considered to explain the formation of these high-turbidity zones. For instance, suppression of turbulence by stratification can give rise to a turbidity maximum (Geyer, 1993) in moderately and highly stratified estuaries. However, in medium and high tidal range estuaries, the prevailing mechanisms causing turbidity maxima are tidal asymmetry and density-induced residual circulation (Dyer, 1986). Tidal asymmetry involves a sequence of erosion, deposition and transport of sediment by short but more intense flood currents and longer but less intense ebb currents, with net particle movement landward towards the head of the estuary, up to a tide decay zone, where river flow becomes dominant (Allen et al., 1980). Density-induced residual circulation causes trapping of suspended material by downstream transport in the surface layer, settling into the lower layer, and upstream transport up to the convergence of the bottom residual flow (Dyer, 1986).

The Gironde estuary (in southwest France) provides a good example of a well developed turbidity maximum. Descriptions of its dynamics are numerous (Allen et al., 1977, 1980; Castaing and Allen, 1981), and both tidal asymmetry and density-induced residual circulation have been invoked to explain the turbidity maximum formation. Abundant observations in the late 1960s (Allen, 1972) revealed a coincidence of the high-concentration zone and the freshwater-saltwater interface leading to a density-induced turbidity maximum, as explained by Nichols and Poor (1967). Later, Allen et al. (1980) demonstrated that the turbidity maximum could be explained merely by the effect of tidal asymmetry (Postma, 1967) coupled with resuspension processes (Schubel, 1971), which gives rise to a tidal pumping mechanism (Uncles et al., 1984). At present these latter mechanisms are believed to be dominant, even though significant density gradients occur during moderate fluvial discharge periods. Hence the aim of this paper is to elucidate the basic mechanisms responsible for the accumulation of suspended sediment in the Gironde estuary. In other words, the objective is to highlight the roles of tidal asymmetry and residual density-induced circulation in turbidity maximum formation.

2. THE GIRONDE ESTUARY

The Gironde estuary is one of the largest estuaries of the European Atlantic coast (Figure 1). It originates from the confluence of the Garonne and Dordogne Rivers which supply freshwater at a mean annual rate of 1,000 m^3/s (Castaing and Allen, 1981). In winter, the rivers have a mean flow of 2,000 m^3/s, with peak floods of more than 4,000 m^3/s. During summer, the mean freshwater flow decreases to less than 300 m^3/s. Tidal range varies from 2 to 5 m at the mouth and increases to more than 6 m towards the upper estuary due to the convergence of estuarine shape. Tides propagate up to 170 km upstream during low river flow periods (up to Pessac in the Dordogne River and La Reole in the Garonne River, Figure 1).

Turbidity maximum is an important feature of the Gironde. This high-turbidity zone is characterized by suspended solids concentrations (SSC) of about 1 g/l (Allen et al., 1977). During slack water periods, and especially at neap tides, settling of suspended matter is enhanced and thick patches of fluid mud appear on the channel bottom, with concentrations up to 300 g/l. The estimated total sediment mass contained in the turbidity maximum/fluid mud system is believed

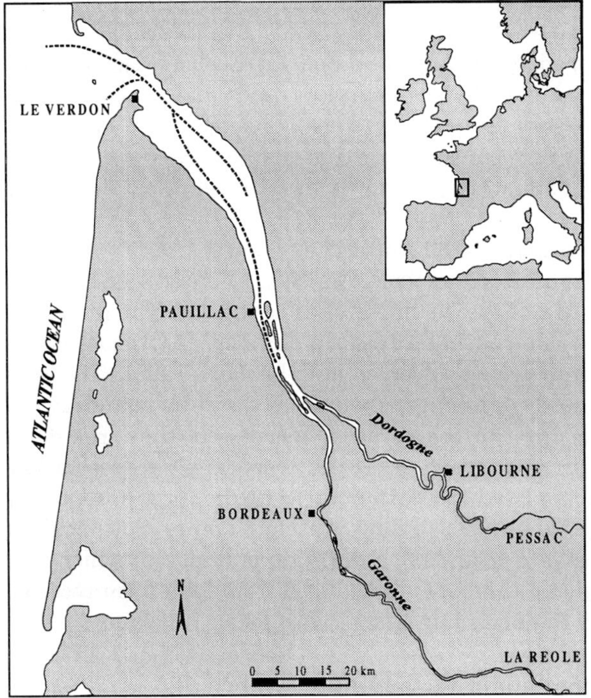

Figure 1. The Gironde estuary (dashed lines represent axes of the main navigation channels).

to reach 5×10^6 tons, which represents two years of riverine solids input (Jouanneau and Latouche, 1981).

3. BRIEF DESCRIPTION OF THE MODELS

Brenon (1997) has demonstrated the use of an evolving modeling strategy to understand estuarine sediment dynamics. This consists in applying the simplest models as a first step, and subsequently increasing their complexity and incorporating mechanisms until a satisfactory representation of reality is obtained. For estuarine sediment transport, this is an effective method. In some cases, depending on the goals, a three-dimensional approach may be essential, but in other cases, if some mechanisms can be roughly represented, a two-dimensional (2D vertical or horizontal) model may be relevant. As noted above, tidal pumping is dominant in the formation of the Gironde estuary turbidity maximum (Allen et al., 1980, Castaing, 1981). Thus, a 2D horizontal model is applied first to reproduce tidal propagation and to take into account lateral variations in the bathymetry. A three-dimensional model is implemented next in order to incorporate density gradients and the vertical structure of mean flow. These two codes are parts of the SAM (Multivariable Advection Simulator) software developed by the Ifremer Coastal Ecology Laboratories (Le Hir et al., 1997). They couple hydrodynamics and mass constituent conservation (advection/dispersion equation). The Navier-Stokes equations is classically simplified by the Boussinesq approximation (for the density distribution) and the hydrostatic approximation (for the vertical distribution of pressure).

3.1. Depth-averaged (2DH) model

The 2D horizontal code solves the Saint-Venant equations that result from vertical integration of the Navier-Stokes equations. The numerical scheme uses a finite difference Alternating Direction Implicit (ADI) method and accounts for tidal wetting and drying (Brenon and Le Hir, 1999). The computational grid covers the southern continental shelf, from the shelf boundary (200 m isobath) up to the usptream limit of the estuary. Because of the large domain covered, the grid is Cartesian but irregular, so as to obtain fine resolution in the estuary (especially in regions where bathymetric gradients are important), as well as to retain a reasonable number of computational points (Sottolichio, 1999).

Along the seaward boundary a real tide is imposed, which is calculated as the sum of 21 harmonic components deduced from long-term records at the shelf (Le Cann, 1990). At the upstream limit, river flow is introduced.

3.2. Three-dimensional (3D) model

The complete Reynolds-averaged Navier-Stokes equations are solved explicitly, with the sea surface level provided by the 2DH model. Thus, the two codes are coupled, especially by friction terms which are computed by the 3D model (see Cugier, 1999 for details). As for the turbulent closure, the mixing length theory is applied to calculate the vertical eddy diffusivity and viscosity. Turbulence damping effects are accounted for by empirical functions dependent on the local gradient Richardson number. An advection-dispersion equation is solved in order to compute the fate of dissolved or particulate matter and the salinity field (Sottolichio, 1999).

In the results presented here, the original computational grid (used with the 2DH model) has been slightly modified. Modifications were done in the eastern part of the grid, where a lower resolution was adopted for 3D calculations. Thus, a lesser number of computational points allowed rapid calculations, with no significant changes in the resulting circulation and sedimentary patterns.

3.3. Sedimentary behavior

Important processes affecting suspended sediment transport are considered by the model, but they have been simplified, as the present goal is to focus on the effect of the circulation pattern on suspended sediment distribution. In particular, the sediment and water column are treated separately, and water-bottom exchanges are expressed following classical empirical formulas. The parameterization of the sedimentary behavior is fairly arbitrary, but the values are in good agreement with the range of values used by other authors in the Gironde estuary (Du Penhoat and Salomon, 1979, Li et al., 1995).

The deposition flux is obtained according Krone (1962). Here, the critical shear stress for deposition is constant and equal to 1 N/m^2. Because little is known about floc sizes in the Gironde estuary, the settling velocity is chosen to be equal to 1 mm/s, which is representative of flocculated particles (van Leussen, 1994; Brenon and Le Hir, 1999).

The erosion flux is based on the Ariathurai-Partheniades equation (Ariathurai et al., 1977), according to which the erosion rate is proportional to the excess shear stress. The critical shear stress for erosion is a function of sediment concentration. As consolidation is not included, an arbitrary value of concentration is allocated for the deposited mud (which is considered to be instantaneously consolidated). The critical shear stress then becomes constant and equal to 0.8 N/m^2 and the deposited mud is instantaneously erodible. The erosion constant is fixed at 0.01 kg/m^2/s, which allows entrainment of eroded matter into the water column.

4. RESULTS

4.1. Two-dimensional horizontal simulation

Suspended sediment dynamics. In the model, tide is the main forcing. It propagates from the adjacent continental shelf, which has been included in the computational grid. The model has been shown to well reproduce tidal asymmetry and phase, and a reasonable flow field (Sottolichio, 1999). Simulations were obtained with a total sediment budget of $5{\times}10^6$ tons. The runs started with the total sediment mass deposited in the navigational channel. After two semi-diurnal tidal cycles, the initial sediment was completely eroded and dispersed in the water column, and a turbidity maximum was formed. Straightforward validation of computed SSC by measurements are difficult as modeled SSC is depth-averaged, and available measurements generally correspond to surface or bottom values. However, comparisons between instantaneous measurements obtained from a helicopter (Romaña, personal communication) and computations are presented for two typical flow regimes. In both cases, model results correspond to the SSC distribution after 300 hours of simulation. For a moderate river flow (1,000 m^3/s), the simulated turbidity maximum is correctly positioned downstream from the islands region (Figures 2a and 2b).

For a low river flow (300 m^3/s on July 1981) two SSC maxima are distinguished: a secondary maximum in the middle part of the estuary, and the main turbidity maximum at the end, which had moved in response to changing river flow (Figure 3a). The model satisfactorily reproduced these features (Figure 3b).

Sediment balance. The above mentioned patterns correspond to a situation observed after about one spring/neap tidal cycle. Longer simulations have been

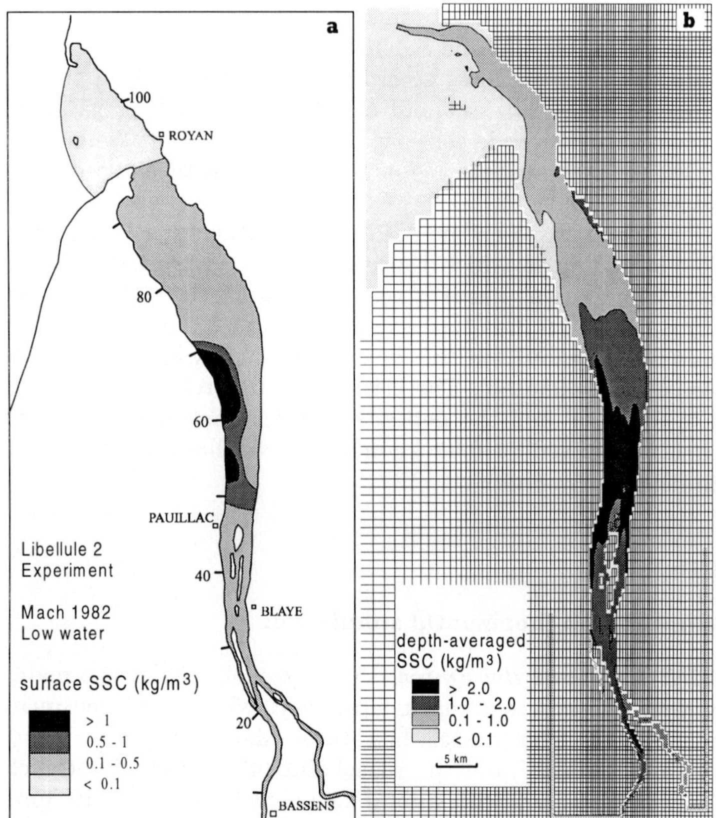

Figure 2. Comparison between observed (a) and 2DH computed (b) turbidity maximum at low tide under mean river flow conditions (1,000 m^3/s).

carried out in order to evaluate the fate of simulated turbid structures over longer periods of time. They have been obtained under a constant river flow of 1,000 m^3/s (annual mean value), and a propagating tide. Figure 4 illustrates the time-evolution of the total sediment mass in the estuary during the first 50 days of simulation (1,200 hours). These lead to a turbidity maximum in the lower reaches as expected from observations. The time-evolution of total suspended solids and the total deposited mud illustrate important water-bottom exchanges correlated with the fortnightly tidal cycle. However, a large decrease of estuarine total sediment mass is seen, even though the total mass in the computational domain remained constant. At the same time, deposited mass is dominant overall, but not in the estuary. Actually, two-thirds of the initial mass has left the estuary and deposited on the inner continental shelf. This feature is conceptually realistic, as a thick mud patch deposit area exists westward of the Gironde

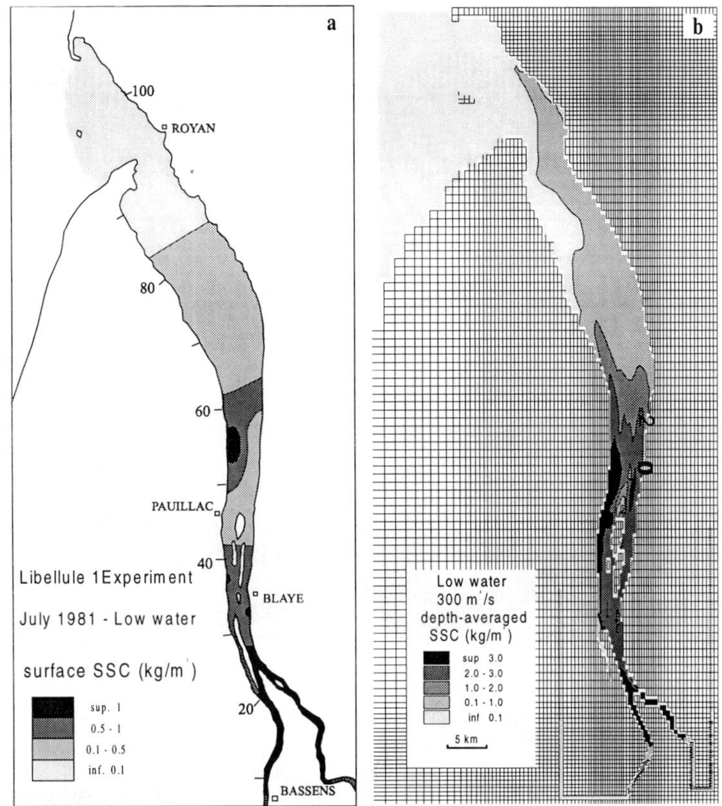

Figure 3. Comparison between observed (a) and 2DH computed (b) turbidity maximum at low tide under low river flow conditions (300 m³/s).

mouth (Castaing, 1981), which is nourished by 50% of the escaped sediment from the Gironde. But the model clearly overestimates the exported mass, which cannot be compensated by riverine input (1 million tons per year, Castaing, 1981).

After 50 days of simulation, the total estuarine mass is still evolving, and the suspension is still dispersing seaward, suggesting that the model had not reached a steady state. Figure 5 illustrates the tide-averaged SSC distribution at two consecutive spring tides. It shows a high-concentration zone, which corresponds to the turbidity maximum at its mean position around the islands region, but with lower SSC values during the second spring tide. Despite the evolution of the total sediment in the computational domain, this can be considered as an equilibrium position of the turbidity maximum, which will remain as long as the river flow remains constant. When a steady state solution is reached, SSC values in the estuary become very low, and the turbidity maximum almost disappears.

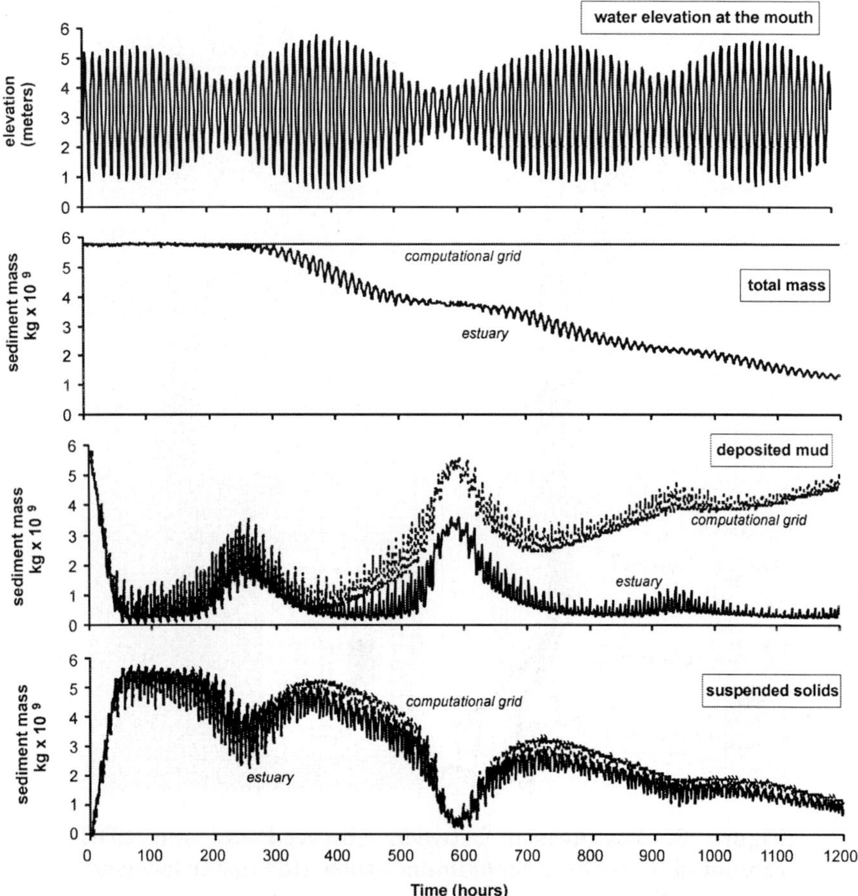

Figure 4. Sediment balance in the Gironde estuary (simulated with the 2H model for a constant river flow of 1,000 m^3/s) over the entire computational grid and estuary-only portion of the grid.

These results tend to confirm the dynamic origin of the turbidity maximum (exclusively due to tidal asymmetry and pumping effects), but point to the need for a mechanism to retain sediment in the estuary. Moreover, one can wonder whether the excessive seaward dispersion is due to inherent mixing within the 2DH model, or if it is due to the absence of density gradient effects. Thus, one needs a full 3D simulation.

4.2. Three-dimensional simulation

Suspended sediment dynamics. Simulations and sediment behavior in the 3D model are the same as those in the 2DH model, which enables a better

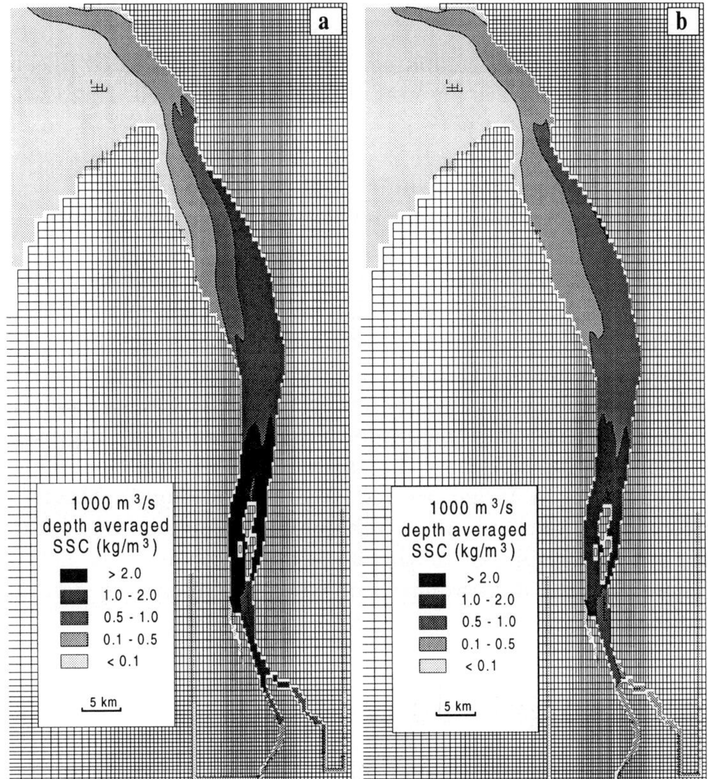

Figure 5. Tide-averaged turbidity maximum location calculated by a 2DH model for two successive spring tides: a) after 400 hours of simulation; b) after 750 hours.

comparison. As expected, the simulated turbidity maximum is correctly positioned for different river flow conditions in the 3D model. However, some differences can be seen in the horizontal distribution of SSC, when compared with 2DH results under mean river flow (Figure 2). In particular, stronger SSC gradients are present in the 3D computations (Figure 4a), which points to some dispersion in the depth-averaged simulation. For instance, SSC maxima are constrained near the left bank along the axis of the navigational channel, which is a more realistic feature, as illustrated by observations (Figure 2a).

Salinity effects on suspended sediment distribution. While the Gironde estuary is considered to be well-mixed, salinity gradients do occur (Allen et al., 1980) both laterally and vertically. In addition, salinity intrusion shifts seasonally in response to changing river flow. These features were simulated with the 3D model, and good simulations of the location of salinity intrusion and salinity gradients were obtained (Sottolichio, 1999). In order to evaluate the effect of salinity-induced vertical stratification on the turbidity maximum, two types of simulations have been compared. The first consist of fully 3D simulations, in which salinity gradients are accounted for and computed. In the second case, water density is a function of SSC only, and is considered to be entirely independent of the salinity, which is still transported. In this way, salinity-induced gravitational effects are excluded, but not those related to high SSC. The sedimentary behavior and boundary conditions are identical in both simulations, under a constant river flow of 1,000 m^3/s.

Differences between with- and without-salinity effect are remarkable. When salinity (Figure 7a) effects are present, the lower limit of the turbidity maximum is sharp, and a strong longitudinal SSC gradient is seen (Figure 7b). When salinity effects are neglected (Figure 7c), the turbidity maximum occupies the same position but there is a large excursion of suspended matter seaward and the lower limit of the turbidity maximum becomes imprecise. Figures 6a and 6b illustrate the same feature, but for the surface SSC distribution. The absence of salinity effects induces a larger escape of SSC, although turbidity maximum contours can still be identified. Moreover, lateral dispersion and induced lateral SSC gradients remain similar to the simulation inclusive of salinity effects, which confirms that 2DH model dispersion can not explain the significant loss of estuarine sediment.

5. DISCUSSION AND CONCLUSION

Allen et al. (1980) suggested that tidal pumping is likely to be the main mechanism responsible for turbidity maximum formation in the Gironde estuary. Sediment patterns are accurately reproduced by the two-dimensional horizontal model, in which only tidal propagation and river flow are considered, and vertical stratification is neglected. This provides a clear evidence of the tidal origin of the turbidity maximum. Further proof is provided by the simulation under low river flow. Two high-turbidity zones were observed in July 1981 and are reproduced by the 2DH model. The lower peak probably corresponds to the steady *erosion maximum* zone, in which the tidal wave reaches its maximum power dissipation at the bottom (Allen et al., 1980). The upper peak corresponds to the turbidity maximum in its summer location. It results from pumping mechanisms, which cause net landward transport of suspended sediment. This transport is effective up a to a convergence zone which moves in response to changing river flow.

The simple formulation of several sedimentary processes as flocculation, consolidation, erosion and deposition lead to only a partial validation of the model. For instance, SSC values obtained by the 3D model seem to be overestimated in relation to observations. Further sediment related calibration is thus necessary. However, despite this crude parameterization, the turbidity maximum geometry and behavior are well simulated. This suggests the importance of bathymetry, which controls local hydrodynamics. Since values of

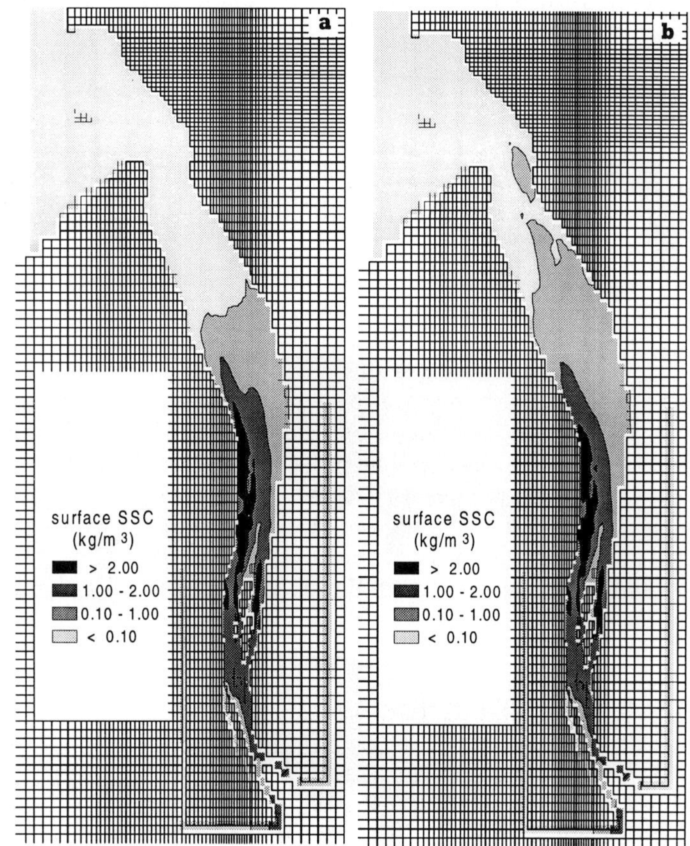

Figure 6. 3D simulated SSC distribution under mean river flow (to be compared to Figure 2): a) fully 3D; b) without salinity effect.

the critical shear stress for erosion and deposition are constant in the models, realistic turbidity maxima locations depend only on tidal currents, and especially on their spatial variability.

Despite good simulations of the location of the turbidity maximum, the depth-averaged model permits too much sediment to escape the estuary. Three-dimensional calculations without salinity-induced circulation better reproduced the observed lateral gradients of SSC (less dispersion), but were not accurate enough to reproduce turbidity maximum mass within the estuary.

Salinity-induced density effects probably do not contribute to the turbidity maximum formation, but they seem to be essential to moderate the escape of suspended matter in the lower parts of the estuary, and to maintain a stable mass of the turbidity maximum. For instance, when the salt effect is suppressed,

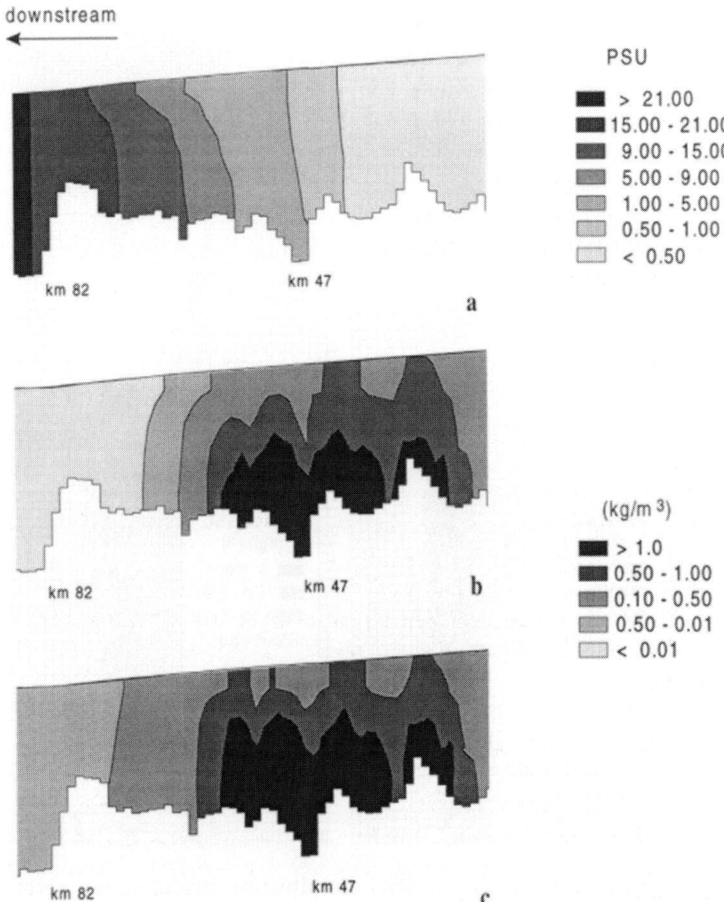

Figure 7. 3D simulation of the turbidity maximum at ebb tide, under moderate river flow. a) salinity; b) SSC with salinity effects; c) SSC without salinity effects.

SSC values are ten times greater at km 82 (Figure 7c). It is thus suggested that gravitational circulation impacts stability to the tidal turbidity maximum in the Gironde estuary. However, this observation remains hypothetical and needs further verification, in particular by means of sediment balances simulated with the 3D model. These sediment balances, coupled with suitable sedimentary parameterization, would provide information on the importance of seaward fluxes over the long time-scale.

6. AKNOWLEDGMENT

This work was carried out as a part of the URM 13 Project, linking IFREMER, CNRS and the University of Bordeaux I. The authors are grateful to Dr. Philippe Cugier and Dr. Cyril Mallet fort their help and support at several stages of the development and verification of the SAM models, and to L.A. Romaña, who conducted the Libellule experiments. The authors also wish to thank the Bordeaux Harbour Authority (PAB) and the French Navy Oceanographic Service (SHOM) for kindly providing bathymetric data of the estuary. This is DGO-EPOC contribution no. 1312.

REFERENCES

Allen, G. P., 1972. Etude des processus sédimentaires dans l'estuaire de la Gironde. *Ph.D. Thesis*, University of Bordeaux I, 314p.

Allen, G. P., Sauzay, G., Castaing, P., and Jouanneau, J. M., 1977. Sediment transport processes in the Gironde estuary. In: *Estuarine Processes*, M. Wiley ed., Academic Press, New York, 63-81.

Allen, G. P., Salomon, J. C., Bassoullet, P., Du Penhoat, Y., and Degranpré, C., 1980. Effects of tides on mixing and suspended sediment transport in macritidal estuaries. *Sedimentary Geology*, 26, 69-90.

Ariathurai, R., MacArthur, R. C., and Krone, R. B., 1977. Mathematical model of estuarial sediment transport. *Technical Report D-77-12*, U.S. Army Engineer Waterways Experiment Station, Vicksburg, MS, 79p + appendices.

Brenon, I., 1997. Modélisation de la dynamique des sédiments fins dans l'estuaire de la Seine. *Ph.D. Thesis*, University of Bretagne Occidentale, 207p.

Brenon, I., and Le Hir, P., 1999. Modelling the turbidity maximum in the Seine estuary (France): Identification of formation processes. *Estuarine Coastal and Shelf Science*, 49(4), 525-544.

Castaing, P., 1981. Le transfert à l'océan des suspensions estuariennes. Cas de la Gironde. *Ph.D. Thesis*, University of Bordeaux I, 535p.

Castaing, P., and Allen, G. P., 1981. Mechanisms of seaward escape of suspended sediment from the Gironde: a macrotidal estuary in France. *Marine Geology*, 40, 101-118.

Cugier, P., 1999. Modélisation du devenir à moyen terme dans l'eau et le sédiment des éléments majeurs (N, P, Si) rejetés par la Seine en Baie de Seine. *Ph.D. Thesis*, University of Caen, France, 249p.

Du Penhoat, Y., and Salomon, J. C., 1979. Simulation numérique du bouchon vaseux en estuaire. Application à la Gironde. *Oceanologica Acta*, 2 (3), 253-260.

Dyer, K. R., 1986. *Coastal and Estuarine Sediments Dynamics*. John Wiley & Sons, Chichester, UK, 342p.

Geyer, W. R., 1993. The importance of suppression of turbulence by stratification on the estuarine turbidity maximum. *Estuaries*, 16(1), 113-125.

Jouanneau, J. M., and Latouche, C., 1981. *The Gironde Estuary. Contributions in Sedimentology*. Springer-Verlag, Stuttgart, Germany, 115p.

Krone, R. B., 1962. Flume studies of the transport in estuarine shoaling processes. *Final Report*, Hydraulics Engineering Laboratory and Sanitary Engineering Research Laboratory, University of California, Berkeley, CA, 118p.

Le Cann, B., 1990. Barotropic tidal dynamics of the Bay of Biscay shelf: observations, numerical modelling and physical interpretation. *Continental Shelf Research*, 10(8), 723-758.

Le Hir, P., Thouvenin, B., Silva Jacinto, R., Brenon, I., Cugier, P., and Bassoullet, P., 1997. Modeling sedimentary processes in the lower estuary. *Final Report*, 'Seine Aval Project' (in French), 116-157.

Li, Z. H., N'Guyen, J. C. Brun-Cottan and Martin, J. M., 1995. Numerical simulation of the turbidity maximum transport in the Gironde estuary. *Oceanologica Acta*, 17(5), 479-500.

Nichols, M. N., and Poor, G., 1967. Sediment transport in a coastal plain estuary. *Journal of the Waterways and Harbors Division*, American Society of Civil Engineers, New York, 23(4), 83-95.

Postma, H., 1967. Sediment transport and sedimentation in the estuarine environment. In: *Estuaries*, G. Lauff ed., American Association for the Advancement of Science, Washington DC, 158-179.

Schubel, J. R., 1971. Sedimentation in the upper reaches of the Chesapeake Bay estuary. Fourteenth International Association of Hydraulic Engineer and Research, Paris, France, 4, 28-10.

Sottolichio, A., 1999. Modelling dynamics of highly-concentrated structures in the Gironde estuary. *Ph.D. Thesis* (in French), University of Bordeaux I, 184p.

Uncles, R. J., Stephens, J. A., and Barton, M. L., 1984. Observations of fine-sediment concentrations and transport in the turbidity maximum region of an estuary. In: *Coastal and Estuarine Studies. Dynamics and Exchanges in Estuaries and the Coastal Zone*, D. Prandle ed., American Geophysical Union, Washington, DC, 255-276.

van Leussen, W., 1994. Estuarine macroflocs and their role on fine-grained transport. *Ph.D. Thesis*, University of Utrecht, The Netherlands, 488p.

Coastal and Estuarine Fine Sediment Processes
W.H. McAnally and A.J. Mehta (Editors)
2001 Elsevier Science B.V.

The role of fecal pellets in sediment settling at an intertidal mudflat, the Danish Wadden Sea

T. J. Andersen

Institute of Geography, University of Copenhagen, Øster Voldgade 10, DK 1350 Copenhagen K, Denmark

Settling tube samples were collected at a very fine-grained intertidal mud flat in the Danish Wadden Sea and the suspended sediment examined under a microscope. The temporal variability in *in situ* settling velocity is discussed mainly in relation to the biological activity at the site. *In situ* settling of suspended sediment is shown to be strongly dependent on suspended sediment concentration as shown before in numerous publications; however, a strong temporal variability has been found as well. It is argued that this temporal variation is mainly due to differences in the aggregation of the bed material at the site, as no differences in hydro-dynamics, primary grain size distributions or organic content which could explain the variation have been observed.

During the study period of 1997, 50-90% of the bed surface material was aggregated into fecal pellets. These pellets were mainly produced by the mud snail *Hydrobia ulvae*, which lives in densities of up to 40,000 indiv./m^2. During the study period of 1998, only ~20% of the surface material consisted of fecal pellets due to lower numbers of *H. ulvae*, and the bed was covered with an algae mat in summer. Microscopic examination of the suspended material showed it to be temporarily consisting of high amounts (> 20%) of fecal pellets during periods of erosion. These pellets have settling velocities that are orders of magnitude higher than their constituent particles, and the presence of pellets enhances the overall settling velocity of the suspended material. The settling velocity of individual, unbroken fecal pellets from *H. ulvae* and the worm *Heteromastus filiformis* were examined in the laboratory, and found to be 3-8 mm/s and 12-33 mm/s, respectively.

The lack of grain size discrimination in the feeding process of *H. ulvae* has the consequence that fast and slowly settling aggregates generally have similar primary grains. Thus, hydraulic sorting with respect to primary grains does not seem to occur.

1. INTRODUCTION

Flocculation/aggregation of fine-grained suspended sediment is an important process in determining the settling velocity of the suspended material, and hence the transport and deposition potential of the material. Here, the term flocculation will be used to describe the aggregation process which takes place in the water column, whereas the term aggregation will be applied to the aggregation of material prior to its suspension. A number of processes responsible for flocculation have been identified. Some of them enhance inter-particle contacts by collisions induced by fluid and particle forces, and also enhance inter-particle bonds by organic coatings or dissolved salt. Comprehensive reviews of the processes responsible for the flocculation are given by van Leussen (1988) and Eisma (1993).

Flocculation occurs in the water column. Bioaggregated material (e.g., fecal pellets, pseudo-feces) on the other hand is material which is mostly aggregated prior to suspension of the material. In some important ways, this material behaves differently from the flocculated material otherwise found in suspension. The most important difference in the sedimentological context is that the material is eroded as aggregates and, thus, aggregation is very different from flocculation in the water column. Flocculation in the water column is governed mainly by the physical and chemical properties of the suspended material and the turbulent intensity (floc growth at low intensities and break up at high intensities). Bioaggregation, on the other hand, is mainly controlled by the feeding processes of the different species responsible for aggregation, and by the nature (mainly shape, size and density) of the excreted material (e.g., fecal pellets and pseudo feces).

As aggregation of the suspended material partly occurs prior to the actual suspension of the material, the settling velocity of the suspended material is partly the result of the shape, size, density and cohesion of the aggregates, as well as the critical bottom shear stress for erosion for different aggregates and particles. Fecal pellets show high settling velocities compared to the constituent primary grains, and thereby enable deposition of fine-grained material in environments where such material is normally not found (Haven and Morales-Alamo, 1972).

Only a few investigations concerned with the settling rates of fecal pellets in suspension have been reported. The works by Haven and Morales-Alamo (1968, 1972) are some of the earliest ones and, recently, suspension of fecal pellets in tidal channels in the Lister Dyb area has been described by Edelvang and Austen (1997). To the author's knowledge no description of the *in situ* settling velocity distributions of fecal pellets has been presented. Since the suspension of fecal pellets is an important contributor to the total suspended load at the study site, it was decided to examine this phenomenon in greater detail. It will be shown that the suspension of fecal pellets enhances the overall settling velocity of the suspended sediment, and that temporal differences in the aggregation of bed material may possibly explain the observed temporal variation in the settling velocity.

2. STUDY SITE

The study site is a fine-grained, intertidal mud flat at Kongsmark near the Danish barrier island Rømø, which forms part of the European Wadden Sea area. The mud flat is situated in the northern part of the Lister Dyb tidal area (Figure 1)

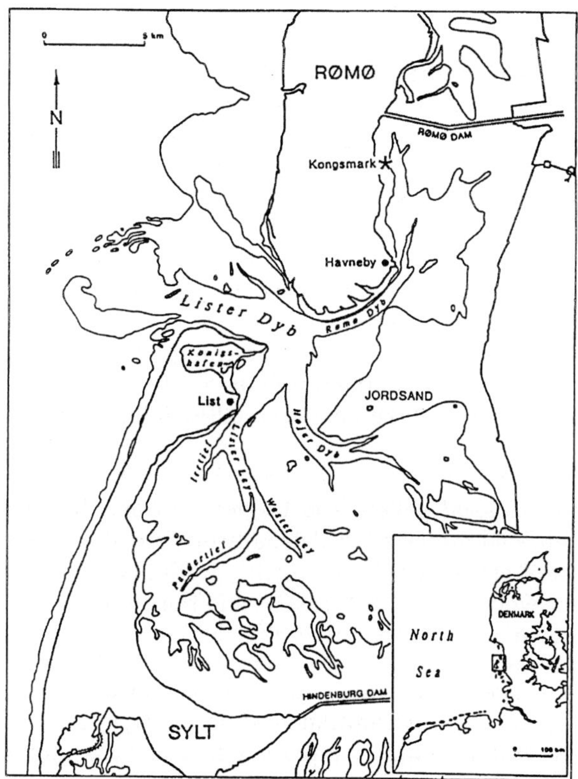

Figure 1. Map of the Lister Dyb tidal area. The study area is situation in the northwestern corner of the area marked ⋆.

where tidal range is approximately 1.8 m, and the maximum current velocity at the mud flat is on the order of 30 cm/s. Samples were taken at a station 550 m from the salt marsh edge of the tidal flat. The station is submerged 6-8 hours during each tide under normal weather conditions, and the typical maximum water depth is 1.2 m. Because of the low current velocities, the bed material is fine-grained with a mean size of 10 μm and a sand content below 5% (Andersen, 1999). The mineralogy of the material has been determined by Edelvang (1996) to be 50% montmorillonite, 20% kaolinite, 30% illite and 1% quartz. The mud flat at the site was formed after the construction of a road dam in 1948, and the average net deposition rate is presently 3.4 $kg/m^2/yr$ (Pejrup et al., 1997). The Lister Dyb area has been the subject of numerous sedimentological investigations in recent years. Results of these investigations are presented in Pejrup et al. (1997), Bayerl et al. (1997), Edelvang (1996), Edelvang and Austen (1997) and others.

3. METHODS

In situ settling velocity distributions of the suspended sediment were examined by Owen tube samples taken from rubber dinghies. The station was visited in the spring, summer and autumn of 1997, and again in spring and summer of 1998. Owen tube model Braystoke SK110 is a 1 m long Perspex tube which is lowered into water in a horizontal position. Upon release by a messenger the tube is closed, trapping a volume of 2 l. The tube is then taken to the surface and placed in a tripod in a vertical position and a stopwatch is started. At approximately logarithmic time-intervals, 200 ml sub-samples are withdrawn from the bottom of the tube and later filtered for determination of suspended sediment. Settling velocities are determined by use of conventional bottom withdrawal technique (e.g., Interagency Committee, 1953). For further description of the technique see Pejrup and Edelvang (1996). The settling tube samples were taken between 0.3 to 0.5 m above the bed during this investigation.

Before filtering, a number of the settling tube samples were wet-sieved at 63 μm in order to separate fecal pellets and fragments from the remaining suspended material. The retained material was visually examined under microscope and the fecal pellet content estimated. Subsequently, retained material and the material passing the sieve were filtered separately, and the amount of fecal pellet material determined. Knowing the content of fecal pellet material in each sub-sample from the settling tube samples, the settling velocity distribution for this material was determined using the bottom withdrawal technique. Visual estimation of the content of fecal pellets was possible as unbroken fecal pellets were clearly different from other aggregates. Broken pellets could be mistaken for other aggregates however, hence the presented fecal pellets contents are minimum values.

Selected filters were given an ultrasonic treatment for two minutes in a 0.002M $Na_4P_2O_7$-solution after weighing. The grain size distributions of the dispersed suspended material were then found using a model Malvern Mastersizer/E laser particle sizer.

Samples of the bed material were analyzed for the content of fecal pellets by a method very similar to the one presented above. A sub-sample was gently wet-sieved at 63 μm and the retained material examined under microscope in order to estimate the fecal pellet content in this material (generally in the order of 90%). The retained material was subsequently given an ultrasonic treatment for 2 min and wet-sieved at 63 μm again in order to retain sand and shells (which was always less than 5%). In this way the sample was split into three parts: fecal pellets, sand and shells, and, lastly, single grains, fecal pellets or other aggregates smaller than 63 μm.

Settling velocities of individual, unbroken fecal pellets of *H. ulvae* and *H. filiformis* were determined in the laboratory by timing of the settling through 36 cm of distilled water at 20 °C.

4. RESULTS AND DISCUSSION

A correlation between the suspended sediment concentration (SSC) and median settling velocity (W_{50}) of the suspended material has been found for various sites in the Danish Wadden Sea area. The correlation can typically be described by a power function of the form

$$W_{50} \sim aC^b \tag{1}$$

where W_{50} = median settling velocity (mm/s), C = suspended sediment concentration (mg/l) and a and b are empirical constants.

In Figure 2 a compilation of some of the regressions found for the Danish Wadden Sea area is presented. It is obvious that large differences between the settling velocities have been observed. There seems to be a marked difference between the settling velocities found in the tidal channels and those for tidal flats, with generally higher settling velocities in the channels. The reason is probably that the suspended sediment in the channels includes sand, and the sand tend to have much higher settling velocities than the fine-grained material. That sand indeed is suspended is confirmed by the results of Edelvang and Larsen (1995). The line based on the results of Pejrup (1988) is calculated on the basis of settling tube samples taken at tidal flats with a sandy or mixed mud bed. The sand content is believed to be the reason why this line shows higher settling velocities than the two lines calculated in the present investigation. These last two lines are based on samples taken at the same fine-grained site, but during the year 1997 and 1998, respectively. As the samples were obtained at the same site it was expected that the samples would be represented well by a single function, but this was not the case.

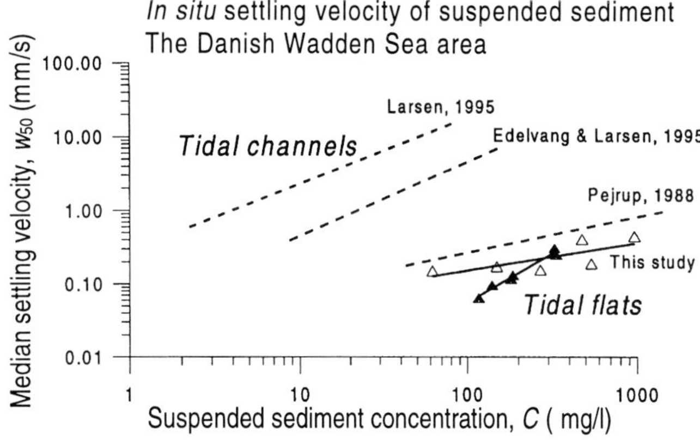

Figure 2. Correlations between suspended sediment concentration and median settling velocity from the Danish Wadden Sea area.

Median settling velocities plotted against suspended sediment concentration for the present investigation are presented in Figure 3 together with the best fitting power functions. It is seen that the correlation between SSC and settling velocity changed from 1997 to 1998. In 1997 the settling index (the exponent of the power function) was rather high (1.34), but settling velocities at low SSC's were generally lower than in 1998, which showed a lower settling index. This shift in the correlation between SSC and settling velocity is not caused by a change in primary grain size composition, as both years showed almost constant and similar grain size composition of the suspended material. Mean grain size was 10 μm and sorting about 2 ϕ, and the organic content of the suspended sediment was 10 to 15%. The mineralogical composition of the suspended material in the two years was not examined, but based on the similarities in grain size composition and organic content for the two years it can be assumed that the suspended sediment very likely was of similar composition. Therefore, the differences in settling velocities must have been due to other factors. Hydrodynamic factors including wave climate and current velocity varied from one field campaign to another. Samples were obtained during the two years during periods with varying but roughly similar wind directions and wind speeds, and as the current velocity pattern varied little between tidal periods it is concluded that differences in the hydrodynamic forcing cannot account for the differences in settling velocities. Comparisons of water temperature (10 to 20 °C) and salinity (25 to 29%) show that these parameters can also not explain the observed differences. Some of the measured parameters are listed in Table 1.

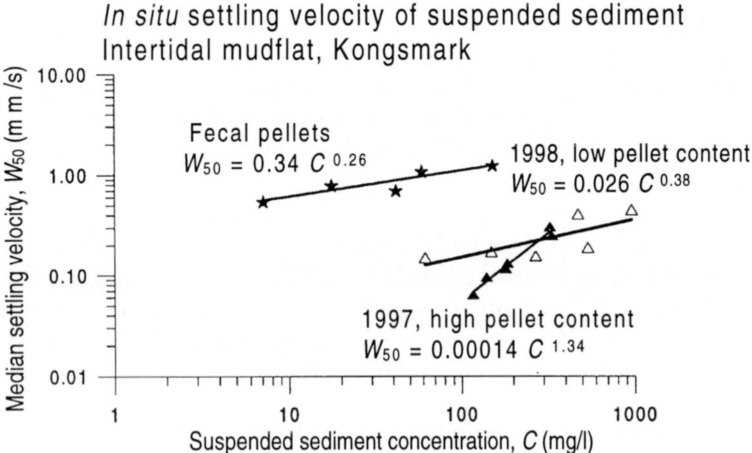

Figure 3. Median settling velocity as a function of SSC. Note the difference between 1997 and 1998, and the high settling velocities of fecal pellets.

Table 1
Main physical parameters for the investigation periods

Date	Wind direction	Wind speed (m/s)	Water temp. (°C)	SSC (mg/l)	D_{50} (μm)	W_{50} (mm/s)
05 May 1997	SW	8 - 12	10	334	19	0.25
05 May 1997	SW	8 - 12	10	140	12	0.097
05 May 1997	SW	8 - 12	10	116	8	0.064
14 May 1997	SW	6 - 8	12	326	21	0.31
28 May 1997	NW	8 - 10	16	181	12	0.12
28 Aug. 1997	SE	6 - 8	18	185	12	0.13
13 April 1998	S	8 - 13	10	958	28	0.45
13 April 1998	S	8 - 13	10	538	18	0.19
02 May 1998	N	4 - 8	10	269	15	0.16
20 July 1998	SE	2 - 3	17	62	13	0.15
11 Aug. 1998	E	8 - 10	20	474	21	0.41
11 Aug. 1998	E	8 - 10	20	149	14	0.17

Microscopic examination of the suspended bed sediment showed that the material was generally highly aggregated. The bed material was especially aggregated, consisting of up to approximately 90% fecal pellets. Based on a comparison with the pellets described by Austen (1995), it was found that they were generally produced by the small mud snail *H. ulvae,* and sometimes in small amounts by the worm *H. filiformis*. This was not surprising as both species have been observed at the site (Austen et al., 1999), and since *H. ulvae* has been observed in very high densities (> 40.000 indiv./m^2), it was expected that the fecal pellets from this species would be dominant. Both *H. ulvae* and *H. filiformis* are deposit feeders which excrete the ingested material as fecal pellets. From investigations at other tidal flats in the same tidal embayment, the lengths of the pellets from these two species have been reported to be 0.1 to 0.3 mm and 0.4 to 0.7 mm and densities 1.03 to 1.14 g/cm^3 and 1.22 to 1.36 g/cm^3 respectively (Austen, 1995). These fecal pellets have settling velocities that are orders of magnitude higher than their constituent particles, and are therefore important in determining the settling velocity of the suspended material, when they form part of the suspended load (e.g., Haven and Morales-Alamo, 1968, 1972; Edelvang and Austen, 1997).

The settling velocity of individual, unbroken fecal pellets from *H. ulvae* and *H. filiformis* were examined in the laboratory, and found to be 3 to 8 mm/s and 12 to 33 mm/s, respectively. These ranges (converted to equivalent fall-diameters) are plotted in Figure 4 together with the settling velocity distributions from a settling tube sample representative of high SSC, and thus high settling velocities. The unbroken pellets show equivalent settling diameters in the coarse range of the *in situ* settling distributions, and pellets from *H. filiformis* generally show settling velocities higher than the observed *in situ* velocities. Pellets from *H. filiformis* were not found in suspension, because their high settling velocities prevent re-entrainment.

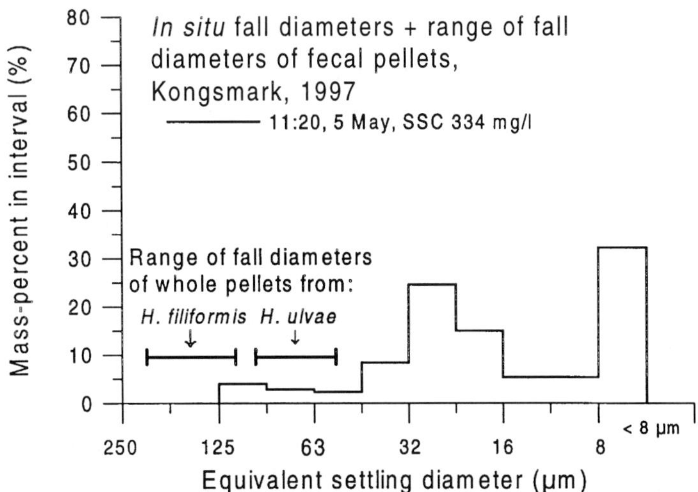

Figure 4. An example of a histogram of equivalent settling diameters of suspended sediment 1998. The ranges of settling diameters of unbroken fecal pellets are also shown. Note log-scale of the horizontal axis.

In Figure 5 the equivalent settling diameters of the pelletal matter and the total suspended sample are shown. This separation is based on wet-sieving at 63 μm of the suspended material in each of the sub-samples of the settling tube analysis. Median settling velocities of the suspended pelletal material are in the range 0.6 to 1.2 mm/s. It is clear that the settling diameters of the pelletal matter (both broken and unbroken pellets) are coarser than the equivalent settling diameters of the total material. Thus, the suspension of fecal pellets causes the overall settling velocity of the suspended material to increase. Unfortunately, wet-sieving of the suspended material was not undertaken for samples from 1997, but as the bed consisted of ~90% fecal pellet matter, it can be expected that the pellet percentages of the suspended sediment were considerably higher than during 1998.

From Figure 5 it is also seen that the flocculated/aggregated sediment distribution is much coarser than the distribution of the dispersed suspended material. It is clear that flocculation in the water column is also very important in determining the velocity distribution, as the suspended pellets can only explain part of the increase in settling velocity compared to the dispersed material. The pellets do have a strong effect however, as they generally possess settling velocities higher than the remaining suspended load. Additionally, the sample presented in Figure 5 is from a period with relatively few pellets in the bed material, and therefore the importance of fecal pellets could potentially be higher.

There was a shift in the degree of aggregation of the bed material at the site from 1997 to 1998 because of a strong decrease in the abundance of *H. ulvae*. The density

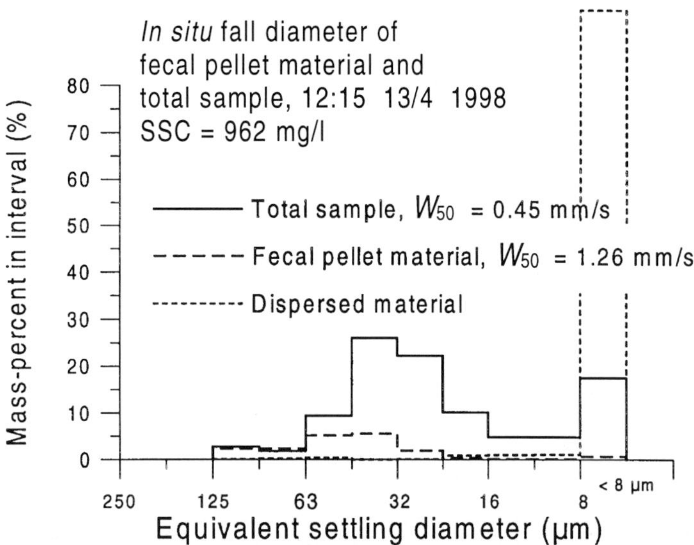

Figure 5. Histograms of settling velocities of suspended sediment, dispersed suspended material and fecal pellets. Fecal pellets formed a substantial part of the fast settling aggregates. Note the log-scale of the horizontal axis.

of *H. ulvae* was not determined in 1998, but based on measurements of the fecal pellet content of the bed material it is believed to be less than 5% of the density in 1997. The result was that a smaller percentage of the bed material was aggregated into fecal pellets. In 1997, the concentration of fecal pellets in the bed material was determined to be 87% by volume (Austen et al., 1999; based on visual examination after a method suggested by Terry and Chilinger, 1955), whereas the concentration in 1998 was 1 - 20% by weight. The method of determining the fecal pellet content was not the same in the two years, but analyses of samples from 1999 have shown contents of up to 70% using the method presented here. It is concluded that although the methods are not directly comparable, the results determined by the two methods probably do not differ by more than approximately 20%.

During 1998 the fecal pellet content of the suspended material was 1 to 20% (whole and broken pellets not passing a 63 μm sieve), with generally higher fecal pellet content at higher sediment concentrations. Also, a tendency for suspension of larger pellets or fragments of pellets at higher sediment concentrations (higher bed shear stresses) was observed. In Figure 3 the correlation between the concentration of fecal pellets (mg/l) and the median settling velocity of pellets is shown. There is a weak increase in settling velocity with increased fecal pellet concentration, but the settling index is low (0.26). The increase was caused by the suspension of larger pellets at higher bed shear stresses, but as the pellets are found only within a certain size range (length 0.1 to 0.3 mm), the increase was limited and the median settling velocity was never higher than the settling velocity of the largest pellets (5 to 8 mm/s).

From Figure 3 it is seen that the median settling velocities of the fecal pellets are up to an order of magnitude higher than the median settling velocity of the total suspended load. Unfortunately, the fecal pellet content of the suspended material was not determined in 1997, but as the bed consisted of at least five times as many fecal pellets as in 1998 (see below), it is strongly believed that the fecal pellet content was considerably larger than in 1998. The low settling velocity at low SSC (low bed shear stress) during 1997 was probably due to the erosion of the matrix of single particles and small aggregates in which the pellets were lying on the bed. At higher bed shear stresses, broken and unbroken fecal pellets with high settling velocities were suspended, thus causing a rapid increase in the settling velocity and thereby a high settling index.

Probably as a consequence of the much lower density of *H. ulvae* in 1998, a dense algae mat was formed at the site during part of 1998. This is believed to be the result of the lower density of *H. ulvae,* as the snail feeds on the diatoms forming algae mats (Austen et al., 1999). The Extra Cellular Substances (EPS) produced by the diatoms enhance interparticle cohesion and create a network effect that "glues" the surface material together (e.g., review by Paterson, 1997). This surface has a high critical bed shear stress, but once broken, large quantities of easily eroded material below the surface can get resuspended. How this material is eroded (as single particles or as aggregates) has not been examined, but it is very likely that erosion as aggregates dominates. Aggregates other than fecal pellets or fragments of pellets were only found in limited quantities in the suspended sediment, and it is therefore inferred that the bed covered with algae mats erodes as very small aggregates. These aggregates were generally smaller than the aggregates retained on the 63 μm sieve, and were therefore not examined under microscope. However, the rather high settling velocities observed at low sediment concentrations during 1998 are believed to be a result of the erosion of material bound by diatoms. This evidence is circumstantial however, and further investigations are required to examine this inference.

The discussion above has shown that the high settling index of the suspended material during 1997 was an effect of the suspension of fecal pellets, whereas the low settling index but higher settling velocities at low SSC during 1998 probably was the result of the erosion and suspension of material from a bed covered by an algae mat.

The small pellets of *H. ulvae* were easily eroded (Austen et al., 1999), but the high settling velocities of the pellets also caused the pellets to deposit easily, and therefore fecal pellets were only found in suspension during periods of high bottom shear stresses (due to onshore winds at the site). The net transport direction of the fecal pellets could not be determined, but it is possible that it was directed offshore simply because of a diffusive process. This inference still remains to be examined however. Offshore-directed transport of fecal pellets has been shown by Minoura and Osaka (1992) in Mutsu Bay, Japan, and also in the tidal channels in the Lister Dyb area (Austen, 1995; Edelvang and Austen, 1997).

4.1 Primary grain composition of floc fractions

The suspended material from selected sub-samples from the settling tube was analyzed for primary grain size in order to determine if size selection occurred during aggregation/flocculation. A representative example is given in Figure 6. Small variations are obviously present, but the overall picture is that there were no significant differences between the fast settling aggregates of the first sub-samples compared to the slow settling material of the last sub-samples. Hydraulic sorting of the flocs and aggregates took place (e.g., larger aggregates were suspended only at high bed shear stresses), but as the different aggregates contained material of similar grain size, no sorting occurred. Furthermore, the organic content of the suspended sediment from selected sub-samples from the settling tube samples was determined. Although some scatter was observed, no significant differences between the organic content of fast and slowly settling aggregates was seen. Thus, the results from the grain size as well as the organic content analyses indicate that no selection with respect to these two parameters occurred during aggregation/flocculation.

The results of the examination of the particle size selection of *H. ulvae* presented by Fenchel et al. (1975) show that at fine-grained deposits no particle size selection can be expected. This inference was examined in the present study by grain size analysis of the fecal pellet material (retained on the 63 μm-sieve) and the remaining material. This was done for both bed sediment and suspended sediment. An example

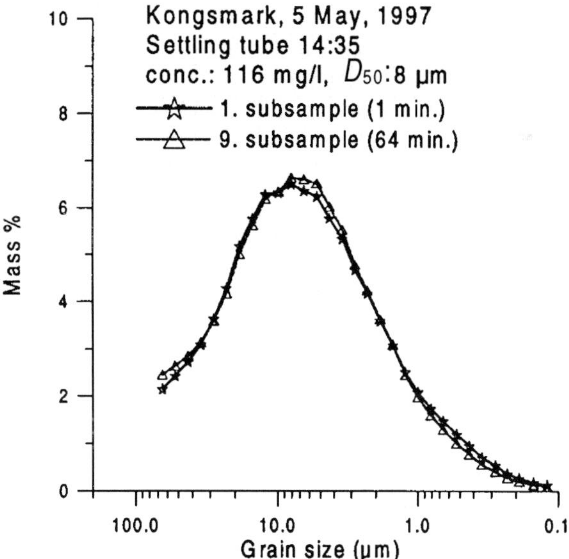

Figure 6. An example of grain size distribution of fast and slowly settling aggregates (large and small flocs/aggregates, respectively).

for the bed material is given in Figure 7. It is obvious that the material incorporated into fecal pellets was of a similar composition as the rest of the bed sediment with respect to grain size. The analysis of the suspended material gave a similar result. Based on this result and the work of Fenchel et al. (1975), it is therefore believed that at least part of the reason for the lack of difference in primary grain size composition of different aggregate size fractions was the lack of particle size selection during one of the important aggregation processes--the feeding activity of *H. ulvae*. It is also possible that the lack of grain size selection of *H. ulvae* feeding on fine-grained deposits is common for many deposit feeders, and that this could partly explain the uniform grain size distributions of fine-grained deposits in the European Wadden Sea area, e.g., as mentioned by Eisma (1993).

The lack of grain size variation between fast and slowly settling aggregates has not been reported before to the author's knowledge, and only very few studies of this kind have been conducted. Puls et al. (1988) reported from the Elbe and Weser estuaries that fast settling flocs generally consisted of coarser material than the smaller and more slowly settling flocs. The reason for this result is probably that the studies were carried out in river mouths with higher current velocities and coarser bed sediments than at the tidal flat site of the present study. With coarser bed material and coarser suspended sediment, flocculation will probably not be as important in determining the *in situ* settling velocity of the suspended material. Additionally, it is possible that bio-aggregation of the sediment is not as important in tidal channels as on tidal flats. Kranck (1975) reported results which in essence

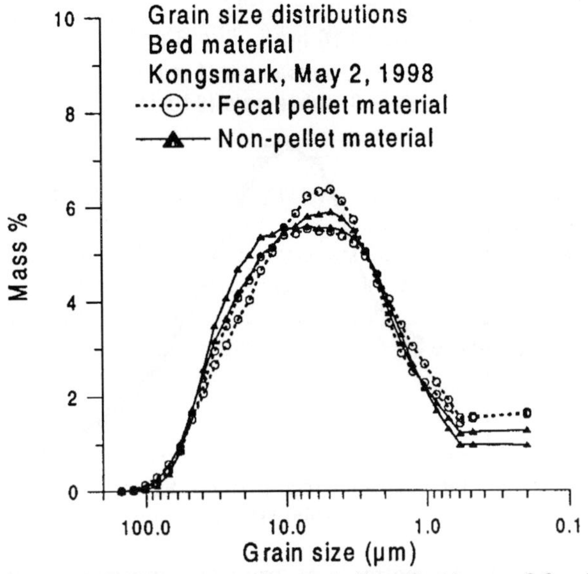

Figure 7. Primary grain size distributions of fecal pellet material compared to non-pellet material.

are similar to the ones described by Puls et al. (1988). This is probably caused by similarities in the environment of the study sites, but perhaps also by floc break-up during the sampling process (e.g., Eisma, 1993).

5. CONCLUSIONS

It is demonstrated that fecal pellets can form a substantial part of the suspended load at intertidal mud flats. Fecal pellet contents of up to 25% of the total suspended load were measured, and as this was during a period when there were only few fecal pellets in the bed material, it is believed that the fecal pellet content of the suspended load can potentially become considerably higher. It is also demonstrated that the fecal pellets had equivalent settling diameters in the coarse range of the size distribution, and therefore enhanced the overall settling velocity of the suspended material.

Differences between the correlations of SSC with W_{50} for the years 1997 and 1998 can be explained by differences in the aggregation of the bed material at the study site. 1997 was characterized by high densities of the deposit feeder *Hydrobia ulvae,* and hence high content of fecal pellets in the bed material. In contrast, 1998 was characterized by lower numbers of *H. ulvae* and fecal pellets.

The lack of grain size selection during the feeding activity of *H. ulvae* at fine-grained deposits partly explains the lack of grain size sorting in the suspended sediment with respect to primary grains. The major part of the sediment seems to be eroded as aggregates, and as no grain size selection takes place during the aggregation processes, no hydraulic sorting with respect to primary grain size occurs.

6. ACKNOWLEDGMENT

The fieldwork was made possible through financial support by the MAST III project INTRMUD (MAS3-CT95-0022). The author would like to thank Heini Larsen and Kirsten Simonsen of the Skalling Laboratory, Esbjerg, for logistical support, and Annette Lützen Møller and Ole Aarup Mikkelsen for help with the field work. Special thanks go to Ingrid Austen and Karen Edelvang for fruitful discussions which initiated the current work.

REFERENCES

Andersen, T. J., 1999. Suspended sediment transport and sediment reworking at an intertidal mud flat, the Danish Wadden Sea. Meddelelser fra Skallinglaboratoriet XXXVII, 1-72.

Austen, I., 1995. Die Bedeutung der Fecal Pellets mariner invertebraten für den Sedimenthaushalt im Sylt-Rømø-Watt. FTZ Bericht 7, Christian-Albrechts-Universität zu Kiel, 1-107.

Austen, I., Andersen, T. J., and Edelvang, K., 1999. The influence of benthic diatoms and invertebrates on the erodibility of an intertidal mud flat, Danish Wadden Sea. *Estuarine, Coastal and Shelf Science*. 49, 99-111.

Bayerl, K., Austen, I., Köster, R., Pejrup, M., and Witte, G., 1997. Dynamic der Sedimente im Lister Tidebecken. In: *Ökosystem Wattenmeer. Austausch-, Transport- und Stoffumwandlungsprozesse*, Ch. Gätje and K. Reise eds., Springer, Berlin, 127-159.

Edelvang, K., 1996. The significance of particle aggregation in an estuarine environment. Case studies from the Lister Dyb tidal area. Geographica Hafniensia, A5, Institute of Geography, University of Copenhagen, 105p.

Edelvang, K., and Larsen, M., 1995. The flocculation of fine-grained sediment in Ho Bugt, The Danish Wadden Sea. Folia Geographica Danica, Tom XXII, Reitzel, Copenhagen, 1-120.

Edelvang, K., and Austen, I., 1997. The temporal variation of flocs and fecal pellets in a tidal channel. *Estuarine, Coastal and Shelf Science*, 44, 361-367.

Eisma, D., 1993. *Suspended Matter in the Aquatic Environment*. Springer-Verlag, Berlin, 315p.

Fenchel, T., Kofoed, L. H., and Lappalainen, A., 1975. Particle size-selection of two deposit feeders: the amphipod *Corophium volutator* and the prosobranch *Hydrobia Ulvae*, *Marine Biology*, 30, 119-128.

Haven, D. S., and Morales-Alamo, R., 1968. Occurrence and transport of faecal pellets in suspension in a tidal estuary. *Sedimentary Geology*, 2, 141-151.

Haven, D. S., and Morales-Alamo, R., 1972. Biodeposition as a factor in sedimentation of fine suspended solids in estuaries. *The Geological Society of America*, Memoir 133, 121-130.

Interagency Committee, 1953. Accuracy of sediment size analyses made by the bottom withdrawal tube method. *Report No. 10*, Subcommittee on Sedimentation of the Federal Interagency River Basin Committee on Water Resources, St. Anthony Falls Hydraulic Laboratory, Minneapolis, MN.

Kranck, K., 1975. Sediment deposition from flocculated suspensions. *Sedimentology*, 22, 111-123.

Larsen, M., 1995. Transport og aflejring af kohæsive sedimenter i Lister Dybs tidevandsområde. *Ph.D. Thesis*, Institute of Geography, University of Copenhagen, 1-162.

Minoura, K., and Osaka, Y., 1992. Sediments and sedimentary processes in Mutsu Bay, Japan: Pelletization as the most important mode in depositing argillaceous sediments. *Marine Geology*, 103, 487-502.

Paterson, D. M., 1997. Biological mediation of sediment erodibility: ecology and physical dynamics. In: *Cohesive Sediments*, N. Burt, R. Parker, and J. Watts eds., John Wiley, Chichester, UK, 215-229.

Pejrup, M., 1988. Flocculated suspended sediment in a micro-tidal environment. *Sedimentary Geology*, 57, 249-256.

Pejrup, M., and Edelvang, K., 1996. Measurements of in situ settling velocities in the Elbe estuary. *Journal of Sea Research*, 36(1-2), 109-113.

Pejrup, M., Larsen, M., and Edelvang, K., 1997. A fine-grained sediment budget for the Sylt-Rømø tidal basin. *Helgoländer Meeresuntersuchungen*, 51, 253-268.

Puls, W., Kuehl, H., and Heymann, K., 1988. Settling velocity of mud flocs: results of field measurements in the Elbe and the Weser estuary. In: *Physical Processes in Estuaries*, J. Dronkers, and W. van Leussen eds., Springer, Berlin, 356-372.

Terry, R. D., and Chilinger, G. V., 1955. Comparison charts for visual estimation of percentage composition. *Journal of Sediment Petrology*, 25, 229-234.

van Leussen, W., 1988. Aggregation of particles, settling velocity of mud flocs. A review. In: *Physical Processes in Estuaries*, J. Dronkers, and W. van Leussen eds., Springer, Berlin, 347-403.

Coastal and Estuarine Fine Sediment Processes
W.H. McAnally and A.J. Mehta (Editors)

Parameters affecting mud floc size on a seasonal time scale: The impact of a phytoplankton bloom in the Dollard estuary, The Netherlands

W.T.B. van der Lee

Institute for Marine and Atmospheric Research Utrecht, Department of Physical Geography, Utrecht University, P.O. Box 80.115, 3508 TC, Utrecht, The Netherlands

Seasonal variation of suspended mud floc sizes in the Dollard estuary in The Netherlands was investigated. Factors like salinity, water temperature and wave height did not seem to have a substantial influence on the seasonal fluctuation of floc size. Suspended sediment concentration and turbulence mainly affected the tidal variation of floc size. Biological processes had a large impact on the seasonal floc size variation, and caused larger flocs. Especially at the end of a plankton bloom, polymer exudates of phytoplankton and bacteria may increase the stickiness of the flocs, resulting in larger floc sizes. Such a process was observed in the Dollard estuary in May 1996. Measurements during this plankton bloom suggest that biologically assisted flocculation does not depend on the absolute quantity of the biopolymer exudate, but on the gluing quality of the polymers. The quantity as well as the quality of the biopolymers may depend on the species composition, concentration and physiological state of phytoplankton, and bacteria.

1. INTRODUCTION

Aggregation of mud particles into flocs is essential for the sedimentation of mud. Mud flocs exhibit a larger settling velocity than the individual particles, which increases sedimentation or even creates the possibility of sedimentation. Floc size and settling velocity are determined by the balance between flocculation and breakup. Breakup is primarily caused by turbulent shear (Dyer, 1989; Eisma, 1986; Hunt, 1986; Luettich et al., 1993; van Leussen, 1994). Flocculation depends on the floc collision frequency and collision efficiency. The collision frequency is mainly determined by the suspended sediment concentration (SSC) and the turbulence intensity. The collision efficiency mainly depends on mud and water properties like salinity, organic coatings and the presence of biopolymers that may 'glue' flocs together. Dyer (1989) used the SSC and turbulence intensity in his conceptual model describing the floc diameter. Both factors show a strong variation on the time scale of a semi-diurnal tide. Over this time scale Dyer's conceptual model can be reasonably applied since the factors that determine the collision efficiency remain relatively constant over this time scale. Over a seasonal time scale however, the collision efficiency may also play an important role.

On a seasonal time scale mud and water properties may show large variations which can affect the collision efficiency of flocs and therefore floc size and settling velocity. The collision efficiency of flocs is determined by the physico-chemical properties of the floc surface. An electrical charge causes repulsion of flocs thereby preventing floc collisions. However, when flocs are close enough to each other, van der Waals or hydrogen bonds will bind the flocs together. This binding can be intensified by polymeric substances on the floc surface, or by separate organic flocs that act as a glue.

Biological production of polymeric substances mainly varies over a larger time scale than the 12.42 hours of a semi-diurnal tide. The impact of biological processes on floc size and settling velocity can therefore only be studied over a larger time scale. Floc sizes were measured every season from October 1995 until December 1996. This paper will show the significant biological impact on floc size, especially at the end of a plankton bloom.

2. BINDING PROCESSES

Changes of floc size and settling velocity on a seasonal time scale are largely determined by changes in the collision efficiency. The binding processes that take place between particles determine this efficiency. The two main binding processes, salt flocculation and biological cohesion of mud flocs, and their possible impact on floc size will be dealt with.

2.1. Salt flocculation

Most fine-grained suspended particles in natural waters exhibit a negative surface charge (Neihof and Loeb, 1972; Loder and Liss, 1985). This negative charge is balanced by a cloud of positively charged ions around it. Together they form the so-called electrical double-layer. When two particles approach each other and the positive ion clouds overlap, electrostatic repulsion between the particles prevents particle collision. Increasing salinity, however, decreases the thickness of the double-layer, due to the lesser gradient between the positive ion concentration around the particle and the ion concentration in the surrounding water. Particles may then approach so close to each other that the van der Waals forces are stronger than the electrostatic repulsion, which leads to flocculation.

In the past this classical concept of salt flocculation was thought to cause flocculation of suspended matter in estuaries at the interface between fresh and salt water. In that case floc sizes and settling velocities should increase at the transition from fresh to saline water. This classical concept of salt flocculation at the limnic-brackish boundary is however disputed by several authors.

Puls and Kuehl (1986) found decreasing settling velocities at the transition from fresh to saline water in the Elbe estuary. They stated that physico-chemical flocculation (salt flocculation) can not become effective due to organic coatings around the particles.

Eisma et al. (1991a) measured particle sizes in five northwest European estuaries *in situ* and by coulter counter and pipette analyses. Sizes measured by coulter counter or pipette became finer at the saltwater contact. *In situ* floc sizes measured with the camera however did not become finer but remained the same. There was however no relation between changes in salinity and the *in situ* floc size. Therefore Eisma et al. (1991a) concluded that particle sizes measured by the coulter counter and pipette were the fragments of flocs broken during sampling and size analysis. The decrease in these sizes with increasing salinity did indicate an increase in the fragility of the suspended flocs, possibly due to mobilization of organic matter in the saline water which weakened the floc structure.

Neihof and Loeb (1972); Hunter and Liss (1979); Hunter and Liss (1982); Loder and Liss (1985) all showed that particles in natural waters have a highly uniform negative surface charge due to organic and/or metallic coatings. The surface charge is therefore independent of the type of suspended material (e.g., clay minerals). Hence, organic coatings can have a considerable effect on the degree of salt flocculation in natural waters. Hunter and Liss (1982) also showed that the surface charge decreases with salinity.

Gibbs (1983) measured the effect of organic coatings on the collision efficiency factor of flocs. Figure 1 shows the collision efficiency factor plotted against salinity. It demonstrates that increasing salinity causes increasing collision efficiency due to the decrease in double-layer thickness and perhaps also due to the decrease in surface charge. In addition, the collision efficiency of coated particles is much less than that of uncoated particles. Coatings therefore slow down the flocculation process. Since particles in natural waters are coated, this may explain why several authors do not observe flocculation at the limnic-brackish boundary.

Eisma et al. (1991b) state that flocculation does not take place primarily by van der Waals forces but by organic matter, whereby long-chain organic molecules, such as natural polymers, play an important role. These molecules extend through the water mantle around the particle and stick together when particles approach each other. Salinity has little effect on this process.

van Leussen (1994) however measured the largest macroflocs at the seaward end of the estuary. This corresponded with an increase in flocculation ability (i.e., collision efficiency) with increasing salinity. Therefore, he stated that salinity remains an important parameter in the flocculation of estuarine sediments, but not in the upper estuary with a few ‰ salinity. There organic coatings may retard the flocculation rate. At higher salinities however the double-layer

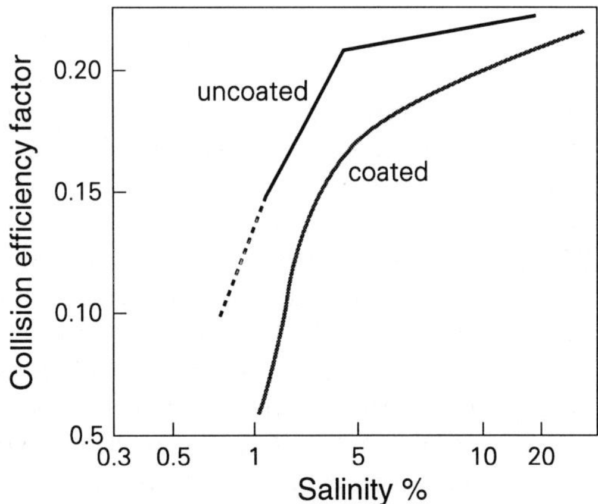

Figure 1. Effect of organic coatings on flocculation of natural sediment particles in blade type reactor tests (Gibbs, 1983). Reprinted with permission from *Environmental Science & Technology* 1983, 17, 237-240. Copyright 1983 American Chemical Society.

thickness may be sufficiently reduced for bridging between particles by polymers to become effective.

It appears that the impact of salinity may differ along the estuary depending on the salinity gradient. Salinity changes may also occur at a specific location due to the tidal excursion, and due to seasonal changes in freshwater input. It remains uncertain whether salinity always has an effect on the flocculation of the suspended sediments.

2.2. Biological cohesion of flocs

Wilkinson et al. (1997) studied the role of natural organic matter (NOM) for the coagulation of colloidal material in freshwater. They found that the chemical nature and structure of NOM is very important for flocculation. NOM mainly produced in soils tended to form coatings around particles whose negative charge diminishes coagulation. NOM of aquatic origin appeared to accelerate the rate of coagulation by a polymer bridging process. Therefore organic matter content alone is not likely to be a good parameter for defining biological cohesion of flocs. A more specific parameter like the biopolymer concentration in the suspended mud seems better, especially for defining the amount of sticky polymers present that may glue flocs together.

As stated above, natural polymeric substances can increase the collision efficiency which causes increased aggregation and therefore larger flocs (Decho, 1990; Eisma, 1986; Ten Brinke, 1993). These polymeric substances are mainly polysaccharides produced by plankton, benthos and bacteria (Harris and

Mitchell, 1973; Avnimelech et al., 1982). Polymer bridging is thought to be an important biological flocculation mechanism. Polymer chains that are on one side attached to a particle and with the other side extending into the surrounding water can become attached to another particle (La Mer and Healy, 1963 in: Harris and Mitchell, 1973).

Another form of biological flocculation could be by transparent exopolymeric particles (TEP). TEP are transparent particles that exist as discrete particles, rather than as cell surface coatings (Alldredge et al., 1993). Discrete particles (TEP) are probably generated abiotically from the extracellular polysaccharides produced by diatoms (Passow et al. 1994). Passow and Alldredge (1995) observed that TEP are an important factor for the flocculation of diatoms during a bloom in a mesocosm. It seems that phytoplankton aggregation and the occurrence of organic flocs may be highly influenced by TEP. It depends however on phytoplankton species composition, abundance and physiological state whether TEP is produced and how high the stickiness of these TEP is. In the natural environment such as the Dollard estuary, TEP are not only capable of flocculating diatoms but also capturing other organic or inorganic particles. In this way TEP could very well be responsible for the formation of large mud flocs.

It seems that biological cohesion of flocs is strongly related to phytoplankton (e.g., diatoms) and possibly also to bacteria. The growth of these organisms obviously depends on seasonal variations, which explains why changes in biological cohesion of flocs mainly take place on a seasonal time scale.

3. FIELD EXPERIMENTS AND METHODOLOGY

3.1. Study area

The study was carried out in the Dollard, which is near the head of the Ems-Dollard estuary (Figure 2). This meso-tidal estuary may be classified as well mixed, even at high discharges (van Leussen 1994). The tidal prism is about $115x10^6$ m^3 and the tidal excursion is about 12 km (De Jonge 1992). Eighty-five percent of the Dollard consists of muddy tidal flats along the main tidal channel, Het Groote Gat. The surface of the tidal flats is smooth and uniform. The largest regular bed forms are wave ripples that occur when and where the surface material contains a sufficient amount of sand. Small dewatering channels are present on the lower flats along the main channel.

3.2. Measurement methods

Floc sizes were measured every season from October 1995 until December 1996. These measurements took place in Het Groote Gat throughout a tidal cycle with an underwater video camera, the Video In Situ (VIS) (van Leussen and Cornelisse, 1993). The VIS was deployed from a research vessel floating with the current, in a quasi-Lagrangian approach. During a VIS operation the measurement section of the VIS, where floc sizes were measured, was about 2.9 m below the water surface. The VIS floated some 30 m away from the research

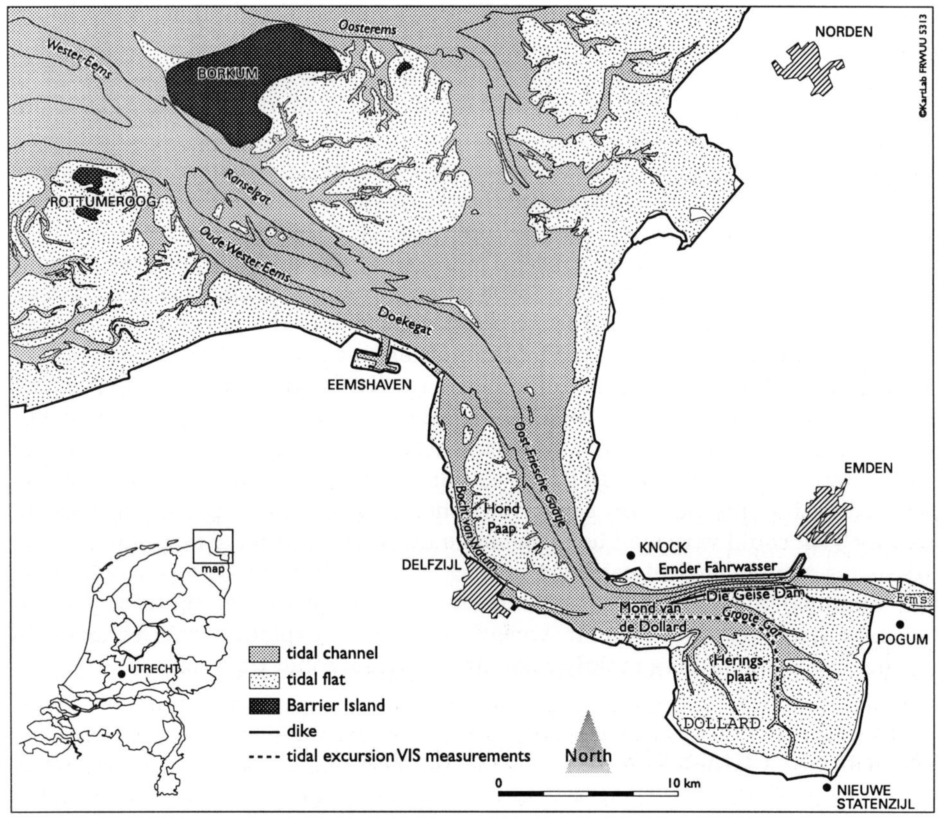

Figure 2. The Ems-Dollard estuary.

vessel, and was connected to the vessel by power and video cables. Floatation with the current minimized the effect of turbulence around the VIS, which otherwise could have caused floc break-up. It also minimized the effect of advection on the measurement. The variation in suspended sediment concentration or floc size was therefore due mainly to local erosion/sedimentation or flocculation/break-up, rather than advection.

Besides floc size measurements with the VIS, velocity and concentration profiles were also measured. Velocities were measured with an OTT-type propeller current meter and concentrations were measured by a pump sampling system followed by filtration through glass fiber filters. The velocity profile measurements required the ship to anchor. After each anchoring the position of the water column in which the preceding VIS measurement took place was calculated from the current velocity data. The ship sailed to this position before starting the next VIS measurement.

Other collected parameters were water level, wave height, water temperature, salinity, chlorophyll-a, and the amount of biopolymers present in the suspended sediment. Water level was determined with a tide gauge of Rijkswaterstaat at a fixed location in the tidal channel north of the Heringsplaat. Wave height was estimated visually. Water temperature and salinity were measured with a temperature and an electrical conductivity sensor. Chlorophyll-a was determined from water samples by colorimetric measurements as part of the regular monitoring program of the National Institute for Coastal and Marine Management. The amount of biopolymers present in the suspended sediment could not be directly determined. As a measure of these biopolymers, the EDTA extracted carbohydrate content of the suspended sediment was used. EDTA or EthyleneDiamineTetraAcetic acid, is a ligand that captures metal ions. Therefore EDTA extracts sugars which bond to sediment by metal-ion interactions. These sugars are in a large part responsible for the gluing effect of the biopolymers. The colloidal carbohydrates attached to the sediment on the filters were extracted with EDTA. The carbohydrates were hydrolyzed with concentrated sulphuric acid, and phenol was added as an organic color developer. Carbohydrate concentration was then determined by colorimetric measurements (Dubois et al., 1956).

3.3. Measurement frequency

A VIS measurement of the velocity and concentration profiles and sample collection, followed by sailing to the preceding water column took about 1 hour and 15 minutes. These measurement series were continued throughout the tidal cycle. The tidal excursion of the VIS and research vessel is shown in Figure 2. These so-called 13-hour tidal measurements took place on 11 and 19 October 1995, 11 and 18 April 1996, 29 May 1996, 1 and 8 August 1996, 14 and 21 October 1996 and 12 December 1996.

3.4. VIS data processing

The VIS consists of a settling tube in an underwater housing (van Leussen and Cornelisse, 1993). The settling flocs in the tube are illuminated by a light sheet and filmed with two video cameras with different magnifications. The video images of the VIS were digitized and processed to obtain floc sizes. A lower detection limit of 35 pixels per floc was chosen, to avoid interpreting noise in the images as flocs. These 35 pixels represent a floc diameter of 85 μm, which means that flocs smaller than 85 μm remain undetected. The second camera took 5 times enlarged images. With this camera smaller flocs could be detected. Since this camera viewed a very limited measurement section, it did not detect enough flocs for a good statistical determination of the floc size distribution. It was however used to estimate the volume percentage of flocs that remained below the detection limit of the main camera. This showed that on average about 9% and a maximum of 15% of the suspended sediment volume remained undetected by the larger camera. Thus there was a certain overestimation of the size of flocs determined with the large camera.

4. RESULTS

The results of the first VIS measurement are shown in Figure 3. Figure 3a shows water level variation throughout the tidal cycle with the corresponding bed level referenced to Dutch ordnance datum (N.A.P.). The bed level changes with time because the measurement ship was moving with the current. During flood tide the ship floated up to the estuary head where the channel was shallower. During ebb tide, the ship floated back towards the estuary mouth where the channel was deeper. The tidal excursion along which the measurements took place is shown in Figure 2. Figure 3b shows the measured floc size distribution expressed as a volume percentage. This means that at about 12:00 h, 76% of the volume of suspended sediment consisted of flocs smaller than 300 μm and 24% larger than 300 μm. When the lines in Figure 3b go down the floc sizes actually increase, and vice versa. Figure 3b shows that the floc sizes varied quite strongly during the tidal cycle. Figure 3c shows the depth-averaged current velocity and SSC at the depth of the size measurements, 2.9 m below the water surface. The SSC strongly correlates with the floc size, which can be explained by the increased floc collision frequency due to a larger abundance of flocs at higher SSC. Another factor that could influence floc size over the tidal time scale is the turbulence intensity that causes increasing collision frequency with increasing current velocity. This may result in larger flocs, but it may also cause floc breakup when the turbulence level is too high.

In general, there are two likely causes of the large flocs observed at high SSC. They were either formed by aggregation in the water column, or the large flocs were eroded from the bed. Observations on 29 May 1996 showed increasing floc size while sediment was settling out of the water column due to a decelerating current. This suggests that at that time flocs were formed by aggregation in the water column, since it is very unlikely that large flocs were eroded from the bed and dispersed into the water column during a period of sediment settling.

The measurements of 11 October 1995 took place about two days after spring tide. Figure 4 shows the measurements on 19 October 1995, about two days after neap tide. These measurements are comparable to the spring tide measurements. There is still a large tidal variation of the floc size and a good correlation with SSC. There also are some differences however. The range of current velocity, SSC as well as the range of the floc size distribution are somewhat smaller during neap tide. The smaller range of floc size distribution during neap tide appears to be coincidental, since this effect was not observed in all measurements.

Daily averaged floc size, the daily averaged range of the floc size and SSC of all measurements are shown in Figure 5. These averages are based on floc volumes. The average floc volume was calculated and then the nominal diameter for a sphere of this volume was determined. Thus it is a volume-weighed average diameter. In this way, the calculated floc sizes are representative of the larger flocs. It is these larger flocs that make up most of the suspended volume, and therefore they are the most important ones for mud transport and sedimentation.

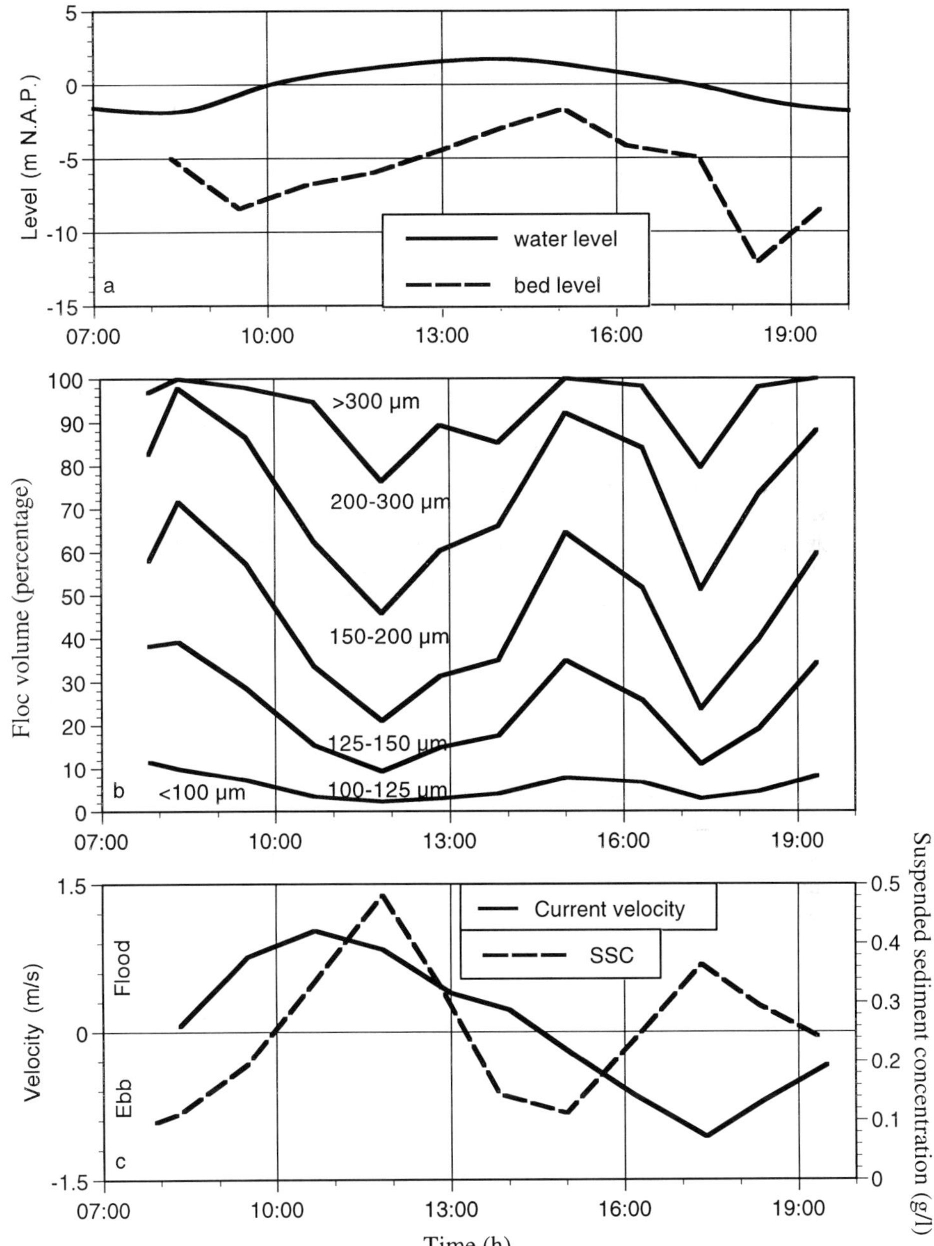

Figure 3. VIS measurements on 11 October 1995.

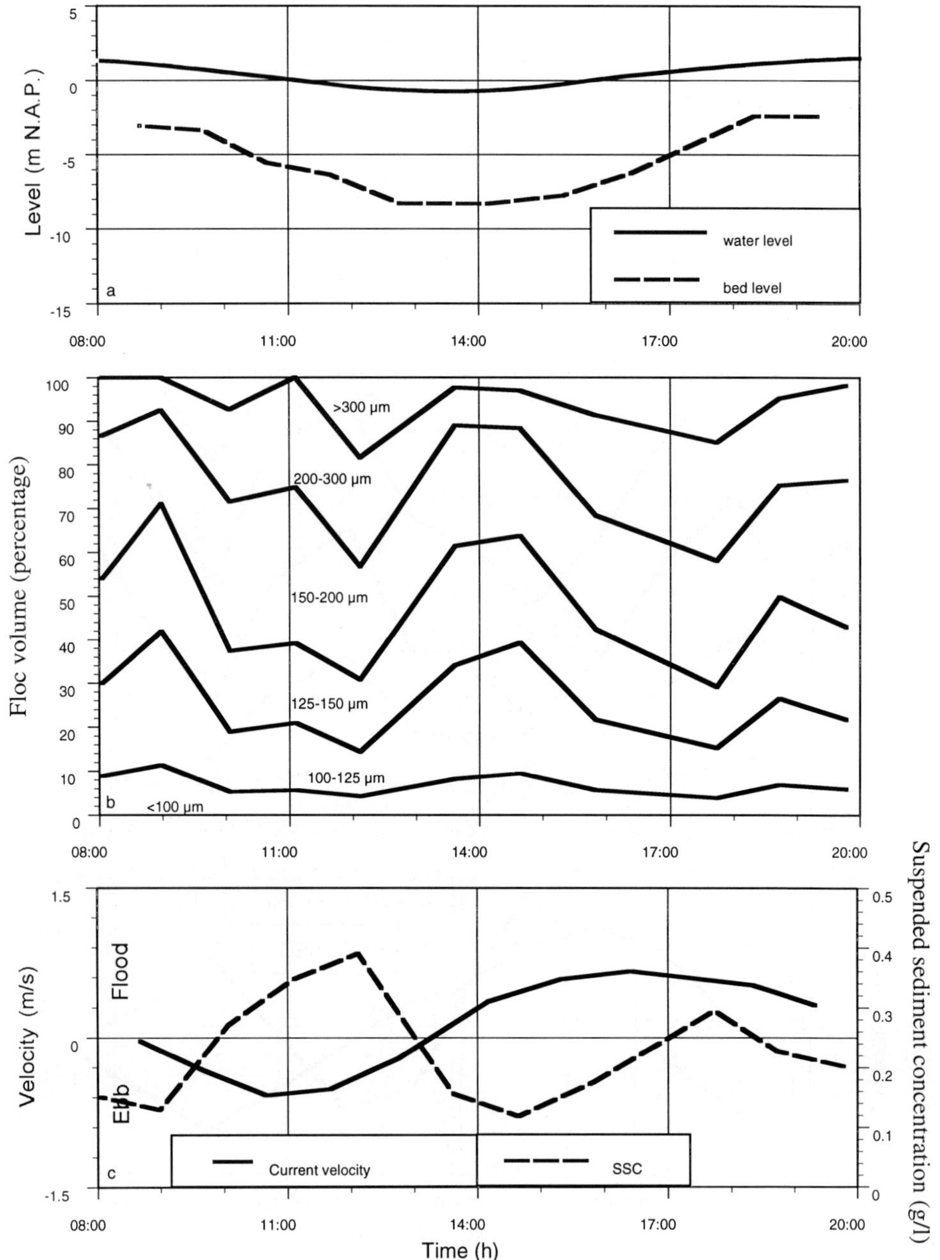

Figure 4. VIS measurements on 19 October 1995.

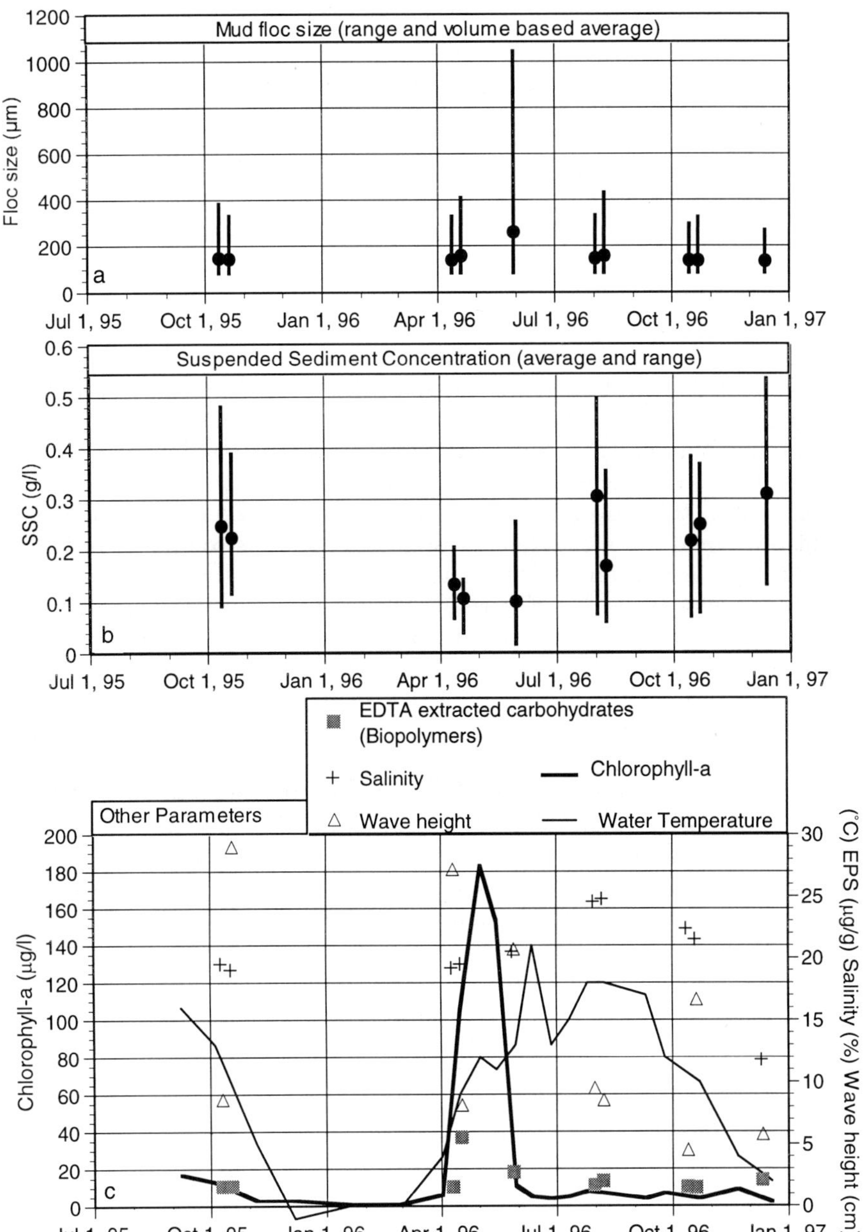

Figure 5. Seasonal variation of mud floc size, suspended sediment concentration, and related parameters. (Chlorophyll-a and water temperature data are from the National Institute for Coastal and Marine Management (RIKZ), The Netherlands).

The use of a mass-weighed average floc diameter would represent the suspended sediment mass. For mud transport and sedimentation mass fluxes rather than volume fluxes, are important. Therefore, floc sizes and settling velocities were determined from a limited number of VIS measurements. The Stokes density of flocs was calculated and then the mass-weighed average diameter was determined. A comparison between the mass-based diameter and the volume-based diameter showed that the volume-based diameter was only about 3% larger than the mass-based diameter. This difference is small enough to justify the use of volume-based floc diameter as representative of most of the suspended material.

Figure 5a shows that the average floc size was approximately the same in all measurements. However, there was one measurement on 29 May 1996, where the average floc size as well as the range of floc sizes were significantly larger than during the rest of the year. Figure 6 shows this measurement. Figure 6b shows that flocs larger than 1000 μm were abundant on this day. Figure 6c shows that there was again a good correspondence with SSC. The floc size variation within this measurement can largely be explained by the SSC variation. But when the measurement of 29 May 1996 is compared to the other measurements, the SSC on 29 May is too low to explain the observed large floc sizes. This can also be seen in Figure 5b, which shows the seasonal variation of SSC. From Figure 5a and 5b, it is clear that SSC did not determine the floc size variation over the yearly time scale as it did over the tidal time scale. There can still be an influence of the SSC, but clearly some other factors must also have determined the floc size variation through the year. There may have been a relation with a plankton bloom in April and May of 1996, as discussed later.

The measurement of 29 May 1996 also shows that flocculation in the water column at least partly determines the floc size, and that floc size is not entirely dependent on erosion of large flocs from the bed. This can be seen from the observed floc sizes during the ebb tide on 29 May 1996. At that time maximum floc sizes did not coincide with the maximum SSC. Maximum floc size occurred about two hours after maximum SSC while the current was decelerating and sediment was settling. Lower turbulence levels may have limited floc breakup at that time, which allowed the growth of large flocs. This suggests that flocs were formed by aggregation in the water column, either by settling-induced collisions, or by collisions due to the remaining turbulence. An increase in floc size due to erosion of large flocs from the bed and dispersion into the water column is very unlikely during this period of sediment settling.

5. DISCUSSION

There are several factors other than SSC that may have influenced the seasonal variation of floc size. Changes in the collision efficiency seem to have played an important role. While these changes typically depend on salinity and

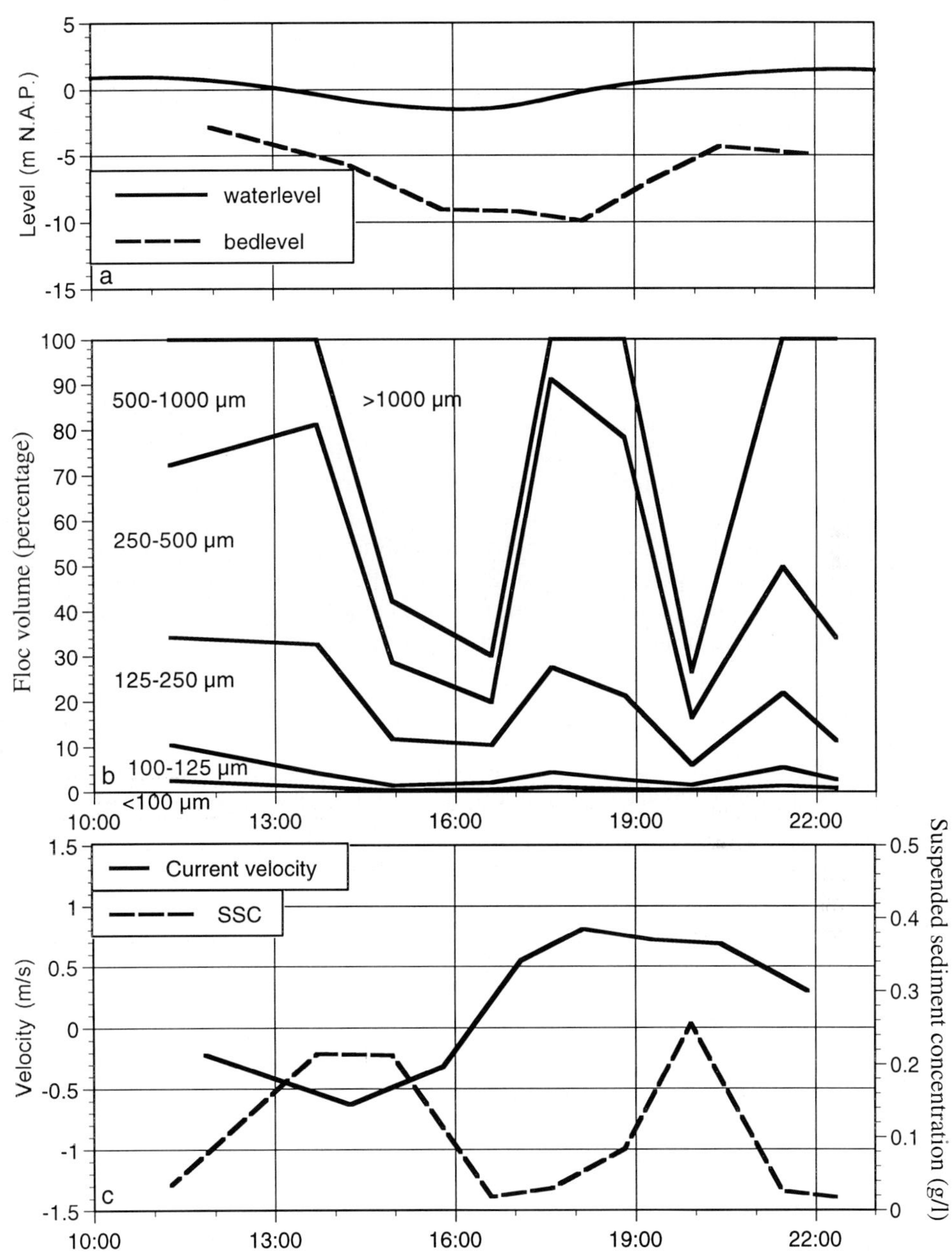

Figure 6. VIS measurements on 29 May 1996.

biological cohesion of flocs, water temperature and turbulence in the water due to wind waves may also have a role.

These factors are plotted in Figure 5c. The possible impact of biological cohesion is expressed in the chlorophyll-a content and the amount of EDTA extracted carbohydrates from the suspended sediment.

Figure 5c shows the following typical results:

- The wave height did not seem to have a large influence on floc sizes in the tidal channel. On 11 and 19 October 1995, 11 and 18 April 1996 and on 14 and 21 October 1996 the wave height varied considerably between the measurements in the same week. However, there was hardly any variation in average floc size (Figure 5a). The range of floc size was lower on 19 October 1995 and 11 April 1996 when waves were high, but the range was also lower on 14 October 1996 when the waves were low. There is no evident correlation between the range of floc sizes and wave height. There is also no clear correlation between range of floc size and neap/spring tide.
- In April and May 1996 there was a large phytoplankton bloom in the Dollard, as can be seen from the Chlorophyll-a content of water. At the end of this bloom the much larger floc sizes of 29 May 1996 were measured. Earlier in the bloom however, on 11 and 18 April, floc sizes were still in the same range as in the other measurements. Thus there could have been a relation between the large flocs and the bloom termination.
- The water temperature did not seem to have a direct influence on floc size. The large average floc size on 29 May 1996 coincided with high water temperature; however, on 1 and 8 August 1996 water temperatures were also high but no large flocs were observed. In addition, the low water temperature on 12 December 1996 is not associated with smaller flocs. The increase in water temperature in spring however may have been an incipient factor for the plankton bloom in May and June.
- Low salinity occurred on 12 December 1996 and high salinity on 1 August 1996. On these days the average SSC was quite high and comparable. The floc sizes were of the same order. This indicates that salinity was not a governing parameter for floc size. That corresponds with the findings of Eisma et al. (1991a), who found no relation between salinity and *in situ* floc size. The lower salinity in winter however is consistent with winter run-off and nutrient fluxes necessary for the onset of the spring plankton bloom.

In summary, waves, water temperature and salinity do not seem to explain the seasonal variation in floc sizes. However, there may be a relation between the plankton bloom in April and May 1996 and the large flocs at the end of this bloom on 29 May 1996.

Diatom blooms are often terminated by mass sedimentation of diatoms. Flocculation of the phytoplankton cells into large flocs (marine snow) enhances the settling velocity of diatoms (Smetacek, 1985). Diatoms are also known to form more sticky flocs during later stages of a bloom (Decho, 1990; Kiørboe et al., 1990). Decho (1990) suggests that exopolymeric substances excreted by diatoms cause this aggregation. These biopolymers may act as a gluing agent and increase floc size.

Mass flocculation of diatoms into macroflocs (marine snow) has regularly been observed by several investigators in the field (Alldredge and Silver, 1988; Kranck and Milligan, 1988; Alldredge and Gottschalk, 1989; Riebesell, 1991) and in mesocosms (Riebesell, 1989; Passow and Alldredge, 1995). The occurrence of large flocs only at the end of the phytoplankton bloom in the Dollard is therefore consistent with the observations of mass flocculation at the end of blooms reported in the literature. The large flocs observed on 29 May 1996 are possibly due to biopolymers excreted by phytoplankton or bacteria.

As a measure for biopolymers, the EDTA-extracted carbohydrate content of the suspended sediment was used. During the plankton bloom the carbohydrate content was higher than during the rest of the year, but the highest amount of biopolymers was measured on 18 April. This was not at the end but closer to about one-third of the plankton bloom duration. Large flocs occurred only at the end of the bloom. Therefore the quantitative amount of EDTA-extracted carbohydrate is probably not a good governing parameter describing the stickiness of suspended particles.

Figure 7 shows the plankton species composition in the Groote Gat channel of the Dollard. In the beginning of the bloom the diatom *Skeletonema costatum* dominated. Then the diatoms almost disappeared and dinoflagellates, mainly *Heterocapsa rotundata*, and micro-flagellates dominated. Subsequently, in May, the diatom *Thalassiosira nordenskioeldii* was most abundant.

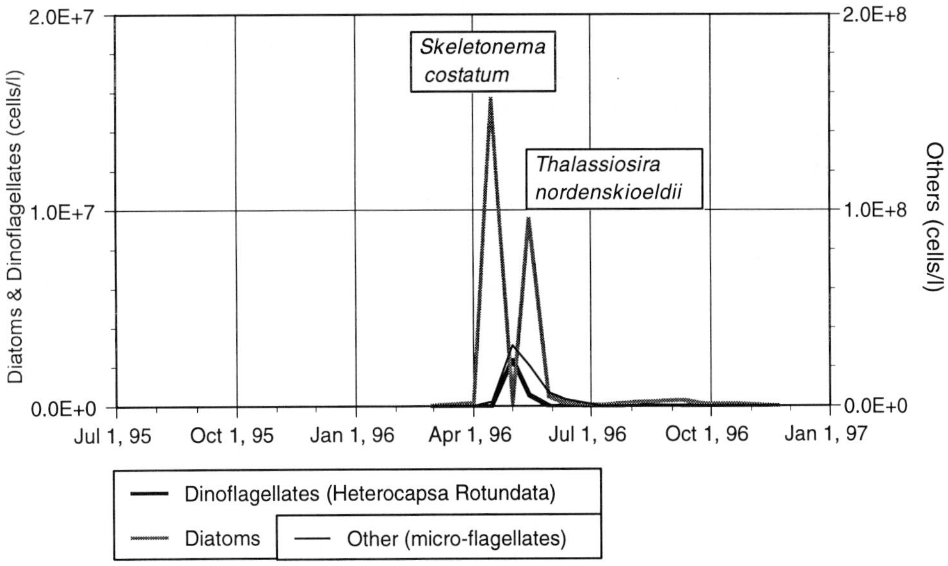

Figure 7. Observed phytoplankton in the Dollard estuary. (Source: National Institute for Coastal and Marine Management, The Netherlands.)

Kiørboe and Hansen (1993) investigated the production of particulate mucus or transparent exopolymeric particles (TEP) by phytoplankton cells and its role in aggregate formation. They found that *Skeletonema costatum* cells are sticky, and coagulation depends on cell-cell sticking and does not involve mucus. This suggests that *Skeletonema costatum* can not contribute to mud floc aggregation. Kiørboe and Hansen (1993) even found that *Skeletonema costatum* at times excretes a solute substance which depresses flocculation. This seems in accordance with the observed "normal" floc sizes on 11 and 18 April 1996 during the *Skeletonema costatum* bloom (Figure 7).

Other investigated diatoms produce TEP and coagulation dependent on TEP-cell rather than cell-cell sticking. *Thalassiosira nordenskioeldii* was not investigated by Kiørboe and Hansen (1993), but *Thalassiosira pseudonana* was among the investigated diatoms, and it produces TEP. Such a production of TEP may have contributed to the growth of flocs at the end of the plankton bloom in the Dollard. Bacterial activity may also favor the formation of large aggregates during the decline of a bloom (Riebesell, 1991).

It is evident that the quantity of biopolymers present in the mud flocs is not a good measure for the stickiness of these flocs. Particulate mucus produced by different species of diatoms differs in chemical composition (Decho, 1990), and appears as well to differ in its surface properties (Kiørboe and Hansen, 1993). This means that the gluing quality of biopolymers present in water is more important than the total quantity. Therefore, biological cohesion of flocs may depend on phytoplankton composition, phytoplankton concentration, the physiological state of phytoplankton and bacterial production or degradation of biopolymers during cell lysis at the decline of a bloom.

It is still not clear how the large flocs observed on 29 May 1996 were formed. The total quantity of EDTA-extracted carbohydrates could not explain it. The large floc sizes did however occur during the decline of the bloom, which is a commonly observed phenomenon in the ocean and coastal seas. Possibly, the biopolymers present on 29 May 1996 had a large impact on the overall stickiness of the suspended sediment, even though the total quantity was less than on 18 April 1996 when large flocs did not appear. Although the exact cause of the intensive flocculation on 29 May 1996 can not be determined, it is likely that it is related to plankton bloom. It seems that mortality associated with the end of bloom enhances the average size of the visualized flocs.

6. CONCLUSIONS

Several parameters that could affect mud floc size in the Dollard were measured on ten measurement days over the period of one year. This is a relatively short observation period for establishing seasonal fluctuations in floc size. Furthermore, it is difficult to establish cause-and-effect relationships from correlations between observations. Despite these limitations the following conclusions can be drawn:

On a tidal time scale, floc sizes in the Dollard were mainly determined by SSC. The average floc size was about 140 µm.

Over a seasonal time scale the observations of floc size and phytoplankton showed that during the decline of a spring plankton bloom floc sizes substantially increased. An average floc size of about 250 µm was observed. Factors like salinity, water temperature and wave height did not seem to have a direct influence on the seasonal fluctuation of floc size.

Large flocs can form from smaller aggregates or they can be derived from bed erosion. The observation of growing flocs while SSC was decreasing on 29 May 1996 shows that at least part of the suspended flocs were formed from smaller flocs or particles in the water column.

Biological cohesion of flocs is probably more dependent on the gluing quality of the biopolymers than on the total quantity. The quantity as well as the quality of the biopolymers depend on the species composition, concentration and physiological state of phytoplankton and bacteria.

Biological processes are important for flocculation of fine sediment during the decline of a plankton bloom. This may have impact on the seasonal variation in fine-grained sediment dynamics.

7. ACKNOWLEDGMENT

This research was funded by The Netherlands Organization for Scientific Research (ALW/NWO) and the EU-sponsored Marine Science and Technology Program (MAST-III) within the INTRMUD project under contract no. MAS3-CT95-022-INTRMUD. I am grateful to the National Institute for Coastal and Marine Management (RIKZ) for the use of VIS, chlorophyll data and water temperature data, and to Bert Wetsteyn of the RIKZ for the phytoplankton data. I am indebted to 'Rijkswaterstaat, Meetdienst Noord' for the use of the RV Regulus and RV Dr. Ir. Johan van Veen. Finally, I would like to thank the laboratory of Physical Geography for logistic support, and The Netherlands Institute for Oecological Research (NIOO) for the use of their laboratory facility.

REFERENCES

Alldredge, A. L., and Silver, M. W., 1988. Characteristics, dynamics and significance of marine snow. *Progress in Oceanography,* 20, 41-82.

Alldredge, A. L., and Gotschalk, C. C., 1989. Direct observations of the mass flocculation of diatom blooms: characteristics, settling velocities and formation of diatom aggregates. *Deep-Sea Research,* 36, 159-171.

Alldredge, A. L., Passow, U., and Logan, B. E., 1993. The abundance and significance of a class of large transparent organic particles in the ocean. *Deep-Sea Research,* 40, 1131-1140.

Avnimelech, Y., Troeger, B. W., and Reed, L. W., 1982. Mutual flocculation of algae and clay: evidence and implications. *Science,* 216, 63-65.

De Jonge, V. N., 1992. Physical processes and dynamics of microphytobenthos in the Ems estuary (The Netherlands). *Ph.D. Thesis*, University of Groningen, The Netherlands.

Decho, A. L., 1990. Microbial exopolymer secretions in ocean environments: their role(s) in food webs and marine processes. *Oceanographic Marine Biology Annual Review,* 28, 73-153.

Dubois, M., Gilles, K.A., Hamilton, J.K., Rebers, P.A., Smith, F., 1956. Colorimetric method for determination of sugars and related substances. *Analytical Chemistry,* 28, 350-356.

Dyer, K. R., 1989. Sediment processes in estuaries: future research requirements. *Journal of Geophysical Research,* 94(C10), 14,327-14,339.

Eisma, D., 1986. Flocculation and de-flocculation of suspended matter in estuaries. *Netherlands Journal of Sea Research,* 20, 183-199.

Eisma, D., Bernard, P., Cadee, G. C., Ittekot, V., Kalf, J., Laane, R., Martin, J. M., Mook, W. G., van Put, A., and Schuhmacher, T., 1991a. Suspended-matter particle size in some West-European estuaries; Part 1: particle-size distribution. *Netherlands Journal of Sea Research,* 28(3), 193-214.

Eisma, D., Bernard, P., Cadee, G. C., Ittekot, V., Kalf, J., Laane, R., Martin, J. M., Mook, W. G., van Put, A., and Schuhmacher, T., 1991b. Suspended-matter particle size in some West-European estuaries; Part 2: a review on floc formation and break-up. *Netherlands Journal of Sea Research,* 28(3), 215-220.

Gibbs, Ronald J., 1983. Effect of natural organic coatings on the coagulation of particles. *Environmental Science & Technology,* 17, 237-240.

Harris, R. H., and Mitchell, R., 1973. The role of polymers in microbial aggregation. *Annual Review of Microbiology*, 27, 27-50.

Hunt, J.R., 1986. Particle aggregate breakup by fluid shear. In: *Estuarine Cohesive Sediment Dynamics. Lecture Notes on Coastal and Estuarine Studies,* A. J. Mehta ed., Springer-Verlag, Berlin, 14, 85-109.

Hunter, K. A., and Liss, P. S., 1979. The surface charge of suspended particles in estuarine and coastal waters. *Nature,* 282, 823-825.

Hunter, K. A., and Liss, P.S., 1982. Organic matter and the surface charge of suspended particles in estuarine waters. *Limnology and Oceanography,* 27, 322-335.

Kiørboe, T., Andersen, K. P., and Dam, H. G., 1990. Coagulation efficiency and aggregate formation in marine phytoplankton. *Marine Biology,* 107, 235-245.

Kiørboe, T., and Hansen, J. L., 1993. Phytoplankton aggregate formation: observations of patterns and mechanisms of cell sticking and the significance of exopolymeric material. *Journal of Plankton Research,* 15, 993-1018.

Kranck, K., and Milligan, T. G., 1988. Macroflocs from diatoms: in situ photography of particles in Bedford Basin, Nova Scotia. *Marine Ecology Progress Series,* 44, 183-189.

Loder, T. C., and Liss, P. S., 1985. Control by organic coatings of the surface charge of estuarine suspended particles. *Limnology and Oceanography,* 30, 418-421.

Luettich, R. A., Wells, J. T., and Kim, S., 1993. In situ variability of large aggregates: preliminary results on the effects of shear. In: *Nearshore and Estuarine Cohesive Sediment Transport,* A.J. Mehta ed., American Geophysical Union, Washington, 42, 447-466.

Neihof, R. A., and Loeb, G. I., 1972. The surface charge of particulate matter in seawater. *Limnology and Oceanography,* 17, 7-16.

Passow, U., Logan, B. E., and Alldredge, A. L., 1994. The role of particulate carbohydrate exudates in the flocculation of diatom blooms. *Deep-Sea Research,* 41, 335-357.

Passow, U., and Alldredge, A. L., 1995. Aggregation of a diatom bloom in a mesocosm: The role of transparent exopolymer particles (TEP). *Deep-Sea Research,* 42, 99-109.

Puls, W., and Kuehl, H., 1986. Field measurements of the settling velocities of estuarine flocs. In: *Proceedings of the Third International Symposium on River Sedimentation,* S.Y. Wang, H.W. Shen and L.Z. Ding eds., Jackson, Mississippi, 525-536.

Riebesell, U., 1989. Comparison of sinking and sedimentation rate measurements in a diatom winter/spring bloom. *Marine Ecology Progress Series,* 54, 109-119.

Riebesell, U., 1991. Particle aggregation during a diatom bloom.2. Biological aspects. *Marine Ecology Progress Series,* 69, 281-291.

Smetacek, V., 1985. Role of sinking in diatom life-history cycles: Ecological, evolutionary and geological significance. *Marine Biology,* 84, 239-251.

Ten Brinke, W. B. M., 1993. The impact of biological factors on the deposition of fine-grained sediment in the Oosterschelde (The Netherlands). Thesis Utrecht University, 252p.

van Leussen, W., 1994. Estuarine macroflocs and their role in Fine-grained sediment transport. Thesis Utrecht University, 484p.

van Leussen, W., and Cornelisse, J. M., 1993. The determination of the sizes and settling velocities of estuarine flocs by an underwater video system. *Netherlands Journal of Sea Research,* 31, 231-241.

Wilkinson, K. J., Negre, J. C., and Buffle, J., 1997. Coagulation of colloidal material in surface waters: the role of natural organic matter. *Journal of Contaminant Hydrology,* 26, 229-243.

Coastal and Estuarine Fine Sediment Processes
W.H. McAnally and A.J. Mehta (Editors)

Salt marsh processes along the coast of Friesland, The Netherlands

B. M. Janssen-Stelder

Institute for Marine and Atmospheric Research Utrecht, Department of Physical Geography, Faculty of Geographical Sciences, Utrecht University, P.O. Box 80.115, 3508 TC, Utrecht, The Netherlands.

Along the coast of Friesland in The Netherlands, hydrodynamic and morphodynamic processes were examined in order to understand the future development of salt water marshes during rising sea level. Measurements show sedimentation dominating over erosion in the pioneer zone of the salt marshes during May to October 1997. The sedimentation rates vary along the coast. Alongshore differences in wave height and suspended sediment concentrations may partly explain the spatial variation in sedimentation rates, which may also be related to differences in large-scale morphology seaward of the pioneer zone. Morphological changes are likely to be particularly rapid during rough weather conditions.

1. INTRODUCTION

Salt marshes tend to develop on gently sloping shores with low wave energy (Dijkema, 1987). The landward edge of the intertidal area of the mainland coast of the Dutch Wadden Sea has extensive salt marshes (Figure 1) that are protected from wave energy by the Wadden barrier islands. Four centuries ago humans began to stimulate further salt marsh development in this area by the construction of small dams and drainage channels in order to reclaim land. But from the beginning of this century salt marshes have ceased to develop in this manner and net erosion has occurred. This change may be attributed to rising mean high tide (MHT) levels. During 1933-1989, MHT levels increased at the rate of 0.23 cm/y as a result of relative sea-level rise. During 1961-1983 there was an acceleration in the rate of rise of MHT levels to 0.44 cm/y. This acceleration is ascribed to an increase in yearly wind speed in this period (Bossinade et al., 1993). Because of the importance of the salt marshes as nature reserve areas combined with their role in affording coastal protection by absorbing wave energy, from 1935 on, government policy has been to protect the remaining salt marshes by the construction of brushwood groins and drainage channels (Dijkema et al., 1988; Dijkema et al., 1995).

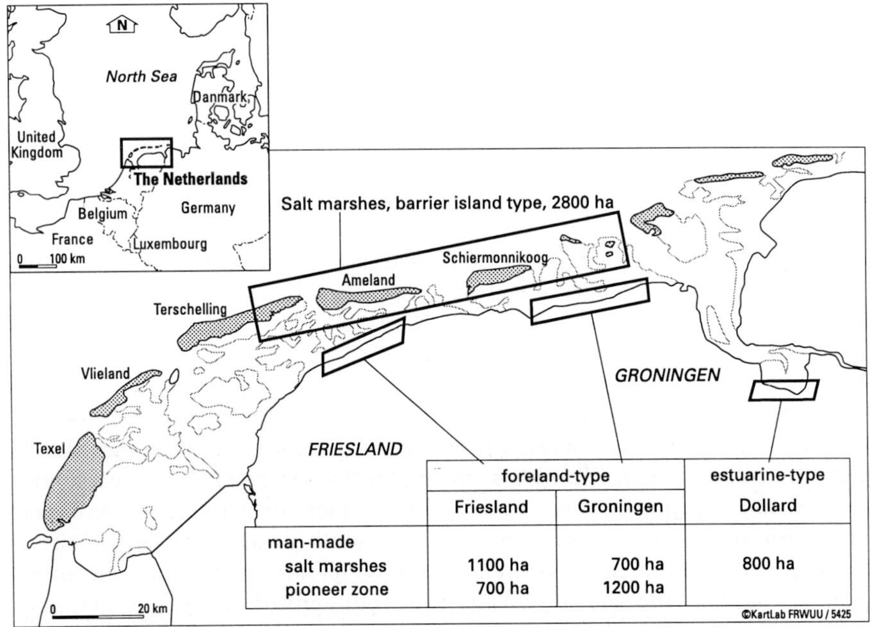

Figure 1. Locations, geomorphological types and areal extent of salt marshes in the Dutch Wadden Sea (after Dijkema, 1987).

Acceleration in sea-level rise may disturb the equilibrium between erosion and sedimentation in the salt marshes. The pioneer zone is the most sensitive area with regard to this equilibrium (Dijkema et al., 1990; Houwing et al., 1995). This transition area between mudflat and salt marsh is flooded twice a day and the vegetative cover is discontinuous (Figure 2). During rising sea level, the pioneer zone (Figure 2) is flooded for a slightly longer period of time each consecutive high water. In the absence of a compensating fine sediment input, shear stresses induced by currents and waves increase because of this longer flooding period, and because of related increase in water depth in the pioneer zone. This increase in shear stress leads to an increase in turbulence in the pioneer zone. A larger amount of seeds and seedlings tend to be washed out due to increased turbulence. Also, because of the increase in wave heights and current velocities, coarser sediment is transported into the pioneer zone (Figure 3). Increase in turbulence and sand content together with decrease in seeds and seedlings cause a decrease in consolidation and bed shear strength in the pioneer zone. Under these circumstances erosion of the pioneer zone occurs by currents and waves. The lower marsh (Figure 2) is also flooded more frequently, but because of the continuous vegetative cover most of the sediment supplied to this area may accumulate. Steepening of the salt marsh profile takes place and a height difference develops because of net erosion in the pioneer zone and net sedimentation

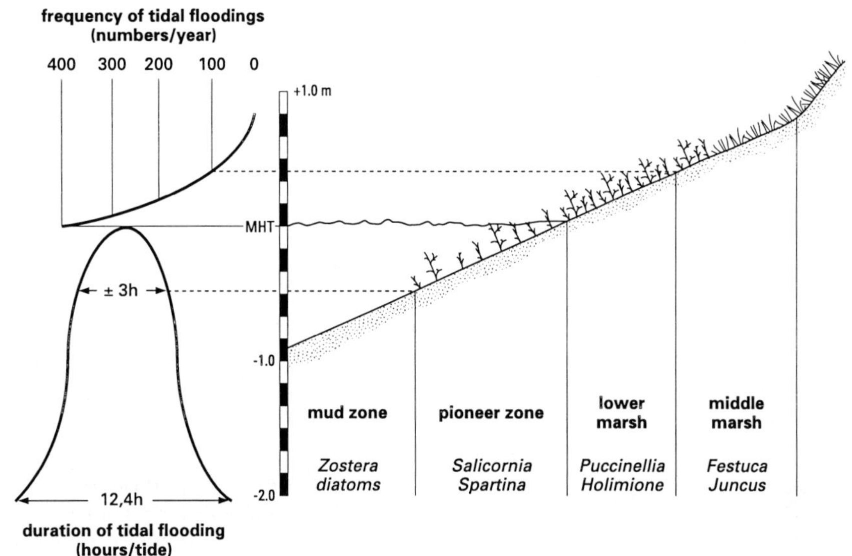

Figure 2: Zonation of the salt marsh in relation to the frequency and duration of tidal flooding (after Erchinger, 1985).

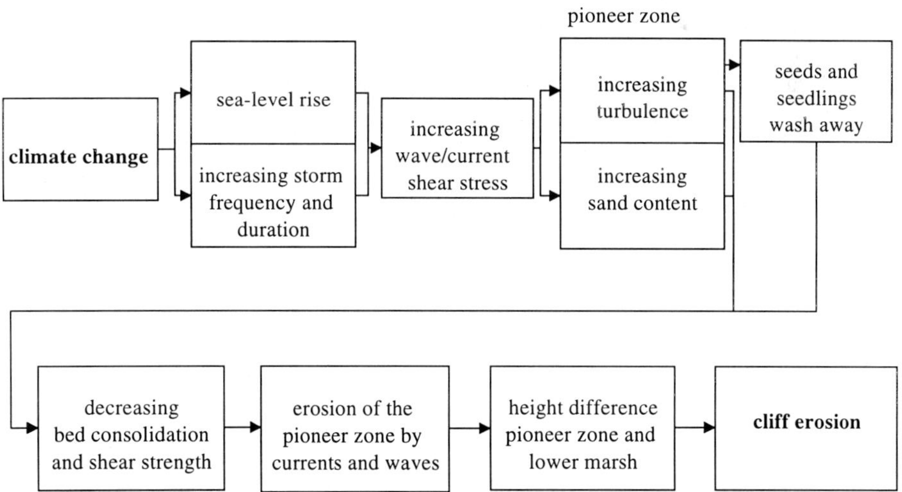

Figure 3. The influence of sea-level rise on salt marsh development.

in the landward lower marsh zone (Figure 2). This will lead to the formation of a cliff, which is sensitive to wave attack, and horizontal erosion of the salt marsh will then occur.

Presently, an extensive study of salt marsh behavior at different temporal and spatial scales is being carried out. In order to predict the effects of sea-level rise on salt marsh development in the future, past (time scale of decades) and present salt marsh developments are being studied. Long term data are available concerning height and extension of the salt marshes, but little is known about the hydrodynamic conditions or other influencing factors such as sediment supply. Historic salt marsh development can be analyzed more properly when the influence of processes on present salt marsh development is better understood. This paper includes results of a study concerning the present coastal behavior (time scale of months) of the Friesland part of the Wadden Sea salt marshes (spatial scale of kilometers). The main question is whether present spatial variability can be explained by studying different governing factors. These include hydrodynamic factors (waves, currents), type of sediment and morphology. During this study salt marsh processes were examined for the present situation. An analysis of recent hydrographic maps was carried out to define the influence of large-scale morphology in front of the pioneer zone on the development of this zone.

2. MEASURING METHODS

A morphodynamic and hydrodynamic analysis of the pioneer zone was carried out in six sedimentation fields along the coast (Figure 4). Morphological data from the pioneer zone were collected during six months by using a leveling instrument.

One measuring frame was placed in the pioneer zone of the sedimentation field 171 (Figure 4) to continuously measure the hydrodynamic conditions and sediment in suspension. The measuring frame was built of steel and was about 7 m high. A solar panel on top of the frame supplied the energy needed to operate the instruments. The instruments, a level switch and a computer system in a waterproof tube were attached to the frame. All instruments were located near the bed. A pressure sensor was used to measure water level fluctuations. An electromagnetic flow meter (EMF) measured horizontal current velocities. A turbidity meter (MEX) gave information about suspended sediment concentrations. Data were collected when the instruments were submerged, every thirty minutes in bursts of ten minutes with a sampling frequency of 2Hz.

Besides the automated frame measurements in field 171, portable instruments were used to determine the hydrodynamic conditions and suspended sediment concentration in the remaining five sedimentation fields along the coast. These instruments included a capacitance wire to measure water level fluctuations at a frequency of 10 Hz and floaters and a compass to determine current velocities and directions. In addition, water samples were taken to measure suspended

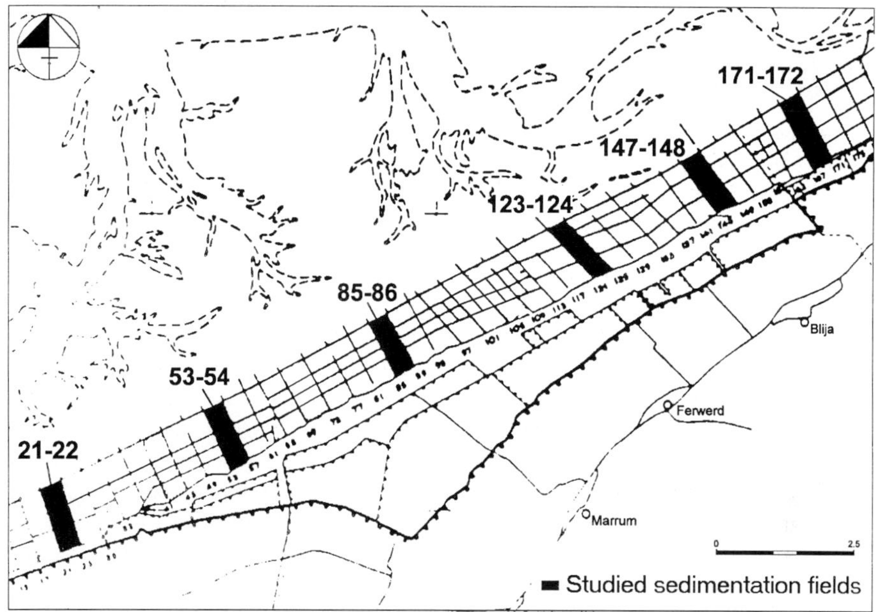

Figure 4. Sedimentation fields studied along the salt marsh coast of Friesland (after Rijkwaterstaat, 1988).

sediment concentrations. The data were registered every 15 minutes during a high water period.

Differences in meso- and large-scale morphology seaward of the pioneer zone were studied by examining hydrographic maps of the Wadden Sea (Marine Service of Hydrography, 1998). Important aspects examined included the distance to the nearest channel bank, the size and orientation of the nearest channel, the distance to the tidal inlet, the slope and the morphology of the area seaward of the sedimentation fields.

3. METHODS OF ANALYSES

From the leveling data, heights above the Dutch Ordinance Level were calculated at intervals of 5 m. These equidistant transects were used to calculate sedimentation rates. The accuracy of the calculated sedimentation rates was 0.5 cm. Suspended sediment concentrations were determined by filtering water samples and weighing the filters. This was done in duplicate to reduce the measurement error. As a measure of wave height, the significant height $H_{1/3}$ was used.

Bed shear stresses induced by waves ($\tau_{b,w}$) and currents ($\tau_{b,c}$) were calculated using $H_{1/3}$ and the mean current velocity following the approach of van Rijn (1990). The bed shear stress due to the combination of currents and waves, which is important for sediment entrainment, was determined by vector addition of the wave related and the current related shear stresses. The assumption was made that the bed shear stress is related to the square of the corresponding instantaneous near-bed velocity. The magnitude of the combined time-averaged bed shear stress was calculated by adding $\tau_{b,c}$ and $\tau_{b,w}$:

$$\tau_{b,cw} = \tau_{b,c} + \tau_{b,w} \tag{1}$$

$$\tau_{b,c} = \frac{1}{8}\rho f_c \bar{U}^2 \tag{2}$$

$$\tau_{b,w} = \frac{1}{4}\rho f_w \hat{U}_\delta^{\,2} \tag{3}$$

where $\tau_{b,c}$ = bed shear stress induced current [Pa], ρ = fluid density [kg/m^3], f_c = Darcy-Weisbach friction coefficient [-], $\bar{U}$ = average velocity over the depth [m/s], $\tau_{b,w}$ = bed shear stress induced by waves [Pa], f_w = wave friction coefficient (constant over a wave cycle), and $\hat{U}_\delta$ = peak value of the horizontal fluid velocity just outside the bottom boundary layer [m/s].

4. RESULTS

4.1. Local morphodynamics in the pioneer zone

The morphology of the six different fields was very similar because of the uniform composition of the sedimentation fields in squares of 200 by 400 m with a north-northwestern orientation. In general, two field pairs were situated between brushwood groins built perpendicular and parallel to the coast (Figure 5).

A straight channel and an opening in the coast-parallel groin separated the field pairs. The pioneer vegetation *Salicornia dolichostachya* was spread up to halfway through the fields, with a coverage of up to 50%. The field to the east of the channel was at a higher level than the western field. This may have been due to the exposure of the sedimentation fields (avg. 332°) to northwesterly winds. Storms from a northwest to westerly direction induce a large set-up of the water level in the Wadden Sea. The brushwood groins of the sedimentation fields are then completely submerged for a period, and current velocities can increase to 1 m/s. However, the groins continue to obstruct the westerly current, and sediment accumulation takes place in the eastern fields. In the eastern fields the vegetation had colonized further seaward than in the western fields due to the higher elevation (Figure 5).

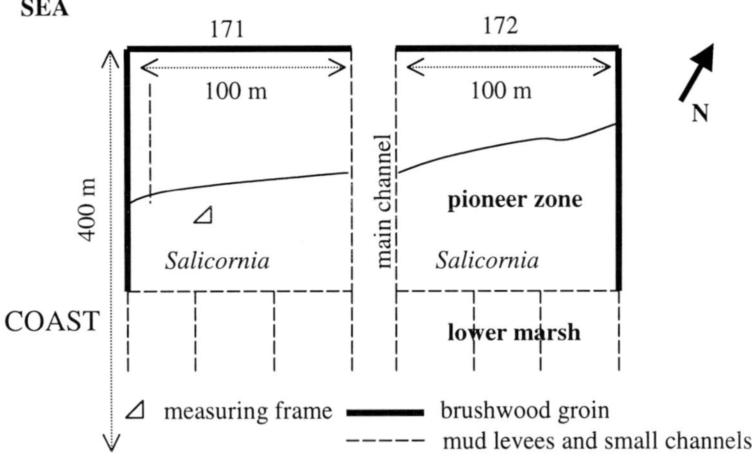

Figure 5. Plan view of sedimentation fields 171 and 172, and the location of the measuring frame.

Sedimentation dominated over erosion in the measurement period of May to October 1997, and net sedimentation was, in general, higher in the eastern than in the western field (Figure 6). During high water, when the entire fields were flooded to an average water depth of about 50 cm, most of the sediment was deposited in the eastern field where the flow was obstructed by the vegetative cover. During ebb tide the water turned clockwise and flowed back to the Wadden Sea through the western field and along the eastern side of the groins, where in many cases a small channel was formed (Figure 5).

The average sedimentation in the pioneer zone was about 3 cm for the measuring period. The very high sedimentation of 6-7 cm in fields 21 and 22, situated in the western part of the area is striking (Figure 6). The height of fields 53-54, also situated in the western part of the area, showed practically no change. Fields 147-148 and 85 exhibited less than average sedimentation. In these fields signs of cliff erosion were found. These observations imply that the sedimentation rates vary along the coast, but without an obvious spatial trend.

4.2. Local hydrodynamics

Hydrodynamics and suspended sediment concentrations were examined during calm weather (1-2 Beaufort) and rough weather (4-6 Beaufort) conditions at different locations along the coast. These conditions were studied in order to determine what conditions are important for the development of the pioneer zone. The spatial variation in wave height, current velocity and suspended sediment concentration were studied in order to explain the variation in sedimentation rate along the coast.

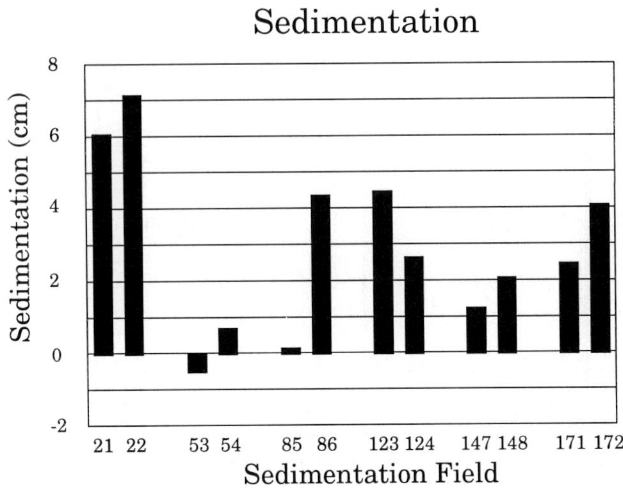

Figure 6. Net sedimentation along the coast of Friesland between May and October 1997.

From the data collected by the measuring frame it appears that wave heights were three times higher during rough weather than during calm periods (Figure 7). During calm weather wave heights did not exceed 5 cm, whereas during rough weather conditions, wave heights increased to about 15 cm. These small wave heights are relevant because of the shallow water depth. The water depth increased from 26 cm during calm weather to 100 cm during rough weather. Current velocities were low but increased with increasing water depth. However, the increase in maximum ebb current velocity from 0.06 m/s during calm weather to 0.15 m/s during rough weather is rather small (Figure 7).

The current velocities were not influenced by the weather but by tides. The sediment in suspension was four to six times greater during rough weather, varying from 0.1 g/l to 0.75 g/l (Figure 8). During rough weather waves decidedly had a great impact on erosion and sedimentation.

The significant wave height ($H_{1/3}$) and the average current velocity were used in (1), (2) and (3) to calculate the shear stresses induced by waves and currents on the bed during different weather conditions. Calculated total shear stresses did not exceed 0.10 Pa during calm weather conditions. However, they were up to 0.30 Pa during rough weather. The increase in shear stress is caused by the waves, which were responsible for 85% of the total shear stress during rough weather. Average bed shear strengths in the study area were 0.15 Pa (Houwing et al., 1995; Kornman and De Deckere, 1998). It appears that only during rough weather did the shear stresses exceed the bed shear strength and the sediment of the pioneer zone brought into suspension.

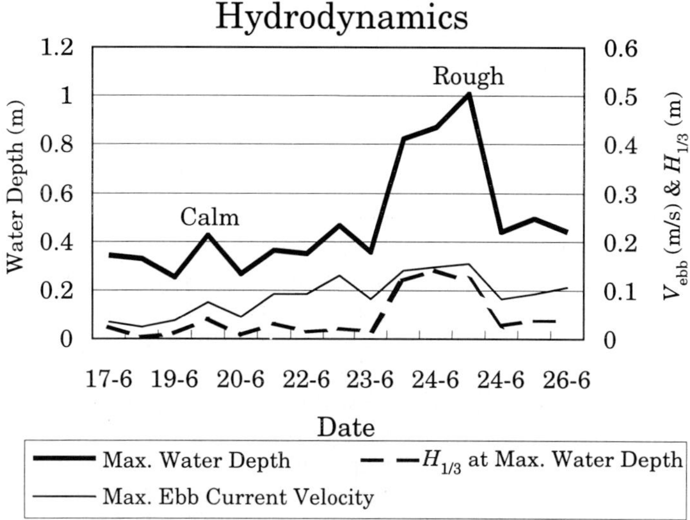

Figure 7. Hydrodynamic parameters during different weather conditions. V_{ebb} is ebb current velocity.

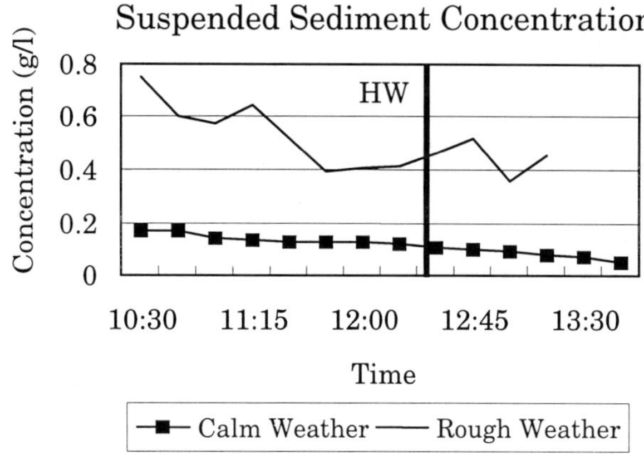

Figure 8: Suspended sediment concentrations in the western part of Friesland during calm and rough weather conditions.

Comparing the measurements of the measuring frame in field 171 with the portable measurements along the coast, wave heights measured in fields 21 and 85 were found to be significantly higher than in field 171 (Figure 9). Possible causes of this difference in wave height along the coast will be discussed later. No significant differences were found in currents along the coast. For the surface suspended sediment concentrations it is difficult to make a spatial analysis, because the measurements were not simultaneous. A minor difference in weather conditions can have a large influence on the amount of sediment in suspension. For example, concentrations measured at one location on the first day of a storm can be much smaller than concentrations measured at another location on the third day of the storm under the same wave and current conditions. During rough weather much variation in suspended sediment concentrations can be found along the coast, but the cause can not be determined from these data.

4.3. Morphology of channels and tidal flats seaward of the pioneer zone

Sedimentation and suspended sediment concentrations varied substantially along the coast during the measuring period. This may have been caused partially by a difference in sediment supply arising from differences in the large-scale morphology in front of the pioneer zone. Important factors are the distance to the nearest channel bank, the size and orientation of the nearest channel, the distance to the tidal inlet, and the slope and the morphology of the area in front of the sedimentation fields. These factors can be examined (Table 1) using hydrographic maps of the Wadden Sea (Marine Service of Hydrography, 1998). Figure 10 is a general view of the hydrography.

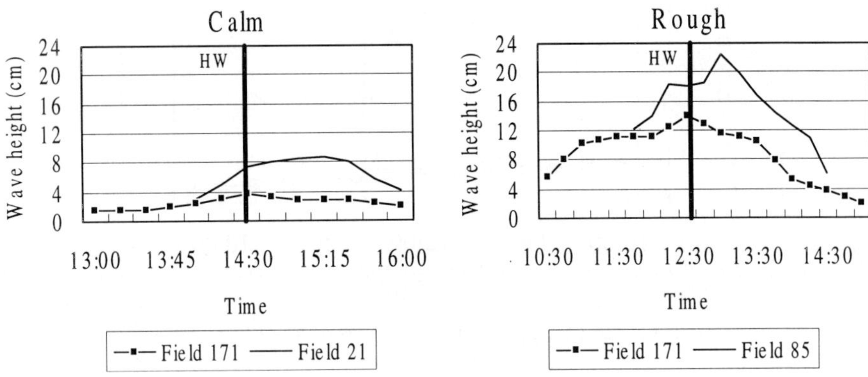

Figure 9: Wave heights in the eastern and western part of Friesland during calm and rough weather conditions.

Table 1:
Factors influencing sediment supply

Sedimentation field	Sedimentation rate in field (cm/0.5y)	Distance to channel bank (m)	Size of nearest channel width (m) /depth (m)	Orientation of nearest channel	Slope of tidal flat in front of field in ° ($\times 10^{-3}$)	Distance to nearest tidal inlet (km)
21-22	6	1,850	200/0.3 = 667	E-W	15	13.9
53-54	0	2,850	250/2.7 = 93	E-W	8	14.2
85-86	1	5,250	750/7.0 = 107	W-E	7	14.7
123-124	3	1,600	350/6.0 = 58	N-S	14	16.1
147-148	1.5	1,750	250/4.0 = 63	W-E	7	17.4
171-172	3	1,000	200/3.4 = 59	N-S	11	18.9

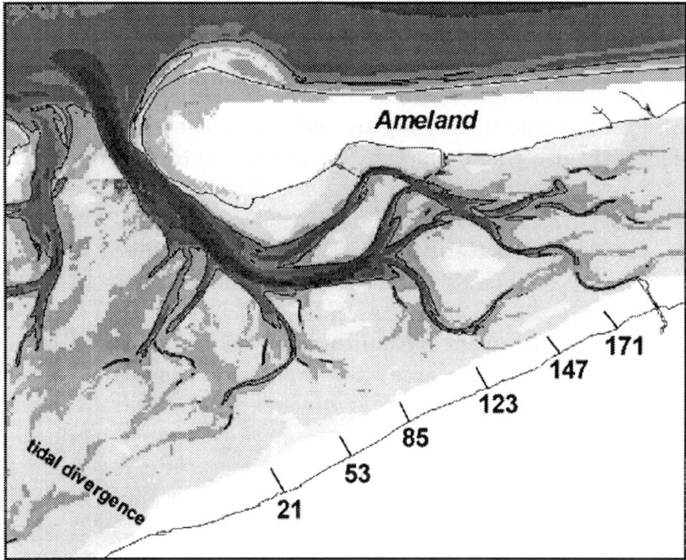

Figure 10. Morphology of channels and tidal flats seaward of the salt marshes.

When a large channel is orientated perpendicular to the coast near the sedimentation fields, large waves and strong currents propagate along this channel. Much sediment from the Wadden Sea can be transported into the salt marshes and is available for deposition. The sedimentation fields act as sediment traps as waves break on the brushwood groins and the groins interrupt currents. When a large tidal flat is situated in front of the sedimentation fields or when channels are orientated parallel to the coast, waves and currents diminish. Sediment settles on the tidal flat or is transported alongshore rather than in the salt marshes during calm weather. During rough weather wave and current energy is dissipated before reaching the sedimentation fields.

The measured sedimentation rates in fields 123-124 and 171-172 are similar (3 cm in six months). The characteristics of the large-scale morphology in front of these fields are also comparable and permit high sediment transport (Table 1 and Figure 10). They are roughly equi-distant from the nearest channel bank, equal with regard to the size and orientation of the nearest channel and have equal slopes of the tidal flat in front of the fields. The sedimentation rate in the field in-between, 147-148, is only half of the sedimentation rate in 123-124 and 171-172. The characteristics of the large-scale morphology also differ. The orientation of the nearest channel is west-east instead of north-south, and the tidal flat in front of the sedimentation fields has a gentler slope (Table 1). Under these circumstances sediment can be expected to be transported through the channel to the east rather than in a landward direction, leading to lower sedimentation rates in fields 147-148. Waves and currents transport less sediment into the fields, so less sediment is available for deposition compared to fields 123-124 and 171-172.

Fields 53-54 are comparable to fields 147-148 with regard to the orientation of the nearest channel (Table 1 and Figure 10). Also, the slope of the tidal flat in front of the sedimentation fields is similar. However, the sedimentation rates in fields 53-54 are lower than in fields 147-148. The long distance to the nearest channel bank may cause this low sedimentation rate. Since waves and currents have to propagate over an extensive shallow tidal flat before the sedimentation fields are reached, a significant amount of sediment is likely to be deposited on the tidal flat in front of the salt marshes of fields 53-54.

Low sedimentation rates were measured in fields 85-86 (Table 1). The distance to the nearest channel bank was great (>5 km). A well-developed channel bank of over a meter in height was situated at an elevation of almost 2 m above the Dutch Ordinance Level. A long distance from the channel bank, the elevation decreased again to about 1.2 m. During flooding tide, waves and currents propagated through the deep channel towards the east and not towards the coast. At high water when the tidal flats were submerged, much of the sediment was deposited on the channel bank. Only a small fraction of the sediment transported within the banks of the large channel could reach the sedimentation fields and accumulate there.

High sedimentation rates were measured in fields 21-22 (Table 1). Only a very shallow channel was present in front of the fields. These sedimentation fields were situated directly landward of the tidal inlet between Ameland and Terschelling. The tidal divergence was located in the vicinity of these fields. At a tidal divergence the current velocities were low and sediment accumulated. During stormy conditions this sediment was resuspended and may have been transported from the tidal divergence into the sedimentation fields. Sedimentation fields 53-54 were about 3 km east of fields 21-22, also behind the tidal inlet, but sedimentation rates were very low in these fields. During stormy conditions the sediment supplied from the tidal divergence probably could not reach this area.

5. DISCUSSION

Despite the similarity in the morphology of the sedimentation fields, the sedimentation rates varied along the coast between 0 cm and 7 cm in six months with an average sedimentation of 3 cm. Spatial variation was also found in wave height and morphology seaward of the salt marshes. The variation in wave height may have been caused by the differences in height and maintenance of the brushwood groins in the different fields. The wave field was interrupted when high, well-maintained groins were present. Wind generated local waves within the sedimentation fields. Waves could partly propagate through the groins when they were low and poorly maintained. In that case, higher waves occurred within the sedimentation fields, as in sedimentation fields 21 and 85.

In fields 21-22, high sedimentation rates and high waves were measured. This area was situated in the vicinity of a tidal divergence. As noted, at a tidal divergence the current velocities are low and sediment tends to accumulate. This sediment can be resuspended and transported into the sedimentation fields during stormy conditions where it is kept in suspension by the high wave turbulence within the sedimentation field. Because a large amount of sediment is transported into the sedimentation fields, at the turn of the tide a substantial amount of the sediment settles and is not transported back to the Wadden Sea during ebbing tide, resulting in high sedimentation rates.

In fields 85-86 the same wave heights were measured as in fields 21-22. The sedimentation rates, however, were low in this area and some cliff erosion was observed. The elevation of the pioneer zone occupied by *Salicornia dolichostachya* was several centimeters lower than in the lower marsh zone occupied by *Puccinellia maritima,* and the sedimentation in the pioneer zone could not keep up with the sedimentation in the more densely vegetated lower marsh zone. Possibly a smaller amount of sediment is supplied to this area than in case of fields 21-22, and during rough weather sediment is eroded within the pioneer zone instead of being transported into the pioneer zone. There is no tidal channel situated near fields 85-86; the nearest channel being located 5 km to the north. This west-east orientated large channel has a well-developed channel bank. Much of the sediment is either transported to the east or is deposited on the channel banks and therefore cannot reach the sedimentation fields.

In the remaining sedimentation fields (53-54, 123-124, 147-148 and 171-172) the wave heights were lower and comparable. As in fields 21-22 and 85-86, in these fields the large-scale morphology in front of these fields appears to have determined the amount of sediment transported into the sedimentation fields, and therefore may have been a contributory factor to the difference in the sedimentation rates.

It is difficult to use suspended sediment concentration as a measure of transport. Sediment concentrations were not measured simultaneously and were therefore not comparable. During similar weather conditions the temporal variation in the concentration was too high to enable a comparison between different measuring locations along the coast. A comparison between different weather conditions is possible for the individual locations. Suspended sediment concentrations were 4 to 6 times higher during rough weather than during calm weather.

During rough weather, waves were three times higher and suspended sediment concentrations were four to six times higher than during calm weather. This is supported by the magnitude of the shear stresses calculated using the van Rijn (1990) approach. The shear stress only exceeded the shear strength of the bed in the pioneer zone during rough weather. The wave-induced shear stress was responsible for 85% of the total shear stress during rough weather. Currents were responsible for transport of sediment that was brought into suspension by waves. However, the morphology changed only when gradients in this transport occurred.

6. CONCLUSIONS

Sedimentation dominated over erosion during the measurement period. Sedimentation rates varied along the coast. In the western part of the study area sedimentation rates were high. In two fields in the center of the study area sedimentation rates were below average and cliff erosion was observed.

Rough weather conditions are likely to cause morphological changes in the sedimentation fields. Such conditions cause wave heights to be three times larger and suspended sediment concentrations to be up to six times larger than during calm weather.

A significant part of the alongshore variation in the height of the six studied sedimentation fields can be explained by alongshore differences in wave height and sediment supply. Sediment supply is an important boundary condition and is strongly related to the morphology of the channels and the tidal flats seaward of the sedimentation fields. Whether or not the sediment is deposited (when the supply is sufficient) depends on the wave climate within the sedimentation fields. This climate is greatly influenced by the height and maintenance of the brushwood groins. When these groins are low and poorly maintained, waves can propagate and keep the sediment in suspension.

7. ACKNOWLEDGMENT

This study was carried out as part of the National Research Program (NOPII) on Modeling the Impact of Climatic Change on the Wadden Sea Ecosystem, and supported by The Netherlands Foundation for Scientific Research (NWO). The author wishes to thank Prof. Pieter Augustinus and Dr. Aart Kroon for their critical comments on the paper.

REFERENCES

Bossinade, J. H., van den Bergs, J., and Dijkema, K. S., 1993. The influence of the wind on yearly averaged high water along the Friesland and Groningen Wadden coast (in Dutch). *Nota GRAN 1993-2009 + IBN-report 049*, Groningen, Texel: Department of Public Works and Water Management, Directorate Groningen + DLO Institute for Forest and Nature Studies, 22p.

Dijkema, K. S., 1987. Geography of salt marshes in Europe. Z. Geomorph. N.F., 31/4. p. 489-499.

Dijkema, K. S., van den Bergs, J., Bossinade, J. H., Bouwsema, P., de Glopper, R. J., and van Meegen, J.W.Th.M., 1988. Effects of brushwood groynes on the sedimentation and extension of the vegetation zones of the Friesland and Groningen reclamation works (in Dutch). Department of Public Works and Water Management, Directorate Groningen, Nota GRAN RIN-rapport 88-66; Service for IJselmeerpolders, Lelystad, RIJP-report 1988-33 Cbw: 1-119.

Dijkema, K. S., Bossinade, J. H., Bouwsema, P., and de Glopper, R. J., 1990. Salt marshes in The Netherlands Wadden Sea: rising high tide levels and accretion enhancement. In: *Expected Effects of Climatic Change on Marine Coastal Ecosystems*, J. J. Beukema, W. J. Wolff and J.J.W.M. Brouns eds., Kluwer, Dordrecht, NL, 173-188.

Dijkema, K. S., van den Bergs, J., Bossinade, J., and Kroeze, T. A. G., 1995. Monitoring reclamation works in Groningen and Friesland: evaluation mid 1992-mid 1994 (in Dutch). *Technical Report IBN-DLO Texel*, 34p.

Erchinger, H. F., 1985. Dünen,Watt und Salzwiesen. *Der Niedersachsische Minister für Ernährung, Landwirtschaft und Forsten,* Hannover, 59p.

Houwing, E. J., van der Waay, Y., Terwindt, J. H. J., Augustinus, P. G. E. F., Dijkema, K. S., and Bossinade, J. H., 1995. Salt marshes and sea-level rise: plant dynamics in relation to accretion enhancement techniques. *IMAU-report R95-27*, 146p.

Kornman, B. A., and De Deckere, E. M. G. T., 1998. The temporal variation in sediment erodibility and suspended sediment dynamics in the Dollard Estuary. In: *Sedimentary Processes in the Intertidal Zone*, K. S. Black, D. M. Paterson, and A. Cramp eds., Geological Society, London, 231-241.

Marine Service of Hydrography, 1998. Hydrographic Map for Coast and Rivers, part 1811 and 1812. Wadden Sea (west) and bordering North Sea coast (in Dutch), 9p.

Rijkswaterstaat, 1988. Survey map of measuring, testing and additional testing fields (in Dutch). Directorate Groningen, Service Delfzijl, GRDD 1988-8091. From: Dijkema et al., 1988. Effects of brushwood groynes on the sedimentation and extension of the vegetation zones of the Friesland and Groningen reclamation works (in Dutch), 130p.

van Rijn, L. C., 1990. *Principles of Fluid Flow and Surface Waves in Rivers, Estuaries, Seas* and *Oceans.* Aqua Publications, Oldemarkt, 335p.

Coastal and Estuarine Fine Sediment Processes
W.H. McAnally and A.J. Mehta (Editors)
2001 Elsevier Science B.V.

Prediction of contaminated sediment transport in the Maurice River-Union Lake, New Jersey, USA

E. J. Hayter[a] and R. Gu[b]

[a]U.S. Environmental Protection Agency, National Exposure Research Laboratory, Ecosystems Research Division, 960 College Station Road, Athens, Georgia 30605, USA

[b]Iowa State University, Department of Civil and Construction Engineering, Ames, Iowa 50011, USA

A sediment and contaminant transport model and its application to the Maurice River-Union Lake system in southern New Jersey, USA is described. The application is meant to characterize and forecast sediment and arsenic (*As*) distributions before and after proposed dredging activities. The model, HSCTM-2D, is a two-dimensional, depth-averaged, finite element code capable of simulating the hydraulics of both steady and unsteady surface water flows, cohesive and cohesionless sediment transport, and the transport and fate of inorganic contaminants. Interactions between dissolved (i.e., desorbed) and particulate (i.e., adsorbed) contaminants and sediments are accounted for by simulating the processes of adsorption and desorption of contaminants to and from sediments, respectively. Four model simulations yielded *As* flushing times ranging from 25 years for the "no action" scenario to four years for dredging of contaminant bed sediments in the Maurice River and Union Lake.

1. TRANSPORT OF CONTAMINANTS IN SURFACE WATERS

Contamination of surface waters by both point and non-point sources is a critical water quality problem. The ability to predict the potential impacts that contaminants such as metals and pesticides have on aquatic ecosystems and possible remediation alternatives are requisite for mitigation. A necessary component of the assessment of environmental effects of contaminants in surface waters is predicting the transport rates and fates of the contaminants. To simulate the transport of contaminants, it is necessary to reproduce not only the governing physical-chemical processes (e.g., adsorption/desorption), but also changes in the various factors (e.g., pH) that govern them. The latter require an ability to predict the hydraulics, water quality, and sediment transport in the water system, because the movement of surface waters, sediments and contaminants are highly coupled. For example, the role of sediments in accumulating contaminants in depositional environments such as reservoirs, lakes, and marinas is well documented (Reese et al., 1978; Abernathy et al., 1984; Medine and McCutcheon, 1989; Brown et al., 1990).

The transport of both cohesionless sediments (i.e., medium size silts and larger particles) and cohesive sediments (i.e., mostly composed of terrigenous clay size particles and fine silts; and sometimes small quantities of biogenic detritus, algae, and organic matter) must be modeled to account for adsorption and desorption of contaminants onto and from sediments of all sizes. Properties of clays that cause the sorption of contaminants are their large specific area, surficial negative electrical charge, and cation exchange capacity. Other processes that affect the association of inorganic contaminants with aquatic sediments include complexation, and organic coatings (Stumm and Morgan, 1981, Dzombak and Morel, 1987). The processes of adsorption and desorption are discussed below. Biological factors, e.g., presence of organic coatings on sorbents (sediments), are beyond the scope of this study.

The bulk of the contaminant load in surface waters is often transported sorbed to cohesive sediments rather than in the desorbed state (Kirby and Parker, 1973). The portion of the contaminant load adsorbed to cohesionless sediments is often less than that sorbed to cohesive sediments. This follows from studies that have shown that increasing metal concentrations typically correlate positively with decreasing sediment particle size, increasing sediment specific area, and increasing concentrations of organic matter and manganese and iron oxides (Horowitz and Elrick, 1987). However, there are a few exceptions to this trend. A study by Brook and Moore (1988) found weak correlation between percent of fine grained sediment (less that 63 microns) and trace metal concentrations in the Clark Fork River in Montana, while coarse grain sediments contributed significantly to bulk metal concentrations. This emphasizes the need to consider the transport of both cohesive and cohesionless sediments in contaminant transport modeling.

Sorption of contaminants generally refers to both adsorption and absorption (Elzerman and Coates, 1987). Adsorption of a sorbate to a sorbent occurs at a surface or interface, while absorption continues beyond the interface and involves incorporation of the sorbate into the interior of the sorbent. Distinction between the two processes is usually not precise, thus explaining the use of the collective term. Sorption of inorganic compounds is a chemical coordination process involving certain reactions between absorbents and the inorganic adsorbate (Dzombak and Morel, 1987). Sorption models for inorganic contaminants should simulate these chemical reactions and account for interactions between electrical surface charges and ion adsorption. Sorption models are empirical in nature due to the complex electrochemical interactions, and are generally applicable to a specific sorbent-sorbate pair (Dzombak and Morel, 1987). Typically, a geochemical-metals speciation model, such as MINTEQA2 (Brown and Allison, 1987), is used to compute metals precipitation and sorption/desorption for the site specific geochemical conditions. As such, sorption of contaminants onto sediments is usually treated as a process in thermodynamic equilibrium that occurs rapidly compared to transformation processes such as hydrolysis, microbial transformation, photolysis, volatilization, and chemical oxidation. These slower processes are usually simulated using a kinetic approach (Baughman and Burns, 1980).

A partition coefficient, K_p, is used to define the distribution of a contaminant between the particulate and dissolved phases. Values for K_p usually correspond to equilibrium conditions at which rates of desorption and sorption are equal. Despite typically high values of K_p, the total mass of contaminants sorbed to suspended sediments is usually lower than that in the dissolved phase because of relatively low concentrations of suspended sediment. In addition, sorbed contaminant mass on bed sediments must be considered. Contaminants that are sorbed on bed sediments either sorbed directly onto deposited sediments or sorbed onto suspended sediments that subsequently settled to the bed.

Factors which control the fate of a contaminant in aquatic environments have been classified according to system, sorbent, and sorbate (contaminant) characteristics (Elzerman and Coates, 1987). System characteristics include: transport of water; transport and fate of sediment; nature of contaminant loading; temperature; pH; ionic strength; concentration gradient of contaminant; and competing sorbates and sorbents. The item listed as "nature of contaminant loading" refers to quantity, location, timing, and mode (i.e., point or non-point source) of contaminant release. Sorbent characteristics include composition; size; shape; pore structure; and surface charges. Sorbate characteristics include molecular structure, which itself is controlled by polarity, size, shape and electric charge. The effects of these factors on contaminant transport are discussed by Karickhoff (1981), Lyman (1985), and Dzombak and Morel (1987).

2. CONTAMINATED SEDIMENT TRANSPORT MODEL

The modeling system HSCTM-2D consists of coupled modules which simulate two-dimensional (2D), depth-averaged surface water flow, cohesive and cohesionless sediment transport, dissolved contaminant transport, and particulate contaminant transport. HSCTM-2D is an unsteady finite element model that uses the Galerkin method of weighted residuals to solve the governing equations of each module to determine the spatial and temporal variations in the field variables. Interactions between dissolved (i.e., desorbed) and participate (i.e., sorbed) contaminants and sediments are accounted for by simulating the processes of sorption and desorption of contaminants to and from sediments, respectively. A brief description of each module is given below. Details are given by Hayter and Mehta (1986).

2.1. Hydrodynamic module

The equations of motion (conservation of momentum) and continuity (conservation of mass), given below, are solved for the flow depths and nodal velocity components.

$$\frac{\partial h}{\partial t}+h\left(\frac{\partial u}{\partial x}+\frac{\partial v}{\partial y}\right)+u\frac{\partial h}{\partial x}+v\frac{\partial h}{\partial y}=0 \tag{1}$$

$$\frac{\partial u}{\partial t}+u\frac{\partial u}{\partial x}+v\frac{\partial u}{\partial y}+g\frac{\partial b}{\partial x}+g\frac{\partial h}{\partial x}-\varepsilon_{xx}\frac{\partial^2 u}{\partial x^2}-\varepsilon_{xy}\frac{\partial^2 u}{\partial y^2}-\tau_x=0 \tag{2}$$

$$\frac{\partial v}{\partial t}+u\frac{\partial v}{\partial x}+v\frac{\partial v}{\partial y}+g\frac{\partial b}{\partial y}+g\frac{\partial h}{\partial y}-\varepsilon_{yx}\frac{\partial^2 v}{\partial x^2}-\varepsilon_{yy}\frac{\partial^2 v}{\partial y^2}-\tau_y=0 \tag{3}$$

where u, v = depth-averaged velocity components in the x- and y-directions [m s^{-1}], respectively; t = time [s]; g = acceleration due to gravity [m s^{-2}]; h = water depth [m]; b = bottom elevation [m]; ρ = water density [kg m^{-3}]; ε_{ij} = turbulent exchange coefficient tensor [m^2 s^{-1}]; and τ = external traction [m s^{-2}]. The turbulent exchange coefficient tensor is equal to the eddy viscosity tensor multiplied by the water density. The external traction includes bottom and surface friction and the Coriolis force. The components of the external traction are given by:

$$\tau_x=\frac{gu\left(u^2+v^2\right)^{0.5}}{C^2h}-\frac{\zeta}{h}W^2\cos\psi-2\Omega v\sin\varphi \tag{4}$$

$$\tau_y=\frac{gv\left(u^2+v^2\right)^{0.5}}{C^2h}-\frac{\zeta}{h}W^2\sin\psi+2\Omega u\sin\varphi \tag{5}$$

where C = Chézy coefficient [$m^{0.5}$ s^{-1}]; ζ = empirical wind shear coefficient; W = wind speed [m s^{-1}]; ψ = wind direction; Ω = angular ψ speed of earth's rotation [s^{-1}]; φ = local latitude. The first term on the right-hand-side represents bottom friction, the second term represents wind shear stress, and the last term represents the Coriolis force. In the application to the Maurice River system, C was spatially varied to obtain optimal agreement between the predicted and measured water surface elevations. Assumptions incorporated in (1), (2) and (3) include: 1) the water is incompressible, 2) vertical variations in velocities are negligible, 3) external (i.e., bottom and surface) friction is uniformly distributed over depth, and 4) pressure is hydrostatic. The predicted flow field at each time step is used in the other modules to solve for the advective and dispersive transport of the represented constituents. This module also includes a drying and wetting routine that adds elements representing floodplains to the grid during higher flows associated with storm events and deletes floodplain elements when the storm hydrograph recedes. Because of the extensive vegetated floodplains along the Maurice River, this feature is particularly vital to be able to accurately represent the transport of sediment and sorbed *As* during storm events when the majority of sediment and *As* loads are typically transported. During the rising limb of a storm event hydrograph, sediment will be transported out onto the floodplains, while during the falling limb of the hydrograph, a sizeable fraction of the sediment load is deposited on the floodplains. Thus, it is necessary to include the floodplains in the grid to simulate this source of contaminated sediments.

2.2. Cohesive sediment transport module

Algorithms that simulate re-entrainment, resuspension, aggregation, dispersion, settling, deposition, bed formation and bed consolidation of up to three sediment sizes (or fractions) are included in this module. Changes in the bed elevation due to erosion, deposition and consolidation are predicted for each active (i.e., wet) element.

The principle of conservation of mass with appropriate source and sink terms describes the advective and dispersive transport of suspended sediment in a turbulent flow field. Thus, the governing equation for cohesive sediment transport is the two-dimensional advection-dispersion equation, given by:

$$\frac{\partial C}{\partial t}+\frac{\partial}{\partial x}(uC)+\frac{\partial}{\partial y}(vC)=\frac{\partial}{\partial x}\left(D_{xx}\frac{\partial C}{\partial x}+D_{xy}\frac{\partial C}{\partial y}\right)+\frac{\partial}{\partial y}\left(D_{yx}\frac{\partial C}{\partial x}+D_{yy}\frac{\partial C}{\partial y}\right)+S \quad (6)$$

with $C = C_s$ [kg m^{-3}] (where C_s = mass concentration of suspended sediment); D_{ij} = effective sediment dispersion coefficient tensor [m^2 s^{-1}]; and S = source/sink term, expressed as:

$$S = \left.\frac{dC}{dt}\right|_e + \left.\frac{dC}{dt}\right|_d + S_L \quad (7)$$

where the first term on the right hand side is the rate of sediment addition (source) due to erosion from the bed, and the second term is the rate of sediment removal (sink) due to deposition of sediment. S_L accounts for removal (sink) of a certain mass of sediment, for example, by dredging in one area (e.g., navigational channel) of a water body, and dumping (source) of sediment as dredge spoil in another location.

The governing equation (6) for two-dimensional, depth-averaged transport of suspended sediment in a turbulent flow field includes dispersive transport terms that account for transport of sediment by processes other than advective transport. Some of these processes include the effects of spatial (i.e., transverse and vertical) velocity variations and turbulent diffusion. Thus, the effective dispersion coefficient tensor in (6) must include the effect of all processes whose scale is less than the grid size of the model, or, what is averaged over time and space (Fischer et al., 1979).

The sediment mass deposition rate, d, is given by:

$$d = -P_d W_s C_s \quad (8)$$

where P_d = dimensionless probability of deposition ($P_d = 1-\tau_b/\tau_{cd}$, where τ_b = bed shear stress [Pa], and τ_{cd} = critical shear stress for deposition [Pa]), and W_s = aggregate settling velocity [m s^{-1}]. Both parameters are calculated as functions of the local suspension concentration and bed shear stress. From laboratory tests on sediment collected in the Maurice River, τ_{cd} was determined to be equal to 0.1 Pa, and W_s was found to vary as C_s to the 3/2 power when C_s was greater than approximately 0.30 kg m^{-3}. For C_s less than 0.30 kg m^{-3}, W_s had an approximately constant value of 0.00015 m s^{-1}. The sink term in (7) is given by (8) divided by the local flow depth (Hayter, 1986).

Deposited sediment is represented as overburden in the bed consolidation module, which consists of the finite strain consolidation model developed by Cargill (1982). The model predicts the change in bed structure (i.e., bulk density) over the bed thickness in response to sediment self-weight and deposited sediments. Constitutive relationships between void ratio and effective stress and between void ratio and permeability needed for the consolidation module were determined by analysis of sediment cores collected in Union Lake and the Maurice River.

For delineating the mode of erosion the bed undergoes when subjected to an excess bed shear, the thickness of deposited sediment that has a bulk density greater than 1,200 kg m^{-3}, as determined in laboratory tests, is simulated to undergo resuspension using the resuspension rate, e, given by Parchure and Mehta (1985):

$$e = e_f \exp\left[\alpha\left(\tau_b - \tau_s\right)^{1/2}\right] \tag{9}$$

where e_f = floc erosion rate [kg $m^{-2}s^{-1}$]; α = factor that is inversely proportional to the absolute temperature [$Pa^{-1/2}$]; τ_b = bed shear stress; and τ_s = bed shear strength [Pa]. The values determined in laboratory testing for e_f, α and τ_s for the Maurice River sediment samples were 1.7x10^{-6} kg $m^{-2}s^{-1}$, 9.0 $Pa^{-1/2}$, and 0.17 Pa, respectively. The source term in (7) is given by (9) divided by the local flow depth.

The portion of the bed with a bulk density less than 1,200 kg m^{-3} is simulated to undergo re-entrainment (i.e., mass erosion). Part of this unconsolidated bed is re-entrained when τ_b is greater than the surface shear strength of the unconsolidated bed, i.e., $\tau_s(Z = 0)$, where Z is the depth below the initial bed surface. The thickness of the unconsolidated bed that is instantly (in one time-step) re-entrained is equal to Z_*, where Z_* is the bed depth at which $\tau_s(Z) = \tau_b$. Z_* is determined from the $\tau_s(Z)$ profile, the latter determined using cores collected in the Maurice River.

Once eroded, cohesive sediment is transported entirely as suspended load by the flow. Such transport is the result of three processes: advection, turbulent diffusion, and longitudinal dispersion. The dispersion algorithm in HSCTM-2D simulates shear flow dispersion only, and thus is most applicable to a vertically well-mixed water body. Following the analysis of Holley et al. (1970), it is assumed that dispersion in such water bodies is associated primarily with the vertical shear. These limitations are consistent with those associated with a depth-averaged model.

The dispersion tensor derived by Fischer (1978) for two-dimensional, depth-averaged bounded shear flow is used in the dispersion algorithm to model shear flow dispersion of cohesive sediments. The four components of this tensor are:

$$\begin{aligned} D_{xx} &= 0.30\frac{u^2 h}{u_f} \\ D_{xy} &= 0.30\frac{uvh}{u_f} \\ D_{yx} &= 0.30\frac{uvh}{u_f} \\ D_{yy} &= 0.30\frac{v^2 h}{u_f} \end{aligned} \tag{10}$$

where u_f is the shear velocity. Values of D_{ij} are calculated at each time step in the model using the specified nodal values of u, v and h. The physical interpretation of the cross product dispersion coefficients D_{xy} and D_{yx} is that a velocity gradient in the x (or y) direction can produce mass (dispersive) transport in the y (or x) direction. The derivation of (10) is given by Hayter et al. (1999).

2.3. Cohesionless sediment transport module

Cohesionless sediment transport occurs when the hydrodynamic forces (i.e., bed shear) acting on sediment particles at the bed surface exceed the resisting forces of interparticle friction and gravity. Thus, estimation of sediment transport requires calculation of the flow induced bed shear stress (or shear velocity). The equation used in this module to calculate the bed shear stress is the following Darcy-Weisbach type relationship:

$$\tau_c = \frac{1}{2}\rho f_c |V| V \tag{11}$$

where ρ = fluid density, V = current speed, and f_c = current friction factor, given by the well known relationship from turbulence theories (Christoffersen, 1982):

$$\left(\frac{2}{f_c}\right)^{1/2} = 2.5 \ln\left(\frac{11.04h}{k_N}\right) \tag{12}$$

in which k_N is Nikuradse's roughness.

Two methods for predicting the non-cohesive sediment transport rate are incorporated in this module. These are the Einstein methodology (Simons and Senturk, 1976) and the Ackers-White algorithm (Ackers and White, 1973). The latter is used when the median diameter, d_{50}, is greater than 0.04 mm. When d_{50} is less than 0.04 mm, both methods are used and the results are compared.

When sediment transport away from a point is not equal to that towards the same point, erosion or deposition will occur causing changes in the bottom elevation. Erosion will result if there is a net transport of sediment away from the point, while deposition will occur if there is a net sediment transport towards it. Using a control volume approach in two horizontal dimensions, Fahien (1983) presented a differential balance of sediment volume flux and accretion/scour. The resulting sediment volume conservation equation is given by:

$$\frac{\partial b}{\partial t} + \frac{\partial q_x}{\partial x} + \frac{\partial q_y}{\partial y} = 0 \tag{13}$$

where b = local bed surface elevation, and q_x, q_y = components of the sediment load transport per unit width in the x- and y-directions [m^2 s^{-1}], respectively, as determined using either the Einstein or Ackers-White equation. Equation (13) is used to calculate the nodal bottom elevation change. The sediment size fractions in the bed are also updated depending on the quantities eroded and deposited from each size fraction. The predicted bathymetric changes at a given time step are used in the hydrodynamic module during the next time step to predict the new flow field. The total sediment transport rate is determined by: 1) calculating the transport rates of both cohesive and cohesionless sediments, 2) multiplying the rates by the respective size fraction in the top bed layer, 3) adding the two rates, and 4) adding the transport rates from upstream sources.

2.4. Contaminant transport module

In HSCTM-2D sorption of dissolved contaminants onto suspended sediments and the bed surface and desorption of particulate contaminants from these surfaces into the water column are the modeled physico-chemical processes that affect the fate of contaminants. The sorption module is a thermodynamic equilibrium model that assumes constant partitioning between the dissolved and particulate phases. It is based on the following assumptions:

a. Local equilibrium between solute and solid phases is instantaneously achieved. Thomann and DiToro (1983) found that equilibrium is usually attained within minutes. Thus, assuming that equilibrium is achieved over one time-step in the model (one-hour time-steps were used for the Maurice River) is very reasonable.
b. The relationship between dissolved and particulate concentrations given by the sorption isotherm is approximately linear for low contaminant concentrations (O'Connor, 1980). Thus, a single partition coefficient can be used.
c. The sorption and desorption processes are completely reversible. Thus, a single partition coefficient can be used to represent both processes.

The equilibrium partition coefficient for a toxicant on sediment is given by:

$$K_p = \frac{C_p}{C_d C_s} \tag{14}$$

where C_p = particulate contaminant concentration [kg m^{-3}] {adsorbed mass of contaminant divided by the total volume}; and C_d = dissolved contaminant concentration [kg m^{-3}] {dissolved mass of contaminant divided by the total volume}. The total contaminant concentration, C_t [kg m^{-3}], is given by:

$$C_t = C_p + C_d = C_d(1 + K_p C_s) \tag{15}$$

For this study, partition coefficients were determined by laboratory analysis of the collected sediment cores.

Dissolved contaminant transport module. The following processes are represented in modeling the transport of dissolved contaminants: sorption, desorption, chemical degradation, and point sources of contaminants. The governing equation for dissolved contaminant transport is (6) with $C = C_d$ and S given by:

$$S = -\sum K_i(K_{pi}C_dC_{si} - C_{pi}) - \sum K_i\gamma_i(1-P)\frac{D_i}{h}K_dC_d + \sum K_i\gamma_i(1-P)\frac{D_i}{h}(C_{pi})_{Bi} + S_d - \lambda C_d \tag{16}$$

where K_i = sorption/desorption rate to approach equilibrium with the ith sediment fraction; K_{pi} = partition coefficient between particulate contaminant associated with ith sediment and dissolved contaminant; C_{si} = suspended mass concentration associated with ith sediment; C_{pi} = particulate contaminant concentration associated with ith sediment; P = porosity of the sediment bed; D_i = diameter of the ith sediment in the bed [m]; γ_i = density of ith sediment in the bed [kg m^{-3}]; $(C_p)_{Bi}$ = particulate contaminant concentration associated with the ith sediment in the bed; λ = bio-chemical degradation rate [s^{-1}]; and S_d = source strength of dissolved contaminant [kg m^{-3}s^{-1}]. The two terms inside the first summation on the right hand side represent the sorption rate of dissolved contaminants and the desorption rate of particulate contaminants onto and from suspended sediments, respectively. The next two terms represent the rate of sorption of dissolved contaminants onto bed sediments (i.e., sink) and the rate of desorption of particulate contaminants from bed sediments into the water column (i.e., source), respectively. S_d was used to represent the flux of *As* contaminated groundwater into a tributary of the Maurice River.

Particulate contaminant transport module. The following processes are accounted for in modeling the transport of particulate contaminants: sorption, desorption, local sources/sinks, and chemical degradation. The transport of contaminants sorbed to sediments is determined separately for each of the three size fractions. The governing equation for particulate contaminant transport is (6) with $C = C_{pi}$ and S given by:

$$S = \sum[-K_i(C_{pi} - K_{pi}C_{si}C_d) + \frac{d}{h}C_{pi} + \frac{e}{h}(C_p)_{Bi} + S_{pi} - \lambda C_{pi}] \tag{17}$$

where S_{pi} = source strength of particulate contaminant associated with the ith sediment fraction [kg m^{-3}s^{-1}]. The first two terms on the right hand side of (17) represent the desorption rate of particulate contaminants and the sorption rate of dissolved contaminants. The third and fourth terms represent the deposition rate and resuspension rate of particulate contaminants, respectively.

3. MAURICE RIVER-UNION LAKE ARSENIC TRANSPORT MODELING

The application of the HSCTM-2D model to the Maurice River is discussed following descriptions of the site and contamination problem.

A chemical plant (ViChem) manufactured organic *As* herbicides and fungicides at a site in Cumberland County, New Jersey from approximately 1949 to the mid

1980s (see Figure 1). The plant is located next to the Blackwater Branch, which flows approximately 2.4 km downstream to the Maurice River (see Figure 2). Approximately 96,300 m^3 of soils at the site were contaminated with *As* above the action level concentration of 20 mg kg^{-1} (Ebasco, 1991). The groundwater in the unconfined aquifer beneath the plant was also contaminated, with *As* concentrations exceeding 300 mg l^{-1}. Contaminated groundwater discharged into the Blackwater Branch downstream of the plant. Approximately six metric tons of *As* per year entered the Blackwater Branch through this pathway. It is estimated that more than 500 metric tons of *As* have entered the Blackwater Branch and subsequently transported downstream to the Maurice River (Ebasco, 1991).

The Maurice River flows into Union Lake, which is located approximately 11 km downstream from the confluence with the Blackwater Branch. Union Lake is an impoundment on the Maurice River, with a surface area of approximately 3.52 km^2 (Figure 2). The Maurice River flows into Delaware Bay approximately 40 km downstream of Union Lake. Previous investigations have found *As* contamination in the Maurice River up to 50 km downstream of the chemical plant.

Both sediments and surface waters in Blackwater Branch, Maurice River, and Union Lake are contaminated with *As*. Surface waters in the Maurice River and

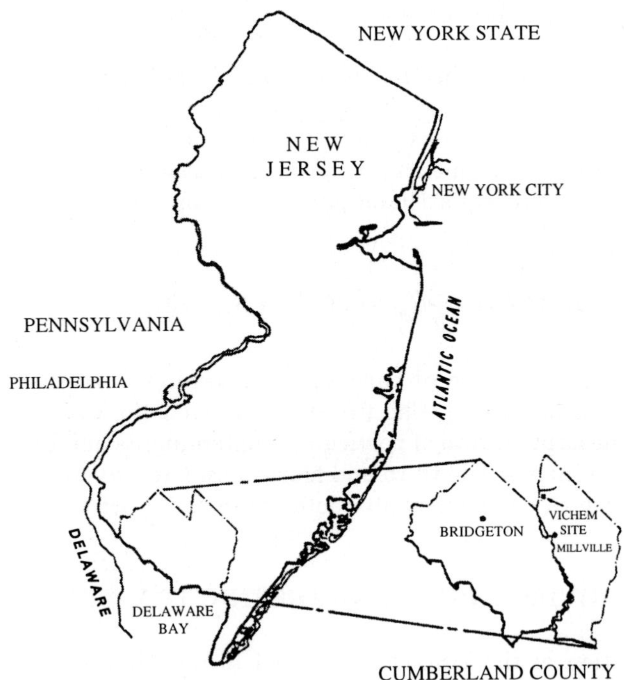

Figure 1. Site of Vichem plant (after Ebasco, 1992).

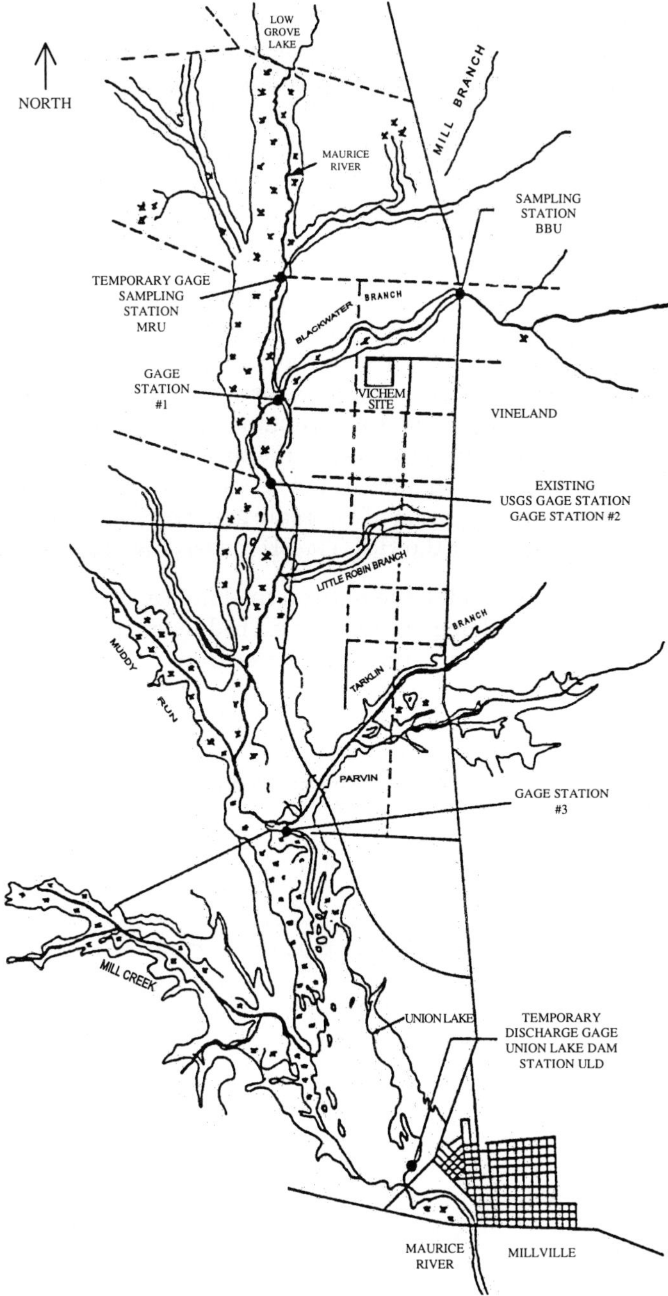

Figure 2. Blackwater Branch, Maurice River, and Union Lake (after Ebasco, 1992). Scale: 1 cm - 1.26 km.

Union Lake have *As* concentrations above 0.050 mg l^{-1}, which is the New Jersey standard for *As* in fresh surface waters. The sediment contamination includes bed sediments in the Blackwater Branch, the Maurice River and Union Lake, and floodplain sediments along the Blackwater Branch. The U.S. EPA determined from a risk assessment that *As* contaminated sediments with concentrations greater than 120 mg kg^{-1} in submerged sediments posed excess carcinogenic risks to humans above the action threshold $1x10^{-5}$ (Ebasco, 1991). The sediment *As* concentrations in this riverine system are more than one order of magnitude higher than this.

The remedial measures specified that remediation would proceed as follows: 1) contaminated sediments in the Blackwater Branch (both in the branch and in the floodplain) would be remediated; 2) a three year natural flushing period would commence to allow the contaminated sediments in the Maurice River to remediate themselves naturally, owing to the concern over potential environmental damage which may be caused by dredging; 3) contaminated sediments in the Maurice River would be remediated, if necessary, following the three year flushing period; and 4) contaminated sediments in Union Lake would be remediated, if necessary, following the flushing period and remediation of the Maurice River sediments.

The remedial strategies that the HSCTM-2D model was used to simulate were:

(1) No Action - used to determine the natural flushing time of the system;
(2) Blackwater Branch Dredging - contaminated bed sediments were replaced with uncontaminated sediments in the Blackwater Branch;
(3) Maurice River Dredging - contaminated bed sediments were replaced with uncontaminated sediments in the Blackwater Branch and the Maurice River;
(4) Union Lake Dredging - contaminated bed sediments were replaced with uncontaminated sediments in the Blackwater Branch, the Maurice River and Union Lake.

The data required to perform the modeling included the following:

- Recorded daily water levels and discharges (determined from the stage-discharge rating curves) over a three-year period at the gaging stations (Stations MRU, BBU, 1, 2, 3, and at the Union Lake Dam) shown in Figure 2, and periodic discharge measurements in the tributaries of the Maurice River. Strong correlations ($r^2 > 0.95$) were found between all the measured tributary flows and the measured stage at Station 2. These correlations were used to construct the discharge hydrographs for the tributaries.
- Cross-sectional surveys at the transects in the Maurice River and tributaries shown in Figure 3, and a bathymetric map of Union Lake.
- Sediment cores at the transects shown in Figure 3 to determine grain size distributions, mineral composition, mean bed density, percent organic material, and total sorbed *As*.
- Pore water samples at selected locations in the system to determine dissolved concentrations of *As*, and the pH and conductivity of the pore water.
- Measurements of flows, suspended sediment concentrations, and dissolved and sorbed *As* concentrations in Union Lake, the Maurice River and its tributaries

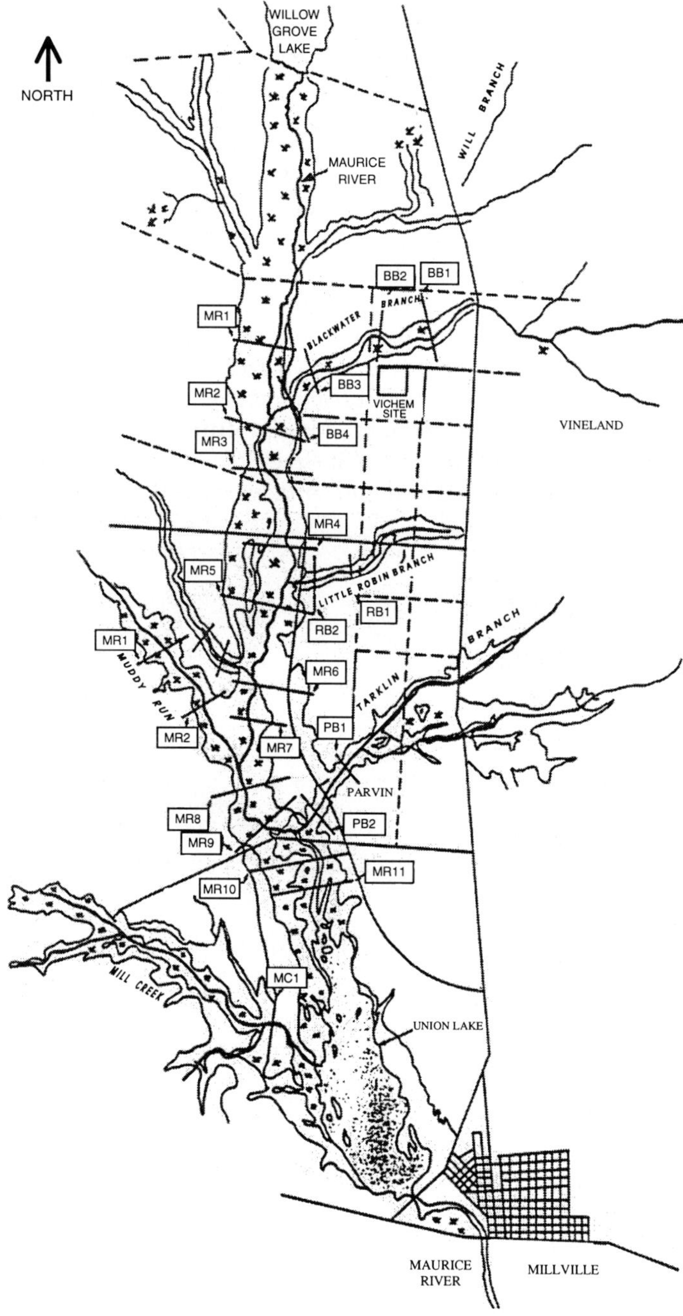

Figure 3. Location of sampling transects - labeled inside small rectangles (after Ebasco, 1992). Scale: 1 cm = 1.26 km.

during both normal and storm flow conditions. At least three storm-events and three baseflow events were measured per year for the three years of field measurements.

Two finite element grids representing this system were developed using the bathymetric surveys of Blackwater Branch, Maurice River, the five tributaries between Blackwater Branch and Union Lake, and Union Lake. Figures 4 and 5 show the finite element grids for the upper and lower regions, respectively. The upper system grid (1494 elements, 4369 nodes; Figure 4) starts upstream of the chemical plant on the Blackwater Branch, and at Garden Road on the Maurice River, and ends at Station No. 2 gaging station. The lower system grid (2773 elements, 7767 nodes; Figure 5) starts just downstream of Station No. 2 and extends to the Union Lake dam. The five tributaries included in these two grids are the Little Robin Branch, Tarkiln Branch, Muddy Run, the unnamed tributary immediately north of Muddy Run, and Mill Creek. To include the floodplains in the grid, 2D triangular and quadrilateral elements were used throughout the grid, as

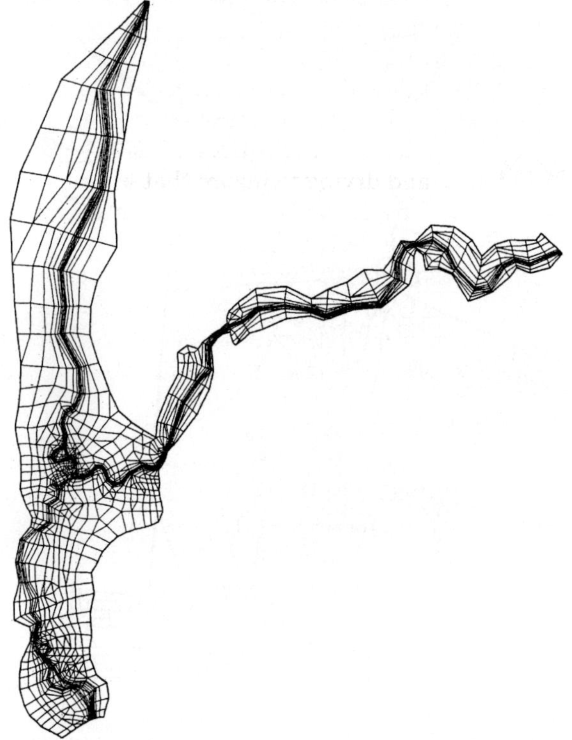

Figure 4. Finite element grid for Blackwater Branch and the Upper Maurice River. Scale: 1 cm = 196.5 m.

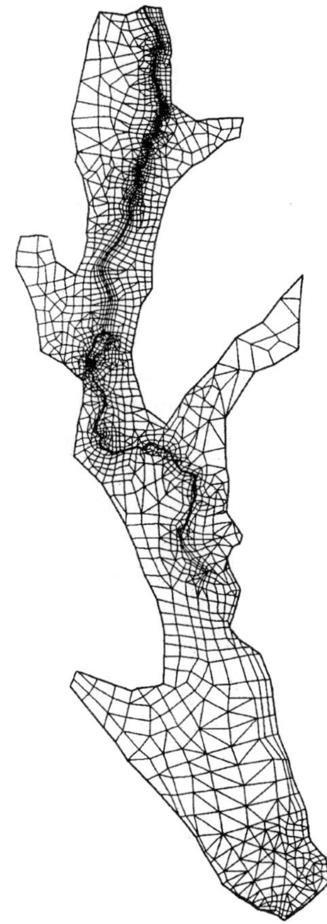

Figure 5. Finite element grid for the Lower Maurice River and Union Lake. Scale: 1 cm = 706.9 m.

seen in Figures 4 and 5. The grid for Union Lake (seen in Figure 5) is more detailed to accurately represent features such as islands. Typically, the rivers and tributaries were represented using a two- to four-element wide grid, while the floodplains were represented using a coarser grid. For example, Figure 6 shows the delineation between the Maurice River-Blackwater Branch (inside the narrow parallel lines) and the adjacent floodplains.

Next, the input data sets for the hydrodynamic module were developed using the three-years of recorded flows at the gaging stations and the calculated tributary flows. The hydrodynamic module was run for the first period of stream flow record (i.e., first year of measurement) to initiate calibration of this module. The second year of measurements, during which a five-year event occurred, was used to refine the calibration. The third year of measurements were used to partially validate the flow module. Satisfactory degrees of calibration and validation were achieved. Particular attention was paid to the model's ability to simulate storm events. This entailed an examination of the simulations of floodplain wetting and drying to insure that water mass was adequately conserved. When water was predicted to flow onto the floodplains during simulation of storm events, a maximum of 10% *of the water that flowed onto the floodplains* was not conserved. The 10% loss occurred during the simulated five-year event. The loss of water mass was less than 5% for the other simulated storm events.

The input data sets for the sediment and contaminant transport modules were developed using the measured sediment and *As* concentrations during the six measured events (three storm events and three baseflow events) per year and the other measured sediment and *As* properties. As previously discussed, the erosional, depositional, and consolidation properties of the cohesive sediment size fractions were determined in laboratory tests. Additional sediment and *As* data that were to be measured, as specified in the work plan, included: bed grain size distributions; mean bed density; mineral composition of bed material; sorbed *As* concentrations on the bed sediments; pore water temperature, pH, conductivity, and *As* concentrations; suspended sediment concentrations and grain size distributions; dissolved *As* concentrations in the surface waters; sorbed *As* concentrations on suspended sediment; *As* desorption rates from sediments; and sedimentation rates in 10 sediment boxes that were installed in the Maurice River and Union Lake.

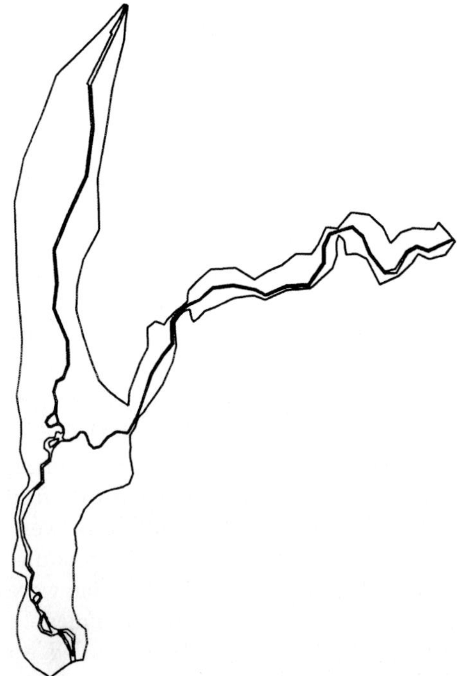

Figure 6. Delineation of the Upper Maurice River and Blackwater Branch inside the narrow parallel lines and the adjacent floodplains outside the parallel lines.

Unfortunately, all of the sediment boxes were lost. Thus, data that would have been used to calibrate the deposition routine were not available. In addition, the *As* desorption rates were not determined as specified in the work plan.

Using these input data sets, and the previously calibrated and partially validated hydrodynamic module, the sediment and contaminant transport modules were run for the three years of record to compare field measurements at the interior sampling stations with the model simulations. Particular attention was paid to the ability of the model to simulate sediment transport during the recorded storm events, when the majority of the sediment load was transported through the system. The measured total suspended solids (TSS) during the monitored events were compared with the model simulations using the experimentally determined erosional, settling and depositional parameters. Over the three years of model simulations, the maximum differences observed between measured and simulated TSS were bounded by -85% to 160%. The average difference was determined to be 35%, with the simulated TSS on average higher than the measured. Because this level of agreement was, considering the previously discussed model and data limitations, deemed satisfactory, the erosional, settling and depositional parameters were not adjusted to achieve a better agreement between the field measurements and the model results. As mentioned above, data that were to be used to calibrate the deposition routines were not collected due to the loss of all ten sedimentation boxes.

Representing a system as complex as the Blackwater Branch-Maurice River-Union Lake system typically involves numerous assumptions. In particular, two physical "components" of this system made it more difficult than most to simulate, the first being the discharge of groundwater from the unconfined aquifer into Blackwater Branch, the second being the manner in which the floodplains flood or are inundated during an extreme rainfall event. The former was represented as a line source along the lower portion of the Blackwater Branch. The latter actually occurs, on at least part of the floodplain, by the water table rising above the ground level on both sides of the river, as opposed to water flowing over the channel banks and onto the floodplains as was simulated by HSCTM-2D. Since HSCTM-2D only simulates surface water systems, it is not possible to explicitly represent the interaction of surface and ground waters as occurs in these two components. This limitation of the model must be considered when interpreting the modeling results.

4. RESULTS OF MODEL SIMULATIONS

Table 1 given below contains estimates of the flushing time of sorbed *As* from the bed sediments in the modeled system under the previously described remediation scenarios. The flushing time is defined as the time if would take for the *As* concentrations to decrease to 120 mg/kg throughout the system. The estimates were determined using hydrographs constructed by repeating the previously described three-year flow, sediment and *As* hydrographs. In addition, these estimates were based on the assumption that the source of *As* to the Blackwater Branch (i.e., contaminated groundwater) was eliminated after three years.

As implied by the title, in the "No Action" simulation the existing system was simulated to determine the natural flushing time. In the "Blackwater Branch Dredging" simulation, only the elements that represented the Blackwater Branch downstream of the chemical plant were "dredged". The latter was simulated by decreasing the sorbed *As* concentration in the top 2.0 m of bed sediments to 120 mg/kg of sediment. It did not seem realistic to assume that all *As* would be removed by the dredging activity. In addition, the simulated dredging did not include excavation of the contaminated floodplain sediments. This also applied to the "Maurice River Dredging" simulation, in which dredging of contaminated bed sediments in the Maurice River as well as in the Blackwater Branch were represented. In the "Union Lake Dredging" simulation, all contaminated sediments in water depths exceeding 2.5 m were dredged, along with the sediments in the Blackwater Branch and Maurice River.

In all four of these multi-year simulations, more than 90% of the sediment mass that was transported into Union Lake deposited, most in the vicinity of the deepest portion of the lake, just upstream of the dam. The remaining sediment was transported in suspension out of the system, i.e., over the dam to the lower reach of the Maurice River. Similarly, more than 85% of the *As* mass that was transported

Table 1
Estimated arsenic flushing times using HSCTM-2D

Remediation action	Flushing time (years)
No Action	25
Blackwater Branch Dredging	12
Maurice River Dredging	8
Union Lake Dredging	4

into Union Lake, either sorbed to suspended sediment or in the dissolved phase, was temporarily stored in the sediments that deposited in Union Lake.

Due to the previously discussed model limitations, the flushing times given in Table 1 for the four remedial scenarios are thought to be "ball-park" estimates. The estimated flushing times are thought to be somewhat lower than the actual flushing times since simulated TSS values were, on average, 35% higher than the measured values. This would tend to decrease the estimated flushing times since the higher simulated TSS indicate more rapid transport of sediment, and therefore bound *As*, through the system. To quantify the impact of the higher simulated TSS values on the flushing time, the no action scenario was re-run using τ_{cd} = 0.16 Pa (as opposed to 0.1 Pa), just less than the value τ_s = 0.17 Pa. Increasing the value of τ_{cd} has the effect of increasing the amount of deposition that occurs. For this simulation, the average difference between the measured and simulated TSS was determined to be -15%, i.e., the simulated TSS was on average 15% lower than the measured, and the estimated flushing time was 33 years.

An approximate estimate for the error associated with the estimated flushing times in Table 1 is $\pm$ 1.5 years. This estimated error was determined from sensitivity analysis performed in which the reported range of values from the scientific literature were used for the *As* desorption rate. As previously mentioned, the desorption rate was not experimentally determined as specified in the project work plan. It is recognized that the reported estimated error is limited since sensitivity analysis was performed for only one parameter. Sensitivity analyses of the measured range of values of, among others, K_p, τ_{cd}, τ_s, α, W_s, and e_f were not performed. The calculated flushing times, even accounting for the model limitations and estimated error, provide remediation project managers with the necessary scientific information to decide which remediation scenario(s) should be implemented.

5. CONCLUSIONS

The application of the HSCTM-2D model to predict estimated *As* flushing times in the Maurice River-Union Lake system demonstrated: 1) the necessity of modeling the transport of both cohesive and non-cohesive sediments when simulating the transport and fate of sorbed contaminants, even when the size fraction of the cohesive sediments in the system was less than 20%, in that the estimated flushing times were much greater when only non-cohesive sediments were modeled, and 2) the limitations imposed on the analysis that result from incomplete data and model-imposed simplifications of the physical system modeled.

6. DISCLAIMER

This paper has been reviewed in accordance with the U.S. Environmental Protection Agency's peer and administrative review policies and approved for publication. Mention of trade names or commercial products does not constitute endorsement or recommendation for use by the U.S. EPA.

REFERENCES

Abernathy, A. R., Larson, G. L., and Matthews, Jr., R. C., 1984. Heavy Metals in the surficial sediments of Pontant Lake, North Carolina. *Water Resources*, 18, 351-354.

Ackers, P., and White, R. W., 1973. Sediment transport: new approach and analysis. *Journal of the Hydraulics Division*, ASCE, 99(11), 2041-2060.

Baughman, G. L., and Burns, L. A., 1980. Transport and transformation of chemicals: a perspective. In: The *Handbook of Environmental Chemistry - Reactions and Processes*, O. Hutzinger ed., Springer-Verlag, Berlin, 2(A), 1-16.

Brook, E. J., and J. N. Moore, 1988. Particle-size and chemical control of As, Cd, Cu, Fe, Mn, Ni, Pb, and Zn in bed sediment from the Clark Fork River, Montana (USA). In: *The Science of the Total Environment*, Elsevier, Amsterdam, 76, 247-266.

Brown, D. S., and Allison, J. D., 1987. MINTEQA2, An equilibrium metal speciation model: User's Manual. *EPA/600/3-87/012*, U.S. Environmental Protection Agency, Environmental Research Laboratory, Athens, Georgia.

Brown, K. P., Hosseinipour, E. Z., Martin, J. L., and Ambrose, R. B., 1990. Application of a water quality assessment modeling system at a superfund site. *Technical Report*, U.S. Environmental Protection Agency, Environmental Research Laboratory, Athens, Georgia.

Cargill, K. W., 1982. Consolidation of soft layers by finite strain analysis. *Miscellaneous Paper GL-82-3*, U.S. Army Engineer Waterways Experiment Station, Vicksburg, Mississippi.

Christoffersen, J. B., 1982. Current depth refraction of dissipative water waves. *Series Paper 30*, Institute of Hydrodynamics and Hydraulic Engineering, Technical University of Denmark, Lyngby.

Dzombak, D. A., and Morel, F. M. M. 1987. Adsorption on inorganic contaminants in aquatic systems. *Journal of Hydraulic Engineering*, ASCE, 113(4), 430-475.

Ebasco. 1991. Final Work Plan for Technical Support Services, Vineland Chemical Company Site, Vineland, New Jersey, Lyndhurst, NJ.

Ebasco. 1992. Final Field Operations Plan for Technical Services, Vineland Chemical Company Site, Vineland, New Jersey, Lyndhurst, NJ.

Elzerman, A. W., and Coates, J. T. 1987. Equilibria and kinetics of sorption of hydrophobic organic compounds on sediments. In: *Sources and Fates of Aquatic Pollutants*, R. A. Hites ed., American Chemical Society, Washington, DC, 263-317.

Fahien, R. W., 1983. *Fundamentals of Transport Phenomena*, McGraw-Hill Book Company, New York.

Fischer, H. B., 1978. On the tensor form of the bulk dispersion coefficient in a bounded skewed shear flow. *Journal of Geophysical Research*, 83(C5), 2373-2375.

Fischer, H. B., Imberger, J., List, E. J., Koh, R. C. Y., and Brooks, N. H., 1979. *Mixing in Inland and* Coastal *Waters.* Academic Press, New York.

Hayter, E. J., 1986. Estuarial sediment bed model. In: *Estuarine Cohesive Sediment Dynamics, Lecture Notes on Coastal and Estuarine Studies,* A. J. Mehta ed., Springer-Verlag, New York, 14, 215-232.

Hayter, E. J., and Mehta, A. J., 1986. Modelling cohesive sediment transport in estuarial waters. *Applied Mathematical Modelling,* 10(4), 294-303.

Hayter, E. J., Bergs, M. A., Gu, R., McCutcheon, S. C., Smith, S. J., and Whiteley, J. H., 1999. HSCTM-2D, a finite element model for depth-averaged hydrodynamics, sediment and contaminant transport. *Technical Report*, U.S. Environmental Protection Agency, Ecosystems Research Division, Athens, Georgia.

Holley, E. R., Harleman, D. R. F., and Fischer, H. B., 1970. Dispersion in homogeneous estuary flow. *Journal of the Hydraulics Division*, ASCE, 96(HY8), 1691-1709.

Horowitz, A. J., and Elrick, K. A., 1987. The relation of stream sediment surface area, grain size, and composition to trace chemistry. *Applied Geochemistry*, 2, 437-451.

Karickhoff, S. W., 1981. Semi-empirical estimation of sorption of hydrophobic pollutants on natural sediments and soils. *Chemosphere*, 10, 833-846.

Kirby, R., and Parker, W. R., 1973. Fluid mud in the Severn estuary and the Bristol channel and its relevance to pollution studies. *Proceeding of the International Chemical Engineers Exeter Symposium*. Exeter, UK, Paper A-4, 1-14.

Lyman, W. J., 1985. Estimation of physical properties. In: *Environmental Exposure from Chemicals,* W. B. Neely and G. E. Blau eds., CRC Press, Boca Raton, Florida, 2, 13-48.

Medine, A. J., and McCutcheon, S. C., 1989. Fate and transport of sediment-associated contaminants. In: *Hazard Assessment of Chemicals*, 6, J. Saxena ed., Hemisphere, New York City.

O'Connor, D. J., 1980. Physics transfer processes. In: *Modeling of Toxic Substances in* Natural *Water Systems*, Manhattan College, New York.

Parchure, T. M., and Mehta, A. J., 1985. Erosion of soft cohesive sediment deposits. *Journal* of *Hydraulic Engineering*, ASCE, 111(10), 1308-1326.

Reese, D. E., Felkey, J. R., and Wai, C. M., 1978. Heavy metal pollution in the sediments of the Coeur d'Alene River, Idaho. *Environmental Geology*, 2, 289-293.

Simons, D. B., and Senturk, F., 1976. *Sediment Transport Technology.* Water Resources Publications, Fort Collins, Colorado.

Stumm, W., and Morgan, J. J., 1981. *Aquatic Chemistry - An Introduction Emphasizing Chemical Equilibria in Natural Waters.* John Wiley, New York, 780p.

Thomann, R. V., and DiToro, D. M., 1983. Physico-chemical model of toxic substances in the Great Lakes. *CR805916 and CR807853*, U.S. EPA Large Lakes Research Station, Grosse Ile, Michigan.

Coastal and Estuarine Fine Sediment Processes
W.H. McAnally and A.J. Mehta (Editors)

Entrance flow control to reduce siltation in tidal basins

T. J. Smith[a], R. Kirby[a] and H. Christiansen[b]

[a]Ravensrodd Consultants Limited, 6 Queen's Drive, Taunton, Somerset, TA1 4XW, United Kingdom

[b]Strom und Hafenbau, Hamburg Port Authority, Dalmannstrasse 1, 20457 Hamburg II, Germany

A means of controlling the flow in the entrance of a tidal basin, which reduces the siltation rate in the basin mouth, is described. The flow is controlled by a structure close to the basin wall that reduces the strength of the mixing layer across the entrance and the entrance eddy. Measurements of velocity and suspended solids concentration in a real tidal basin are presented together with results from hydraulic model studies. Results are also presented from a prototype installation in the Port of Hamburg. These results show siltation rates were reduced by 50% as a result of entrance flow control.

1. INTRODUCTION

The wide-open entrances to tidal basins are the sites of a long standing, and until now irreconcilable, conflict of interest. In order to permit safe navigation, the entrance needs to be as wide as possible but at the same time, the wider the entrance, the greater the influx of sediment from the tidal waters beyond. Sediment accumulates preferentially in the entrances to such basins and may give rise to a severe problem to port operators. For example, in the Port of Hamburg between 55% and 80% of all material deposited within basins lies in entrance shoals. When fine-grained sediments are involved and these are contaminated, the cost of removal and treatment can be very high.

On the flood tide, sediment is transported into an open basin by a number of mechanisms including:

- tidal filling
- entrainment through the mixing layer across the entrance
- salinity driven, densimetric exchange of water between the basin and the river
- gravity currents generated by near-bed, high concentrations of suspended sediment

The entrance shoal is a well-known, characteristic feature of tidal basins and is shown clearly in Figure 1 for some of the basins in the Port of Hamburg. The existence of this shoal has been linked to the occurrence of a major eddy in the basin entrance generated by the shear layer between outer river water and inner basin water. A schematic diagram of the flow structure in the entrance region of a tidal basin on the flood tide is shown in Figure 2. The characteristic eddy in the basin entrance with associated mixing layer are shown. The structure of the flow in the entrance eddy is shown in Figure 3a. The primary azimuthal flow gives rise to a perturbation in the free surface, which provides a hydrostatic force to balance the radial centrifugal force generated by the eddy rotation.

In the near bed boundary layer, the azimuthal velocity and hence the centrifugal force are reduced, becoming zero at the bed while the radial hydrostatic force remains constant. This gives rise to a radial inflow towards the eddy's center at the bed. Conservation of mass then requires an equal and opposite radial outflow at the surface with an upwelling near the center of the eddy and a downwelling at the periphery. This secondary circulation (also shown in Figure 3a) is believed to play an important role in sediment transport and siltation in the basin entrance. This is illustrated schematically in Figure 3b.

The existence and dynamics of the entrance eddy in tidal basins has been demonstrated in basins of simple geometry using tidal flumes and mathematical

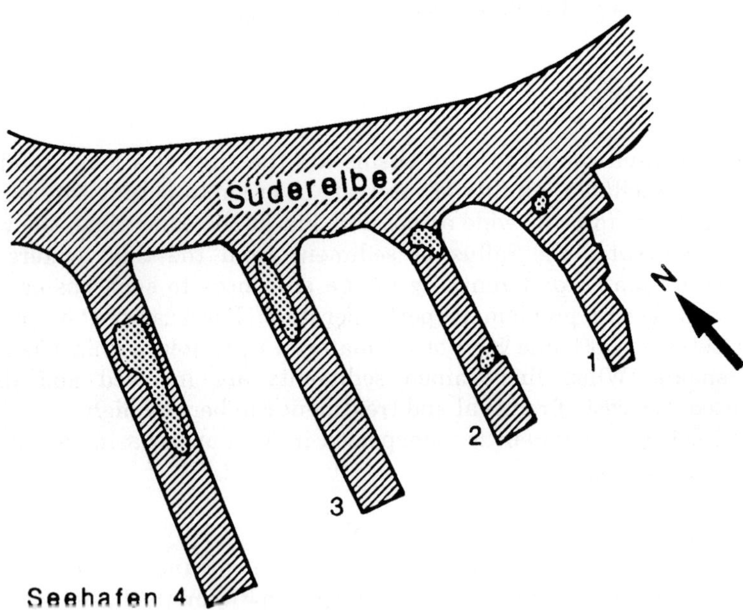

Figure 1. Areas of high siltation, >0.5 m yr^{-1}, (within enclosed dotted areas) in the Seehafen basins, Port of Hamburg, Germany.

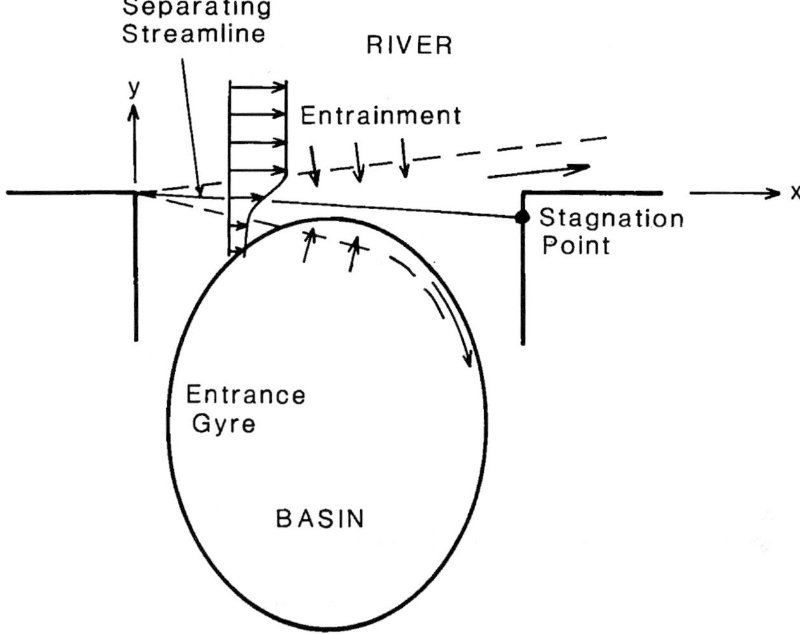

Figure 2. Schematic illustration of the mixing layer across the basin entrance.

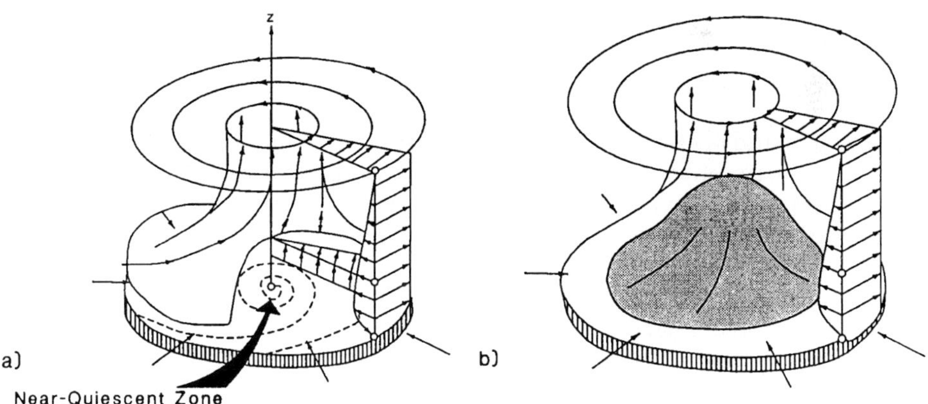

Figure 3. a) Primary and secondary circulation in the entrance gyre. b) Eddy generated sediment build-up.

models and for real basin geometries using hydraulic scale models. However none of these was able to incorporate sediment and hence conclusively demonstrate the interaction of the mixing layer, the entrance eddy and siltation in the basin entrance. In this paper, simultaneous measurements of velocity and suspended solids concentration are described which allowed sediment flux to be determined in the Parkhafen basin in the Port of Hamburg. These measurements demonstrate the secondary currents in the entrance eddy focusing suspended solids at its center. Hence, the causal link between the entrance eddy and the entrance shoal is demonstrated in a full scale tidal basin.

If siltation in the basin is to be reduced, one, some or all of the processes responsible for introducing sediment into the basin must be reduced in magnitude or eliminated completely. This paper considers basins in locations where the salinity difference between inside and outside the basin is sufficiently small that any salinity driven densimetric exchange is negligible. Under these circumstances, it will be shown that the control of the flow in the entrance region of a tidal basin using a device known as a Current Deflecting Wall (Christiansen and Kirby 1991; Kirby, 1994; Winterwerp et al., 1994; Christiansen, 1997) has the effect of reducing siltation in the basin entrance by:

- substantially reducing the strength of the mixing layer across the basin entrance and repositioning that which remains such that sediment entrained across it is less likely to enter the basin
- substantially reducing the strength or eliminating entirely the entrance eddy, thereby removing its effect of concentrating siltation in the basin entrance
- deflecting near bed gravity currents away from the entrance so that they are less likely to penetrate the basin
- filling the basin with water from the upper 75% of the water column where the lowest turbidity is naturally found, thereby reducing the volume of sediment entering the basin through tidal filling

The design and operation of the Current Deflecting Wall (CDW) is described in the context of its effect on the sediment transport processes listed above. The effectiveness of the device is then demonstrated by reference to data from the Köhlfleet in the Port of Hamburg, where a full-scale prototype CDW has been in place since 1990.

In the remainder of this paper, the terms upstream and downstream are used to denote direction relative to the average flow over many tides in the main channel. Hence, the ebb tide flows in the downstream direction while the flood tide flows in the upstream direction. The downstream corner of a tidal basin is that nearest the sea while the upstream corner is that nearest the non-tidal river.

2. MEASUREMENTS OF SEDIMENT TRANSPORT

2.1. Parkhafen, Hamburg

The Parkhafen in Hamburg, Germany, is the port's major basin for container traffic. During 1997, observations were made in the Parkhafen to determine the velocity field and suspended solids distribution in the entrance to the basin throughout a tidal cycle. The objective of the experiments was to directly measure the fluxes of suspended solids into and out of the basin. This was part of a study into the feasibility of entrance flow control in the basin as a means of reducing siltation.

The Port of Hamburg (Figure 4) is located on the Elbe river some 100 km upstream from its confluence with the North Sea. The Elbe is tidal upstream of Hamburg but the saline intrusion stops well short of the port. Hence, the water density in the river and the harbor basins are always equal. Some densimetric currents may occur due to near-bed, localized layers of fluid mud which are known to form (Christiansen and Kirby, 1991). The characteristic features of the Elbe River in the vicinity of Hamburg are given Table 1, while those of the Parkhafen are given in Table 2. The contrast in peak suspended solids concentrations between the river and the basin is due to the concentration effect of the entrance eddy.

2.2. Instrumentation

The measurements were made using a 600 kHz Acoustic Doppler Current Profiler (ADCP) with 20° beam angles. By "range-gating" the emitted pulses, it is

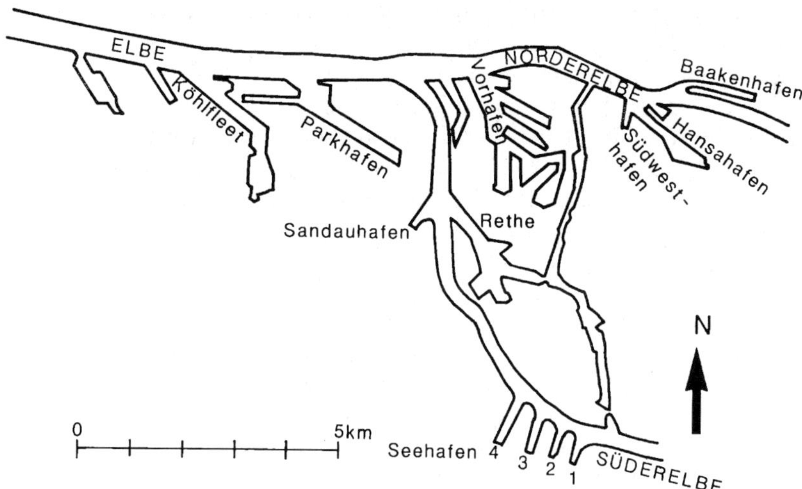

Figure 4. Outline of the Port of Hamburg highlighting the location of the major basins.

Table 1
Characteristics of the Elbe at Hamburg

River discharge (m^3s^{-1})		Tidal range (m)		Suspended solids concentration (mgl^{-1})		Mean depth (m)	Mean width* (m)
High	Low	Spring	Neap	High	Low		
2,400	200	4.0	3.0	100	25	16.8	500

*Outside the Parkhafen

Table 2
Characteristics of the Parkhafen

Mean volume ($m^3 \times 10^{-3}$)	Tidal volume ($m^3 \times 10^{-3}$)	Annual siltation (m^3)		Basin entrance (m)		Suspended solids concentration (mgl^{-1})	
		Basin	Entrance	Width	Depth	High	Low
24,960	4,768	350,000	224,000	515	17	250	25

possible to obtain vertical profiles of the water column with measurements at intervals of depth as small as 0.1 meter. These vertical profiles are virtually instantaneous. Determining the position of the measuring point relative to the ADCP requires that the variation of the speed of sound between the transducer and the measuring point be known. This is obtained by frequent measurements of the vertical profiles of temperature and salinity. The ADCP data processing algorithm then automatically corrects for water density variations along the acoustic path using these temperature and salinity profiles.

It has been recognized for some time that the magnitude of the backscattered energy from the acoustic pulses is related to the concentration of the suspended particles from which it has been reflected. If this relationship can be isolated through a knowledge of the other processes affecting acoustic attenuation and appropriate calibration, an ADCP can provide a non-intrusive means of simultaneously measuring suspended solids concentrations and velocity at the same location. This allows the 3-D mapping of sediment flux in real time.

Provided that sufficient care is taken in deployment, calibration (using frequent pumped or bottle water samples) and data analysis which corrects for known errors, the ADCP is capable of providing estimates of suspended solids concentrations of an accuracy similar to or better than that of conventional siltmeters (both white-light transmissometers and infra-red backscatter devices) up to 2,000 mgl^{-1}. A comprehensive description of the ADCP, its use to measure suspended sediment concentrations and the accuracy obtainable are given in Land et al. (1997).

2.3. Measurements

ADCP surveys were undertaken in the entrance to the Parkhafen during the ebb tide on 20th October 1997, the flood tide on 21st October 1997 and over a full tidal cycle on 7th November 1997. Tidal data are given in Table 3.

The monitoring vessel made repeated traverses within the basin starting at the entrance and working back into the basin. The vessel's track for the traverse between 05:39 PM (LW+1:37) and 06:07 PM (LW+2:05) on 21 October is shown in Figure 5a. The total elapsed time for these observations was 28 minutes.

2.4. Results

Figure 5a also shows the velocity field from the ADCP data, while the synchronous suspended solids distribution is shown in Figure 5b. These data clearly show that the suspended solids are focused towards the center of the eddies which have formed in the basin entrance. This is the case for both the primary and the secondary eddies. If the suspended solids field shown in Figure 5b is compared with the silt accumulation pattern for the Parkhafen shown in Figure 6, the significance of the entrance eddy in controlling sedimentation in the basin entrance becomes apparent. Figure 5a also shows the strong shear layer across the basin entrance. It will be demonstrated in the following section that this shear layer is responsible for enhancing the sediment transport rate into the basin over that from tidal filling alone. Indirect evidence for this is the elevated levels of suspended solids in the core of the primary eddy. These are over 5 times the concentration in the river water and 10 times the average concentration in the basin.

These results clearly indicate that the combination of the shear layer across the basin entrance, together with the primary eddy just inside the basin, play a significant role in determining the rate of accretion in the entrance to tidal basins. Hence, a means of controlling the flow such that the strength of the shear layer is reduced and the primary eddy eliminated should result in beneficial changes to the sedimentation regime in the basin entrance. The potential magnitude of these benefits is now investigated.

Table 3
Tidal data for the Parkhafen observations

Date	Low water	High water	Low water	Tidal range (m)
20 October 1997	-	08:08	15:23	3.64
21 October 1997	16:02	21:14	-	3.54
07 November 1997	03:04	08:20	15:35	3.22

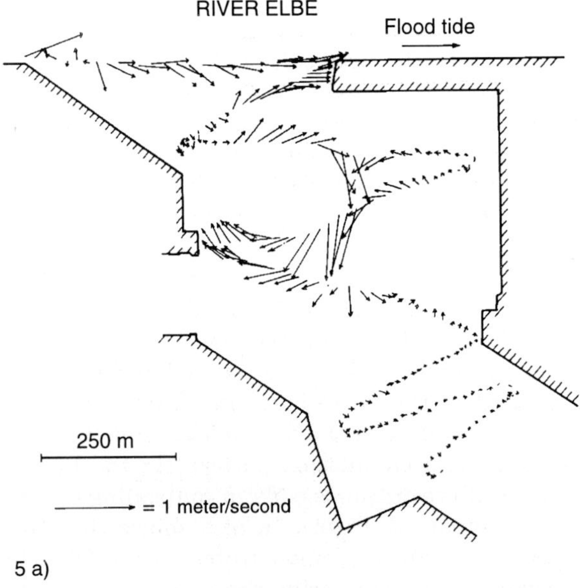

5 a)

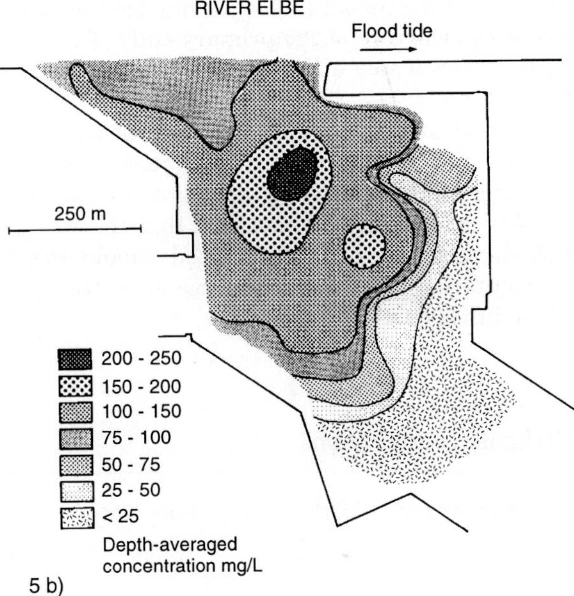

5 b)

Figure 5. Velocity and suspended solids distribution in the entrance to the Parkhafen, Hamburg between 17:39 and 18:07 on 21 October 1997.

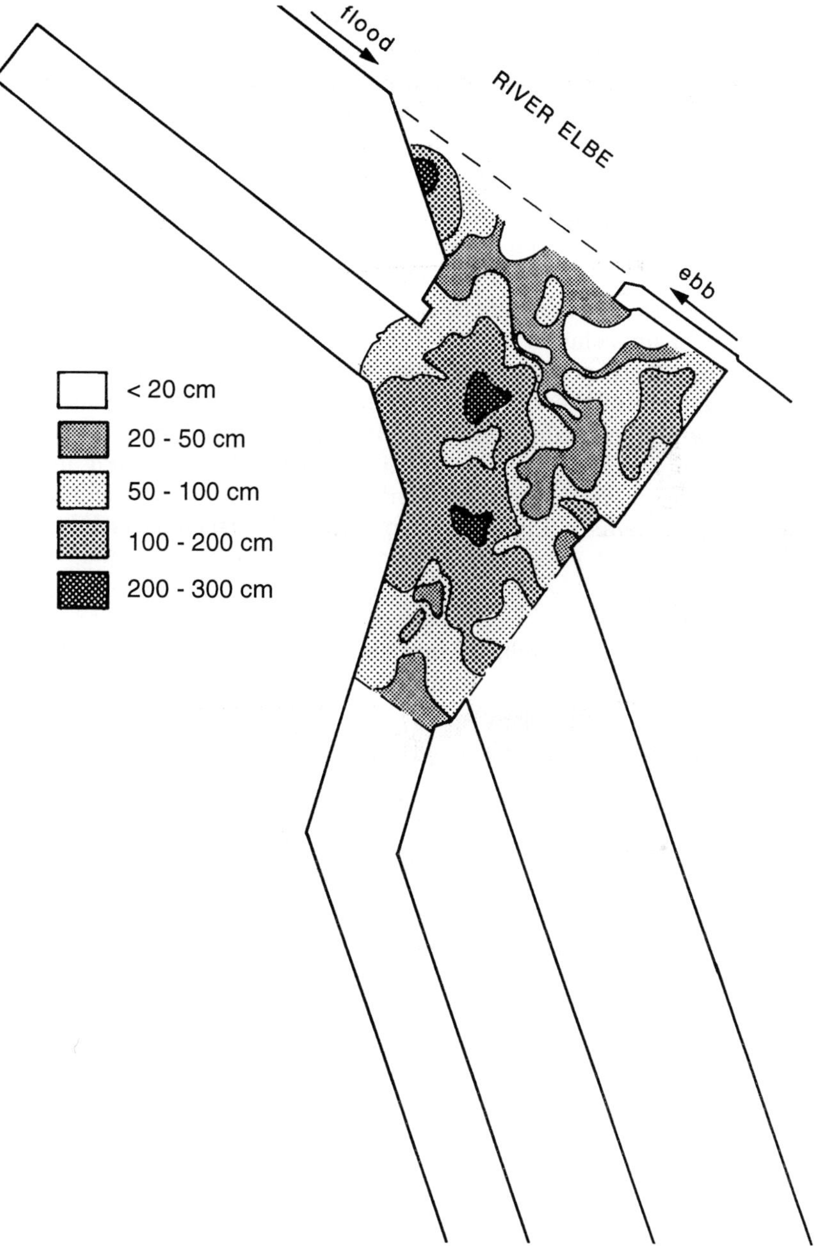

Figure 6. Average siltation pattern in the Parkhafen between 1977 and 1990. Contours show siltation rates per year.

3. SEDIMENT FLUXES INTO A TIDAL BASIN

Consider a basin with tidal volume V, length L, width b and mean depth h in a tidal flow of period T. The angle of the basin to the main stream is α, measured clockwise from the downstream river bank. Thus, for a basin at right angles to the river, $\alpha = 270^{\circ}$.

Let the mean flood speed in the tidal river at the location of the basin be U, the sediment concentration in the river be C and the difference in sediment concentration between river water and basin water just either side of the mixing zone be Δc. The concentration deficit, Δc can be either positive or negative. The position of the basin relative to the turbidity maximum at high water, the horizontal extent and gradients within the turbidity maximum, the tidal excursion, the nature of the suspended sediment and the hydrodynamics of the basin entrance will all contribute to the variation of Δc throughout a tide. Here, it is assumed that Δc represents the mean value over the flood tide and, in line with the majority of basins, is taken to be positive, i.e., on average over a flood tide, the river water contains a higher concentration of suspended solids than the basin water.

3.1. Tidal filling

The mass of sediment introduced into the basin during the flood tide is VC. Hence, the sediment flux into the basin due to tidal filling, F_t, is $2VC/T$. If the cross-sectional area of the basin entrance is A (= bh), then this flux can be written in terms of the tidal filling speed, u_t, as:

$$F_t = f_t AUC \tag{1}$$

in which

$$f_t = \frac{u_t}{U} \tag{2a}$$

$$u_t = \frac{2V}{TA} \tag{2b}$$

3.2. Entrainment across the mixing layer

The mixing layer which forms across the entrance to the basin is shown schematically in Figure 2. Within the mixing layer there is a dividing streamline which separates that part of the mixing layer which returns to the river from that part which flows into the basin. Conservation of mass requires that this streamline be displaced into the basin terminating in a stagnation point at the upstream basin wall (Figure 2). Conservation of momentum shows that for full-scale, non-tidal basins, the stagnation point is only on the order of a few meters inside the basin. For tidal basins, the stagnation point moves deeper into the basin as the tidal volume increases.

The exchange of estuary water and basin water across the mixing layer on the rising tide reduces as the tidal inflow into the basin increases due to the deflection of the mixing layer deeper into the basin. If the tidal inflow is sufficiently strong that the whole of the mixing layer is contained within the basin then the exchange of water with the river across the mixing layer is zero. The net suspended solids flux into the basin due to entrainment across the mixing layer, F_e, can be written as the sum of the entrainment flux in the absence of tidal inflow, F_{eo}, less the effect of the displacement of the mixing layer into the basin by the tidal inflow. This can be approximated by:

$$F_e = F_{eo} - \varepsilon F_t; \qquad F_{eo} > \varepsilon F_t \tag{3a}$$

$$F_e = 0; \qquad F_{eo} \leq \varepsilon F_t \tag{3b}$$

in which $\varepsilon \in [0,1]$ is a constant for a particular basin.

The entrainment flux under conditions of no tidal inflow can be estimated from a simple turbulent diffusion model as:

$$F_{eo} = AD\nabla c \tag{4}$$

If the width of the mixing layer is δ, then the suspended solids concentration gradient can be approximated by:

$$\nabla c = \frac{\Delta c}{\delta} \tag{5}$$

The turbulent diffusivity in the mixing layer, D, can be determined from the turbulent effective viscosity η according to:

$$D = \frac{\eta}{S_c} \tag{6}$$

in which S_c is the turbulent Schmidt Number. The effective viscosity can be obtained using the mixing length approximation:

$$\eta = \lambda^2 \nabla \boldsymbol{u} \tag{7}$$

in which λ is the mixing length and $\boldsymbol{u}$ is the local velocity vector.

Subject to later verification, the mixing length can be written as:

$$\lambda = \frac{1}{2\kappa\delta} \tag{8}$$

while the local velocity gradient can be approximated by:

$$\nabla \boldsymbol{u} = \frac{\Delta u}{\delta} \tag{9}$$

Here, κ is von Karman's constant (= 0.4) and Δu is the velocity difference across the mixing layer. An initial justification for the use of (8) comes from Rodi (1984), who used an error integral velocity distribution in the mixing layer as a fit to experimental data to obtain $\delta = 4.7\,\lambda$. Equation (8) implies $\delta = 5.0\,\lambda$.

If the development of the mixing layer is due to the lateral diffusion of momentum deficit, then the width of the mixing layer should increase according to (Smith, 1982):

$$\delta = 2\sqrt{\frac{\eta x}{u_m}} \tag{10}$$

where x is the distance from the downstream corner of the basin measured parallel to the flow in the river, positive upstream, and u_m is the average velocity in the mixing layer. Substitution of (7), (8) and (9) into (10) and differentiating with respect to x leads to the rate of spread as:

$$(2\phi)^{-1}\frac{\partial \delta}{\partial x} = \kappa^2 \tag{11}$$

$$\phi = \frac{1}{2}\frac{\Delta u}{u_m} \tag{12}$$

Various authors have reported experimental values for $(2\phi)^{-1}\partial\delta/\partial x$ covering both plane mixing layers (i.e., infinite depth) and those in shallow water. Table 4 compares these experimental results with (11).

Table 4
Spreading rate of mixing layer

Source	Equation (11)	Townsend (1976) [1]	Chu and Babarutsi (1988) [2]	Brown and Roshko (1974) [3]
Rate of spread	0.16	0.14	0.18	0.09

Notes:
1. Based on unpublished data for plane mixing layers.
2. Initial spreading rate in shallow water mixing layers with negligible bottom friction. Increasing the friction coefficient reduces the initial spreading rate.
3. Based on their own data together with that of 10 other authors.

The approximation given by (11) is close to the results of Townsend (1976) and Chu and Babarutsi (1988). The rate of spread reported by Brown and Roshko (1974) is much smaller than that suggested by the other experimental data, but is based on a large number of experiments by different authors. The explanation for this discrepancy probably lies in the definition used by Brown and Roshko (1974) for the width of the mixing layer. This was:

$$\delta = \frac{\Delta u}{(\nabla \boldsymbol{u})_{max}} \tag{13}$$

It is straightforward to show that $(\nabla \boldsymbol{u})_{max} > (\nabla \boldsymbol{u})_{mean}$, and hence the width of the mixing layer according to the Brown and Roshko (1974) definition should be less than that of the other authors. Hence, (11) is taken to be an acceptable approximation of the mixing layer's development.

The magnitude of the sediment flux entrained across the mixing layer can now be approximated by:

$$F_{eo} = f_e AUC \tag{14}$$

with

$$f_e = \frac{1}{4}\kappa^2 S_c^{-1}\left(\frac{\Delta u}{U}\right)\left(\frac{\Delta c}{C}\right) \tag{15}$$

The turbulent Schmidt Number in neutrally stratified flows has a value around 0.9 (Smith and Takhar, 1979).

The velocity deficit across the mixing layer, $\Delta u/U$, can be estimated from an analysis of the vorticity in the primary eddy. If the vorticity vector is ω ($= \nabla \times \boldsymbol{u}$) then the depth averaged, vertical component, $\omega \boldsymbol{k}$, is given by:

$$\frac{\partial \omega}{\partial t} + \nabla \cdot \boldsymbol{u}\omega = -\left\{\nabla \times \left(\frac{\tau}{\rho h}\right)\right\} \cdot \boldsymbol{k} \tag{16}$$

in which $\boldsymbol{k}$ is the unit vector in the vertical direction, positive upwards, τ is the bed shear stress and ρ is the water density.

If the bed shear stress is described by the quadratic friction approximation, $\tau = \rho C_D \boldsymbol{u} \mid \boldsymbol{u} \mid$ then:

$$-\left\{\nabla \times \left(\frac{\tau}{\rho h}\right)\right\} = \underbrace{-C_D |\boldsymbol{u}| \frac{\omega}{h}}_{\text{dissipation}} + \underbrace{C_D \boldsymbol{u} \times \nabla\left(\frac{|\boldsymbol{u}|}{h}\right)}_{\text{generation}} \tag{17}$$

If it is assumed that, to a first order approximation, vorticity is generated by the shear layer across the entrance and is dissipated by bed friction throughout the area of the basin occupied by the primary eddy and that, as a simplification, all parameters are represented by their average values within each of these regions, then (16) and (17) can be integrated over the area of shear layer and the primary eddy to give:

$$\frac{\Delta u}{U} = 1 - \left\{ \frac{4\delta}{h} + 1 \right\}^{-1/2} \tag{18}$$

Strictly, this derivation applies to basins whose length to width ratio is unity, as no account is taken of energy transfer nor dissipation beyond the primary eddy.

The width of the mixing layer is obtained by integrating (11). Substitution of this estimate into (18) reveals that, to this level of approximation and for $L/b = 1$, the velocity deficit ratio varies only with the width to depth ratio of the basin, b/h. Given that full-scale basins typically have a width to depth ratio in the range 20 to 40, the velocity deficit, $\Delta u/U$, from (11) and (18) is given in Table 5 for values of b/h in this range.

Experiments in square model harbors ($L/b = 1$) at various width to depth ratios reported by Booij (1986) give $\Delta u/U$ in the range 0.78 to 0.83. Full-scale basins typically have a width to depth ratio in the range 20 to 40.

It remains to estimate the suspended solids concentration deficit across the mixing layer, $\Delta c/C$. This can be estimated from a mass balance applied to the basin. Let the mass of sediment entering the basin during each flood tide be M_{in}. A proportion, $2p$, of the total sediment mass in the basin will deposit during each full tide (i.e., LW to LW). A simple mass balance then shows that an equilibrium state is achieved within the basin in which the mass of sediment being deposited per tide equals the net mass of sediment entering the basin per tide. The net mass entering the basin is that due to tidal filling and entrainment across the mixing layer less that in the tidal prism which leaves the basin on the ebb tide. In this equilibrium state, the total mass of sediment in the basin is M_E where:

$$M_E = \frac{M_{in}\Phi}{\{1 - \Phi(1-p)\}} \tag{19}$$

Table 5
Velocity deficit across the mixing layer in the basin entrance

b/h	20	25	30	35	40
Velocity deficit	0.75	0.79	0.81	0.83	0.84

$$\Phi = 1 - \left(\frac{V}{\Lambda} + p\right) \tag{20}$$

in which Λ is the volume of the basin over which the incoming sediment is mixed per tide. In low aspect ratio basins ($L/b < 2$), Λ can be taken to be the total volume of the basin at HW. However, for high aspect ratio basins ($L/b \geq 2$) where the suspended solids do not penetrate the full length of the basin, then $\Lambda = 2V$.

It is interesting to note that (19) implies that the residence time for sediment in the basin, T_R, is $T_R = T\{1 - \Phi(1\text{-}p)\}/\Phi$. This gives a typical residence time between 1 and 11 tidal periods.

The mass of sediment entering the basin during the flood tide is given by (1), (3) and (14) as:

$$M_{in} = \frac{1}{2}\{(1\text{-}\varepsilon)f_t + f_e\}TAUC \tag{21}$$

Using (2) and (15) for f_t and f_e respectively, and noting that $\Delta c/C$ is given by 1-$[M_E/\Lambda C]$, (20) can be rearranged to give:

$$\frac{\Delta c}{C} = 1 - \frac{\left(1 - \varepsilon + \frac{1}{2}\beta\right)\frac{\Gamma V}{\Lambda}}{1 - \varepsilon + \frac{1}{2}\beta\frac{\Gamma V}{\Lambda}} \tag{22}$$

with $\beta = \frac{1}{4}\kappa^2 S_c^{-1}(\Delta u/U)\mathrm{TUA}/V$ and $\Gamma = \Phi/\{1 - \Phi(1\text{-}p)\}$.

The value of p can be estimated from the effective settling velocity of the sediment in the basin, w_f. This is less than the still water settling velocity as it represents the net value, averaged over the water column, of downward settling (including any concentration related effects such as flocculation and hindered settling), upward turbulent diffusion and hydrodynamic re-entrainment. A detailed discussion of these processes can be found in Smith (1986) and Smith and Kirby (1989).

It should be noted that the above analysis assumes a uniform distribution of sediment within the entrance region of the basin. However, it was shown earlier that the entrance eddy tends to concentrate the suspended solids near its center, thereby reducing the concentration below the area mean at its edge. This means that the suspended solids deficit predicted by (22) is conservative.

The constant ε can be estimated from the requirement that F_e is zero (i.e., $f_e = \varepsilon f_t$) when the mixing layer is just contained within the basin. This occurs when:

$$u_t = \phi\kappa^2 U \tag{23}$$

Hence,

$$\varepsilon = \frac{1}{4}(\phi S_c)^{-1}\left(\frac{\Delta u}{U}\right)\left(\frac{\Delta c}{C}\right) \tag{24}$$

It is shown later that river discharge has a significant effect on the annual variation of sedimentation in a given basin. This is due to changes in the river sediment concentration and flow speeds that are linked to discharge. Hence, while the sediment flux depends strongly on river discharge, the coefficients f_t, f_e and ε do not.

3.3. Application to the Parkhafen

Neap tide conditions for the Parkhafen were used to evaluate the coefficients f_t from (2), f_e from (11), (15), (18) and (22) and ε from (18), (22), (23) and (24). This gave $f_t = 0.038$, $f_e = 0.015$ and $\varepsilon = 0.19$.

Experimental values of f_e and ε were obtained from a 1:75 scale, undistorted, hydraulic model study of the Parkhafen under the same tidal conditions. These were (Delft Hydraulics, 1992, personal communication): $f_t = 0.038$, $f_e = 0.013$ and $\varepsilon = 0.14$ (present profile for the Athabaskahöft) and $\varepsilon = 0.16$ (proposed modification to the Athabaskahöft profile).

The Athabaskahöft is the upstream corner of the basin (see Figure 6). This affects the proportion of the mixing layer captured by the basin, and hence has an influence on the coefficient ε. While the difference between 0.14 and 0.16 may appear slight, the proposed modification to the Athabaskahöft profile saves 2% of the annual sediment input due to tidal filling (~ 7,000 m^3). These results also indicate that the proportion of sediment entering the Parkhafen due to entrainment across the mixing layer is only 17% of the total. However, this still represents 60,000 m^3 of siltation per year.

Delft Hydraulics (1992, personal communication) have determined the coefficients f_e and ε empirically for a large number of basins. In general, they have found $f_e \in [0.01, 0.03]$ and $\varepsilon \in [0.1, 0.25]$. Hence, the agreement between the analytical model and the hydraulic model is reasonable, given the level of approximation in the analytical model. This is taken as sufficient justification to use the analytical model to estimate the effects of entrance flow control on sediment input, and hence sedimentation, in tidal basins in the absence of salinity driven densimetric water exchange.

4. ENTRANCE FLOW CONTROL

4.1. Principles

If the sediment input into the basin is to be reduced, it is necessary to reduce the strength of the mixing layer, eliminate the entrance eddy and deflect any near-bed, high concentration suspensions sufficiently far away from the basin entrance that the combined effects of tidal filling and flow under their own weight does not draw them into the basin.

While eliminating the mixing layer generated at the downstream corner, the CDW generates its own wake which passes across the basin entrance. By suitable design of the shape of the forward portion of the CDW, the strength of the wake can be minimized and the separation point chosen so that the wake does not penetrate the basin. An "overcatch" achieved by setting the wall slightly further offshore than is strictly necessary to permit basin filling helps with this as a small outflow is generated between the device and the upstream corner of the basin. This outflow deflects the wake towards the main stream. The addition of a slot part way along the wall can also be used to control the flow over the device itself and hence the strength and position of the wake produced. Thus, a CDW reduces the magnitude of the mixing layer across the entrance to the basin and also reduces the effect of that which remains by repositioning it away from the basin.

The CDW also has another effect. The stream exiting the trailing edge of the wall generates counter-clockwise vorticity whose magnitude is of the order of $2U/b$. The velocity shear across the basin entrance generates a clockwise vorticity of the same magnitude [see (17)]. Hence the effect of the device is to produce a vorticity equal and opposite to that in the entrance eddy. The net average vorticity in the basin entrance is zero. Consequently the entrance eddy is eliminated. The shape of the rearward portion of the CDW and its length are important design parameters in maximizing the extent to which the device reduces the entrance eddy. In practice the complex geometries of real basin entrances do not always allow the entrance eddy to be completely eliminated, but substantial reductions in its strength are always achieved.

A further design feature of the CDW is the partial depth sill which extends from the base of the leading edge of the wall to the river bank. This is shown schematically in Figure 7. The objective of the sill, which is typically up to 25% of the water depth in height, is to intercept the higher concentration water in the lower portion of the water column, so that the basin fills with the lower concentration water from the upper part of the flow thereby reducing the mass of sediment introduced into the basin due to tidal filling. The sill also intercepts, and diverts away from the basin, the near-bed, high concentration suspensions which flow under a combination of tidal currents and their own weight. High concentration layers are known to form in most estuaries, even those considered high energy environments such as the Severn estuary in the UK (Kirby and Parker, 1983). Surveys have shown them to be present in the Port of Hamburg (Christiansen, 1987; Christiansen and Kirby, 1991). These near-bed sediment layers are capable of transporting significant volumes of sediment. However, as they are ephemeral, their existence and horizontal extent is difficult to predict. The small vertical density gradients generated by the non-uniform suspended solids concentrations lead to the local Froude number being sufficient to ensure that the flow will divert horizontally around the sill rather than flow over it.

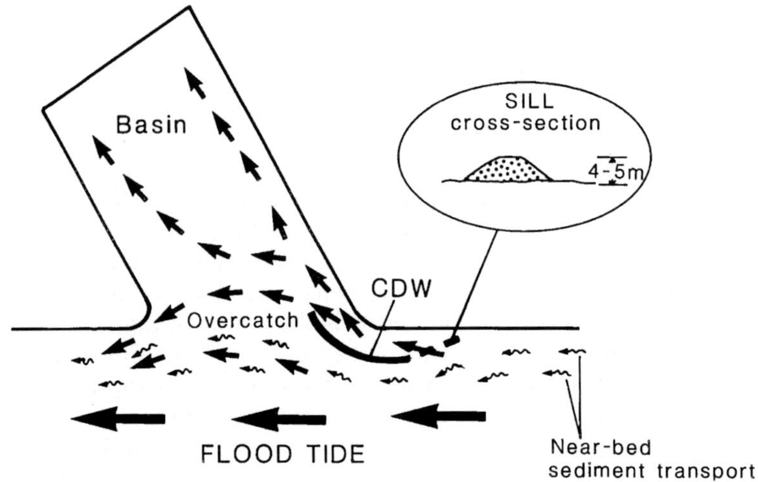

Figure 7. Schematic diagram of the Current Deflecting Wall.

4.2. Impact on the flow regime

A comprehensive study into entrance flow control using the CDW was undertaken in the large tidal flume at Delft Hydraulics (Crowder, 1996; Crowder et al., 1999a; Crowder et al., 1999b). Figure 8 shows particle image velocimetry results for one configuration of basin and CDW. The absence of a vortex in the basin entrance is clear, while the flow remains attached to the device giving rise to very little wake or entrainment of sediment through the wake into the basin. The use of a slot in the CDW in this situation should also be noted. The slot in the device ensures that the boundary layer remains attached leading to a very small wake. This in turn minimizes the entrainment of sediment across the wake, and hence further reduces sediment input into the basin. Two-part CDWs are considerably more difficult to optimize than a single wall. However, in certain circumstances, particularly high-angle basins with complex entrance geometries, there is sufficient incremental return to justify the additional design and construction costs of a slotted wall.

A further example of the dramatic effect of a CDW on the flow regime in a basin entrance is evident in Figure 9 (Winterwerp et al., 1994). This shows two photographs of the entrance region of the Parkhafen, Hamburg, in a 1:75 scale undistorted hydraulic model before and after entrance flow control using a CDW. Again a two-part wall is illustrated. In this example, the entrance eddy is not completely eliminated, but is substantially reduced in magnitude thereby reducing the entrainment of sediment into the basin and its concentration in the entrance shoal. The beneficial effect of the device to the maneuvering of vessels entering and leaving the basin through the reduction in the strong velocity shear across the entrance is also clearly shown.

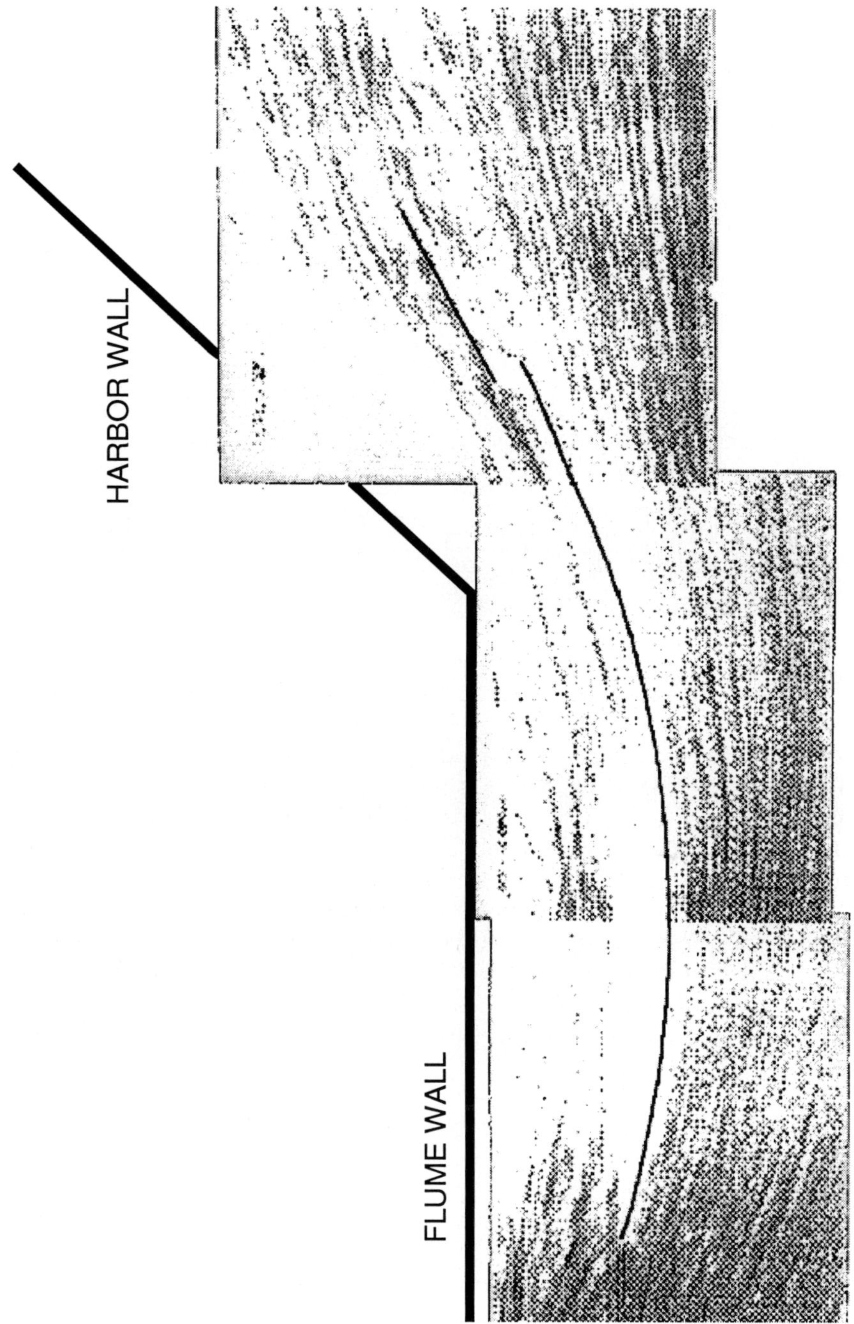

Figure 8. Particle image velocimetry data for a Current Deflecting Wall in a simulated harbor basin off a tidal flume.

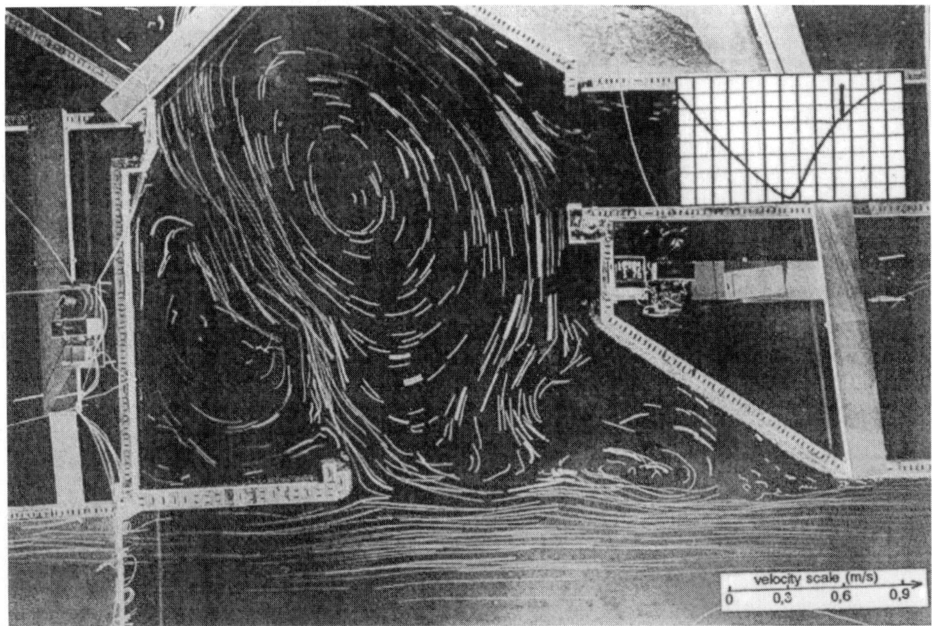

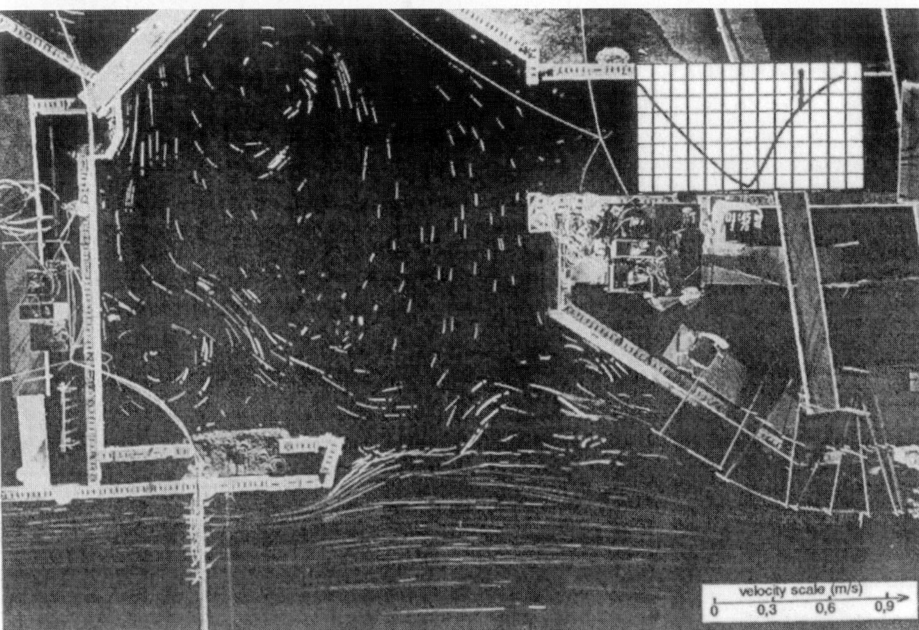

Figure 9. Particle image velocimetry data at mid-flood tide in a hydraulic model of the Parkhafen, Hamburg: a) without a CDW, b) with a CDW.

5. APPLICATION IN THE KÖHLFLEET, HAMBURG

5.1. Potential benefits

The location of the Köhlfleet is shown in Figure 4. It is a major basin within the Port of Hamburg handling bulk goods and oil. Prior to 1991, the average annual maintenance dredging demand in the basin was 255,000 m^3. Due to the contaminated nature of the sediment arriving at Hamburg, it is forbidden to dispose of material dredged from the Köhlfleet entrance shoal back into the Elbe. Instead, it has had to be treated, at substantial cost, before disposal to landfill.

The partial savings from the use of entrance flow control to minimize sediment entrainment into the basin can be estimated from the analytical model developed herein. The coefficients f_t, f_e and ε can be calculated for the Köhlfleet and are found to be $f_t = 0.025$, $f_e = 0.018$ and $\varepsilon = 0.16$.

In the absence of entrance flow control, sediment input into the basin is 1.56 F_t while with entrance flow control it could be as low as 1.0 F_t provided that the optimum design of CDW is used. This represents a 36% saving in sediment input to the basin and hence in maintenance dredging demand. Additional savings are likely to accrue from the reduced probability of near-bed high concentration suspensions penetrating into the basin due to the action of the sill incorporated into the CDW.

5.2. Actual savings

During late 1990, a CDW was constructed in the entrance to the Köhlfleet. The installation is shown in Figure 10. The initial prototype was constructed using vertical "I" section piles with wooden inserts to form the panels. These wooden panels were damaged by ice in late 1992 and were removed. This meant that, for the whole of 1993, there was no CDW in the basin. In 1994, a permanent replacement structure was built using prefabricated concrete panels. Full details of the permanent construction can be found in Christiansen (1997).

The average annual siltation pattern in the basin, established from maintenance dredging returns, with and without the CDW is shown in Figure 11. The annual siltation rate prior to 1991 contrasts sharply with that after the installation of the CDW. Two results are clear from this figure. Since entrance flow control has been introduced:

- siltation is distributed more evenly across the area of the basin entrance
- the maximum siltation rate is significantly less

The former is important as it means that, for a given sediment input into the basin, the navigable depth decays more slowly than with the usual distribution pattern. This means that maintenance dredging is required less frequently and hence is cheaper per unit as mobilization costs are reduced.

Figure 10. The Current Deflecting Wall in Köhlfleet, Hamburg. The River Elbe is along the top of the photograph.

Correlation of the annual siltation rate in each of the Hamburg basins with various environmental parameters since 1982 has shown that the discharge in the Elbe is the major driver of annual variations in the siltation rate of a given basin. The effect of the CDW on the average siltation in the Köhlfleet as a function of discharge in the Elbe is illustrated in Figure 12. Two curves are shown, one for years prior to 1991 and a second for the period 1991 to 1997 inclusive. The siltation rate for a given river discharge is significantly less with the CDW than without it. It will be recalled that for the whole of 1993 the prototype CDW was removed prior to its permanent reconstruction. Figure 12 provides clear evidence of the significant benefit of entrance flow control to reduce siltation in the Köhlfleet over the whole range of river discharge conditions likely to be encountered in the Elbe. The significance of the 1993 result is that it lies on the same curve as the pre-1991 data. This shows that the siltation reduction since 1991 is due to the CDW alone, and not some other unknown, coincidental change in another environmental parameter.

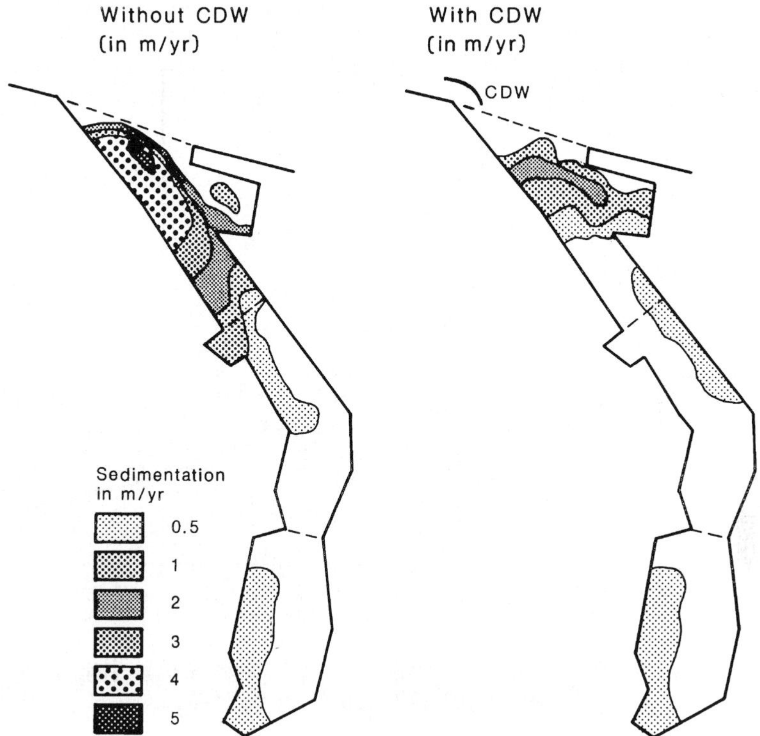

Figure 11. Siltation patterns in Köhlfleet, Hamburg, before and after the installation of the Current Deflecting Wall.

The benefit in reduced overall siltation is even greater than that suggested by the data in Figures 11 and 12. The correlation of annual siltation with river discharge for a given basin is used by the Strom und Hafenbau as a predictive tool for planning maintenance dredging in that basin. The correlations for each basin have proved to be highly accurate. For each basin, the correlation obtained from data prior to 1991 has been used to predict the siltation rate for the years 1991 to 1997 using the known river discharges for those years. For basins other than Köhlfleet, the pre-1991 correlations have provided accurate predictions of the maintenance dredging demand in each basin for the years 1991 to 1997 both close to and at some distance from Köhlfleet. Hence, the pre-1991 correlation for the Köhlfleet together with the actual river discharges for the years 1991 to 1997 should provide an accurate estimate of what the annual siltation in that basin would have been from 1991 to 1997 in the absence of the CDW.

Using this technique, the average maintenance dredging demand in the absence of entrance flow control would have been 280,000 m^3 per year. This is 9.8% higher than the average for the period prior to 1991 due to a lower weighted

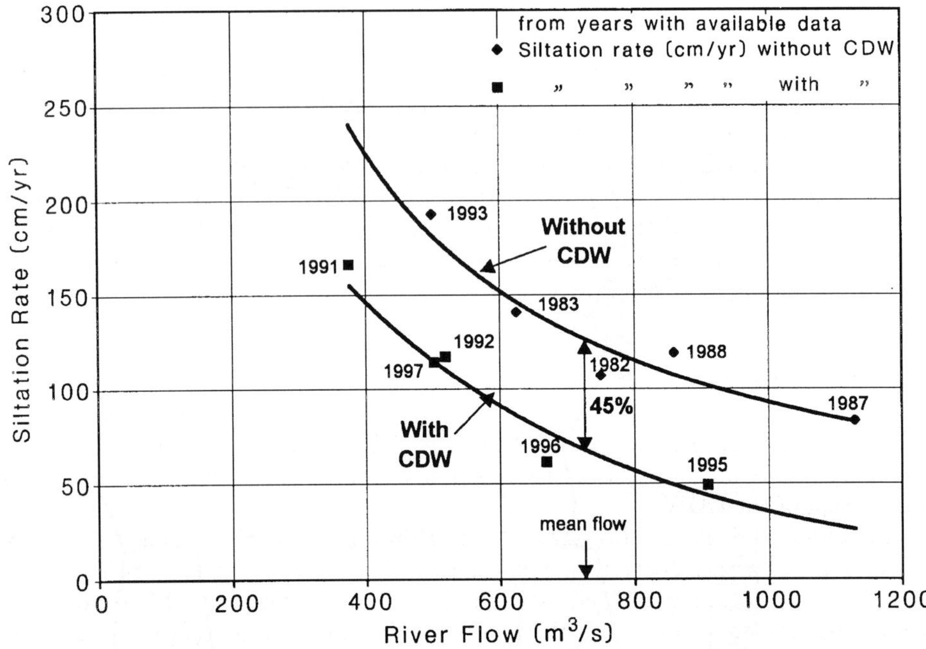

Figure 12. Correlation between siltation rate and river discharge for the Köhlfleet basin with and without a Current Deflecting Wall.

average river discharge. The actual maintenance dredging requirement in the Köhlfleet between 1991 and 1997 was 140,000 m^3 per year. This represents a 50% saving due to the use of a CDW.

The applicability of the pre-1991 correlations for siltation rate in basins other than the Köhlfleet during the period 1991 to 1997 indicates that the sediment, which would have deposited in the Köhlfleet in the absence of flow control, has not found its way into other basins in the port. Hence, the reduction in siltation rate in the Köhlfleet achieved by means of the CDW is a true saving as it has not been achieved at the expense of incremental siltation in the other basins of the port. This is predictable as the effect on the main stream sediment concentration of the sediment which remains in the main stream, rather than entering the basin, is negligible due to its dilution over the tidal excursion.

It is postulated that the additional actual savings over and above the predicted 36% reduction in siltation rate are due to the effect of the CDW on the movement of near-bed, high concentration layers. If the sediment input from these layers is $F_d = f_d AUC$ and the additional savings are due to the sill and the reduced mixing layer eliminating this input, then this leads to a value of the coefficient $f_d = 0.011$. This implies that, prior to the construction of the CDW, 22% of the sediment entering the basin came from these near-bed layers.

6. CONCLUSIONS

It has been shown that the large scale sedimentation observed in the entrance to a tidal basin is due to the capture of the sediment entering the basin by the primary eddy in the basin entrance which is driven by the mixing layer across the entrance. It has also been shown that by modifying and controlling the flow in the basin entrance using a Current Deflecting Wall, the volume of sediment entering the basin and the proportion of that sediment settling in the entrance can both be significantly reduced. Thus, the conflict of interests at basin entrances has largely been overcome.

7. ACKNOWLEDGMENT

The authors are grateful to the Strom und Hafenbau, Freie und Hansestadt Hamburg for permission to publish this paper and for their long term support for the concept of the CDW.

The measurements in the Large Tidal Flume at Delft Hydraulics were made possible by a grant from the European Union under Phase II of its Large Installations Programme. The authors are grateful for this financial support and would also like to thank the staff of Delft Hydraulics and all other collaborators for their input to the LIP-II study. The authors are also grateful to Dr. J. C. Winterwerp and the staff at Delft Hydraulics for undertaking the hydraulic model study of the Parkhafen.

Thanks are also due to Dredging Research Limited who undertook the ADCP survey of the Parkhafen and to the staff of the Franzius Institute, Hannover, for carrying out the hydraulic model study of the Köhlfleet.

REFERENCES

Booij, R., 1986. Measurements of the exchange between river and harbour. *Report No. 9-86*, Department of Civil Engineering, Delft University of Technology, Delft, The Netherlands (in Dutch).

Brown, G. L., and Roshko, A., 1974. On density effects and large structure in turbulent mixing layers. *Journal of Fluid Mechanics*, 64(4), 775-816.

Christiansen, H., 1987. New insights on mud formation and sedimentation processes in tidal harbours. *International Conference on Coastal and Port Engineering in Developing Countries*, Beijing, China, 2, 1332-1340.

Christiansen, H., 1997. Erfahrungen mit der Strömungslenkwand. *HANSA International Maritime Journal*, Vol. 12, 70-73.

Christiansen, H. and Kirby, R., 1991. Fluid mud intrusion and evaluation of a passive device to reduce mud deposition. *Proceedings of CEDA-PIANC Conference "Accessible Harbours"*, Amsterdam, E1-E14.

Chu, V. H., and Babarutsi, S., 1988. Confinement and bed-friction effects in shallow turbulent mixing layers. *Journal of Hydraulic Engineering*, 114(10), 1257-1274.

Crowder, R. A., 1996, Study on the reduction of siltation in harbours by means of a Current Deflection Wall. *Report to the European Union,* Department of Civil and Environmental Engineering, University of Bradford, Bradford, UK, 130+xii p.

Crowder, R. A., Smith, T. J., Christiansen, H., Winterwerp, J. C., Kirby, R., Falconer, R. A., and Schwarze, H., 1999a. Flow control in the entrance to river harbours, embayments and irrigation canals. *Report No. LIP-01-99*, Ravensrodd Consultants Limited, Taunton, UK, 21p.

Crowder, R. A., Falconer, R. A., and Kirby, R., 1999b. Model study to reduce siltation in harbours using Current Deflecting Walls. *Fourth International Conference on Computer Modelling of Sea and Coastal Regions, Coastal Engineering '99*, 151-158.

Kirby, R., 1994. Sedimentation engineering techniques for environmentally friendly dredging. *Underwater Technology*, 20(2), 16-24.

Kirby, R., Ibrahim, A. A., Lee, S-C., Kassim, M. K. Hj., and Christiansen, H., 1997. Environmentally-friendly sediment management systems. *Proceedings of the AEDA 2nd Asian and Australian Ports and Harbours Conference*, Ho Chi Minh City, Vietnam, 114-129.

Kirby, R., and Parker, W. R., 1983. The distribution and behaviour of fine sediment in the Severn Estuary and Inner Bristol Channel. *Canadian Journal of Fisheries and Aquatic Sciences*, 40(1), 83-95.

Land, J. M., Kirby, R., and Massey, J. B., 1997. Developments in the combined use of acoustic doppler current profilers and profiling siltmeters for suspended solids monitoring. In: *Cohesive Sediments*, N. Burt, W. R. Parker and J. Watts eds., John Wiley, Chichester, UK, 187-196.

Rodi, W., 1984, Turbulence models and their application in hydraulics - A state of the art review. *International Association for Hydraulic Research*, Delft, The Netherlands.

Smith, T. J., 1982. On the representation of Reynolds stress in estuaries and shallow coastal seas. *Journal of Physical Oceanography*, 12, 914-921.

Smith, T. J., 1986. Formation and break-up of fine particle aggregates in stirred tanks. *Institution of Chemical Engineers Symposium Series*, 108, 209-219.

Smith, T. J., and Kirby, R., 1989. Generation, stabilization and dissipation of layered fine sediment suspensions. *Journal of Coastal Research,* SI5, 63-73.

Smith, T. J., and Takhar, H. S., 1979. The effect of stratification on the turbulent transport of mass and momentum. *Proceedings of the XVIIIth Congress of the International Association for Hydraulic Research,* Cagliari, 3, 79-86.

Townsend, A. A., 1976, *Structure of Turbulent Shear Flow.* Cambridge University Press, Cambridge, UK.

Winterwerp, J. C., Christiansen, H., Eysink, W. D., Kirby, R., and Smith, T. J., 1994. The Current Deflecting Wall: a device to minimise harbour siltation. *Dock and Harbour Authority*, 74(March/April), 243-246.

An examination of mud slurry discharge through pipes

P. Jinchai[a], J. Jiang[b] and A. J. Mehta[c]

[a]Hydrography Department, Royal Thai Navy, Bangkok, Thailand

[b]ASL Environmental Sciences Inc., 1986 Mills Road, Sidney, V8L 5Y3, BC, Canada

[c]Department of Civil and Coastal Engineering, University of Florida, Gainesville, Florida, 32611 USA

The dependence of the discharge of mud slurries in pipes on cohesion and water content is examined by introducing a slurry cation exchange capacity. Based on tests in which forty-two clay-water mixtures were pumped through a horizontal pipe viscometer, it is shown that this exchange capacity can serve as an approximate determinant of slurry discharge in the viscous flow range. This observation may find application in assessing the pumping requirement for the transportation of dredged cohesive mud at *in situ* water content.

1. INTRODUCTION

Contamination of ambient waters during dredging of bottom mud often occurs when the material to be removed is diluted to enable transportation of the resulting slurry by hydraulic means. In order to minimize this source of contamination, the technology for pumping mud through pipes at *in situ* density or water content has been investigated in recent years. This approach has been shown to hold promise for moderately dense mud slurries, which are characteristically transported in the viscous flow range (Parchure and Sturdivant, 1997). In that regard, *a priori* assessment of the pumping requirements can be facilitated if a measure of the transportability of mud could be acquired by testing for its relevant properties as a precursor to dredging. Here we have examined an empirical approach as a basis for such testing by relating the pump-pressure driven pipe discharge of clayey mud slurries to an easily determinable parameter related to slurry cohesion and water content.

2. MUD SLURRY FLOW

Mud slurry considered is non-settling and homogeneous, and one which flows steadily in the viscous range as a thick suspension. The rheology of such a slurry, which is characteristically non-Newtonian, can be simply described as a power-law fluid following the Sisko (1958) relation for the apparent viscosity η, which is

$$\eta = c\dot{\gamma}^{n-1} + \eta_\infty \tag{1}$$

where c is called the consistency of the non-Newtonian fluid, and n characterizes its flow behavior. Thus, $n<1$ refers to a slurry which exhibits pseudoplasticity or shear-thinning, i.e., its viscosity decreases with increasing shear rate $\dot{\gamma}$. Conversely, $n>1$ denotes a dilatant or shear-thickening flow behavior, i.e., increasing viscosity with increasing $\dot{\gamma}$. When $n=1$ the slurry becomes Newtonian with a constant viscosity equal to η_∞, since in this case the consistency, c, is nil. At high, theoretically infinite, shear rate a pseudoplastic also becomes Newtonian with a viscosity η_∞. A limitation of (1) is that for all $n<1$, $\eta \to \infty$ at the centerline, where $\dot{\gamma}=0$.

In general, knowing η, the steady discharge of a slurry can be readily calculated from the balance of pressure and shear forces in a pipe of radius R, which in turn amounts to a linear variation of the shear stress from zero at the pipe centerline to τ_w at the pipe wall according to

$$\eta\dot{\gamma} = -\frac{r}{R}\tau_w \tag{2}$$

where r is the radial coordinate with origin at the pipe centerline (Heywood, 1991).

Now, combining (1) and (2) leads to

$$\left[\frac{c}{R^n}\left(\frac{\partial u}{\partial \zeta}\right)^{n-1} + \frac{\eta_\infty}{R}\right]\frac{\partial u}{\partial \zeta} + (\zeta - 1)\tau_w = 0 \tag{3}$$

where u is the local flow velocity such that $\dot{\gamma} = \partial u / \partial r$, and the coordinate $\zeta = 1\text{-}(r/R)$. In (3), $\tau_w = R\Delta p / 2L$, where the pressure drop Δp occurs over pipe length L. The corresponding discharge Q is then obtained from

$$Q = 2\pi R^2 \int_0^1 u(\zeta)(1-\zeta)d\zeta \tag{4}$$

At high shear rates, as the flow approaches Newtonian behavior, by setting $c = 0$ in (3) one obtains the well-known Poiseuille formula for pipe discharge, $Q = \pi \Delta p R^4 / 8\eta_\infty L$. In general, for a given pipe and pressure drop, provided the power-law coefficients c, n and η_∞ are known, the discharge can be calculated from (4). Further, if these coefficients can be empirically related to a representative property of the slurry, then measuring this property can lead to a method for estimating discharge. This development is described next.

When the slurry is composed of clay and water and is therefore cohesive, a candidate parameter characterizing slurry behavior is the cation exchange capacity (CEC) of the clay, a measure of cohesion. CEC is determined in the laboratory using standard soil analysis techniques (ASTM, 1993). Inasmuch as cohesion has a strong influence on slurry rheology, CEC can be expected to correlate with the power-law coefficients, if not fully then at least to a significant extent. To determine if such a correlation exists, forty-two clay mixtures flocculated in tap water were tested. Three clays were chosen: a kaolinite (K), an attapulgite (A) and a bentonite (B), all with nominal median dispersed diameters on the order of 1 μm. For these mixtures a composite cation exchange capacity, CEC_s, can be defined as

$$CEC_s = f_K\, CEC_K + f_A\, CEC_A + f_B\, CEC_B \tag{5}$$

where f is the weight fraction, and the subscripts identify the three clays. Note that given the water content f_w, i.e., weight of water divided by weight of solids, we have $f_K + f_A + f_B + f_w$. The CEC values (in milliequivalents per 100 g of sediment) were: 6 for K, 28 for A and 105 for B (Jinchai, 1998). For the selected slurries the water content ranged from 86% for a moderately dense slurry to a high 423% for a dilute slurry, and CEC_s from a very low 1.9 meq/100 g to 10.4 meq/100 g (Table 1). These parametric values cover a reasonably representative range of slurry types.

3. EXPERIMENTAL RESULTS

The power-law coefficients were determined by measuring slurry viscosities over a range of shear rates (0.063 to 20.4 Hz) in a Brookfield (Model LVT) concentric cylinder viscometer. In order to cover a wider range of shear rate, in each case an additional value of the viscosity at a shear rate that was an order of magnitude higher than that attainable in this viscometer was measured in a horizontal pipe viscometer. This apparatus consisted of a 3.1 m long, 2.54 cm i.d. PVC pipe placed horizontally, with one end attached to a piston-diaphragm pump. Over the central 2.50 m length of the pipe the pressure drop was measured by two flush-diaphragm gage-pressure sensors. Each slurry was fed through a hopper at the pump end, and the discharge was measured at the other end. A correction to pressure gradient due to end-effects characteristic of this apparatus was applied by measuring the pressure drop in pipes of different lengths and radii (Heywood, 1991).

Viscosity data for all slurries at room temperature are given in Table 1, and a typical example (slurry no. 15) is shown in Figure 1. In this plot, the data point corresponding to $\dot{\gamma} = 981$ Hz was obtained in the horizontal pipe viscometer at a pressure drop $\Delta p = 95.4$ kPa. The best-fit line corresponding to (1) yields $c = 3.14$, $n = 0.22$ and $\eta_\infty = 0.234$ Pa. This value of n indicates that this slurry, as indeed all others tested, exhibited a shear-thinning behavior. It remains to examine how these

Table 1
Properties of mud slurries tested

Slurry no.	Slurry composition (%)	f_w (%)	CEC_s (meq/100 g)	c	n	η_∞ (Pa.s)
1	100%K	210	1.9	2.35	0.30	0.189
2	100%K	167	2.3	3.43	0.32	0.140
3	100%K	139	2.5	9.02	0.20	0.436
4	100%K	117	2.8	5.08	0.29	0.589
5	100%K	100	3.0	6.99	0.22	0.663
6	100%K	86	3.2	12.99	0.06	0.675
7	75%K+25%A	210	3.7	1.44	-0.07	0.338
8	75%K+25%A	169	4.3	7.69	0.17	0.547
9	75%K+25%A	139	4.8	9.98	-0.01	0.750
10	75%K+25%A	117	5.3	25.96	-0.28	0.904
11	50%K+50%A	210	5.5	4.30	0.22	0.238
12	50%K+50%A	169	6.3	13.21	-0.02	0.270
13	50%K+50%A	153	6.7	21.65	-0.12	0.295
14	50%K+50%A	139	7.1	76.08	-0.25	0.453
15	25%K+75%A	289	5.8	3.14	0.22	0.234
16	25%K+75%A	253	6.4	9.38	0.13	0.382
17	25%K+75%A	215	7.1	22.05	-0.13	0.560
18	25%K+75%A	189	7.8	34.08	-0.20	0.620
19	100%A	409	5.5	1.23	0.39	0.194
20	100%A	333	6.5	3.37	0.07	0.305
21	100%A	280	7.4	13.24	0.07	0.385
22	100%A	239	8.3	21.32	-0.24	0.557
23	90%K+10%B	273	4.3	2.98	-0.16	0.189
24	90%K+10%B	211	5.1	9.25	0.14	0.178
25	90%K+10%B	169	5.9	14.28	0.12	0.259
26	90%K+10%B	140	6.6	42.83	-0.04	0.380
27	65%K+25%A+10%B	231	6.5	9.04	0.18	0.218
28	65%K+25%A+10%B	204	7.0	17.50	0.09	0.277
29	65%K+25%A+10%B	182	7.6	21.05	0.06	0.308
30	65%K+25%A+10%B	163	8.1	57.83	-0.06	0.750
31	40%K+50%A+10%B	299	6.7	7.82	0.18	0.129
32	40%K+50%A+10%B	257	7.5	12.17	0.15	0.072
33	40%K+50%A+10%B	224	8.3	10.46	0.14	0.138
34	40%K+50%A+10%B	197	9.1	36.00	-0.02	0.284
35	15%K+75%A+10%B	423	6.2	1.82	0.21	0.177
36	15%K+75%A+10%B	345	7.3	6.17	0.24	0.145
37	15%K+75%A+10%B	290	8.3	8.71	0.16	0.126
38	15%K+75%A+10%B	248	9.3	22.13	0.03	0.362
39	90%A+10%B	415	6.9	4.12	0.25	0.199
40	90%A+10%B	339	8.1	14.83	0.05	0.107
41	90%A+10%B	284	9.3	17.32	0.03	0.373
42	90%A+10%B	244	10.4	89.59	-0.20	0.771

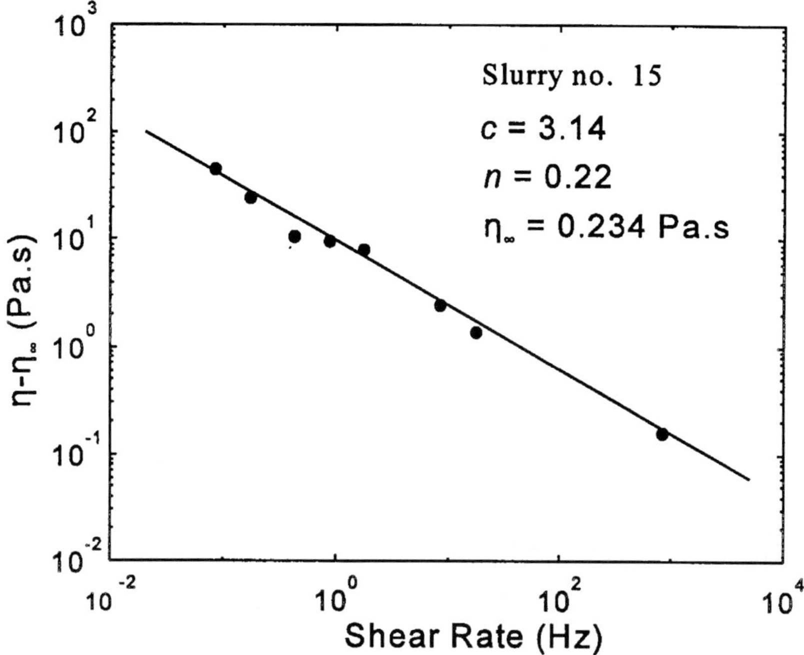

Figure 1. Dependence of excess apparent viscosity on shear rate for Slurry No. 15.

three coefficients vary with CEC_s. In Figure 2, the logarithm of c is plotted against CEC_s for all slurries. Since CEC_s increases with increasing cohesion and decreasing water content, notwithstanding the evident smearing of data points, the observed mean trend of increasing consistency with CEC_s can be expected. Note that, as CEC_s approaches zero, log$c$ would tend to $-\infty$ as the flow becomes increasingly Newtonian. The mean line, a crude measure of the trend within the measured range of CEC_s, is given by:

$$\log c = 0.13\,CEC_s + 0.22 \tag{6}$$

Plots of n and η_∞ against CEC_s in Figures 3 and 4, respectively, also show considerable data scatter. The respective mean trends given by:

$$n = -0.033\,CEC_s + 0.28 \tag{7}$$

and

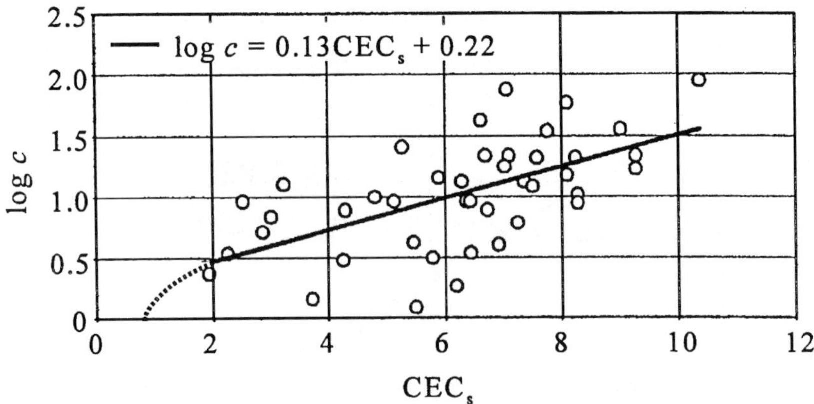

Figure 2. Variation of the logarithm of consistency c with CEC_s.

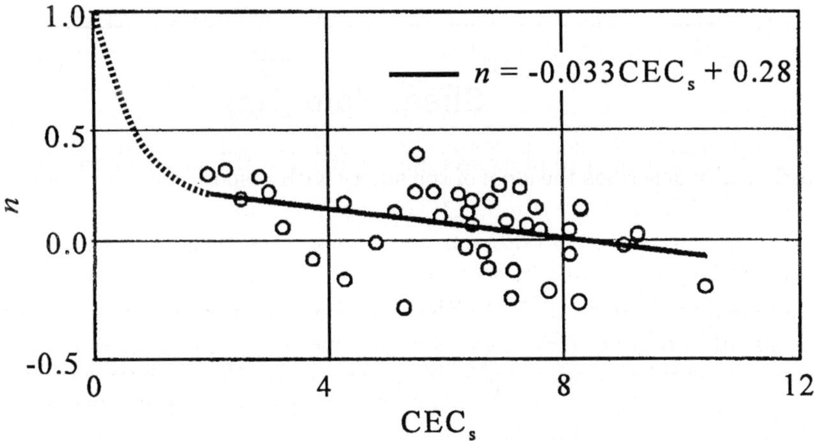

Figure 3. Variation of the flow behavior coefficient n with CEC_s.

$$\eta_\infty = 0.05CEC_s + 0.001 \tag{8}$$

As in the case of (6), relations (7) and (8) are applicable only within the overall range of CEC_s for which they were determined. In any event, (7) implies that as CEC_s increased n decreased, which means that the flow diverged from a Newtonian response towards increasing pseudoplasticity. From (8) we observe that η_∞ expectedly increased with increasing CEC_s along with consistency. When CEC_s is zero, the fluid viscosity is that of water (0.001 Pa.s). Observe also that, although the

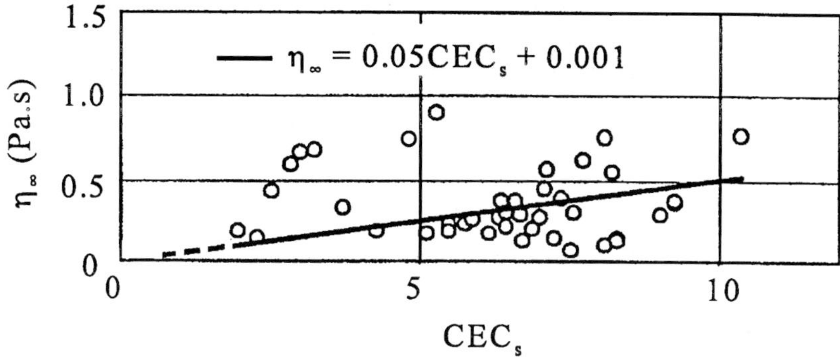

Figure 4. Variation of the viscosity η_∞ with CEC_s.

mean trend-line in Figure 4 is notionally assumed to be linear, further data may reveal a more complex variation of η_∞ with the CEC_s.

As to data scatter in Figures 2, 3 and 4, the main source of experimental error was from the discharge measurement. The capacity of the slurry hopper was only 22 liters, which meant that it took merely a few seconds for the material to pass through the pipe. It proved to be difficult to measure this passage of time accurately with a stop-watch used for the purpose. Apart from the ensuring error, we suspect that an important additional reason for data scatter is that CEC_s, which characteristically describes the *state* of the slurry with respect to cohesion and water content, cannot fully capture the effects that arise during the *flow* of slurry. Thus, while CEC_s appears to be a significant parameter correlating with the power-law coefficients, at best it can only be an approximate measure. Its potential utility lies in the comparative ease with which it can be determined using standard experimental methods and (5).

In order to demonstrate the significance of the method, ten slurries were pumped at pressure drops lower than those used to establish the power-law for those slurries. The corresponding data on pressure drop, mean shear rate, wall shear stress, measured discharge and the characteristic pipe Reynolds number, $Re=2\rho Q/\pi R\eta_\infty$, are given in Table 2. For example for slurry no. 1, at Δp = 68.5 kPa the measured discharge was 0.00135 m^3/s. The corresponding Reynolds number value of 448 indicates that the flow was non-turbulent due to the high viscosity of the slurry (0.209 Pa.s at the prevailing mean shear rate of 836.6 Hz). For all the other slurries in Table 2, the discharge was measured under similar viscous flow conditions. The range of pressure drops was narrow (61.7 to 82.7 kPa). Hence, despite differences in the slurry composition, the range of measured discharges was also narrow (0.00128 to 0.00137 m^3/s).

In Figure 5, the pressure drop from Table 2 and the power-law coefficients from Table 1 for slurry no. 1 have been used to solve (3) numerically, together with the no-slip boundary condition at the wall, $u(\zeta=0) = 0$ (Jinchai, 1998). Although the velocity profile could not be measured within the narrow and opaque pipe, the

Table 2
Comparison between measured and calculated discharges

Slurry no.	Pressure drop, Δp (kPa)	Shear rate, $\dot{\gamma}$ (Hz)	Wall shear stress, τ_w (Pa)	Measured discharge, Q (m^3/s)	Reynolds number, Re	Calculated discharge, Q (m^3/s)	Error (%)
1	68.5	836.6	175.3	1.35×10^{-3}	448	1.31×10^{-3}	-3
7	70.0	807.6	179.2	1.30×10^{-3}	240	0.85×10^{-3}	-35
11	82.7	844.9	211.7	1.36×10^{-3}	354	1.27×10^{-3}	-7
15	67.7	794.5	173.3	1.28×10^{-3}	322	1.08×10^{-3}	-16
19	61.7	842.0	158.0	1.36×10^{-3}	395	1.15×10^{-3}	-15
23	71.7	823.2	183.5	1.33×10^{-3}	424	1.55×10^{-3}	+16
27	67.6	825.4	172.7	1.33×10^{-3}	375	1.01×10^{-3}	-24
31	73.9	800.9	189.3	1.29×10^{-3}	589	1.92×10^{-3}	+49
35	78.4	821.6	200.7	1.32×10^{-3}	420	1.74×10^{-3}	+32
39	74.9	850.0	191.8	1.37×10^{-3}	388	1.34×10^{-3}	-2

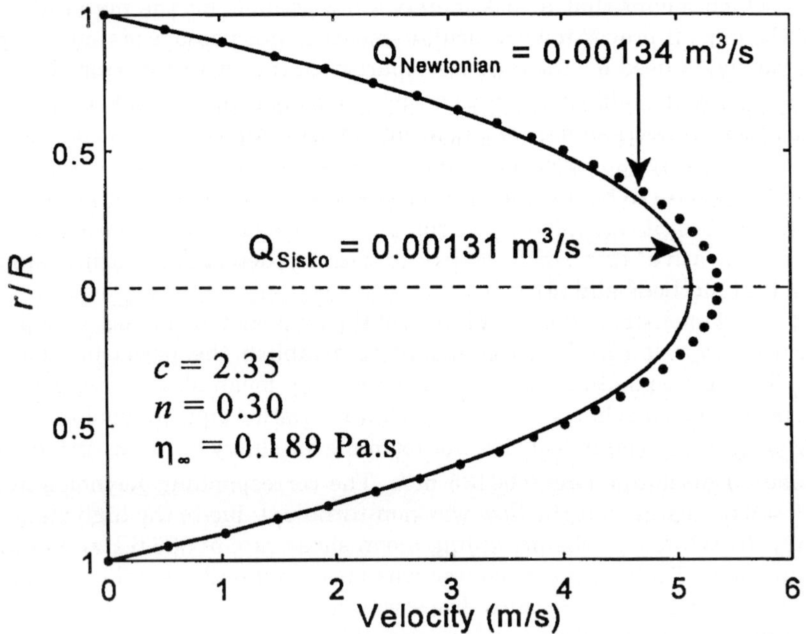

Figure 5. Calculated Sisko power-law velocity profile and discharge for Slurry No. 1, and comparison with Newtonian flow.

corresponding measured discharge was 0.00135 m^3/s (Table 2). The percent error in Table 2 indicates that the calculated discharge (0.00131 m^3/s) was 3% lower the measured value. For slurry nos. 7, 27, 31 and 35 the error was particularly high, ranging from -35 to +49%. Interestingly enough, the mean error for the ten slurries taken together was only -5%, which may imply that the calculations in the aggregate agreed with the measurements.

Also plotted in Figure 5 for the purpose of comparison is the Newtonian discharge that would result at zero consistency. This discharge (0.00134 m^3/s) is only about 2% higher than calculated (0.00131 m^3/s) for the power-law fluid, because at the high flow shear rate in the test the power-law behavior of the slurry was close to Newtonian.

It is essential to point out that in 13 of the 42 slurry rheometric tests (Table 1), the n coefficient was less than zero. Referring to (3), for all $n<0$ a singularity occurs at the centerline ($\zeta=1$), where $\partial u/\partial\zeta = 0$. Thus for calculating the discharge, e.g., as reported for slurry no. 7 in Table 2, the centerline velocity $u(\zeta=1)$ was determined by interpolation between the u values obtained at $r/R = 0.01$.

4. CONCLUSIONS

If the relationships between the power-law parameters and CEC_s prove to be of general validity, the presented approach may be used as follows: 1) for sediment at the dredging site determine CEC_s from (5), 2) calculate the power-law coefficients from the given mean trend-line equations [(6), (7) and (8)], and 3) knowing or setting the discharge requirements for pumping the dredged sediment, back-calculate the pressure drop required to achieve this discharge by iteratively solving for Δp from (3) and (4). The pump must deliver sufficient pressure to sustain this drop.

It is evident that the results of this method cannot be used under prototype conditions without rheometric calibration using site-specific sediment. Furthermore, where the mud is not predominantly clayey, the use of CEC_s must be supplemented with other parameters correlating with slurry rheology. On the other hand, inasmuch as the flow behavior of most natural muds tends to be pseudoplastic within the range of water contents in the natural submerged environment, (3) and (4) should be applicable. Finally, it should be pointed out that to further examine the validity of the use of (6), (7) and (8) it will be essential to run additional tests at pressure drops that are lower than those in Table 2. At these low pressures the non-Newtonian character of flow would be far more apparent than at the pressures tested.

6. ACKNOWLEDGMENT

Support from the Royal Thai Navy and the U. S. Army Engineer Waterways Experiment Station, Vicksburg, Mississippi (contract DACW39-96-M-2100) is acknowledged.

REFERENCES

ASTM, 1993. Standard test methods for operating performance of particulate cation-exchange materials. *Annual Book of ASTM Standards*, American Society for Testing and Materials, Philadelphia, PA, 11(11.02),772-776.

Heywood, N. I., 1991. Rheological characterisation of non-settling slurries. In: *Slurry Handling Design of Solid-Liquid Systems*, N. P. Brown and N. I. Heywood eds., Elsevier, Amsterdam, The Netherlands, 53-87.

Jinchai, P., 1998. An experimental study of mud slurry flows in pipes. *M.S. Thesis*, University of Florida, Gainesville, FL.

Parchure, T. M., and Sturdivant, C. N., 1997. Development of a portable innovative contaminated sediment dredge. *Report No. CPAR-CHL-97-2*, U. S. Army Engineer Waterways Experiment Station, Vicksburg, MS.

Sisko, A. W., 1958. The flow of lubricating greases. *Industrial Engineering Chemistry*, 50, 1789-1792.

Coastal and Estuarine Fine Sediment Processes
W.H. McAnally and A.J. Mehta (Editors)

Beach dynamics related to the Ambalapuzha mudbank along the southwest coast of India

A. C. Narayana[a], P. Manojkumar[a] and R. Tatavarti[b]

[a]Department of Marine Geology and Geophysics, Cochin University of Science and Technology, Lakeside Campus, Cochin 682 016, India

[b]Naval Physical and Oceanographic Laboratory, Thrikkakara, Cochin 682 021, India

The unique Ambalapuzha mudbank in Kerala along the southwest coast of India occurs as a small patch in the pre-monsoon (February – May) season, and increases in both the alongshore and the cross-shore directions as the monsoon (June – September) progresses. This mudbank was observed to be dynamic in character, both with respect to its position and geometry. Beach profile changes were monitored monthly for a period of one year, based on which the volume of beach sediment was computed. Simultaneously the dimensions of the mudbank were monitored in relation to monthly beach profile changes. We observed that the volume of sediment was larger in all seasons along the beach bounded by the mudbank. Sediment accretion and erosion were observed along the beach to the north of the mudbank zone during the monsoon, while erosion was observed south of the mudbank zone. Grain size studies revealed the beach sediment to be medium sized sand. Beach sediment to the north of the mudbank zone, where accretion took place, was observed to be relatively coarse, while that to the south of the mudbank zone, where erosion takes place, was finer.

1. INTRODUCTION

Deposition and removal of non-cohesive beach sediment depend on the shoreline configuration, source and sink of sediment and its size, and the hydrodynamics of the nearshore region. Littoral sediment transport, which contributes to the deposition/erosion processes, is generally strong in the longshore direction. The overall direction of sediment transport and its mechanics determine the areas of coastal erosion and accumulation. However, the erosional and depositional patterns along the beaches of the southwest coast of India are also influenced by factors such as the intensity of the monsoon and the occurrence of mudbanks. Mudbanks are unique coastal features encompassing calm regions of sea with comparatively high suspended sediment concentrations and low wave action. Mudbanks along the southwest coast of India act as natural protectors of the coast, even during the stormy monsoon seasons, thus preventing coastal erosion (Figure 1).

The wind and current systems along the coast play an important role in the dynamics of mudbanks of the southwest coast of India. The important feature of the wind system in the Indian seas is a seasonal reversal of direction associated with two monsoons. Along the west coast of India, during the southwest monsoon winds blow southwards from May to September and attain a northerly direction during the northeastern monsoon. Thus, the seasonally reversing wind pattern influences the southward littoral drift during the southwest monsoon, while a northward drift occurs during the northeastern monsoon. By the middle of May, the southwest monsoonal winds of oceanic origin are established. These winds continue to increase gradually until June when there is a 'burst' or sudden strengthening of the southwest winds. During July and August, the winds have their highest strength, and in September the wind force decreases ahead of the fall transition which lasts through October and November (Sharma, 1978). The two monsoons are: the southwest monsoon from June to September, and the northeast monsoon from October to January. During the southwest monsoonal period the west coast experiences heavy rainfall, with an average annual value of about 300 cm. The northeast monsoon is generally weak along the west coast and the average annual rainfall is about 60 cm. Because of the lesser effects of the

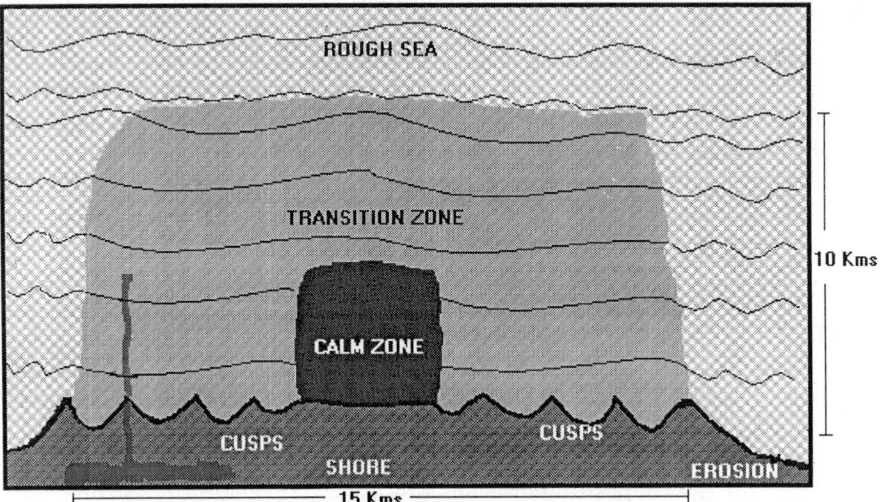

Figure 1. Schematic view of the mudbank of Ambalapuzha along the southwest coast of India (based on an aerial survey).

northeast monsoon on the southwest coast, in this paper we refer to the period from February to May as the pre-monsoon season, June to September as the monsoon season, and October to January as the post-monsoon. Consequent to the establishment of the southwest monsoon in June, the wave activity along the coast increases and the beach erosional/depositional processes are reactivated.

Mudbanks are known to be associated with high biological productivity and prevention of coastal erosion (Damodaran and Hridayanathan, 1966; Gopinathan and Qasim, 1974; Kurup, 1977). Unlike the very few perennial 'mudbanks' found close to the river mouths along muddy coasts of the world, the mudbanks along the southwest coast of India are unique in the sense that they occur away from river mouths and generally appear at the onset of the monsoon and disappear after monsoon. The mudbank at the Ambalapuzha coast expands to about 10-15 km in the alongshore direction and 5-6 km cross-shore during the monsoon. As these features hug the coastline, they influence sediment transport and depositional patterns on the beaches. They comprise of cohesive sediment as a blanket on the sea floor and also as suspended matter.

As it is important to understand how sediments from various sources on the shoreface of the beach are reworked and redistributed by the nearshore hydrodynamic processes (Liu and Zarillo, 1990, 1993), the present study is focussed on the role of mudbanks in the transportation and deposition of non-cohesive sediments along the beaches. Here, mudbank dynamics, spatial and temporal variations of beach sediment volume, and sediment grain size in relation to beach erosional and depositional patterns along the Ambalapuzha coast are discussed. This coast is of a prograding type with a wide beach in its central and northern parts, while in the southern part it is narrow and erosional.

Erosion and deposition of sediment are examined based on the spatial variability of the beach profiles on monthly and seasonal time scales. Characteristic features, including seasonal changes in the shape of the profile, have been examined quantitatively in a number of studies (Inman and Frautschy, 1966; Aubrey, 1979; Weishar and Wood, 1983; Aubrey and Ross, 1985; Larson and Kraus, 1994). In addition to being fundamental to understanding the morphodynamics of beaches, the spatial and temporal behaviors of the beach profile have direct application in coastal engineering projects involving beach nourishment and in the siting of coastal structures (Larson and Kraus, 1994).

2. METHOD

The physical dimensions of the Ambalapuzha mudbank were recorded monthly for about a year from March 1995 to March 1996. Beach profiles were monitored monthly during this period at 38 alongshore stations covering a distance of 20 km from 9° 15' – 9° 25' N latitude (Figure 2). An interval of 500 m between two consecutive profile lines along the shore was maintained, and beach elevation measurements were carried out at each profile. Elevation changes were recorded using a theodolite generally at 5 m intervals in the cross-shore direction. At

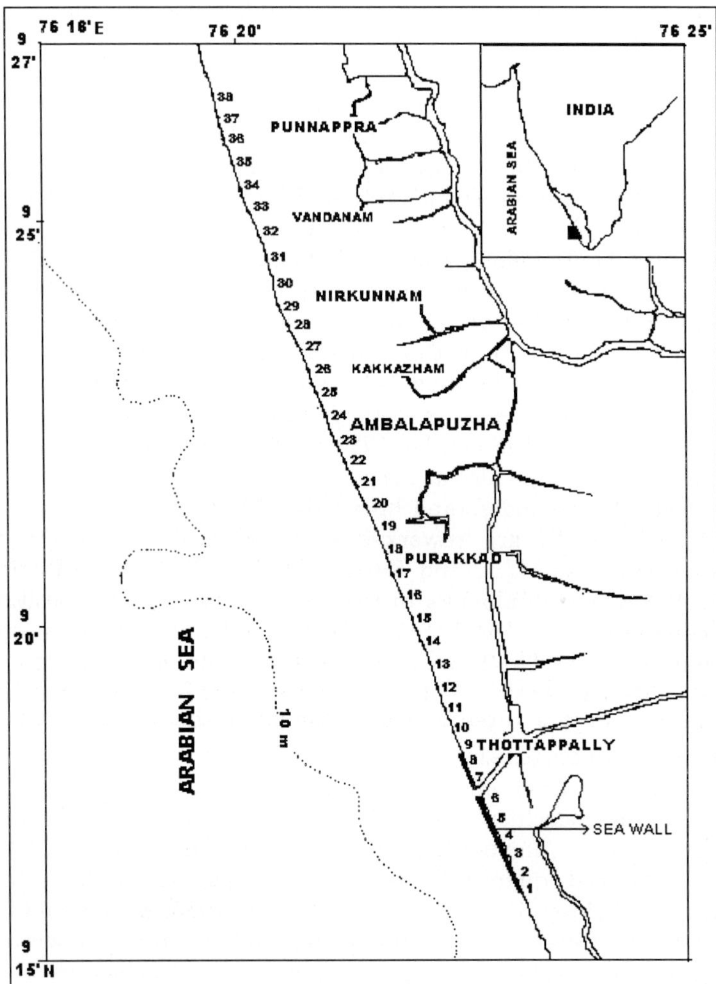

Figure 2. Map of the study area with beach profile locations.

places where sharp changes in the beach slope occurred, such changes were recorded at appropriately smaller (<5 m) cross-shore intervals. Monthly and seasonal variations of sediment volume were calculated for the period from March 1995 to January 1996 using the beach profile data. The linear volume of the beach was calculated by multiplying the area of the profile of unit width (i.e., 1 m) of the beach. Sediment samples from the beach shoreface were collected and mean grain sizes of samples were calculated using the conventional method of graphic measures described by Folk (1974). The samples were collected in all three seasons, i.e., pre-monsoon (February – May), monsoon (June – September) and post-monsoon (October – January).

3. RESULTS

During the non-monsoon period the shoreline was generally straight, devoid of any beach cusps. Gravity waves were predominant and observed to be characteristically breaking slightly offshore (approximately 200 m from the shore). Wave reflections were observed to be rather weak. The significant wave heights during our observations were on the order of 1.5 – 2.5 m. The surf zone comprised of plunging and collapsing breakers. During the monsoon period, the mudbank region and its surrounding sea were observed to have significantly strong shoreline reflections of waves. Surging and spilling breakers were observed in the transition zone, while plunging breakers occurred in the rough zone.

The Ambalapuzha beach is divided into three zones: (i) the mudbank zone, (ii) the zone north of mudbank bordering the northern periphery of the mudbank, and (iii) the zone south of the mudbank, bordering its southern periphery. The beach was wider in the mudbank zone and the adjacent northern zone, whereas in the southern zone there was no beach for the most part of the year.

The position of the mudbank along the coast and its spatial extent in different months of the study period are shown in Figure 3. Figure 4 shows average beach profile variations in the three zones of Ambalapuzha beach during different

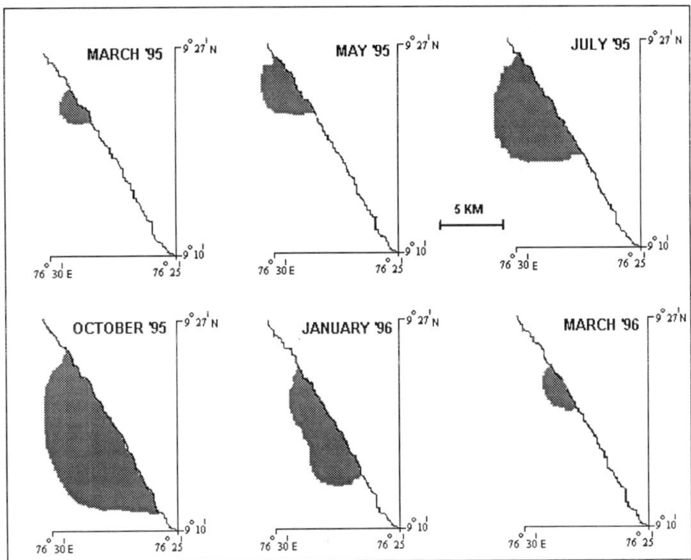

Figure 3. Spatial extent of mudbank in different months.

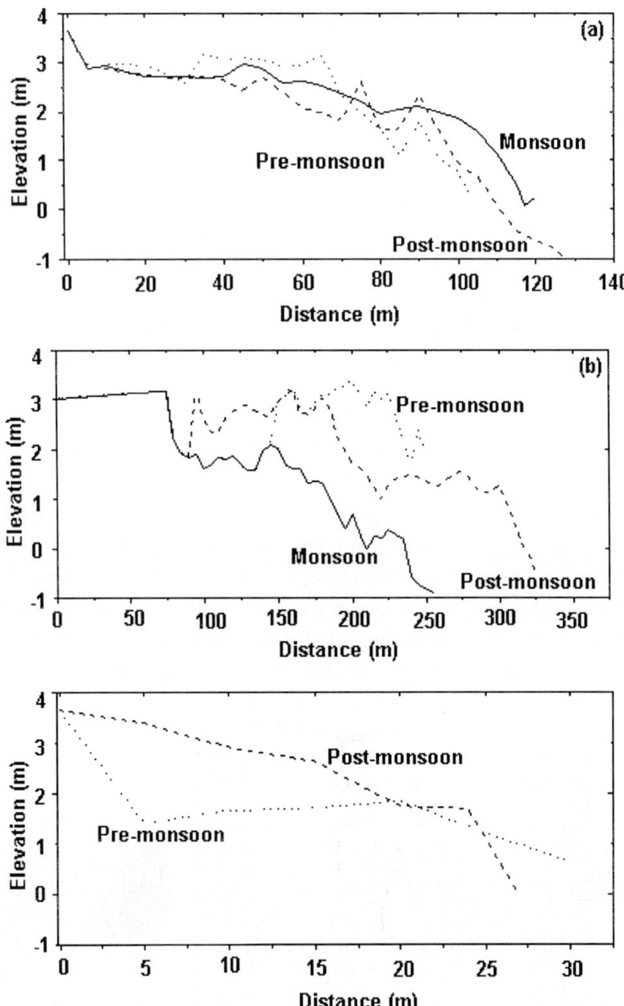

Figure 4. Typical beach profile variations in three different zones of the study area: (a) north of mudbank (b) mudbank (c) south of mudbank.

seasons. Sediment volume variations north of the mudbank, in the mudbank area and south of mudbank for one year period are shown in Figure 5.

The season-averaged mean grain size of beach shoreface sediment of the three zones of the study area in different seasons is shown in Figure 6. In general, the sediment is medium sized (0.23-0.36 mm) sand. Observations on mudbank extension, beach volume change and mean grain size variations in different seasons are discussed below.

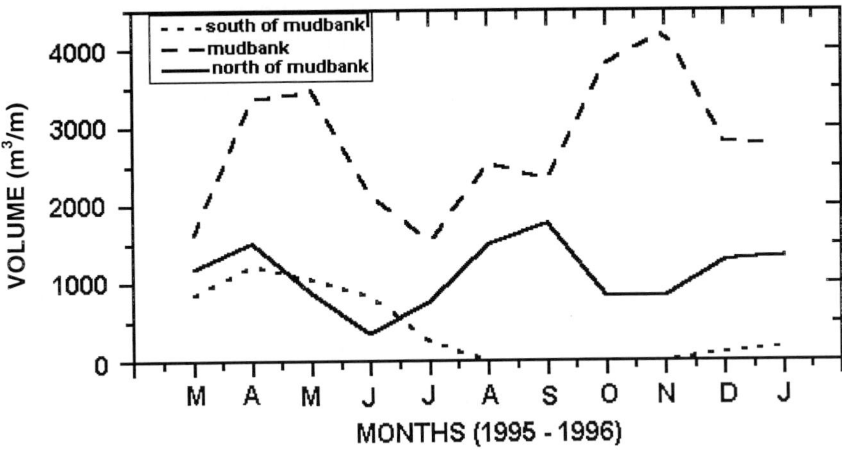

Figure 5. Beach sediment volume variations from March '95 to January '96.

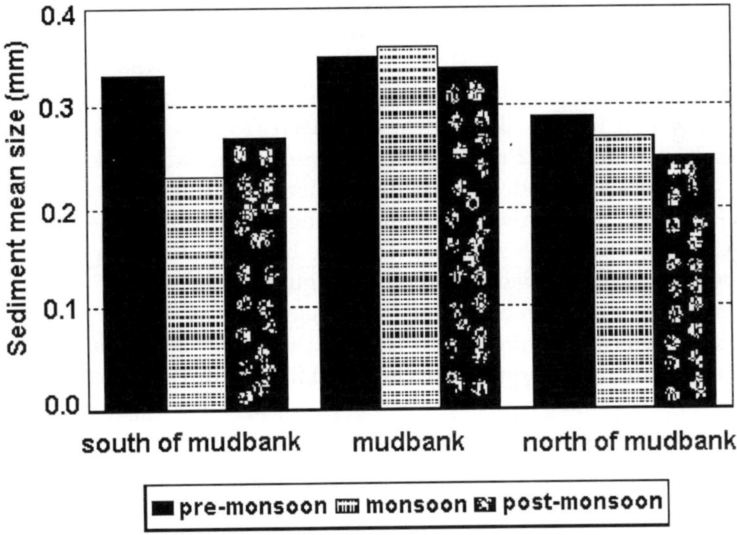

Figure 6. Variations in season-averaged mean size of beach shoreface sediment.

3.1. Pre-monsoon season

It is observed that the alongshore extent of the mudbank was about 1.5 km in March 1995 and shifted northwards about 1.5 km with an increase in its extent by about 0.5 km in April (Figure 3). In May 1995 the alongshore extent increased to about 4.5 km towards south with an offshore extent of 3 km. Our observations reveal that the mudbank occurs as a small patch in the pre-monsoon season.

During the pre-monsoon season, the beach width was higher in the mudbank zone than in the north and south zones (Figure 4). The average beach volume in the south zone was higher (1,754 m^3/m), compared to the north zone in the pre-monsoon season (Figure 5). The beach linear volume was higher (2,110 m^3/m) in the mudbank zone compared to the other two zones. During the pre-monsoon season the beach sediment in the mudbank area had a high mean size (0.35 mm), whereas the sediment in the north zone had a lower mean size (0.29 mm) (Figure 6). The season-averaged mean size of beach sediment in the south zone was 0.33 mm.

3.2. Monsoon

It was observed that the mudbank shrank by about 1 km in its alongshore length while retaining its width in the offshore direction in June 1995, i.e., in the initial phase of the southwest monsoon. From July onwards the mudbank started increasing in its length towards the south and reached its maximum extent by the end of the monsoon (September) (see Figure 3). Its extent was 10 km alongshore and 5 km in the cross-shore direction.

Beach width increased in the mudbank zone as well as north of mudbank during the monsoon (Figure 4). However, beach accretion was more in the north zone compared to the mudbank zone. There was no beach at many profile stations in the south zone during the monsoon (Figure 4). In this season the highest beach volume (2,142 m^3/m) was observed in the mudbank zone (Figure 5). The south zone experienced a complete erosion of the beach by the end of the monsoon. However, a beach was present during the initial phase of the monsoon in the south zone. The north zone gained more beach material in this season, and the average linear volume was 1,103 m^3/m (Figure 5).

The grain size data revealed that the mean size of beach sediment of the mudbank zone was 0.36 mm, while that of the south zone was 0.23 mm. The mean size of sediment in the north zone was 0.27 mm during monsoon (Figure 6).

3.3. Post-monsoon season

By October, i.e., the initial phase of the post-monsoon season, the mudbank had reached a length of about 10 km alongshore, and about 5 km width cross-shore. This is the period when the southwest monsoon withdraws and the northeast monsoon sets in. By the end of November the mudbank started to shrink, and by December it was about 6.5 km in length and 4 km in width. Shrinking continued in January 1996, with the alongshore and cross-shore extents being about 5 km and 3.5 km, respectively (Figure 3).

The beach accreted in all the three zones during the post-monsoon season (Figure 4). Beach width increased to about 100 m in the mudbank zone compared to the monsoon season, while in the north zone beach width increased by about 10 m. The beach was replenished to a width of about 25 m in the south zone during the post-monsoon season (Figure 4). In this season the highest beach linear volume (3,411 m^3/m) was observed in the mudbank zone (Figure 5). Beach volume slightly decreased at the beginning of this season, then increased towards the end of the season in the north zone. Beach accretion was observed in the south zone and its volume was 73 m^3/m (Figure 5).

During the post-monsoon season as well, the highest sediment grain size (0.34 mm) was observed in the mudbank zone, while the lowest grain size (0.27 mm) occurred in the south zone (Figure 6).

4. DISCUSSION

Field observations revealed that the mudbank along the Ambalapuzha coast exists almost throughout the year, even though its spatial extent and suspended sediment concentration are low in the non-monsoon season compared to their monsoonal values. This was in contradiction to the general belief that the mudbanks of Kerala (southwest) coast of India appear only with the onset of southwest monsoon, i.e., in June (Varma and Kurup, 1969) or during pre-monsoon swells (Du Cane et al., 1938) and disappear with the withdrawal of southwest monsoon.

The mudbank had shown its maximum alongshore extent in October and minimum in March 1995 (Figure 3). These observations reveal that just before the onset of the southwest monsoon the mudbank moved slightly northward, whereas during monsoon it extended towards the south, and by the end of the northeast monsoon it had begun to shrink. However, the expansion of the mudbank in the cross-shore and alongshore directions during May was probably due to pre-monsoonal showers, as they activate mudbanks by churning the bottom sediment.

Numerous studies have suggested that mudbanks migrate from their place of origin, particularly along the French Guyana coast (Froidefond et al., 1988) and the southwest coast of India (Gopinathan and Qasim, 1974). However, our studies reveal that mudbanks along the southwest coast of India do not migrate but only expand in their dimensions as the monsoon progresses. They occur at more or less the same locations every year, except in a few cases where they shift their place of origin by a few hundred meters. It appears that the movement and expansion of the mudbank at the Ambalapuzha coast is due to meteorological and hydrodynamic conditions.

In the Arabian Sea, of the two monsoons, the southwest monsoon endures over a longer period of the year and is strong and steady. During this period, the coastal current sets-in in the clockwise direction due to the coastal configuration, and in the counter-clock wise direction from November to January during the peak of the northeast monsoon. As a result, the longshore current is stronger and towards the south during monsoon. Even in the mudbank region the longshore current is stronger than the cross-shore current (Tatavarti et al., 1996). Hence, mudbank extension is towards the south during the monsoon. From February to April, when the northeast monsoon weakens, the circulation has a character which differs from the other transitional period, and the mudbank starts shrinking during this period.

The highest volume of sediment was found to have accumulated in the mudbank zone in all the months (Figure 5). This volume was noticed in the mudbank zone during May, October and November of 1995, whereas in the north zone higher sediment volume was observed during August and September (Figure 5).

The mudbank zone had a maximum sediment volume when compared to the other two zones in all three seasons – pre-monsoon, monsoon and post-monsoon. In the south zone a considerable quantity of sediment accumulated only in the pre-monsoon season, when compared to the mudbank zone. During the monsoon, in the north zone an increase in sediment volume was observed. The study further suggests that during the pre-monsoon season sediment exchange had been mainly between the mudbank zone and the south zone. It is interesting to note that a large quantity of sediment was transported to, and deposited in, the north zone during monsoon, whereas intense erosion took place in the south zone during the same period (Figure 5). However, no significant deposition was observed on the beach within the mudbank zone during monsoon. The variation in sediment volume on either side of the mudbank zone may be attributed to the variation in the longshore current and wave energy conditions, while the mudbank acts as a shore-connected barrier or breakwater. Figure 7 is a schematic presentation of the same.

In all three seasons the mean size of beach sediment of the mudbank zone was slightly higher (0.34-0.36 mm) than that of the northern (0.25-0.29 mm) and the southern (0.23-0.33 mm) zones (Figure 6). Both the highest and the lowest mean sizes were observed in the monsoon--the highest size being in the mudbank zone and the lowest in the south zone. The beach sediment of the north zone was coarser in the monsoon and finer in the pre- and post-monsoon seasons when compared to the other two zones. In the south zone beach sediment was finer size in monsoon when erosion occurred, and in the other two seasons (i.e., pre- and post-monsoon seasons), when accretion was dominant, the sediment was coarser. The relation between shoreline erosion and beach grain size has been established by various workers (Anwar et al., 1979; Frihy and Komar, 1993), who have found the finest grains in areas of erosion. Conversely, areas of shoreline accretion are characterized by coarser sands. These relationships result from grain sorting as the waves and longshore currents first erode the sand from the beach face, transport the sand alongshore, and finally deposit it in areas of accretion (Frihy and Komar, 1991).

Our earlier investigations of the surficial sediments (on the sea bed) have showed that the nearshore area was blanketed with clayey silts at all times. In the pre-monsoon season the sediment blanketing the inner-shelf did not contain much sand. However, in the monsoon about 20% of sand sized sediment covered

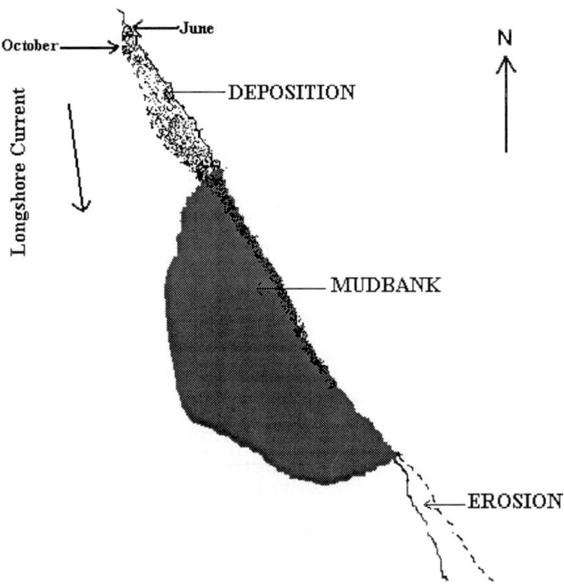

Figure 7. Schematic view of mudbank as a shore-connected barrier, depicting sediment deposition in the north and erosion in the south.

the bottom up to 5 m water depth (Manojkumar et al., 1998). We now speculate that the higher sand content in the nearshore region during the monsoon as reported by Manojkumar et al. (1998) may be due to supply of eroded beach sediment by the prevalent longshore current.

5. CONCLUSIONS

The mudbank at the Ambalapuzha coast was activated with the onset of the southwest monsoon. Initially 'a small patch', it began enlarging towards the south. By the end of the monsoon it attained its maximum size in both the cross-shore and the alongshore directions. The extended size remained practically unchanged during the post-monsoon period but began to shrink at the end of post-monsoon period (i.e., from January). These observations suggest that mudbank dimensions are primarily controlled by the prevailing meteorological conditions and coastal currents.

Beach volume remained almost constant in the mudbank zone in all seasons. However, accretion was observed in the north zone and erosion in the south zone during monsoon. This suggests that the mudbank acts as a shore-connected barrier and obstructs sediment transport from north to south during monsoon.

The sediment size reflected the erosional and accretional pattern of the beach. Coarse sediment was found in the north zone, i.e., zone of accretion, and finer sizes occurred in the south zone, i.e., zone of erosion during monsoon. In the pre- and post-monsoon seasons, the grain size was finer in the north zone and coarser in south zone, respectively, reflecting erosion and deposition during these seasons.

We conclude that the mudbank acts as a shore connected-barrier and plays a significant role in the erosional and depositional processes of the beach, particularly during the monsoon.

5. ACKNOWLEDGMENT

This work was supported by the Department of Science and Technology, Government of India, under the SERC Programme (ESS/CA/A1-14/94).

REFERENCES

Anwar, Y. M., Gindy, A. R., El Askary, M. A., and El Fishawi, N. M., 1979. Beach accretion and erosion, Bullurus-Gamasa coast, Egypt. *Marine Geology*, 30(3/4), M1-M7.

Aubrey, D. G., 1979. Seasonal patterns of onshore/offshore sediment movement. *Journal of Geophysical Research*, 84(C10), 6347-6354.

Aubrey, D. G., and Ross, R. M., 1985. The quantitative description of beach cycles. *Marine Geology*, 69(1/2), 155-170.

Damodaran, R., and Hridayanathan, C. 1966. Studies on the mudbanks of the Kerala coast. *Bulletin of the Department of Marine Biological Oceanography of the University of Kerala*, 2, 61-68.

Du Cane, C. G., Bristow, R. C., Brown, C. J., Keen, B. A., and Russel, E. W., 1938. Report of the special committee on the movement of mudbanks. Cochin Government Press, Cochin, India.

Folk, R. L., 1974. *Petrology of Sedimentary Rocks*, 2nd edition, Hemphill, Austin, TX.

Frihy, O. E., and Komar, P. D., 1991. Patterns of beach-sand sorting and shoreline erosion on the Nile Delta. *Journal of Sedimentary Petrology*, 61(4), 544-550.

Frihy, O. E., and Komar, P. D., 1993. Long-term shoreline changes and the concentration of heavy minerals in the beach sands of the Nile Delta, Egypt. *Marine Geology*, 115(3/4), 253-261.

Froidefond, J. M., Pujos, M., and Andre, X., 1988. Migration of mud banks and changing coastline in French Guiana. *Marine Geology*, 84(1/2), 19-30.

Gopinathan, C. K., and Qasim, S. Z., 1974. Mudbanks of Kerala - their formation and characteristics. *Indian Journal of Marine Sciences*, 3, 105-114.

Inman, D. L., and Frautschy, J. D., 1966. Littoral processes and development of shorelines. *Proceedings of the Coastal Engineering Specialty Conference*, ASCE, New York, 511-536.

Kurup, P. G., 1977. Studies on the physical aspects of the mudbanks along the Kerala coast with special reference to the Purakad mudbank. *Bulletin of Marine Science (University of Cochin)*, 8, 1-72

Larson, M., and Kraus, N. C., 1994. Temporal and spatial scales of beach profile changes, Duck, North Carolina, *Marine Geology*, 117(1/2), 75-94.

Liu, J. T., and Zarillo, G. A., 1990. Shoreline dynamics: evidence from bathymetry an surficial sediments. *Marine Geology*, 94(1/2), 37-53.

Liu, J. T., and Zarillo, G. A., 1993. Simulation of grain-size abundances on a barred upper shoreface. *Marine Geology*, 109(3/4), 237-251.

Manojkumar, P., Narayana, A. C., and Tatavarti, R., 1998. Mudbank dynamics: Physical properties of sediments. *Journal of the Geological Society of India*, 51(6), 793-798.

Sharma, G. S., 1978. Upwelling off the southwest coast of India. *Indian Journal of Marine Sciences*, 7(4), 209-218.

Tatavarti R., Narayana, A. C., Ravisankar, M., and Manojkumar, P., 1996. Mudbank dynamics: Field evidence of edge waves and far infra gravity waves. *Current Science*, 70(9), 837-843.

Varma, P. U., and Kurup, P. G., 1969. Formation of the Chakara (mudbank) on the Kerala coast. *Current Science*, 38, 559-560.

Weishar, L. L., and Wood, W. L., 1983. An evaluation of offshore and beach changes on a tideless coast. *Journal of Sedimentary Petrology*, 53(3), 847-858.